Comparative Protozoology

O. Roger Anderson

Comparative Protozoology
Ecology, Physiology, Life History

With 114 Illustrations

Springer-Verlag
Berlin Heidelberg New York
London Paris Tokyo

O. Roger Anderson
Department of Biological Oceanography
Lamont-Doherty Geological Observatory of
  Columbia University
Palisades, New York 10964, USA

Library of Congress Cataloging-in-Publication Data
Anderson, O. Roger
  Comparative protozoology.
  Bibliography: p.
  Includes index.
  1. Protozoology. I. Title.
  QL366.A53 1987   593.1   87-20686

© 1988 by Springer-Verlag New York Inc.
All rights reserved. This work may not be translated or copied in whole or in part without the written permission of the publisher (Springer-Verlag, 175 Fifth Avenue, New York, New York 10010, USA), except for brief excerpts in connection with reviews or scholarly analysis. Use in connection with any form of information storage and retrieval, electronic adaptation, computer software, or by similar or dissimilar methodology now known or hereafter developed is forbidden.
The use of general descriptive names, trade names, trademarks, etc. in this publication, even if the former are not especially identified, is not to be taken as a sign that such names, as understood by the Trade Marks and Merchandise Marks Act, may accordingly be used freely by anyone.
While the advice and information in this book are believed to be true and accurate at the date of going to press, neither the authors nor the editors nor the publisher can accept any legal responsibility for any errors or omissions that may be made. The publisher makes no warranty, express or implied, with respect to the material contained herein.

Typeset by Publishers Service, Bozeman, Montana.
Printed and bound by Arcata Graphics/Halliday, West Hanover, Massachusetts.
Printed in the United States of America.

9 8 7 6 5 4 3 2 1

ISBN 3-540-18082-6 Springer-Verlag Berlin Heidelberg New York
ISBN 0-387-18082-6 Springer-Verlag New York Berlin Heidelberg

Preface

The protozoa are an eclectic assemblage of organisms encompassing a wide range of single-celled and multiple-celled colonial organisms lacking tissue organization, but exhibiting remarkably refined biological behavior. In some modern classifications, they are classified as a subkingdom among the Protista (eukaryotic single-celled organisms). Although they are not considered a formal category by some taxonomists and some biologists consider the name inappropriate (inferring that they are the first unicellular animals, although some photosynthesize), it is still convenient to consider this group of organisms as an informal collection under the heading of protozoa. Their cosmopolitan distribution, significant ecological role in mineral recycling and enhancement of carbon flow through lower trophic levels of food webs, and remarkable cellular adaptations to enhance survival in diverse environments make them significant organisms for biological investigation.

In some cases, biologists are introduced to this group in first level courses or in invertebrate zoology, but never develop a full appreciation for the diverse and biologically sophisticated characteristics of these organisms. This book is intended as a survey of broad concepts in protozoan biology with an emphasis on comparative data. The focus is on the zoological aspects of the group. Topics more closely related to plantlike characteristics, as presented in books on phycology, are not considered in detail here. A sound background in modern biology and an introduction to cellular biology will be helpful in understanding Chapters 15 and 16, which include a substantial amount of information on biochemistry. Where possible, basic information has been included to help orient the reader who may not be thoroughly grounded in modern biological concepts. The field of protozoology has grown to such proportions, and so many organisms have been investigated, that it is not possible to treat all aspects with equal weight nor to represent the work in all of its diversity. Therefore, major foci have been identified with the intent of introducing the reader to some organizing conceptual perspectives that can serve as guides to further inquiry in specific fields. Where appropriate, sufficient detailed information is presented to establish a clear understanding of the range and diversity of knowledge that has been acquired, in addition to the broad ideas.

There are three major sections: (I) Morphology and Ecology, (II) Functional Microanatomy, and (III) Physiology and Life Processes. Section I provides a broad overview of the diversity of organisms included among the protozoa and provides a conceptual perspective on their ecology. This section is intended as a conceptual organizing framework for more specific topics considered in subsequent chapters. Where possible, comparative data are presented. Each chapter on morphology of a major group (flagellates, amoebae and related organisms, and ciliates) is followed by a chapter on ecological concepts. These chapters are designed to introduce basic concepts which are readily illustrated by examples within each group of protozoa, and also establish a theoretical perspective that is appropriate for broader applications. The chapters in Section I culminate in an integrative view of protozoan ecology presented in Chapter 8. The parasitic protozoa are considered in a separate chapter (Chapter 9) to highlight some of their particular biological characteristics in an ecological perspective, and to emphasize the importance of eradicating this serious group of disease organisms. Sections II and III consider the fine structure, physiology, and life processes of major groups of protozoa. To enhance comparative perspectives, these chapters are organized around major biological concepts rather than according to taxonomic categories. Wherever feasible, however, each chapter has subsections for each major taxonomic group (ie, flagellates, amoebae and their relatives, and ciliates). To aid the reader who prefers to follow as much information as possible about a particular group, the following organizational chart is presented as a guide. The page numbers for each of the major groups are listed in the chart by columns. Rows indicate biological concepts.

	Flagellates	Amoebae etc.	Ciliates
Morphology	16–34	56–73	94–107
Ecology	35–55	74–93	108–130
Microanatomy	197–230	231–252	253–271
Metabolism	285–291	291–294	294–305
	300–304	302–305	339–342
	342–345	345–346	345–348
Nutrition	308–309	309–311	311–314
	314–323	323–327	328–333
Motility	357–360	351–353	355–357
	362–366	366–373	360–363
Life Cycles	386–388	377–379	379–392
	393–398	398–407	408–417
Genetics	418–419	419–420	420–423

Other specific topics such as respiration, osmoregulation, and reproduction are presented in a much more integrated comparative format. Therefore, information about a particular group of protozoa is obtained by examining the subsections within each chapter.

I am particularly indebted to colleagues who have provided illustrations. They are cited in the figure legend accompanying their contribution. I also extend hearty thanks to my students in biology at Columbia University Teachers College, who have sustained my intrinsic interest in protozoa. To many colleagues who over the years have given formal and informal advice on my research and in preparing lectures, I extend my warm thanks. I trust that they will not be offended by omission of their names here, as it would be too voluminous for a modest preface. I especially thank the Biological Oceanography Division of the National Science Foundation for financial support of my research on biomineralizing marine protozoa, and for making it possible for me to continue as an active participant in this intellectually lively and challenging field.

O. Roger Anderson
Columbia University

Contents

Preface v

Section I Morphology and Ecology

Chapter 1 The Protozoa in Broad Perspective 3

    Protozoan Form and Diversity 3
    Comparative Morphology and Taxonomy 5
    Phylogenetic Relationships 8
    Taxonomic Summary 12
    Summative Perspective 15

Chapter 2 The Flagellates (Phylum: Sarcomastigophora;
    Subphylum: Mastigophora) 16

    General Morphology 16
    Phytoflagellates 21
    Summative Perspective for Phytoflagellates 26
    Zooflagellates 28
    Summative Perspective for Zooflagellates 32

Chapter 3 Some Perspectives on the Habitats and Comparative
    Ecology of Flagellates 35

    Organismic Environmental Interactions 35
    Physico-chemical Environmental Factors 39
    Biotic Factors 45
    Diversity 47
    Summative Perspective 53

Chapter 4 Amoebae and Their Relatives (Phylum: Sarcomastigophora;
    Subphylum: Sarcodina) 56

    General Morphology 56
    Amoebae in Broad Perspective 57
    Naked Amoebae (Class: Lobosea) 58

	Slime Molds (Class: Mycetozoea)	62
	Testate Amoebae (Classes: Lobosea and Filosea)	63
	Foraminifera (Class: Granuloreticulosea)	65
	Radiolaria (Class: Polycystinea)	70
	Heliozoans (Class: Heliozoea)	72
	Summative Perspective	73
Chapter 5	Some Comparative Perspectives on the Ecology and Habitats of Sarcodina	74
	General Ecological Concepts	74
	Naked and Testate Amoebae	77
	Benthic Foraminifera	86
	Planktonic Foraminifera and Radiolaria	90
	Summative Perspective	93
Chapter 6	Ciliates (Phylum: Ciliophora)	94
	General Morphology	94
	Class: Nassophorea	100
	Class: Oligohymenophorea	103
	Class: Colpodea	104
	Class: Karyorelictea	105
	Class: Spirotrichea	105
	Class: Protostomatea	106
	Class: Phyllopharyngea	106
	Summary of Ciliate Morphology	107
Chapter 7	Perspectives on Ciliate Ecology	108
	Basic Concepts	108
	Population Growth and Interspecies Competition	112
	Structure of Natural Communities	120
	Summative Comment	129
Chapter 8	General Perspectives on Protozoan Ecology	131
	Protozoan Communities	131
	Protozoan Distribution and Colonization Rate	136
	Succession and Community Structure	138
	Summative Comment	149
Chapter 9	Human Parasitic Protozoa	150
	The Concept of Protozoan Parasitism	150
	Amoebae	152
	Flagellates	157
	Malarias: *Plasmodium* (Apicomplexa)	169
	Other Parasites of Medical and Economic Importance	173
	Summative Perspective	173

Section II Functional Microanatomy

Chapter 10 General Principles of Cell Fine Structure 177

- Introduction 177
- Plasma Membrane and Cytoskeletal Structures 177
- Cytoplasmic Organelles 182
- Nucleus 185
- Summative Perspective 187

Chapter 11 Protozoan Fine Structure and Functional Microanatomy 189

- A Concept of Protozoan Fine Structure and Function 189
- Common Features and Specialized Function of Organelles 191
- Summative Perspective 195

Chapter 12 Comparative Microanatomy of Flagellates 197

- Flagellate Microstructure 197
- Flagellum and Haptonema 197
- Cytoplasmic Fine Structure 205
- Summative Perspective 229

Chapter 13 Fine Structure of Amoebae and Their Relatives 231

- Comparative Fine Structural Features 231
- Major Organelles 238
- Cytoplasmic Specialized Structures 242
- Extracellular Surface Structures 244
- Summative Perspective 252

Chapter 14 Fine Structure of Ciliates 253

- General Cytoplasmic Features 253
- Cortical Fine Structure 253
- Specialized Cytoplasmic Structures 263
- Summative Perspective 271

Section III Physiology and Life Processes

Chapter 15 Basic Biochemistry and Physiology 275

- Cellular Dynamics 275
- Intracellular Regulation 282
- Specialized Metabolic Pathways in Flagellates 285
- Some Specialized Metabolic Pathways in Amoebae 291
- Some Metabolic Pathways in the Ciliate *Tetrahymena* 294
- Other Ciliates 298
- Some Selected Topics on Comparative Protozoan Metabolism .. 299
- Summative Perspective 306

xii   Contents

| Chapter 16 | Nutrition | 307 |

- Basic Perspectives ... 307
- Flagellate Feeding Behavior ... 308
- Sarcodina Feeding Strategies ... 309
- Ciliate Feeding Strategies ... 311
- Flagellate Nutrition ... 314
- Nutrition of Amoebae ... 323
- Feeding and Nutrition in Ciliates ... 328
- Symbiosis ... 333
- Summative Perspective ... 336

| Chapter 17 | Respiration and Osmoregulation | 338 |

- Energy Budgets and Growth Efficiency ... 338
- Osmoregulation ... 340
- Intracellular Ionic Balance ... 345
- Summative Perspective ... 348

| Chapter 18 | Motility | 350 |

- Sensation and Coordination ... 350
- Ciliary Motion ... 355
- Flagellar Motion ... 357
- Mechanisms of Ciliary and Flagellar Motion ... 358
- Contractile and Locomotory Behavior ... 360
- Amoeboid Locomotion ... 366
- Shuttle Streaming in Slime Molds ... 371
- Summative Perspective ... 373

| Chapter 19 | Reproduction | 375 |

- Reproductive Strategies ... 375
- Sexual Reproduction ... 376
- Asexual Reproduction ... 386
- Summative Perspective ... 392

| Chapter 20 | Life Cycles and Genetic Continuity | 393 |

- Conceptual Perspective ... 393
- Flagellates ... 393
- Amoebae and Related Organisms ... 398
- Ciliates ... 408
- Some Perspectives on Protozoan Genetics ... 417
- Genetic Studies of Volvocidans ... 418
- Genetics of a Myxamoeba ... 419
- Genetic Studies of Ciliates ... 420
- Summative Perspective ... 423

References ... 425
Subject Index ... 461
Genera and Species Index ... 473

Section I Morphology and Ecology

1
The Protozoa in Broad Perspective

Protozoan Form and Diversity

Knowledge of the diversity of life, within a context of unifying biological concepts, is a valuable source of insight in advancing biological theory. A thorough understanding of biological diversity including variations in form, function, ecological roles, modes of reproduction, and genetic control of life processes is a rich source of information whereby creative and often practical insights are generated into the nature of life. Hence, a complete understanding of the broad range of living things (particularly some of the smallest and also most intricately organized forms of life) is an asset in advancing biological theory and devising innovative approaches to modern biological research. The diversity of protozoa provides a unique opportunity to explore some of these major biological themes in a comparative format. Protozoa span a spectrum of sizes from microscopic forms, barely larger than many bacteria, to those clearly visible with the unaided eye, and comparable in size to some of the smallest metazoa. The protozoa comprise a vast ensemble of organisms with wide diversity in morphology and environmental adaptations. One of the major unifying principles, however, amidst this diversity is the central defining idea of a protozoan; ie, a one-celled or colonial (polycellular) organism possessing all of the elegantly coordinated processes of life, usually (though not exclusively) animallike, within the boundaries of a single cell. Even the colonial forms, designated here as polycellular, may be considered fundamentally a loosely coupled set of single cells with most functions of the colony duplicated within each of the member unit cells. There are no tissues as observed in more advanced organisms. Hence, the concept of the protozoa as single-celled but highly complex organisms is probably justified. Functionally, and to a large extent morphologically, there is considerable autonomy and uniformity in cellular activity. There are, however, notable exceptions, especially among some colonial species. For example, in *Volvox* and its relatives some of the cells in the colony differentiate into reproductive cells. However, these are not organized into a tissue. Furthermore, in some colonial radiolaria (marine protozoa) the gelatinous spheroids attain diameters of up to several centimeters. Cells at the periphery extend rhizopodia into the environment to snare prey, while those deeper within the colony distribute nutrients throughout the colony using a network of rhizopodial strands. They can also coordinate movements to repair damage to the gelatinous hollow sphere enclosing the colony. For example, a rupture in the colony is healed by the coordinated action of the peripheral, contractile rhizopodia. The breach is closed and new gelatin is secreted at the site of healing, producing a fully reformed colony with characteristic shape for the species. Species specific shapes vary from spheroids to cylindroids. Although these integrated processes are noteworthy for such "primitive" organisms, there are no clearly differentiated groups of cells within the organism. Hence, there are no true tissues and the whole colony can still be considered as a polycellular aggregate of unit cells. Nonetheless, it is clear that the diversity of function and widely divergent adaptive responses of protozoa often necessitate qualifications in their definition. The qualification of the animallike properties of protozoa arises from the somewhat perplexing situation that some are photosynthetic in varying degrees. Thus, distinctions between some protozoa and the algae become blurred, and indeed the pigmented species are claimed by protozoologists and also most botanists, who include them among the algae. Some pigmented species have fine structural features or physiological properties that suggest close affinities with nonpigmented species, thus justifying their inclusion among the protozoa. This is the case with *Euglena gracilis*, for example, which has fine structural similarities to the colorless flagellate *Astasia* and some parasitic flagellates (trichomonads). In general, most of the pigmented forms have other animallike characteristics that justify their assignment to the protozoa. In some cases, in addition to being partially or substantially photosynthetic, these pigmented species also assimilate organic nutrients from the environment, thus exhibiting an animallike form of nutrition. Other pigmented species are only marginally photosynthetic and cannot maintain themselves without the presence of substantial sources of dissolved or particulate organic nutrients in the environment. A clear tendency toward animal nutrition occurs in those photosynthetic species that also ingest particulate food by phagocytosis. A remarkable trophic dualism is found in some species exemplified by *Euglena*. They are pigmented (possessing chlorophyll, among other pigments) when temperature, light, and inorganic nutrients are favorable for autotrophy. However, they become bleached (lacking chlorophyll) and assume a completely heterotrophic nutrition (assimilating dissolved organic molecules) when subjected to darkness, elevated temperatures, or other heterotrophic-inducing factors in the environment. This facultative capacity for varied nutrition, while often perplexing to biologists who prefer to easily categorize species as plantlike or animallike, further illustrates the intriguing physiological diversity and behavioral complexity of many protozoa.

Most of the fundamental processes of life that we normally associate with the clearly specialized structures of metazoa, composed of numerous cells organized into tissues and organs, are accomplished within the membrane-enclosed compartments of these "one-celled" organisms. The complexity of microanatomical structures and physiological mechanisms that have evolved in these small organisms to permit their efficient exploitation of far-ranging environments makes them among the most aesthetically attractive and intellectually intriguing groups to study. Herein lies also a rich opportunity to inquire into some of the most fundamental principles of biology including plasticity and coordination of life processes at the cellular level, and integration of form and function as occurs among the rather closely united cells in some colonial protozoa.

Although unified by the fundamental concept of single-celled organization, protozoa constitute an eclectic assemblage of phylogenetically, distantly related organisms. In addition to large size variations among species, they have adapted to highly diverse environments. Their habitats include all major terrestrial and aquatic environments, spanning temperature extremes from polar sea-ice to the tepid tropical forests of the equator; and range from the darkened ocean depths to sun-splashed mountain peaks. Their ability to exploit diverse niches is further documented by their presence in unexpected environments such as the edge of the insect eye, carapace of microcrustacea, thermal spring effluent, mountain snow banks, and the intestines of insects and advanced metazoa. Some are parasites within algal cells, other protozoa, or mammalian erythrocytes and macrophages. Moreover, recent explorations of the ocean floor have disclosed the presence of large, rhizopod-bearing protozoa known as Xenophyophores (Tendal et al 1982). These testate protozoa occur abundantly at depths as great as 8,260 m, and are equally diverse in the Atlantic (18 species) as in other oceans including the Ogasawara Trench of the Western Pacific.

Comparative Morphology and Taxonomy

The heterogeneous morphology and complexity of physiological processes characterizing the protozoa yield substantial data for taxonomic distinctions, but also provide innumerable perplexities in defining common characteristics that unify the group phylogenetically (through evolutionary lineages) or conceptually based on structural-functional characteristics. Consequently, some modern taxonomic schemes do not recognize a formal category for the protozoa. They are considered members of the kingdom Protista following the five kingdom classification scheme (Whittaker 1969; Margulis 1974); ie: (1) Monera, the bacteria; (2) Protista, including the algae (nontissue-bearing unicellular or multicellular photosynthetic organisms), and the "protozoa"; (3) Plantae, multicellular photosynthetic organisms with cellulose cell walls and at least a tissue-level organization; (4) Fungi, nonphotosynthetic multicellular simple organisms with chitinous (nitrogen-containing organic) cell walls including the molds, mushrooms, and their relatives; and (5) Animalia, the nonphotosynthetic (heterotrophic) metazoa. In some classifications, the category of Protozoa is retained as a subkingdom under the Protista.

A committee of the Society of Protozoologists chaired by Levine (Levine et al 1980) substantially reorganized the classification of protozoa and addressed the issue of their complex taxonomic status:

"The Protozoa are essentially single-celled eukaryotic organisms. They are not a natural group, but have been placed together as a matter of convenience. In the classical, broad classification of living things, which divides them into plants and animals, the Protozoa comprise a phylum of animals. More modern classifications, however, divide living things into 4 to many kingdoms.... In this, the Protozoa might be considered a subkingdom of the kingdom Protista."

Although they do not make a final conclusion on the matter, their classification divides the protozoa into phyla consistent with subkingdom status. The problem of determining what constitutes natural affinities can be perplexing when classically accepted ideas of broad natural boundaries become blurred, and unless some clear fossil evidence is available to reconstruct evolutionary relationships (which is minimal for many protozoa), one must rely heavily on evidence from living forms. This requires, among other sources of insight, a clear knowledge of fundamental physiological and biochemical characteristics of the organisms, and where possible, substantial information on the ontogenetic development of individuals as a means of finding possible common points in evolutionary history. Some comprehensive and very readable reviews of current thought in protozoan taxonomy provide useful background information on these problems and issues (eg, Corliss 1960, 1972; Baker 1977; Whittaker 1977; Page 1976a; Lee et al 1985a). In contrast to the problems arising from their taxonomic diversity, protozoa exhibit a remarkable degree of morphological intergradation among major groups, which, in a broad synthetic perspective, provides a conceptual basis for their scholarly treatment as a biological group. A diagram of some morphological relationships (Figure 1.1) is presented as a conceptual organizing view and does not necessarily reflect current understanding of phylogenetic relationships or taxonomic affinities.

The flagella-bearing species, formally classified in the subphylum Mastigophora, or informally known simply as flagellates, exhibit wide morphological variation. The single-celled, flagellated organism typically associated with this group (Figure 1.1a) is but one morphological form. Others are polyflagellated (eg, Figures 1.1 b,c). Some flagellates transform into amoeboid organisms either still possessing the flagellum (Figure 1.1f), or lacking it after resorption. Others are entirely amoeboid, without flagella, and only their cytoplasmic organization confirms their affinity with the Mastigophora. The overall morphology and locomotion of these nonflagellated Mastigophora is very similar to the "true amoebae" (Figures 1.1g, h), and related organisms with fingerlike to weblike cytoplasmic extensions (pseudopodia). These amoeboid organisms may be typically attached to the substratum (eg, many of the "naked amoebae" and "testate amoebae"), or floating with a halo of fine pseudopodia (eg, planktonic Foraminifera). They are classified within the subphylum Sarcodina. Other members of this group, known as amoebo-flagellates, produce flagellated swarmers (Figure 1.1g) as part of their life cycle. The flagellated stages may represent a primitive characteristic that has persisted into modern time. If this is correct, it further reinforces the hypothesis that amoebae may have arisen from flagellated

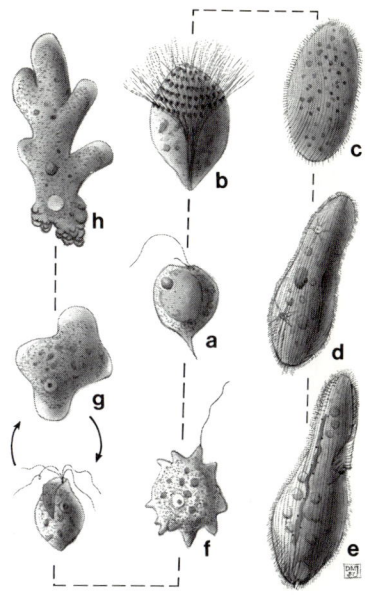

FIGURE 1.1. Some representative species of protozoa exhibiting intergradational morphological features. (a–c, and f) flagellates; (d and e) ciliates; (g and h) amoebae. (a) *Spumella hovassei*, (b) *Calonympha grassii*, (c) *Opalina ranarum*, (d) *Paramecium* sp, (e) *Blepharisma* sp, (f) *Mastigella polymastix*, (g) *Tetramitus rostratus*, and (h) *Amoeba* sp. The dashed lines do not imply phylogenetic relationships, but only indicate patterns of intergradation in morphology to illustrate the remarkable variation and intergradation in form among major groups of some protozoa. Artwork by Dean Jacobson.

ancestors. Indeed, based on these similarities and possible phylogenetic affinities among amoebae and flagellates, the subphyla of Sarcodina and Mastigophora are grouped into the phylum Sarcomastigophora by some modern taxonomists (eg, Levine et al 1980).

The morphological diversity of the flagellates, as conceptualized in Figure 1.1, also extends in another direction toward polyflagellate forms. These include elegant species dwelling in the intestines of termites and wood roaches. The species illustrated in Figure 1.1b are placed in the order Hypermastigida. They are characterized by numerous flagella arranged in whorls or patches on the cell surface. Further flagellar elaboration, covering the cell surface, characterizes the subphylum Opalinata (Figure 1.1c). These organisms are easily confused with ciliates (Figures 1.1 d,e) owing to the densely distributed rows of flagella on the cell. But, they are grouped with the flagellates in modern classifications based on their cytoplasmic organization, fine structure of the basal body of the flagellum, and in some species, according to differentiating mode of cell division. These opalinids have symmetrical binary fission, separating into two identical daughter cells by cleavage through the midline of the body. Other species of the Sarcomastigophora with reduced size and cellular complexity may superficially resemble the parasitic protozoa placed in the phyla Apicomplexa and Microspora, but fine cytoplasmic details clearly separate the phyla. The ciliates (Figures 1.1 d,e), phylum Ciliophora, are a more homogeneous group taxonomically and perhaps phylogenetically (Corliss 1972) as compared to the Sarcomastigophora. They possess cilia at least during one stage of their life cycle, exhibit highly diversified structures for food gathering, and some have highly specialized locomotory organelles depending on, among other factors, their habitat and mode of nutrition. For example, some astomatous species (lacking an oral apparatus) may superficially resemble opalinids; but the majority of the species of ciliates possess diverse and sometimes complex structures for gathering a wide range of food particles. While this cursory overview of major morphological similarities of some protozoa along an integradational scheme exhibits the unity of major features, fine details of form and function permit much clearer separation of the species into increasingly well-defined taxonomic categories. As our knowledge of form, physiological function, and molecular organization increases, taxonomic categories will undoubtedly continue to be refined to better reflect our emerging understandings of the "natural" affinities among the protozoa. Although the early evolutionary history of the protozoa is obscure, the close morphological and fine structural similarities of the flagellates with other groups, particularly the amoebae, suggest that a common ancestral stock may have risen from flagellated organisms. Indeed, it has been proposed that the more phylogenetically advanced metazoa may have arisen by a lineage from an early flagellated organism. In accordance with the heterotroph hypothesis of ancestral origins, some theorists propose that the flagellates arose from a nonpigmented, phagotrophic form (eg, Margulis 1970). Plastid-containing species could have arisen by phagocytic engulfment of small algae that eventually became symbiotic with the protozoan host. Through increasing interdependence between host and symbiont during evolution, the alga may have become structurally reduced to only a plastid enclosed by peripheral membranes (eg, Gibbs 1978).

The origins of the Sarcodina are also obscure. In one view, some species of amoebae may have arisen from colorless ancestral flagellates through gradual loss of flagella accompanying increasing amoeboid properties. Other species of amoebae, however, could have arisen separately along one or more parallel lines of evolution (Sleigh 1979) in a polyphyletic evolutionary model. The ornate siliceous skeletons of radiolaria (Sarcomastigophora) and their complex cytoplasmic organization (Anderson 1983) suggest that these organisms may have evolved from an ancient ancestor along quite separate phylogenetic lines compared to other sarcodinads. Their within-group diversity also implies a rather complex polyphyletic evolution. Ciliates may also have several lines of phylogenetic development, and some could have arisen from polyflagellated ancestors such as those giving rise to the opalinids (Figure 1.1c). Given the complexities of protozoan form and function, it is likely that a polyphyletic model of their evolution is most appropriate with several parallel lines of development rather than a monophyletic (single, branching-tree) model.

Phylogenetic Relationships

A polyphyletic pathway model of protozoan evolution and the origin of some higher organisms is presented in Figure 1.2 (Sleigh 1979). In addition to general morphological features indicating affinities among the major protozoan groups

Phylogenetic Relationships 9

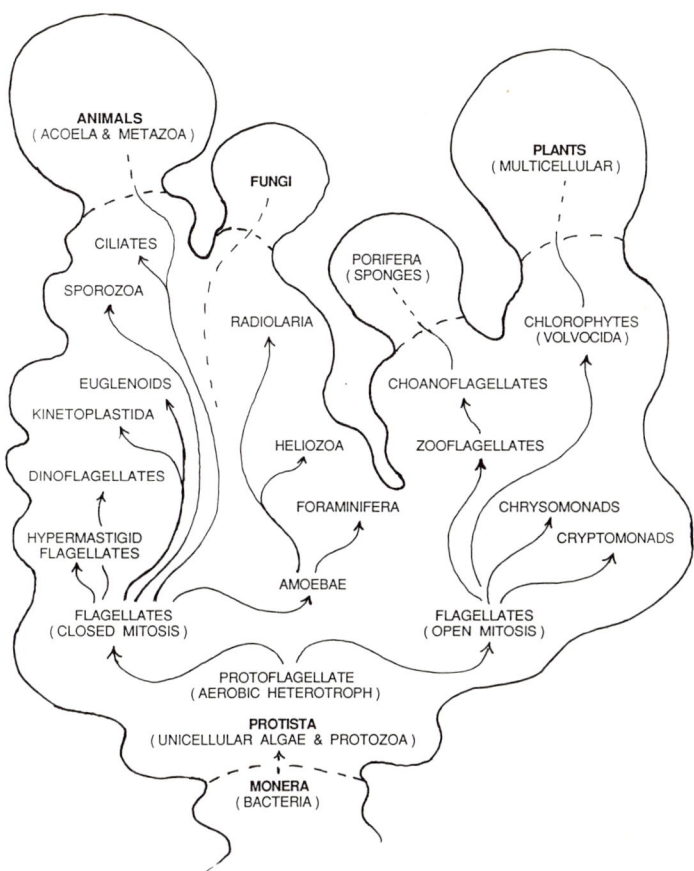

FIGURE 1.2. A phylogenetic map of some members of the five kingdoms, with special emphasis on major protozoan groups. The protozoa are presumed to have arisen from a primitive protoflagellate with a major bifurcation along two lines based on whether the form of mitosis is closed or open. Based on data from Whittaker (1969), Margulis (1974), and Sleigh (1979).

(phytoflagellates, zooflagellates, amoeboid protozoa, and the ciliates), some fine-detailed cytoplasmic characteristics are incorporated into this scheme. For example, there are at least two markedly different ways that mitosis occurs in protozoa. In "open mitosis", the nuclear membrane is resorbed during prophase, and at telophase, the dividing chromosomes move to opposite poles of the cell, forming two daughter chromosomal centers that are subsequently enclosed within a newly synthesized nuclear envelope. This form is typically observed in higher organisms. In "closed mitosis", the nuclear membrane remains intact. The chromosomes separate within the enlarging nuclear envelope, which becomes considerably elongated during late metaphase. In the last stages, the elongated nuclear envelope is constricted at the midregion, and is eventually pinched in two, forming two daughter nuclei. Some intermediate forms are also observed where the nuclear envelope is only partially resorbed. Spindle fibers pass through the fenestrations in the nuclear envelope and connect with chromosomes. In addition to nuclear fine structure, the organization and microstructure of other organelles are also used in clarifying ancestral origins and reconstructing phylogenetic lineages. The number of membranous thylakoids within the plastids and the organization of the enclosing membranes in pigmented flagellates are also used in constructing the polyphyletic model proposed by Sleigh. The major radiation of protozoa according to this model occurred by multiple lines arising from flagellated ancestors with closed mitosis. The cryptomonads, chrysomonads, and some zooflagellates arose from precursors with open mitosis. The choanoflagellates, with collared cells, may have given rise to the sponges (Porifera). Higher land plants arose, according to this scheme, by a major, early, separate pathway through the green algae with cellulose walls. This rather thoroughly documented view represents one of the more recent perspectives on the complex problem of deciphering protozoan phylogeny and its relationship to the origins of higher organisms. There are, however, varied interpretations of protozoan phylogenies, and much remains to be done toward elucidating their long evolutionary history. Some of the issues are discussed critically in reviews by Sleigh (1979) and Corliss (1960, 1972).

In comparison to other forms of life, the protozoa are in a central position encompassing many of the features that link them superficially or more substantially to the other kingdoms of living things (Figure 1.3). In this diagram, biological characteristics of some major groups of protozoa are mapped onto a two-coordinate system. The horizontal axis represents mode of nutrition extending from autotrophic to heterotrophic; ie, from photosynthetic, plantlike nutrition to food-gathering or organic nutrient-absorbing, animallike modes of nutrition. The vertical axis extends from prokaryotic to eukaryotic cellular organization. The dashed lines at the base of the vertical axis signify that none of the protozoa are truly prokaryotic, but only approach the structural features characteristic of prokaryotes (eg, absence of mitochondria). The midpoint of the vertical axis denotes the mesokaryotes, those protozoa with mitochondria and a nuclear envelope surrounding permanently condensed chromosomes resembling those in some prokaryotes. The presence of a nuclear envelope, and in some cases, plastids in the cytoplasm of these protozoa separates them from the prokaryotes and suggests that they are a "mesoform" in between prokaryotes and eukaryotes. As with any conceptual scheme, the amount of information represented is limited by the defining dimensions, and this diagram is intended only as a map of some salient features of protozoan groups and their diversity. It is clear from this conceptualization that the flagellates are among the most diversified of protozoa, with locations in major segments of the coordinate system. Some of the parasitic flagellates lack mitochondria. Their parasitic life cycle, reduced cytoplasmic organization, and specialized organelles subserving oxidative metabolic functions in lieu of mitochondria make them very different from most eukaryotes. Their habits and cellular organization are more reminiscent of prokaryotes.

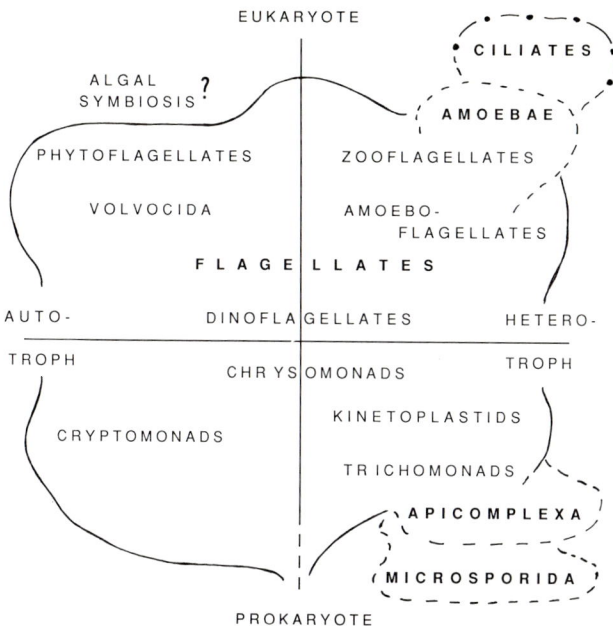

FIGURE 1.3. A conceptual diagram mapped on a coordinate system illustrating the relative diversity of some major groups of protozoa. The horizontal dimension represents variation on a continuum from those species that are largely autotrophic (as with most pigmented phytoflagellates) to those that are heterotrophic. The vertical dimension represents the degree of cellular organization from very "primitive" forms with reduced cytoplasmic complexity approaching that of the prokaryotes to those with advanced eukaryotic features. The central location and extensive breadth of the flagellates (enclosed by a solid line) may indicate their breadth of adaptation and possibly their central position in phylogenetic origins of other groups such as the amoebae and ciliates localized largely in the upper right-hand quadrant. Some parasites, that have become highly dependent on unique environments rich in nutrient sources, possess less elaborate cytoplasmic organization and therefore, in some cases, are plotted in the lower right-hand quadrant. Note that the dashed line at the base of the vertical dimension denotes that none of the protozoa are true prokaryotes.

These probably did not evolve directly from prokaryotes, but may have become reduced in cytoplasmic organization through adaptation over long periods of evolution. These organisms are mapped in the lower right quadrant of the diagram denoting their heterotrophic mode of nutrition, in addition to a "simplified" cell organization. At least one amoeba (*Pelomyxa*) is reported to lack mitochondria, and, although clearly possessing a eukaryotic nucleus, is mapped in this quadrant at a level closer to the midpoint of the vertical axis. The remaining major groups of pigmented and colorless flagellates occupy large regions of the coordinate system. The dinoflagellates occur largely along the horizontal axis, spanning from autotrophic to heterotrophic forms, Eukaryotic, nonpigmented flagellates are mapped in the, upper right-hand quadrant, while pigmented forms occur in the upper, left-hand quadrant. The amoebae and their relatives (Sarcodina), with the exception of the few noted in the lower, right-hand quadrant, are mapped in the upper, right-hand quadrant, as they are eukaryotic, heterotrophic organisms. One small group of Sarcodina labeled with a question mark is included in the upper, left-hand quadrant. The symbiont-bearing Foraminifera and radiolaria are included here. Also, some ciliates and benthic Foraminifera capable of extracting chloroplasts from algal prey and sequestering them in their cytoplasm are included here. This myxotrophic group of protozoa can be conceptualized as having heterotrophic and autotrophic functions owing to their algal inclusions. The question mark indicates the tentativeness of this assignment. Some algal, symbiont-bearing species of ciliates (eg, *Paramecium bursaria*) may also be placed here, but such fine distinctions are not quite valid within the broad scope of the diagram. Most ciliates are mapped in the upper right-hand quadrant, as they are clearly eukaryotic and heterotrophic.

In summary perspective, it is helpful to remember that the protozoa, though undoubtedly of ancient origin, have had a long evolutionary history. Therefore, most extant species represent forms that have adapted over long periods of geological history and are not necessarily characteristic of the ancient ancestral stock from which this amazingly diverse group of organisms evolved. Many species have become adapted to modern environments that may have persisted for long evolutionary periods, and in some cases, others have clearly become adapted to very specialized environments. Thus, it is wise to use due care in ascribing a primitive status to extant species of protozoa. Within the broad range of protozoan species living today, one may assume that there are varying degrees of phylogenetic advances. And, moreover, protozoa may represent broadly the kind of early ancestral stock from which more phylogenetically advanced organisms could have arisen. In most cases, however, extant protozoan species are also elegantly advanced organisms. Therefore, a thorough biological analysis of protozoa should include adequate attention to their advanced properties, in addition to possible insights derived from probing their "primitive status" relative to more advanced metazoans. Proper care should be given to understanding their biology in broad perspective including: (1) their physiological ecology, with an analysis of adaptive features in relation to their environments; (2) the biochemical basis for their strategies in adapting to varied environments; (3) molecular genetics, especially of those species that have been studied intensively biochemically and physiologically; (4) the cellular biological bases of functional morphology and control of morphogenesis; and (5) phylogenetic analyses to clarify taxonomic categories.

Taxonomic Summary

A current taxonomic summary of major phyla and subphyla is presented as an organizing perspective for the more detailed comparative analyses among species to be presented in the remaining chapters of section I. The descriptions are according to those published by the Committee appointed by the Society of Protozoologists (Levine et al 1980) and more fully addressed in the *Illustrated Guide to the Protozoa* edited by Lee, Hutner, and Bovee (Lee et al 1985a).

Kingdom Protista

SUBKINGDOM PROTOZOA

Essentially single-celled eukaryotic organisms exhibiting highly organized animallike, but sometimes also autotrophic, biological functions either totally contained within a single cell or among groups of cells forming a colony.

Phylum: Sarcomastigophora

Flagellated and amoeboid organisms having but one form of a nucleus (no micronucleus) either single or multinucleated, and in some flagellates bearing plastids.

Subphylum: Mastigophora. One to many flagella typically present in the active trophic stage. Reproduction by symmetrical binary fission; sexual reproduction known in some groups. Subdivided further into pigmented species (phytoflagellates) and colorless species (zooflagellates).

Subphylum: Opalinata. Surface covered by flagella (undulipodia), no cytostome (food canal), single or multiple nuclei all of one form (no micronucleus), symmetrical binary fission common. Live as commensals in the alimentary tract; observed commonly in amphibians and fish.

Subphylum: Sarcodina. An eclectic assemblage of organisms producing pseudopodia (varying among groups as fingerlike, netlike, raylike or filament-shaped cytoplasmic extensions), known informally as the amoebae and related organisms. Flagella when present usually restricted to developmental or other temporary stages; body naked or covered with organic coat, test, scales, or internal skeleton; reproduction by fission; sexuality when present associated with flagellated or more rarely amoeboid gametes; most species free-living in terrestrial, freshwater and marine environments. Representatives include the naked amoebae (represented by the familiar *Amoeba proteus*); testate amoebae ("shell-covered" amoebae); radiolaria (silica-secreting floating marine organisms with a pore-bearing organic wall, and stiffened radial pseudopodia called axopods); and Heliozoa (axopod-bearing species without an organic wall).

Phylum: Labyrinthomorpha

Active feeding stages with spindle-shaped or spherical, non-amoeboid cells within an ectoplasmic (external) network. In some genera, amoeboid cells move within network by gliding. Unique cell-surface organelles (sagenogenetosomes) associated with ectoplasmic network. Motile spores with unequal flagella produced by most species. Absorb nutrients (saprobic) or parasitic on algae. Live mostly in marine and estuarine waters.

Phylum: Apicomplexa

Parasitic species bearing characteristic cytoplasmic organelles known as the apical complex at one end of cell consisting of (1) one or more polar rings; (2) a conical array of spirally arranged cytoplasmic microtubules; (3) rhoptries (strandlike bodies with swollen ends); and (4) micronemes, rod-shaped granules lying longitudinally in the anterior half of the cell. Some of the most serious human parasites are included here (eg, malarial parasites of man and other animals).

Phylum: Microspora

Intracellular parasites that infect many kinds of animals forming spores of unicellular origin, with a single threadlike sporoplasm discharged from the spores through the channel or tubular polar filament. Development takes place totally within host cells and dispersal is by the small (1 µm) sporoplasm (the "planont") which is injected into the host cell through the polar filament by the internal pressure of the spore.

Phylum: Myxozoa

Parasites of tissues and organ cavities of cold-blooded vertebrates or annelids, forming spores of multicellular origin containing 1 or 2 sporoplasms and 1 to 6, but usually 2 polar capsules (each containing a coiled polar filament). Several serious diseases of fish are caused by myxosporida, which invade major organs and cartilage.

Phylum: Ciliophora

Ciliate-bearing during at least part of life cycle. Two forms of nuclei commonly observed: macronucleus (one or more) and micronucleus. Many are phagotrophic, actively ingesting food particles.

Subphylum: Postciliodesmatophora. Ciliates with cilia grouped in units of two (dikinetids) on cell surface and lacking parasomal sacs (minute depressions within the outer cell membrane). Other features involve fine structural details of the overlapping arrangement of cytoplasmic microtubular ribbons attached to the base of the cilia, and the arrangement of the ciliary rows within the oral cavity (1 to 3 rows with complex cytoplasmic constituents). The familiar genus *Stentor* is included here.

Subphylum: Rhabdophora. Ciliates with individual cilia (monokinetids) on the cell surface, except around the cytostome (oral apparatus), where typically a crown of dikinetids occur. The surface pellicular structure of the cell does not have well-developed alveoli (membrane enclosed flattened, polyhedral spaces lying immediately beneath the plasma membrane).

Subphylum: Cyrtophora. Ciliates with varied forms of ciliary patterns on the surface of the cell, including monokinetids or dikinetids (depending on the species). The cortical alveolar system and underlying fibrillar system are well developed. The commonly known *Paramecium* species are included here.

Summative Perspective

The wide range of morphologies and habitats of protozoa reflect a long and complex evolutionary history. Hence, modern species of protozoa, though small relative to other forms of life, are equally sophisticated in many of their life processes. Although we think they have arisen from some of the very earliest forms of life, they have clearly become advanced in many of their biological functions during geological history. Some of them produce exquisite mineralized products such as surface scales, or shells, and exhibit other forms of highly coordinated and specialized functions clearly indicating an advanced evolutionary status. The flagellates appear to represent an ancestral group from which many other protozoa and higher forms of life may have evolved. The wide range of flagellate morphology and physiology as summarized in Figure 1.3 strengthens the likelihood that this group could be ancestral to a broad range of other protozoan species. Phylogenetic relationships, and hence taxonomic affinities among protozoa are determined by fine structural and physiological analyses, and preferably also by molecular genetic analyses. Modern taxonomic categories are increasingly established by similarities in the protein or nucleic acid composition of cells. The base composition of RNA and DNA is believed to be a fundamental indicator of biological similarity among organisms. Since these informational molecules determine basic properties of living systems, they are probably one of the most conservative indicators of biological identity. Further research combining fine structural and molecular genetic analyses promises to provide substantial information on the taxonomic affinities and phylogenetic history of protozoa.

2
The Flagellates
(Phylum: Sarcomastigophora; Subphylum: Mastigophora)

General Morphology

An efficient means of motility is a decided asset to most living organisms, especially those that reproduce rapidly and must move about efficiently to invade new territory and reduce competition for resources. If the environment is nutritionally limited (oligotrophic), then the pressure toward survival by efficient and sustained locomotion is all the more critical. Both autotrophic and heterotrophic organisms are benefited by efficient locomotion, either to move to regions of appropriate illumination, or to increase the probability of finding external sources of food. Aqueous environments have been efficiently exploited by the flagella-bearing protozoa whose undulating, whiplike organelles provide an efficient and dexterous mode of moving through fluids. Each flagellum or undulipodium is a thin cylindrical organelle approximately 150 µm long and 0.2 µm in diameter. The surface membrane surrounding the flagellum is continuous with the cell plasma membrane at the point where the flagellum is attached to the cell. The internal structure (see Figure 10.2) is complex consisting of a central pair of cytoplasmic tubules surrounded by nine doublets. Each doublet consists of a pair of tubules connected along their length. A basal body (kinetosome), usually of somewhat denser refractile properties and staining more intensely than the distal part of the flagellum, anchors it within the cytoplasm. In some species, the basal body also has one or more fine fibrils attached to it that are clearly visible only with the electron microscope. Flagellar fine structure is presented more fully in Chapters 10 and 12. Whiplike undulations or wagging motions are produced by complex movements of the internal tubules and/or by the action of fibrils attached to the base of the flagellum within the cell cytoplasm. The presence of a flagellum is one of the key characteristics defining members of the subphylum Mastigophora. Some lack flagella during phases of their life cycle, and in the case of *Dientamoeba*, a fully amoeboid organism formerly assigned with the amoebae, no flagellum is present; but, the fine cytoplasmic details clearly indicate its affinity with the flagellates. Two major classes of flagellates are recognized: (1) the phytoflagellates (Phytomastigophorea) with pigmented plastids, and frequently possessing heterotrophic modes of nutrition in addition to autotrophy;

and (2) the zooflagellates (Zoomastigophorea) lacking pigmented plastids and obligately heterotrophic. In addition to flagellar locomotion, a few of the phytoflagellates are capable of amoeboid locomotion by flattening out and creeping upon substrata. This is more commonly observed among some of the zooflagellates. There are colonial forms of phyto- and zooflagellates, either free-swimming or attached to surfaces. Sessile (attached) colonies use flagellar motion to move currents of water around the cells; thus enhancing supplies of nutrients and keeping a fresh supply of fluid in the vicinity of the colony. Though anchored to a substratum, the colonies are able to create constant encounter with new portions of the environment by moving it within their domain, rather than moving themselves through the fluid space.

The major characteristics of flagellates are depicted in Figures 2.1 through 2.3. Flagellar number varies from one (Figures 2.1c) to two or more (Figures 2.1 b,d,f,g), often in multiples of two. The flagellum may emerge from basal bodies anchored immediately beneath the apex of the cell (eg, Figures 2.1a,b,e,g) or from basal bodies situated beneath the surface membrane lining a flagellar reservoir (Figures 2.1c and d). The position of attachment of the flagellum within the pouch is of taxonomic significance. In other species exemplified by some zoomastigorphorans, one active (undulating) flagellum may be directed largely forward (Figure 2.1e) with one trailing. In other species, the posteriorly directed flagellum is linked to the surface of the plasma membrane (Figure 2.1j), producing a thin undulating membrane when the flagellum is active and lifts the membranous fold. Thus, the membrane is not a permanent structure, but is elevated when the flagellum pulls it upward during undulation. The remaining flagella (in this case, three; Figure 2.1j) may be directed forward and exhibit whiplike motions. In other species, one flagellum may trail and the other is wrapped around the cell body (as in the dinoflagellates), usually within a specialized groove (girdle) near the midline of the cell (Figure 2.1h). Flagellar motion varies among species. Activity includes massive armlike strokes, vigorous whiplike undulations, or fine wavelike ripples moving along the flagellar axis. In some cases, the flagellum is held rather rigidly in the anterior direction with fine wavelets of motion traversing its length. Water is propelled along the flagellar axis. This propulsion of water results in movement of the cell through its aqueous medium. The trajectory of the cell may be in large spiral paths, or in rather direct paths with gyrations around one axis. In some species, the cell exhibits a tumbling motion.

The cytoplasm as viewed with a light microscope typically contains one nucleus; although, multinucleated forms occur as for example among the hypermastigida (Figure 1.2b), and Opalinata (Figures 1.1 and 2.3c). The surrounding cytoplasm contains vacuoles of varying size and internal composition, one or more plastids (when present), mitochondria (respiratory organelles), and Golgi (secretory bodies) barely visible when stained and viewed with high power light optics. In plastid-bearing species, the pigment composition, internal, membranous-lamellar structures, number and arrangement of pyrenoids (refractile proteinaceus bodies), and the number and kind of surrounding cytoplasmic

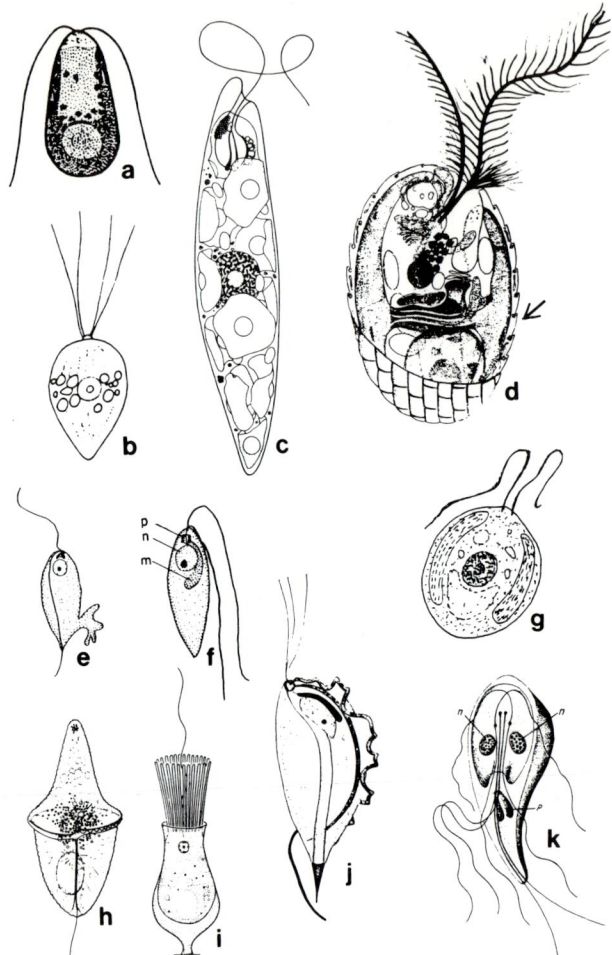

FIGURE 2.1. Comparative morphology of some flagellates: (a) *Dunaliella salina* × 750; (b) *Polytomella caeca* × 750; (c) *Euglena gracilis* × 1800; (d) *Chroomonas mesostigmatica* × 4800, arrow designates thecal plate; (e) *Cercomonas* sp × 1600; (f) *Proteromonas lacertae* × 2200 (p = parabasal body, n = nucleus, m = mitochondrion); (g) *Ochromonas* sp × 2500; (h) *Gymnodinium dogieli* × 270; (i) *Salpingoeca amphoroideum* × 430; (j) *Tritichomonas muris* × 3600; (k) *Giardia intestinalis* × 2400. Figures (a–f) and (j) from Lee et al (1985a), courtesy of the Society of Protozoologists. Figures (h) and (i) from Grell (1973); with permission.

General Morphology 19

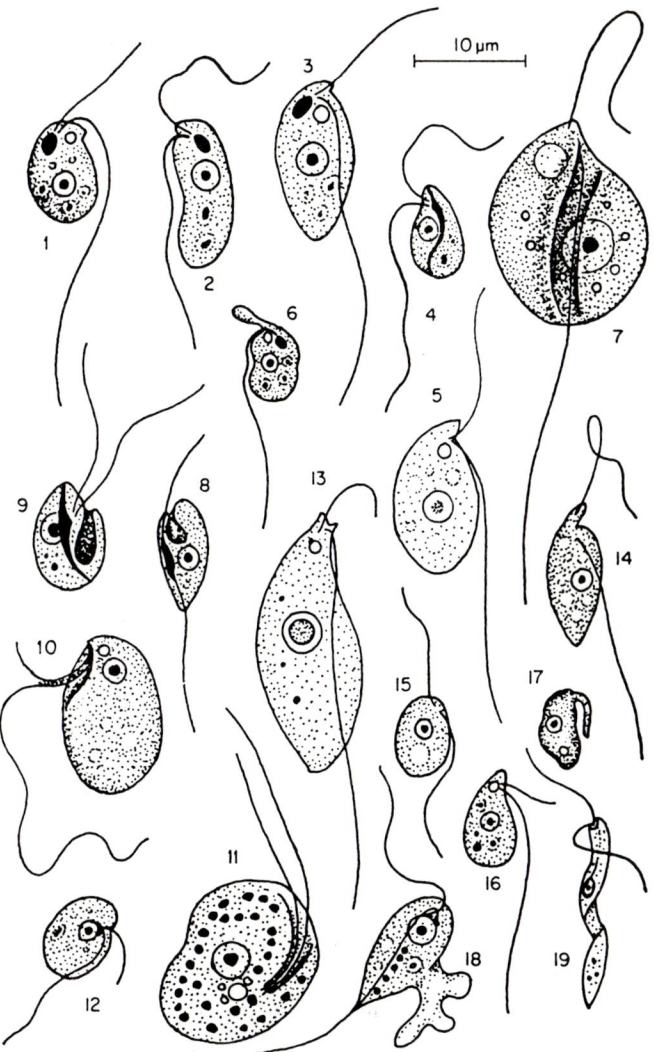

FIGURE 2.2. Family Bodonidae and related flagellates (× 1500): (1) *Bodo saltans*; (2) *B caudatus*; (3)–(5) different versions of *Bodo edax*; (6) *Rhynchomonas nasuta*; (7) *Colponema loxodes**; (8) *C agitans*; (9) *Phyllomitus amylophagus*; (10) *P undulans**; (11) *Ichthyobodo necator* showing multiple kinetoplasts; (12) *Bodomorpha minima*; (13) *Parabodo nitrophilus**; (14) *Cruzella marina*; (15) *Heteromita globosa*; (16) *Amastigomonas debruynei*; (17) *Pleuromonas jaculans**; (18) *Cercobodo* (= *Cercomonas*) sp; (19) *Spiromonas angusta*. (*Unstained specimen. Presence of kinetoplast uncertain. From Vickerman (1976); with permission.)

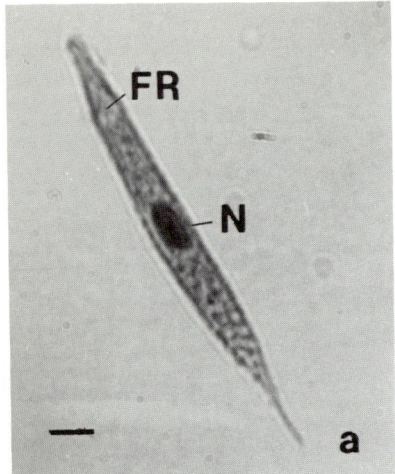
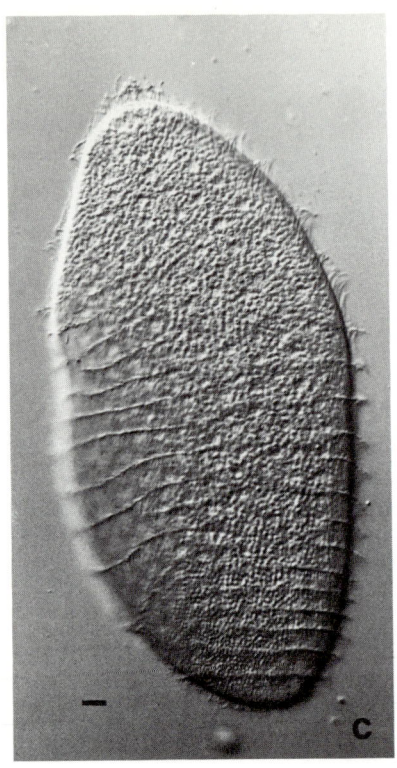
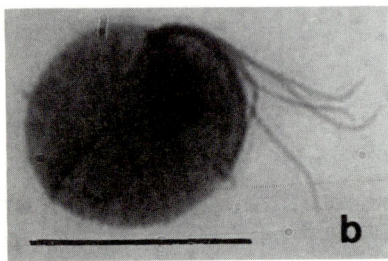

FIGURE 2.3. Comparative light micrographs of some diverse flagellates (a) *Euglena acus*, N = nucleus, FR = flagellar reservoir, (b) *Trichomonas vaginalis*, and (c) *Opalina ranarum*. Scales = 10 μm. (c) Courtesy of Dr. David Patterson, University of Bristol, England.

membranes are important distinguishing features among species. Although most nonpigmented flagellates show no evidence of even a residual plastid, some species lacking chlorophyll, but clearly related to pigmented forms, possess colorless plastids (leukoplasts) indicating their probable close phylogenetic affinity with pigmented precursors. This cytoplasmic variability is clear evidence of possibly complex evolutionary pathways among the flagellates. In some of their pathways, they appear to have progressed from heterotrophic to autotrophic forms. In other cases, they seem to have made transitions from autotrophic to heterotrophic forms with loss of photosynthetic capacity and retention of only a colorless plastid. Storage substances including oil droplets, starch and other carbohydrate deposits (either large grains or granular masses), or various organic deposits are sometimes clearly visible within the cytoplasm. The type of storage product is often of taxonomic significance and may be related to the kind of cellular nutrition. Some autotrophic species store carbohydrates largely as massive starch grains or occasionally as small polymers (leucosin) in a liquid form. Sometimes lipid droplets are present; particularly with increasing cell maturation. Heterotrophic species, though not exclusively, may store oil droplets, waxy substances, or other lipids within the cytoplasm. Some possess masses of glycogen (animal starch) scattered throughout the cytoplasm. It is not possible, however, to make rigid categorizations of food storage products between autotrophic and heterotrophic species. The physiological state of the organism, other variations in metabolic characteristics, and stage of growth often produce modifications. Where appropriate, the major reserve substance for each organism will be presented as part of the comparative descriptions of flagellate species.

Some species possess a cell covering in addition to the peripheral plasma membrane; eg, (1) elastic, complex-ridged pellicles; (2) loricas (organic or partially to completely mineralized walls); or (3) thin proteinaceous or cellulose shroud-like coverings. These external enclosures can have a very elegant complex geometric design. Some silicoflagellates produce delicate tests resembling baskets woven of siliceous rods. Euglenoids and some cyst-forming chrysomonads produce flasklike mineralized enclosures around the cell either during active growth or at the time of encystment. In other species, the external covering is a rather inconspicuous layer, as observed by light optics, that immediately surrounds and follows the contour of the cell surface.

As a general perspective on comparative morphology of this group, major features of several contrasting types of flagellates are presented with a fair amount of detail. This hopefully will provide a conceptual orientation for the more concise taxonomic summaries and comparative ecological topics that follow.

Phytoflagellates

The organisms included here possess pigmented plastids, or when colorless, exhibit features closely resembling those of the pigmented forms, including in some groups leukoplasts of identical structure to the pigmented kind.

Cryptomonadida

Chroomonas mesostigmatica (Figure 2.1d), a representative of the cryptomonads (order Cryptomonadida), is a small barrel-shaped organism, slightly flattened on one side, ca 10 μm long with two flagella emergent from a nearly apical pore of the canallike gullet. Each flagellum is anchored in the lateral (dorsal) wall of the gullet. The two flagella of almost equal length bear lateral hairs or mastigonemes composed of protein filaments. The longer carries a bilateral array of mastigonemes ca 1.5 μm long; each tipped with a thin filament. A tuft of identical mastigonemes arises from a platform near the base of the flagellum where it emerges from the pore of the gullet. The shorter flagellum bears a single file of mastigonemes; each tipped with a pair of filaments of unequal length. The gullet has two pouchlike branches near the base. The more dorsal one immediately beneath the flagellar attachment site is lined by ejectile organelles (trichocysts) that propel a thin barbed filament (possibly as a defense in this autotrophic species and not as a prey capturing device). The gullet is not active as a food engulfing organelle. Additional trichocysts are distributed over the surface of the cell, which also is covered by several rows of laterally offset rectangular plates bearing prominent ridges. These are displayed more clearly in the view of the sectioned edges (Figure 2.1d, arrow). The single, bilobed plastid is H-shaped when viewed from one side, owing to a thin isthmus joining the two longitudinally oriented massive lobes of the plastid. The pyrenoid, anterior to the isthmus, may have several layers of starch on the surface. An eyespot is attached to a recurved segment of the dorsal lobe of the plastid and lies just anterior to the pyrenoid near the trichocyst-lined branch of the gullet. A prominent nucleus with a nucleolus is posterior. Healthy cultures have a blue-green color. The normal storage product is starch, but physiological changes and maturation of the cell are often accompanied by increases in lipid that may replace the starch. Locomotion of cryptomonads is by smooth swimming, with the action of the rapidly undulating long flagellum and rather stiff shorter flagellum causing the cell to gyrate about its long axis.

Variations in morphology of cryptomonads include single chloroplasts, as the boat-shaped one of *Hemiselmis* sp or absence of plastids as in the colorless, phagotrophic *Chilomonas*. The shape of the gullet varies in width and prominence of its branched pouches. The position of the gullet opening also varies. It is much more ventrally located (toward the flattened side of the cell) in most species compared to the apical position in *Chroomonas*. The nucleus, however, is typically posterior. Chloroplasts are pigmented olive-green or brownish, or sometimes red or bluish owing to a combination of chlorophylls *a* and *c*, yellow-pigmented carotenes, and a unique system of red and blue pigmented proteins (biliproteins) (phycoerythrin and phycocyanin, respectively). The colorless *Chilomonas* has a leukoplast, though devoid of light trapping pigments, that is structurally similar to those of pigmented cryptomonad species.

Dinoflagellida

The dinoflagellates (order Dinoflagellida), are commonly found in freshwater and marine environments, and constitute an important source of food for small organisms. Some form symbiotic associations with marine invertebrates. Other species (eg, *Gymnodinium breve*) sometimes produce toxic blooms that kill fish and cause serious economic loss to coastal marine fishermen.

The genus *Gymnodinium* (Figure 2.1h) illustrates the key features of the thecate (wall-bearing) dinoflagellates. The cell is typically surrounded by a cellulose covering containing a distinct cingulum (circumferential furrow) with one of the two flagella, a filamentous or ribbonlike undulating type whose action causes the characteristic spiraling motion of the organism. The second is attached close to the point of emergence of the first and lies in a thin, longitudinal fissure known as the sulcus. The two conical portions of the cell wall (hypotheca with broad dome-shaped surface and epitheca with more pointed end) are characteristic of this genus. Wide variations in size, shape, and relative position of the two "conical" halves occur across the many species of dinoflagellates. Other species exhibit remarkably elaborate and ornate thecae composed of closely joined plates (reminescent of jigsaw puzzle pieces); and in some cases with elegant branching or spirelike projections. *Gymnodinium* spp are autotrophic or heterotrophic. The nucleus is situated typically in the lower half of the cell.

Some phagotrophic, colorless species possess a specialized feeding organelle (peduncle) resembling a thickened flagellum, but of very different cytoplasmic organization. The peduncle is thrust out from a pore in the cell wall and aids in the apprehension of food particles. Among dinoflagellates, the thecal plates, when present forming the wall, are secreted and contained within a system of closely knit flattened vesicles (thecal vesicles) that are part of the perialgal membrane system surrounding the cell. Nonthecate species lack the elegant ornamentation characteristic of thecate forms. Thecate species have highly diverse thecal plate patterns that are species characteristic. Pigmented species usually have chlorophylls a and c_2, beta-carotene, and the xanthophylls peridinin, neoperidinin, dinoxanthin, and neodinoxanthin. Starches and oils are stored as reserve substances.

Euglenida

The euglenoid flagellates (order Euglenida), typified by *Euglena gracilis* (Figure 2.1c) and *Euglena acus* (Figure 2.3a), are a diverse group of pigmented or colorless flagellates, with the flagella arising within an anterior canallike invagination known as the flagellar pouch, with a pyriform pocket known as a "reservoir". The typical number of flagella is two, but as is characteristic of *Euglena*, only one is long and emergent. The other is reduced to a short stub with its distal end closely appressed to the base of the long flagellum, sometimes giving the appearance of a branched base to the longer flagellum. A swollen region (paraflagellar body or rod) is located near the base of the longer flagellum near the point of contact by the short flagellum. This swollen region is probably the light sensing part of the cell. The pigmented "eyespot" lying immediately opposite the paraflagellar swelling within the cytoplasm adjacent to the membrane of the flagellar pouch is probably not a light sensing organelle, but may aid in light sensing, perhaps by modulating the intensity of light reaching the sensory portion of the flagellum. *Euglena gracilis* has a tapered elongate form with a centrally located nucleus surrounded by 6 to 12 circular, shieldlike plastids; each bearing a prominent central pyrenoid, covered on either side by a watch-glass shaped mass of storage carbohydrate (not starch but a closely related carbohydrate, paramylon). The surface of the cell is enclosed within a pellicle composed of a set of elastic, interlinked, ribbonlike plates arranged in a spiral pattern around the cell. Swimming occurs through a gyrating motion, with the anterior end tracing a wide circle. Euglenoid movement (alternating phases of contraction and elongation sometimes accompanied by twisting movements) is pronounced when swimming ceases.

Euglenoid flagellates in general possess the characteristic flagellar pouch opening toward the anterior end, although the shape, position (sometimes it is more lateral as in *Entosiphon*), and configuration of the opening vary among species. A contractile vacuole near the reservoir occurs in all freshwater species, including *Euglena gracilis*. Cell shape varies markedly among genera. Forms vary from decidedly spindle-shaped, long, thin, tapering as in *Euglena acus* (Figure 2.3) or its colorless counterpart, *cyclidopsis acus*, to flasklike with a swollen posterior end (*Astasia klebsii*, colorless and osmotrophic). Others are spherical to ovoid with a mineralized lorica (*Trachelomonas grandis*, containing pigmented plastids resembling *Euglena*, but with inwardly projecting cylindrical pyrenoids), or laterally compressed with surface ridges (eg, *Phacus triqueter*).

Modes of nutrition vary, including photoautotrophic or facultative heterotrophic green euglenoids with chlorophylls *a* and *b* (some osmotrophic, but none phagotrophic), and obligate heterotrophic colorless species being osmotrophic and/or phagotrophic. The stored carbohydrate is paramylon, a polymer resembling starch but with a beta-1:3-linked glucose chain. It gives no color reaction with iodine in KI. Mature cells in aging cultures are charged with brown or orange droplets containing lipids, carotenoids, and cyclic metaphosphates.

Chrysomonadida

The chrysomonads (order Chrysomonadida) represented by *Ochromonas* sp (Figure 2.1g) are basically cells with two flagella of unequal length. The longer, bearing bilateral rows of mastigonemes, is directed anteriorly, and the shorter smooth one is often slightly trailing or directed laterally and may have a whiplash tip. There is no flagellar pouch. The cells, naked or covered by a thin cellulose shroud, are globular, ellipsoidal to pyriform, and slightly to extremely flattened. One to several contractile vacuoles are present. Plastids (pigmented yellow-brown or yellow-green) are arranged parietally near the edge of the cell, and one or more large vacuoles occur near the posterior part of the cell beneath a prominent nucleus. Some species of *Ochromonas* are only partially photosynthetic (eg, *O malhamensis*) and must obtain organic nutrients by osmotrophy or phagotrophy. Others, as with *O danica*, are more substantially to completely photoautotrophic and can survive with limited external organic sources of nutrition if sufficient mineral nutrients for photosynthesis are provided. Leucosin within vacuoles and lipid as cytoplasmic droplets are stored as reserves. Chrysomonad plastids contain chlorophylls *a* and *c* (c_1 and c_2), beta-carotene, and several xanthophylls. Some chrysomonads produce siliceous surface scales, or loricas enclosing the cell.

Prymnesiida

The elaborate calcareous or organic scales of the Prymnesiads (eg, Figure 12.10) enclosing the cell body are typical of this diverse group of golden-brown pigmented flagellates in the Order Prymnesiida. A filiform organelle (haptonema), emerging at a point near the origin of the flagella, is characteristic of the group. It may be rather short and twisted, or in other groups it is extensible, thrusting out as a long, thin filament many times the length of the flagella. Though otherwise similar in morphology to many chrysomonads, this group is physiologically very distinct. Cells are spherical, oval or flattened and usually bear two saucer-shaped parietal plastids. Most species are photoautotrophic, but many have also been shown to be phagotrophic. Pigmented species contain chlorophylls a and c (c_1 and c_2) and various carotenoids, with fucoxanthin predominant.

Volvocida

The volvocidans (order Volvocida), including the commonly recognized colonial forms of *Volvox* spp, the solitary *Chlamydomonas* spp, and *Dunaliella* spp (Figure 2.1a), possess usually two or sometimes four naked flagella per cell. The plastids may be varied in shape, but usually are cup-shaped, with the open end oriented toward the flagellar pole of the cell and enclosing the nucleus, which is partially obscured by the peripheral walls of the plastid when observed through light microscopy. A large pyrenoid, surrounded by starch grains, is typically immersed in the basal part of the plastid. A red eyespot occurs in the anterior half of the plastid of most species and consists of one to several layers of carotenoid-containing lipid globules. Two contractile vacuoles are commonly found at the base of the flagella, although there are variations. The cell is naked or enclosed by a firm wall. The wall is composed of protein in *Chlamydomonas*. In some genera, the flagella emerge from an apical papilla as in *Polytomella* (Figure 2.1b). *Polytomella caeca*, bearing four flagella, is a colorless heterotroph with ovoid cells, pointed bluntly posteriorly. *Chlamydomonas* with two flagella, and *Carteria* with four flagella and a prominent wall are typical of the pigmented single-celled members of this group. Colonial forms including *Volvox* and those containing fewer cells (eg, *Pandorina* and *Gonium*) are among some of the most highly organized polycellular protozoa with specialized masses of cells for reproduction. Carbohydrate is stored as starch, either as a thick coat surrounding the pyrenoid or as scattered grains within the plastid. Chloroplasts are usually grass green and contain chlorophylls a and b, alpha- and beta-carotenes, and xanthophylls.

Prasinomonadida

The prasinomonads (order Prasinomonadida) are green flagellates, often bearing multiple flagella (four or eight), with small scales covering the cell body and flagellar surface. Some species secrete a cell wall composed of pectin (a gelatinous product) not cellulose. They are largely free-living, but some species form significant symbiotic associations with marine invertebrates including flatworms; ie, *Tetraselmis convolutae* (Parke and Manton 1967), and radiolaria (Anderson 1983). The genus *Tetraselmis* illustrates some of the typical features of motile monads in this group. Three phases in the life history encompass

(1) flagellated cells, usually with four flagella inserted in one row at the bottom of an anterior trough, covered with scales; (2) nonmotile vegetative stages; and (3) a thick-walled cyst. The cell is often covered by a rigid theca or two or more thecae and may be nonmotile. Chloroplasts are cup-shaped, with a prominent pyrenoid invaded by cytoplasmic protrusions. Some prasinomonads have a lobed chloroplast with one or more large pyrenoids, typically occupying a large part of the cell. Cells are bilaterally symmetrical, or with at least two planes of symmetry. Pigments include chlorophylls *a* and *b*, alpha- and beta-carotene, lutein, and other xanthophylls.

Silicoflagellida

The silicoflagellates (order Silicoflagellida) are the last group considered among the phytoflagellates. They possess a single flagellum located near one of the radial spines of the skeleton, and are enclosed by a distinctive external siliceous skeleton of unique organization resembling an inverted basket formed of fused hollow rods. One extant genus (*Dictyocha*) with two species is widely recognized, although fossil evidence suggests many more diverse species in ancient times (approximately 12 genera and 50 species). Extant species have numerous small discoid chloroplasts (1 to 2 μm diameter), enclosed in frothy masses of cytoplasm that form a peripheral sphere separate from the central mass containing the nucleus. Pigments include chlorophylls *a* and *c*, carotenes, and xanthophylls.

Summative Perspective for Phytoflagellates

A summary of major morphological features among the phytoflagellates is presented in Table 2.1. In general perspective, the lateral gullet of the Cryptomonadida, sulcal and annular grooves in the Dinoflagellida, and the flagellar reservoir of the Euglenida are similar cellular features associated with flagellar insertion. A much reduced flagellar reservoir associated with the thecal slit occurs in some Prasinomonads such as *Tetraselmis cordiformis*. Contractile vacuoles, (or pusules in dinoflagellates) when present in freshwater species occur near the flagellar bases and empty into the flagellar reservoir or depression (sulcus in the case of dinoflagellates) through a membrane covered pore. In other orders lacking a flagellar depression, the contractile vacuole is situated near the flagella but discharges at a site on the surface of the cell. Cup-shaped plastids containing chlorophylls *a* and *b* and enclosing the nucleus, totally or partially, occur in the Volvocida and Prasinomonadida. Although species in both orders have large pyrenoids at the base of the plastid, some prasinomonads also have smaller pyrenoids scattered throughout the plastid. A prominent eyespot is present in the anterior edge of the volvocidan plastid, whereas, in the prasinomonads, the eyespot may be at the periphery near the base of the plastid. The Chrysomonadida and some Prymnesiida possess two or more somewhat concave parietal plastids

TABLE 2.1. Summary descriptions of major orders of phytoflagellates.

Order	Flagella	Plastids	Cell Shape	Cell Surface
Cryptomonadida	Unequal (2) with hairs arising subapically in a ventral groove or gullet.	Boat-shaped, bilobed (H-shaped), parietal, green, brown, red, blue, or absent; pyrenoid varied.	Ovoid and flattened; barrel- to bean-shaped; gullet opens on ventral (flattened) side.	Smooth, or with rectangular or hexagonal pattern periplast plates.
Dinoflagellida	Heterodynamic (2) inserted apically or laterally, one ribbon-like in cingulum other extended or trailing.	Discoidal or elongate, few to many, golden brown, or absent.	Varied, bi-conical or with elaborate ornamentation; mid-line or spiral cingulum (groove) & shorter, longitudinal sulcus.	Smooth or reticulate composed of thecal plates.
Euglenida	Two (rarely more) often only 1 emergent from apical pouch.	Discoidal, ribbon-shaped, or plate-like, several to many; grass-green or absent.	Elongated, ovoid, or leaf-shaped; with spiral or bilateral symmetry; sometimes with cell flattening.	Spiral ridges, or smooth mineralized lorica.
Chrysomonadida	Two inserted apically or laterally; 1 with hairs directed anteriorly, other smooth and trailing.	Curved-shield shaped, two to several; yellow-brown to yellow-green, or colorless or lacking.	Spheroidal to ovoid or tear-drop shaped; some colonial in branched gelatinous sheath.	Naked, smooth, or covered with small siliceous scales.
Prymnesiida	Two equal or unequal, smooth inserted at apex or laterally with 3rd unique appendage (haptonema) between them.	Two prominent, saucer-shaped, parietal; golden brown.	Spherical, ovoid, or flattened, often curved like a saddle.	Calcareous or organic scales of elaborate design.
Volvocida	Two or 4, apical, smooth; sometimes attached to papilla.	One, varied in form but usually cup-shaped with single basal pyrenoid; grass-green.	Ovoid, pear-shaped, or elongate forming colonies within flattened or spheroidal gelatinous sheath.	Smooth with thin or thick organic wall; infrequently mineralized.
Prasinomonadida	One, 2, 4, or 8 (covered with rows of finely patterned scales) emerge sometimes from depression in surface.	One, cup-shaped or lobed, thin, but vividly pigmented green to olive; one or more large pyrenoids, one often basal, others peripheral.	Two or more planes of symmetry; flattened-ovoidal, reniform or chordate.	Fine scales in 1 to several layers.
Silicoflagellida	One emerges from tip of spine on stellate cell.	Many, small, discoidal golden- or greenish-brown.	Star-shaped with pseudopodia emerging from tips of spines.	Complex, silicified stellate framework.

that are often so closely spaced as to give the appearance of a cup-shaped plastid; however, upon closer inspection when the cell is properly oriented in the field of illumination, the space between the plastids can usually be discerned. In other cases, the plastids are sufficiently small and widely spaced around the perimeter of the cell to be seen clearly as separate saucer-shaped bodies. The typically numerous shieldlike plastids of many euglenoids as exhibited by *Euglena gracilis* are distinctive and usually exhibit a clearly discernable pyrenoid. Marked variations occur, as for example, in *E viridis*, with a complex of radially arranged plastids emanating from a paramylon-coated, centrally located pyrenoid. The surface structure of some Cryptomonadida, Dinoflagellida, and Euglenida commonly consists of elaborate, peripheral plate- or strip-like components. The periplast of the cryptomonads, with embedded rectangular or hexagonal plates, or the complex, sometimes interdigitating thecal plates of some dinoflagellates, and the interlocking pellicular ridges of euglenoids provide remarkably different forms of surface enclosing structures with similar organizational plans. Scales or loose surface coverings occur among the Chrysomonadida, Prymnesiida, and many Prasinomonadida. In comparison, a variety of organic enclosures (walls) or mineralized loricas surrounding the cell occur among the Euglenida, Chrysomonadida, Volvocida, and Prasinomonadida. Facultative heterotrophy of some pigmented members, especially in the Euglenida and Chrysomonadida, provides adaptive flexibility in exploiting diverse environments. Colorless species are often clearly related to pigmented forms by the presence of leukoplasts of similar form, though lacking photosynthetic light-trapping pigments. Substantial modifications of the gullet region may occur in some heterotrophic species; for example, among the Euglenida. *Heteronema acus* and *Peranema trichophorum* have two rodlike bodies formed of microtubules that can be thrust out of a pocket, near the opening of the flagellar pouch, to aid prey capture. By contrast, *Entosiphon sulcatum* possesses a tube ("siphon"), also composed of microtubules, that forms part of the feeding apparatus and opens near the pore of the flagellar pouch. To provide a more comprehensive view of the phytoflagellates, a taxonomic synopsis of major groups is presented in Table 2.1.

Zooflagellates

The obligate heterotrophic flagellates, without close pigmented relatives, are placed in the class Zoomastigophorea. These encompass a range of free-living species that obtain nutrition either by assimilation of soluble organic material by osmotrophy (absorption through the plasma membrane), or by ingestion of particulate matter by phagotrophy (engulfing prey in food vacuoles). Some parasitize humans and other animals, causing much hardship and suffering. Only some major representatives of the free-living, symbiotic, commensal forms, and a few of the animal parasitic forms are discussed here. The human parasitic species are treated in Chapter 9.

A considerable variability in form occurs among these species, related in part to their mode of nutrition and habitat. Some are fairly simple biflagellated monads either with smooth surfaces or pseudopodial projections (Figure 2.1e). Some transform into ameboid forms. Others are fairly reduced in size and may have one or two flagella (Figure 2.1f) or possess several flagella, one of which is anchored to the surface of the cell forming an undulating membrane (Figure 2.1j). Some major representatives are comparatively described.

Choanoflagellida

The choanoflagellates (order Choanoflagellida) are intriguing representatives of free-living zooflagellates that have developed unique structures to enhance food particle capture. The apical surface of the cell is fringed with a corona of fine cytoplasmic protrusions (microvilli) that are individually too thin to resolve with usual light optics, and thus appear as a conelike collar surrounding the solitary apical flagellum (Figure 2.1i). The typically ovoid or spherical cells are enclosed by a very thin filamentous shroud (invisible by light optics), or within a thickened wall forming a lorica, or surrounded by a basketlike framework composed of siliceous rods (costae or ribs). Species with siliceous loricae are either marine or found in brackish water. The loricate cells are frequently sessile, attached to the substratum by a thin basal extension of the stalked theca, or adhering directly to algae and other surfaces by the posterior surface of the theca. Both planktonic and sessile forms occur, and some species are colonially united by the basal filaments of the thecae into a branching or subspheroidal aggregate. The nucleus is typically centrally located and surrounded by small vacuoles and mitochondria. The form of the periplast (cell covering) is used in taxonomic categorization. The beating action of the flagellum brings a constant stream of water into contact with the fringe of microvilli that trap suitable food particles and direct them to the surface of the cell, where they are engulfed in food vacuoles. The choanoflagellates have been hypothesized to be possible phylogenetic precursors of more advanced forms of life, including the sponges which bear similar "collar" cells in their food trapping cavities. It is also possible that other lines of evolution arising from primitive choanoflagellates may have given rise to some of the metazoa. The presence of flagellated or ciliated cells in a variety of tissues of higher forms of life have given credence to this concept.

Cercomonadida

Cercomonas (Figure 2.1e) is a small ameboid flagellate (order Cercomonadida) with two flagella, one directed anteriorly and the other trailing. The cell body is pyriform to cylindrical, with an ameboid posterior surface whereby it engulfs small food particles. Thus, in comparison to the choanoflagellates that have developed a unique feeding mechanism coordinating flagellar-driven flow of particle-laden fluid across the fringe of microvilli, *Cercomonas* relies on a more simple,

Proteromonadida

A trailing flagellum is also characteristic of the order Proteromonadida. These parasitic flagellates commonly found in guts of amphibians and reptiles are pyriform, with two flagella (eg, *Proteromonas lacertae*; Figure 2.1f) or four flagella (*Karotomorpha* otherwise resembling *Proteromonas*). A single curved ("bean-shaped") mitochondrion originating anteriorly near the flagellar bases lies adjacent to the centrally located nucleus. Some form cysts.

Trichomonadida

The trichomonads (order Trichomonadida) are multiflagellated monads typically found in the alimentary tracts of their hosts. The type genus *Trichomonas* has four anterior free flagella (Figure 2.3b) and one attached posteriorly directed flagellum forming an undulating membrane in adults. *Tritrichomonas* (Figure 2.2j) has three anterior free flagella as indicated by the prefix "tri", and one undulating membrane formed by the attached posteriorly directed flagellum. The anteriorly located nucleus is associated with a very delicate curved body, the pelta, attached to a fine rod (axostyle) that extends through the long axis of the cell. The undulating membrane is underlaid by a supportive structure, the costa. In some genera (eg, *Metacoronympha*), multiple anterior flagella are arranged in a spiral containing about 150 flagellar units ("karyomastigotes"). Binucleated ameboid forms occur in the genus *Gigantomonas*, and *Dientamoeba fragilis* is permanently ameboid (eg, Figure 12.13d). Only the characteristic bilobed parabasal bodies near the nucleus, reminiscent of those in other flagella-bearing members of the order, clearly indicate its affinity with the Trichomonadida. Additional details of the human parasites in this group are presented in Chapter 9.

Diplomonadida

The diplomonads (order Diplomonadida) dwell typically in a host's alimentary tract, and some are serious human parasites. The free-living ones occur in organic rich waters. The cells are double in appearance (diplozoic), as implied by the name diplomonad. There are two symmetrically united karyomastigonts (Figure 2.1k), each possessing a nucleus and one to four flagella. The double nuclei, resembling two eyes, give a characteristic facial image to the cells when viewed with a light microscope. The free-living *Trepomonas* and *Hexamita* (Figure 3.3, 5 through 9) possess four flagella, but differ in the arrangement and disposition of the flagella. In *Trepomonas*, two oral grooves, one on each side of the body, originate near the base of the nuclei and extend posteriorly on the flattened cell surface. Three flagella lie within each groove, and the fourth is free and directed laterally. *Hexamita*, with a truncated spheroidal body, rounded anteriorly, flattened posteriorly, and somewhat compressed dorsoventrally, has only one flagellum within each tubelike flagellar pocket. The other three on each side are free.

Hypermastigida

The tendency toward multiple flagella, noted, for example, in the genus *Metacoronympha*, reaches a supreme level in the order Hypermastigida. These organisms (Figure 1.1b) living in the intestines of termites, cockroaches, and woodroaches have unique mastigont systems, each with multiple flagella anchored in branched or multiple basal bodies (kinetosomes). The flagella are distributed on the surface of the cell as complete or partial circles, as a plate or set of plates, or in longitudinal or spiral rows that meet in a specialized zone anteriorly known as the rostrum. There is typically one nucleus located medially or anteriorly. An exception, *Rhizonympha*, is a large plastic ameboidlike organism attached by pseudopods to the lining of the gut of the termite *Anacanthotermes ochraceus*. It contains 300 to 400 karyomastigonts within the ameboid body, each with its own nucleus and tuft of flagella. The hypermastigida are of interest not only because of their unique multiple flagellar systems, but also due to their interesting physiological adaptations for dwelling in the oxygen-deficient environment of the insect gut. Many species of wood-eating insects harboring the flagellates are fully dependent on them to break down the wood into nutritionally useful products.

Kinetoplastida

Members of the order Kinetoplastida (eg, Figure 2.2) encompass free-living and parasitic species. The free-living species are significant as bactivores in the natural environment, and are important in the early stages of decomposition and mineralization of organic matter in aquatic environments. Parasitic *Leishmania* and *Trypanosoma*, causing serious human diseases including kala-azar and Chagas' disease, are treated more fully in Chapter 9. The order receives its name due to the presence of a unique organelle (kinetoplast), a large mitochondrionlike body with a localized deposit of DNA in a swollen portion typically occurring near the flagellar bases. The kinetoplast, though sometimes extending as a tubular body within the long axis of the cell, is not easily visualized in some species without the aid of stains (eg, Feulgen, Janus green, or Romanovsky stains such as Giemsa). The presence of a kinetoplast is a defining taxonomic characteristic and flagellates assigned to this group must be shown to have kinetoplasts; otherwise they belong elsewhere.

Free-living representatives are exemplified here by two species: (1) *Bodo saltans*, an aquatic flagellate with a trailing flagellum used somewhat like a skate as it propels itself along surfaces; and (2) *Rhynchomonas nasuta*, a bacterivorus species with one trailing flagellum and a short anterior flagellum flanked on either side by lobes emanating from the point where the cytopharynx (food gathering canal) opens anteriorly. Bodonids typified by *Bodo saltans* (Figure 2.21) have one large kinetoplast, with the DNA-rich lobe near the flagellar bases and the tubular body extending along the dorsal side of the body. The ventral side bears a cytostome leading into a feeding canal (cytopharynx). The trailing flagellum along the ventral surface and the anteriorly directed motile flagellum arise from a flagellar pocket at the anterior of the cell, immediately above or lateral to a short rostrum (ridge) bearing the cytostome. *Bodo* species are commonly found in bacteria-rich fresh or brackish waters high in organics. *Rhynchomonas nasuta* (Figure 2.2, 6) is more truncated and bears a pair of characteristic lobes on the anterior proboscis, which move with the shorter flagellum as it vibrates and propels the organism. A cytopharynx receiving food particles (typically bacteria, selectively collected by the action of the lobes) opens near the flagellar pocket.

Cryptobia is a parasite in the gut or reproductive organs of invertebrates. One flagellum is directed anteriorly as with the bodonids, but the recurrent flagellum is attached to the cell surface. Digenetic cryptobias (alternating between two hosts) are freshwater parasites of a few marine teleosts and their leeches. The life cycle in the leech gut includes several stages. Each cell division produces changes in length and shape, eventually yielding small slender "metacyclic" forms that invade the proboscis sheath of the leech. They are transmitted to a new fish host when the leech feeds. Invasion of fish tissue can result in high mortality and has caused heavy losses among Pacific salmon.

Opalinata

The subphylum Opalinata is a group of clearly definable multiflagellated organisms exemplified by *Opalina* (Figures 1.1c, 2.3c), bearing numerous, closely spaced flagella covering the surface of the rather flattened cell. There is no oral groove or cytopharynx and ingestion occurs by modified pinocytosis (through micro-invaginations of the plasma membrane). Two to several nuclei are present and their number, and presence or absence in pairs, are used to distinguish among genera. Opalinids are all commensals in the alimentary tract of amphibians or occasionally in fish.

Summative Perspective for Zooflagellates

The zooflagellates, characterized by an obligate heterotrophic mode of nutrition, have developed specialized structures to enhance food gathering. These include: (1) a corona of microvilli with basally situated phagocytic sites on the cell membrane; (2) ameboid pseudopodia to engulf prey; and (3) specialized organelles for ingestion as exemplified by the cytopharynx of the bodonids. This reaches a rather elegant level, though on a small size dimension, in *Rhynchomonas* (comparative internal details, Figure 12.15). The coordinated action of its specialized nasal lobes, aiding the selection and engulfment of food particles directed into the cytopharynx by flagellar action, is a remarkable adaptation for a single-celled

Summative Perspective for Zooflagellates 33

TABLE 2.2. Summary descriptions of major orders of zooflagellates including opalinids.

Order	Flagella	Cell Inclusions	Cell Shape
Cercomonadida	Two heterodynamic, trailing.	Nucleus anteriorly located; drawn out into apical cone covered by cap of microtubules. Extrusomes often associated with tracks of microtubules.	Pyriform to cylindrical, ameboid posterior gathers food particles.
Trichomonadida	Four to 6 per mastigont system; 1 recurrent, associated with undulating membrane when present.	Nucleus anteriorly located, associated with curved pelta that gives rise to posteriorly directed axostyle. Fiber-like costa underlies undulating membrane.	Pyriform, ovate, or slightly lunulate. Some species entirely ameboid or sac-like.
Diplomonadida	One or 2 karyomastigonts each with 1–4 flagella.	Typically two nuclei arranged to each side of the midline of bilaterally symmetrical body. Axostyles present. No Golgi or mitochondria.	Cell with mirrored symmetry of nuclei and flagellar systems. Recurrent flagella enclosed within longitudinal grooves.
Hypermastigida	Flagella numerous arranged in mastigont systems either as patches or in spiral bands.	Typically, single nucleus. Peltoaxostylar lamellae arise dorso-laterally from kinetosomes and extend upward toward surface, then posteriorly.	Spheroidal, pyriform to broadly spindle-shaped. Those with spiral of flagella terminate in an apical cone-like rostrum. Some species with apical tufts of flagella.
Kinetoplastida	Typically 2, 1 trailing, and may be associated with an undulating membrane.	Single elongate, tubular mitochondrion associated with one or more kinetoplasts containing DNA. In some, dorsal fibrillar bands.	Single or colonial forms, cells reniform to spindle-shaped. Apical flagellar pouch.
Opalinida	Flagella numerous covering cell surface, resembling rows of cilia.	Nuclei 1 to many; ingestion by pinocytosis, no cytopharynx.	Flattened, leaf-like or somewhat cylindrical. Cell surface ridged with pellicle bands.

organism. Various forms of oral grooves or canals are observed in the Diplomonadida, often serving as guides for the posteriorly directed multiple flagella. A tendency to develop multiple flagella manifests itself in several orders, including the anteriorly directed multiple flagella of trichomonads, which reach a rather subclimactic proportion in the apical karyomastigonts of *Metacoronympha* compared to the elegantly organized mastigonts of the Hypermastigiida living in insect guts. Dependence upon external sources of nutrition have led the Zoomastigophorea to develop some unusual commensal and symbiotic relationships, including the interdependence of xylophagic (wood eating) insects and their flagellate symbionts. Parasitism both in humans and other animals, as encountered particularly among the Diplomonadida and Kinetoplastida, is also an adaptation to enhance survival through exploitation of and specialization in inhabiting unique nutrient-rich, intra-organismic environments. A summary of major features of some groups of zooflagellates is presented in Table 2.2.

3
Some Perspectives on the Habitats and Comparative Ecology of Flagellates

Organismic Environmental Interactions

The establishment and growth of a species in a habitat depends on a combination of biotic and physico-chemical environmental parameters. These include, among other abiotic factors, an adequate source of energy (carbon source or its production by autotrophy), essential nutrients, proper temperature, pH, tonicity, oxygen level, and redox potential. The stability of a community of organisms also depends on balanced growth among the component species and an equitable partitioning of resources. If one species becomes so dominant as to utilize much of the primary resources, the complex web of sustaining interactions may be seriously jeopardized, causing the total community to collapse. As an instance, an increasingly dominant member dependent on another for auxotrophic substances (a vitamin, for example), may compete too heavily for basic resources, leading to the attrition or exclusion of the vitamin producer. As a result, the total community may decline or completely collapse. Often, however, interactions are balanced by the competing needs of each member, so that eventually a relatively steady association develops. The many biotic and abiotic interrrelated variables accounting for population growth and development of protozoan communities are formalized in the principles of microbial ecology. Some of these fundamental principles, pertinent to a wide variety of protozoa, are introduced at this time with special reference to the flagellates. They are not necessarily limited to flagellate biology, but in many cases illustrations are easily made, and hence fundamental concepts to be used in subsequent discussions of protozoan ecology can be established early on.

Habitat

Flagellates occur in a wide variety of habitats, including soil and aquatic (freshwater and marine) environments. Soil-dwelling protozoa have been documented from almost every type of soil and in every kind of environment, from the peat-rich soil of bogs to the dry sands of the Sahara desert (eg, Bamforth 1973, 1985a,b; Stout 1973; Alexander 1977; Foissner, 1987). In general, flagellates are among the most dominant in absolute numbers, and species diversity of the soil-dwelling protozoa. They reach densities of 3000 to 200,000 per gram of soil. Among the many genera described, some of the more widely encounterd are *Allantion, Bodo, Cercobodo, Cercomonas, Entosiphon, Heteromita, Monas, Oikomonas, Sainouran, Spiromonas, Spongomonas,* and *Tetramitus* (the flagellate stage of an amoeba). The pigmented forms are best represented by *Euglena*. In general, protozoa are found in greatest abundance near the soil surface, especially in the upper 15 cm. They are infrequent in subsoils, but occasionally isolates can be obtained at depths of a meter or more. Protozoa do not constitute a major part of soil biomass, but in some highly productive regions such as forest litter and moist rich soils in cool climates the protozoan biomass may reach 20 g/m^2 of soil surface area. More commonly, the value is less than 5 g/m^2.

By far the most widely recognized, and more thoroughly studied flagellates from an ecological perspective, are freshwater and marine species. Some are stenohaline (sensitive to tonicity and requiring a narrow salinity range) others are euryhaline (tolerating wide variations in salinity). The coccolithophorids, for example, are exclusively marine planktonic, whereas *Dunaliella salina* is euryhaline, dwelling typically in brackish waters such as estuaries where large salinity fluctuations occur. Flagellates living as commensals or parasites within plants or animals often require specialized environments, as they have become adapted during their evolution to the limited conditions imposed by their host. Some, for example, can survive only in anaerobic and organically enriched environments such as the guts of insects or vertebrates, and some of these are described more fully in Chapter 9. Protozoa are commonly thought of as soil or aquatic dwelling organisms, but some terrestrial protozoa are parasitic on plants. Plant-infecting trypanosomatids, as exemplified by *Phytomonas* spp, are transmitted by insect vectors and are readily established in milkweeds (Asclepiadaceae) within several weeks after feeding by the infected vector (eg, Hanson et al 1966; McGhee and McGhee 1971). Other terrestrial flagellates are obligate autotrophs and require adequate illumination for growth. They are limited to surface growth in terrestrial environments, eg, chloromonads inhabiting snow surfaces (Hoham 1980), and those found in moist surface films of soil.

Aquatic autotrophic species are limited to the photic zone where light quality and intensity is within the photosynthetic compensation range of the organism. The photosynthetic compensation point is the light intensity level where photosynthesis is just sufficient to maintain respiratory metabolism. Photoautotrophs that are able to adjust their respiration to very low levels and/or make highly efficient use of available light energy have low light intensity compensation values. There is a wide variation in compensation points as exemplified for some dinoflagellates (Table 3.1) with a range of < 1 to 35 µEinsteins/m^2/sec. This illustrates the remarkable efficiency of some species (eg, *Amphidinium carterae*) to utilize even the dimmest light compared to surface illumination with a daily mean intensity of approximately 500 µE/m^2/sec. In the case of *A. carterae*, compensation is achieved by a very low respiration, thus conserving as much as possible of the small amount of photosynthates produced in the dysphotic environment.

TABLE 3.1. Compensation intensity and photosynthetic efficiency of some flagellates.

	Compensation Intensity	Photosynthetic Efficiency
Gonyaulax tamarensis	35	2.0
Gonyaulax polyedra	30	3.1
Olisthodiscus luteus	9	1.9
Pyrocystis fusiformis	10	1.2
Amphidinium carterae	2	2.1

Based on data from Dunstan (1973), Rivkin et al (1982, 1984), and Langdon (1987). Compensation intensity is expressed as µEinstein's/m^2/sec and photosynthetic efficiency as (µg C/Chl a^{-1} h^{-1}/µE m^{-2} sec^{-1}) × 10^{-2}.

Heterotrophic flagellates, although sometimes inhibited by light of a sufficient intensity, are less influenced by variations in light than the photoautotrophic forms. Thus, many heterotrophic flagellates are found in locations where organic carbon sources are abundant, independent of light intensity, although some grow best in low light intensities or in darkness. Growth of pigmented and colorless euglenoids (eg, pigmented *Euglena gracilis* var *bacillaris* and nonpigmented *Astasia longa*) is less in light than in darkness, although chlorophyll-containing *E gracilis* is less inhibited than bleached mutants (SM-L1) (Cook 1968). Some euglenoids (eg, *E sanguinea* and *E haematodes*) possess a red pigment that migrates to the surface of the cell in bright light, conferring a distinctively red color to the cell, but is withdrawn to the inner cytoplasm under lower illumination. The exposed peripheral chloroplasts become visible and the cell assumes a typical distinctive green hue. The red pigment may serve as a protective light-absorbing screen against excessive illumination, as also apparently occurs among the red pigmented snow algae exposed to intensive ultraviolet illumination in mountain peaks (Klausener 1908). Since much of the emphasis in this book is on the clearly zoological aspects of protozoa, the photosynthetic biology of these organisms is not considered in detail. Additional basic information on algal photosynthesis can be found in some standard sources (eg, Stewart 1974).

Many of the effects of physico-chemical variables within the environment can be understood in relation to the protozoan mode of nutrition. Therefore, a convenient classification diagram (Figure 3.1) prepared by Levandowsky and Hutner (1979) is introduced as a context for discussing flagellate ecology. Flagellates in group I are those inhabiting relatively pure, low organic, oligotrophic environments. They utilize mineral nutrients and light energy for carbon fixation and tend to be inhibited by low pH coupled with the presence of organic acids. At low pH, the organic acids such as acetate and short chain acids are less ionized, and these uncharged molecules more easily penetrate the cell plasma membrane. Increased cytoplasmic concentrations can reach levels that "poison" the cell. Group II flagellates are facultative autotrophs capable of photosynthesizing in the presence of light, or subsisting by heterotrophy in darkness. They can be osmotrophic, assimilating dissolved organics across the plasma membrane, or phagotrophic, engulfing particulate matter. Some of these species are known as acetate flagellates, as they can dwell in organic rich (eutrophic) and polluted environ-

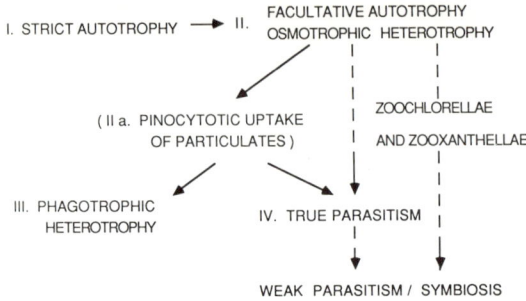

FIGURE 3.1. Categorization of protozoa according to nutritional mode. (I) Strict autotrophs require only mineral nutrients and utilize sunlight as an energy source to fix carbon dioxide in the biosynthesis of large carbohydrate molecules. (II) Facultative autotrophs, utilizing photosynthesis and osmotrophy, inhabit eutrophic and sometimes polluted water as exemplified by the "acetate flagellates." (IIa) An intermediate group tending toward obligate heterotrophy utilizes pinocytotic uptake of particulates and may have given rise phylogentically to (III) phagotrophic species (eg, *Peranema*) and (IV) parasitic species that have become specialized to utilize the unique nutrient sources within body fluids of a specific host. Other possible pathways leading to weak parasitism or symbiosis (dashed lines) may have arisen from the osmotrophic species (II). Incorporation of algal cells in protective vacuoles could have led to algal symbioses. From Levandowsky and Hutner (1979); with permission.

ments where oxygen concentrations and pH may be low due in part to high bacterial activity, and where strong reducing conditions exist. The acetate flagellates (eg, *Euglena*, *Polytomella*, *Chlamydomonas*, and others) are characterized by their ability to utilize acetate, short-chain organic acids, and sometimes alcohol and other organics as carbon sources at low pH values normally lethal to other forms. They are common in stagnant ponds and other shallow, polluted environments. Indeed, in a study of a series of connected ponds and their tributary stream, Cowles and Schwitalia (1923) found that euglenoids were most abundant only in those locales that were low in pH and organically enriched. Even these low-pH tolerant species have limits to tolerance to organic acids, and therefore are excluded from highly eutrophic environments of low pH.

Some acetate flagellates also tend toward facultative anaerobiosis. Others, as exemplified by *Ochromonas malhamensis*, are only weakly photosynthetic and depend to a large extent on external sources of carbon. Species that exhibit autotrophic and heterotrophic modes of nutrition are known as myxotrophic organisms. Group IIa is included in between group II and III, as it represents intermediate forms of flagellates assimilating dissolved organics or colloidal-sized organic particles by pinocytosis. This is a form of "drawing-in" dissolved substances by engulfing them in small vesicles formed at the base of fine conical depressions in the surface of the cell. Other flagellates (group III) are completely heterotrophic. Some obtain nourishment by pinocytosis and/or phagocytosis. These organisms have become evolutionarily fixed as heterotrophs and some have become specialized to engulf whole cells as sources of nutrition. They depend on ingested organisms not only for growth promoting substances (auxotrophy), but also as a source of requisite major organic compounds such as amino acids that they may be unable to synthesize from simpler compounds. These organic substances are often required in the proper proportions characteristic of their particular prey. This may explain their obligate nutritional dependence on one prey species. Some of these live in relatively clean environments and prey on other protozoa or algae. A culminating level of dependence is reached in group IV, including those that are parasites or symbionts. They have become so adapted to a specialized environment within a host that they usually cannot live without the particular physico-chemical environmental factors provided by the host. Flagellates, especially some dinoflagellates, prasinomonads, chrysomonads, and prymesiomonads are commonly observed symbionts in invertebrates.

Physico-chemical Environmental Factors

Metabolism is influenced markedly by temperature, and some protozoa are limited to a rather narrow range of temperatures for growth and reproduction. At the upper extreme, thermophilic species inhabiting hot springs are subjected to temperatures in excess of 40°C. Their thermo-tolerant physiology and relatively heat stable proteins permit them to maintain cellular integrity and metabolize substrates at elevated temperatures where most organisms die. Tolerance of elevated temperatures approaching the upper limit appears to be increased if sufficient nutrients are available (Hutner et al 1957). Death in some species grown at elevated temperatures is due to starvation induced by rapid growth in nutrient-limited environments. A comparative survey of upper temperature limits for protozoa in a variety of environments is presented in Table 3.2. This includes one flagellate, *Chilomonas paramecium*, that incidentally is also not inhibited by light (Cook 1968).

Cryophilic or cryotolerant species dwell at low temperatures in snow and ice packs, for example, where temperatures typically approach or fall below the freezing point of water. Snow flagellates (including chlamydomonads, euglenoids, chrysophytes, dinoflagellates, and cryptomonads) usually develop blooms when the temperature increases above freezing and melt water forms. Some of these species are limited to growth in a narrow temperature range 1 to 10°C and form resting or environmentally resistant stages (eg, palmella or cysts) when the temperature is below or above their optimal range (Hoham 1980). Mosser et al (1977) reported that *Chlamydomonas nivalis* photosynthesized optimally at 10 or 20°C, but remained photosynthetically active at temperatures as low as 0 to −3°C. *Dunaliella* spp, although not cryophilic organisms, are able to withstand subfreezing temperatures due to the presence of glycerol, which acts as an antifreeze agent. This prevents ice crystal formation that disrupts the cytoplasm of other protista during freezing and thawing.

40   3. Habitats and Comparative Ecology

TABLE 3.2. Some representative temperatures in the natural environment where flagellates occur.*

Organism	Temperature (°C)
Mallomonas (alpina?)	6.5
Trachelomonas volvocina	6.5
Dinobryon divergens	7.7
Dinobryon bavaricum	8.8
Gymnodinium helveticum	9.3
Peridinium willei	10.5
Rhodomonas lacustris	10.6
Trachelomonas hispida	10.7
Cryptomonas erosa	12.4
Dinobryon sociale	13.2
Ceratium hirundinella	14.4
Mallomonas (acarioides?)	17.4
Chilomonas paramecium[†]	36.0

*Based on data from Findenegg (1943) and Nisbet (1984).
[†] Thermal spring species.

The pH of the environment imposes a limitation on growth of some species, not only for the direct effects of the amount of hydrogen or hydroxyl ions in the environment, but also due to changes in the ionic composition of molecules and the solubility of inorganic elements that may be essential to their physiology. For example, in environments rich in short-chain organic acids (eg, acetate, propionate, butyrate, etc), the pH determines the proportion of nonionized to ionized molecules (Table 3.3). As the pH decreases below pH 7.0, the proportion of nonionized molecules increases. The proportion of ionized to nonionized organic acid molecules as related to pH is given by the following relationship:

$$\text{pH} = \text{pK} + \log \frac{[\text{R COO}^-]}{[\text{R COOH}]}. \qquad 3.1$$

The pK, a constant for a given compound, is equivalent to the pH of the solution containing equal proportions of ionized and nonionized forms of the organic acid. As an operational definition of pK, consider an acid with pK = 4.6. If the solution containing the acid is adjusted to a pH of 4.6 by adding hydroxyl or hydrogen ions, then there will be equal amounts of the ionized and nonionized form of the acid in the solution. It is important to assess the amount of nonionized organic acids, since the uncharged small organic molecules diffuse more readily through the plasma membrane of some cells than do the charged forms. This is due in part to the low permeability of the lipid layer of the membrane to charged (polar) molecules. The acetate flagellates cited above are able to subsist in low pH environments rich in organic acids, as they have plasma membranes that are relatively impermeable to the organic acids and use membrane transport mechanisms to regulate uptake. Specialized molecules within the plasma membrane bind and translocate the acid molecules into the cytoplasm. Hence, only sufficient amounts of the organics are accumulated as needed for metabolism. Through this

TABLE 3.3. Ratio of ionized to nonionized forms of organic acids and ammonia as a function of pH.

pH	[COO$^-$]/[COOH]	[NH$_3$]/[NH$_4^+$]
10	1.75×10^5	5.70
9	1.75×10^4	5.70×10^{-1}
8	1.75×10^3	5.70×10^{-2}
7	1.75×10^2	5.70×10^{-3}
6	1.75×10^1	5.70×10^{-4}
5	1.75	5.70×10^{-5}
4	1.75×10^{-1}	5.70×10^{-6}
3	1.75×10^{-2}	5.70×10^{-7}
2	1.75×10^{-3}	5.70×10^{-8}
1	1.75×10^{-4}	5.70×10^{-9}

specialized adaptation, the acetate flagellates are able to utilize the abundant but potentially lethal organic acids as a source of carbon for energy production and biosynthesis of organic compounds. Polluted aquatic systems, rich in bacteria, are especially likely to have low pH and abundant organics owing to the release of organic acids and other metabolic end products by the bacteria.

Two major sources of nitrogen for flagellates in the environment are nitrate (NO_3^-) and ammonia (NH_3). Nitrate is accumulated especially by photoautotrophic species through an active transport process whereby specialized membrane molecules bind to and translocate the nitrate across the membrane. Within the cell, the nitrogen can be assimilated only in a reduced form. Nitrate is reduced by a succession of steps probably corresponding to the following sequence:

+5	+3	+1	−1	−3
NO_3^-	NO_2^-	$N_2O_2^{2-}$	NH_2OH	NH_4^+ .

A more detailed treatment of the biochemistry of this process is presented by Morris (1974). There appear to be only two enzymes involved in the entire sequence: nitrate reductase, which catalyzes the reduction of nitrate to nitrite, and nitrite reductase, which catalyzes the reduction of nitrite to ammonium. The ammonium can be utilized in production of amino acids, nucleic acids, and other nitrogen-containing organic molecules. In general, when both nitrate and ammonium are present, ammonium is preferentially assimilated. Usually, colorless flagellates cannot utilize nitrate, with the exception of *Polytoma ocellatum* and some euglenoids including *Euglena gracilis* group, *Trachelomonas abrupta*, *T pertyi*, and *Phacus pyrum* (Provasoli 1958).

It has been known for some time that flagellates accumulate ammonium nitrogen directly from the environment, and more recently it appears that the most likely form for assimilation is NH_3, by the following pathway:

glutamate + NH_3 + ATP → glutamine + ADP + Pi.

Too much ammonia in the growth medium, however, is lethal for most flagellates, as this nonionized form (NH_3) diffuses more readily through the plasma membrane producing poisonous intracellular concentrations. Indeed, above pH 7 presence of ammonia in the growth medium (Table 3.3) usually causes cell lysis (Provasoli 1958). Marine and brackish water flagellates are especially sensitive to ammonium added to alkaline medium. It is unlikely, however, that ammonium reaches these lethal concentrations in the natural environment except in habitats polluted by animal wastes. These environments are likely also to be low in pH favoring the presence of ammonium rather than ammonia.

Redox Potential

The redox potential of the environment is determined in part by the amount of oxygen, and more particularly by the concentration of other oxidizing or reducing compounds. It is a significant environmental parameter correlated with the presence or absence of some protozoan species. Clearly, in a reducing environment, where, for example, oxygen concentrations are low and reducing compounds such as H_2S are abundant, only anaerobic tolerant organisms are likely to be present. Moreover, the redox potential of the environment influences the availability and oxidation state of some mineral nutrients (eg, iron) required by many protozoa, especially flagellates. The redox potential is a measure of the extent to which reducing conditions are present as measured by the electron donating capacity of the environment. It is measured in volts and decreases with increasing proportions of reducing compounds relative to oxidizing compounds. For example, in environments rich in sulfides (S^{2-}), hydrogen, or other electron donating groups, the redox potential is likely to be low, indicating absence of oxidizing compounds. The presence or absence of oxygen alone produces only minor variations in the redox potential within the range of environmentally significant, detectable values. For example, decreasing the partial pressure of oxygen from 100% saturation to 1% saturation lowers the potential only about 0.03 volts from an initial value of about 0.80 at 25°C. The redox potential (E_h) is expressed mathematically in relation to the proportional concentration of oxidized to reduced forms of a compound as follows:

$$E_h = E_o + \frac{RT}{F} \cdot \ln \frac{[\text{oxid}]}{[\text{red}]} , \qquad 3.2$$

where T is absolute temperature, R is the universal gas constant, and F is Faraday's constant. E_o is the standard redox potential and is equivalent to the E_h, where there are equal amounts of oxidized and reduced forms of a compound. Since the last term is a logarithmic expression, and has negative values when the ratio of oxidized to reduced substance is a fraction (less than 1 but greater than 0), E_h will decrease as the proportion of oxidized to reduced substances decreases. As the redox potential is influenced by pH, it is a common practice to express measurements in the biosphere as adjusted to pH = 7, designated E_7 ($E_7 = E_h$ at pH 7). More detailed information on the concept of redox potential and biosphere measurements can be found in standard textbooks of aquatic biology (eg, Hutchinson 1957). Well-mixed, or sufficiently oxygenated, clear water ecosystems may have little variation in E_h with depth of the water column. A representative value for a pond in midsummer stagnation is $E_h = $ ca 0.5. In general, if a body of water is not well-mixed, and biological activity is high throughout the water column, redox potential will decrease with depth, especially if bacteria releasing organic wastes and H_2S are abundant. As an example, in a poorly mixed lake, the redox potential declines from $E_h = 0.140$ volt at 8m depth to $E_h = 0.075$ volt at 20.5m. The H_2S content concomitantly rises from 1.2 to 2.1 mg per liter. The decrease in E_h with depth is especially marked in organic-rich sediments (Figure 3.2) and in general reaches a minimum at ca 5 cm. Below 5 cm the redox potential may increase slightly, indicating less reducing compounds and lower biological anaerobic decomposition. The magnitude of the

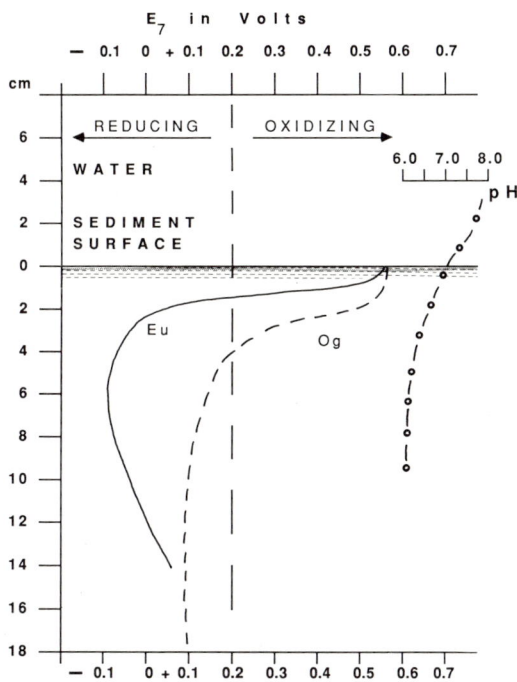

FIGURE 3.2. Gradient in redox potential, expressed as E_7 units in volts, within the sediments of an eutrophic (Eu) and oligotrophic (Og) lake. The pH gradient in the surface water and underlying sediment of an eutrophic marine pool is also plotted. The redox potential in the sediment of an eutrophic lake (Eu) decreased markedly in the first 2 cm depth and reaches a minimum at about 5 cm. The decline is less marked in an oligotrophic freshwater lake (Og) and also has a minimum at about 5 cm, but remains fairly constant near this value with depth. As a general rule, E_7 values greater than 0.2 V are classified as oxidizing and those less than 0.2 V are reducing. The pH of an eutrophic marine pool declines steadily in the surface water from near 8.0 at 4 cm to ca 7.0 at the sediment interface. In the sediment, the pH declines further approaching pH 6.0 with depth. Redox data from Ruttner (1953), and pH data from Fenchel (1969) and Rheinheimer (1985).

change with sediment depth varies for different bodies of water, as illustrated in Figure 3.2. Thus, sediment dwelling colorless flagellates and other protozoa are likely to encounter increased reducing conditions when penetrating several millimeters into the surface.

The solubility of various forms of iron as a function of pH and E_h is presented in Table 3.4. When pH and E_h are low, soluble ferrous iron increases substantially. The critical threshold for conversion of ferric to ferrous iron is in the range of $E_7 = 0.30$ to 0.20. An E_h of 0.20 is generally accepted as a critical value where iron becomes available due to solubilization from hydroxides. Within a redox gradient, the threshold levels of E_7 for oxidation or reduction of other biologically important compounds has been determined: NO_3-NO_2 (0.45 to 0.40 volt); NO_2-NH_4^+ (0.40 to 0.35); and SO_4-S (0.10 to 0.06). For clarity, consider the example of nitrate reduction. When the redox potential falls below 0.45 to 0.40, the nitrate ions increasingly tend to become reduced toward nitrite. A similar explanation can be given for the other compounds in the series. In organically enriched waters, the presence of chelating agents (particulates capable of binding iron ions) may enhance the amount of available iron, even at normally unfavorable E_h. The surface adsorbed ions are biologically available. Some flagellates such as various euglenoids and cryptomonads favor "polluted" acid waters when Fe and the C:N ratio are high (Provasoli 1958).

A variety of classification schemes have been devised to indicate the degree of organic enrichment (saproby) and concomitant oxygen depletion typically found in "polluted" eutrophic environments. One of the more widely used is the "Saprobien System" devised by Kolkwitz and Marsson (1908). This system categorizes water masses according to physico-chemical conditions and the kind of organisms dwelling in it. The most highly polluted regions, called the *polysaprobic* zone, are characterized by high organic content and bacterial activity

TABLE 3.4. Solubility of various forms of iron (all computations in mg–atom per m³).

	pH 8.5	pH 8.0	pH 7.5	pH 7.0	pH 6.5	pH 6.0
[a]Fe+++	$10^{-15.4}$*	$10^{-13.9}$	$10^{-12.4}$	$10^{-10.9}$	$10^{-9.4}$	$10^{-7.9}$
[a]FeOH++	$10^{-9.3}$	$10^{-8.3}$	$10^{-7.3}$	$10^{-6.3}$	$10^{-5.3}$	$10^{-4.3}$
[a]$H_2FeO_3^-$	$10^{-4.8}$	$10^{-5.3}$	$10^{-5.8}$	$10^{-6.3}$	$10^{-6.8}$	$10^{-7.3}$
[a]Fe++						
at $E_h = 0.8$	$10^{-16.0}$	$10^{-14.5}$	$10^{-13.0}$	$10^{-11.5}$	$10^{-10.0}$	$10^{-8.5}$
0.6	$10^{-12.5}$	$10^{-11.0}$	$10^{-9.5}$	$10^{-8.0}$	$10^{-6.5}$	$10^{-5.0}$
0.4	$10^{-9.0}$	$10^{-7.5}$	$10^{-6.0}$	$10^{-4.5}$	$10^{-3.0}$	$10^{-1.5}$
0.3	$10^{-7.25}$	$10^{-5.75}$	$10^{-4.25}$	$10^{-2.75}$	$10^{-1.25}$	2
0.25	$10^{-6.4}$	$10^{-4.9}$	$10^{-3.4}$	$10^{-1.9}$	**0.4**[†]	13
0.2	$10^{-5.5}$	$10^{-4.0}$	$10^{-2.5}$	**0.1**	**3**	**100**
0.0	10^{-2}	**0.3**	**10**	**300**	**10,000**	**300,000**

*Quantities less than 0.1 mg-atom per m³ or 5.6 mg per m³ are not likely to be analytically detectable and are entered in the table as negative powers of 10. Note that the solubility increases markedly at pH below 7.0 and at lower E_h values.
[†]The analytically detectable amounts of ferrous iron are given to one significant figure in boldface type.
From Hutchinson (1957).

with low oxygen concentrations, or anoxia. Anaerobic and low pH tolerant protozoa are most abundant. The *mesosaprobic* zone of intermediate quality is divided into two categories: *alpha-mesosaprobic* with moderate dissolved oxygen and bacterial activity, and *beta-mesosaprobic* where dissolved oxygen concentration is usually greater than 50% and bacterial activity is diminished due to less organic matter. *Oligosaprobic* zones are unpolluted, rich in dissolved oxygen, typically clear, and often supporting diverse autotrophs in moderate, but not high numbers. Bacterial activity and biological oxygen demand is low.

Biotic Factors

Productivity

The amount of carbon fixed by organisms, as assessed by radioactive carbon uptake, for example, is a direct measure of the primary productivity often utilized in ecological research. In Table 3.1, the maximum rate of CO_2 carbon fixed into organic compounds by some dinoflagellates is shown as µg C/µg chlorophyll/hr. For this group it varies from 7.71 (*Amphidinium carterae*) to 3.1 (*Pyrocystis noctiluca*). In general, productivity can also be assessed by the increase in number of individuals per unit of time. This is particularly useful for nonphotosynthetic organisms, as the total increase in carbon, associated with an increase in population, can be calculated based on the mean amount of carbon per individual. Cells undergoing binary fission produce two daughter cells at each division, and thus the total population density increases exponentially as a power of two. For example, the number of individuals produced at each successive day by binary fission of a single cell doubling every 24 hr, would be: day 1 = 2, day 2 = 4, day 3 = 8, etc. Thus, the increase from a single cell is 2^n where n is the number of days. This exponential function can be expressed linearly by using logarithms as in the following formula for the "specific growth coefficient" (μ) or sometimes (r).

$$\mu = \frac{\ln N_t - \ln N_o}{t_n - t_o}, \qquad 3.3$$

where ($t_n - t_o$) is the length of time, N_o is the number of cells at the beginning of the time period, and N_t is the number of cells at the end of the time interval. If logarithms to the base 10 are used instead of natural logarithms, then μ is obtained by multiplying by 2.3. For example, suppose we were to assess the specific growth coefficient for a hypothetical population of flagellates using natural logarithms in the formula. Assume that on day 0 (t_o) the number of cells was $N_o = 20$ and on day 4 the number of cells was $N_t = 5,120$. The time interval is 4 days. We take the natural log of 5,120 ($ln\ 5,120$ = ca 8.5409) and subtract the natural log of 20 ($ln\ 20$ = ca 2.9957) = 5.5452. This is divided by 4 to obtain $\mu = 1.3863$. The specific growth coefficient expressed as natural logarithms can be converted to doublings per unit time by dividing μ by 0.693. Thus, 1.3863/0.693 = 2.00. In this simplified example, the cells were doubling twice

3. Habitats and Comparative Ecology

in 24 hr or doubling once every 12 hr. Note that it is necessary to carry the computations out to the fourth decimal place to obtain adequate accuracy. Another way of expressing μ is to divide ln 2.0 by the time required for doubling of the population. The merit of using μ to express growth is that it is a linear function and therefore the growth rates of two or more species can be compared proportionately. Thus, for example, a species with μ = 1.5 has a growth rate that is twice that of a species with μ = 0.75. As explained above, the specific growth rate can also be directly converted to doublings/unit time, which is a convenient and easily understood expression of cell growth rate. The specific growth rate for some flagellates categorized by nutritional habit (autotrophic or heterotrophic) is presented in Table 3.5.

In addition to the productivity of a species, the total biomass of a population (single species) or of a community (group of species interacting within a specified

TABLE 3.5. Growth of flagellates categorized by nutritional habit.

Species	Medium	Natural Growth Units	Doubling Time (Days)
Heterotrophic Growth			
Peranema trichophorum	Organics	0.07	10.00
Haematococcus pluvialis	Acetate (0.5%)	0.34	2.0
Euglena gracilis	Acetate (0.5%)	0.58	1.20
Euglena gracilis var *bacillaris*	Acetate (0.5%)	0.92	0.75
Chlamydomonas reinhardtii	Acetate (0.2%)	0.92	0.75
Gyrodinium cohnii	Glucose (0.6%)	1.13	0.60
G cohnii	Acetate (0.2%)	1.36	0.51
Chilomonas paramecium	Acetate (0.2%)	1.61	0.43
Astasia longa	Acetate (0.03%)	1.66	0.42
Paraphysomonas imperforata	Bacteria	1.70	0.40
Actinomonas mirabilis	Bacteria	6.00	0.11
Autotrophic Growth			
Chlamydomonas pulsatilla	CO_2, inorganic	0.27	2.60
Haematococcus pluvialis	CO_2, acetate (0.05%)	0.34	2.0
Euglena gracilis	CO_2, inorganic	0.58	1.20
Emiliana huxleyi	CO_2, inorganic	0.63	1.10
Amphidinium carterae	CO_2, inorganic	0.87	0.80
Dunaliella tertiolecta	CO_2, inorganic	0.87	0.80
Ochromonas danica	CO_2, glucose (0.5%)	0.91	0.76
Chlamydomonas reinhardtii	CO_2, acetate (0.2%)	0.92	0.75
Euglena gracilis var *bacillaris*	CO_2, inorganic	0.97	0.72
Pavlova (Monochrysis) lutheri	CO_2, inorganic	1.16	0.60
Chlamydomonas pulsatillia	CO_2, acetate (0.1%)	1.21	0.57
Chlamydomonas mundana	CO_2, inorganic	2.01	0.34
Chroomonas sp	CO_2, inorganic	2.31	0.30
Chlamydomonas mundana	CO_2, acetate (0.34%)	3.30	0.21

From Droop (1974), Fenchel (1982), and Davis and Sieburth (1984). Note that some species in the autotrophic group are listed more than once depending on the source of nutrients. Natural growth units symbolized by μ or r are calculated using Equation 3.3.

region of the environment) is a useful indication of the accumulated biological activity. Biomass is typically expressed as the total dry weight of the individuals in a population or community. This can be assessed by taking several random and representative samples of the biota and determining a mean dry weight which is extrapolated to the total biota by proportional computation. Alternatively, if the mean dry weight per organism is known for each species, the total biomass can be obtained indirectly by multiplying the number of individuals by the mean dry weight per individual. In an efficiently operating ecosystem with vigorous biota, biomass should increase with increasing productivity.

Diversity

Organisms dwelling together in a habitat exhibit variations both in the number of individuals within each species and in the number of different species. The greater the number of different species, and the more equivalent the number of individuals (density) within each species, the greater the diversity. Thus, a community consisting of 10 different species should be more diverse than one with only two species. Moreover, if the population density of each of the species is about the same, so that no one or few species tend to dominate the community, the diversity should also be greater. This intuitive concept can be expressed quantitatively by a diversity index symbolized either by H or D.

$$H = - \Sigma \, p_i \log_2 p_i, \qquad 3.4$$

where p_i is the proportion of individuals of species i to the total number of individuals. The product of $p_i \times \log_2 p_i$ is summed for all species and multiplied by -1. It is usually more convenient to use $\log_{10}$. The $\log_{10}$ is converted to the equivalent value of $\log_2$ by multiplying by 3.3219. The new expression is $H = - \Sigma \, p_i (3.219) \log_{10} p_i$.

This function has a maximum value when all proportions of species (p_i) are equivalent. It decreases as any one or a few species tend to be dominant. Consider, for example, a simple collection of three species with population numbers of $A = 200$, $B = 200$, and $C = 200$. H has a value of 1.58. If, however, the population numbers were $A = 100$, $B = 300$, and $C = 200$, maintaining the same total number (600), then $H = 1.41$. Although the total number of individuals has not changed, the tendency for one or more species to be more dominant than the others produces a decrease in the value of the diversity coefficient. This coefficient is a useful index of species diversity, but we must also consider other factors in making interpretations. The numerical diversity of a community may not adequately reflect the ecological role of its members. For example, a community with nearly equal proportions of all species, but where one is a heavy consumer of nutrients or produces a substantial amount of growth stimulating substances, is a very different community from another with identical diversity index, but where biological characteristics are in better balance. In other words, we should not place complete reliance on the numerical value of H without also considering other biological factors contributed by each of the member species in a community. Other distorting effects can also arise when we attempt to correlate H with other ecological parameters. Since a shift of species composition toward one or more dominants can cause a marked change in the value of H, and given the varied possible roles of the individuals in the community, it is sometimes meaningless to correlate H values with other parameters. As with any statistic, a careful conceptual analysis of the ecological problem should be made first to determine how the index of diversity can be legitimately used in making further quantitative assessments.

Other conceptual and interpretive problems in employing the H equation have been discussed by Hurlbert (1971), who points out that there is little direct correspondence between information theory underlying the H equation and ecological interactions studied by the biologist. He proposes different measures. One of these is especially straightforward and represents the probability of interspecific encounter (*PIE*); ie, the proportion of potential encounters among individuals that are *inter*specific as opposed to *intra*specific. This expression is a more direct measure of interactions of individuals in a community than the H equation, and therefore is likely to have greater potential for making quantitative relationships with other ecological variables. The probability of interspecific encounters (P) is assessed as follows:

$$P = \left[\frac{N}{N-1} \right] \left[1 - \sum_i^s \left(\frac{N_i}{N} \right)^2 \right], \qquad 3.5$$

where N_i = number of individuals of the *ith* species in the community (or collection), N = total number of individuals in the community, and s = number of species. High values of the P equation indicate highly diverse interactions among members of a community, with likelihood that an encounter with a member of a given species being followed by a second encounter of the same species is low. In other words, once one has encountered a member of a given species, the next encounter is likely to be with yet a different species. With a low P value, however, the very opposite is true. Most frequent encounters are likely to be with the same species. Hence, there is less diversity.

Within a given environment, the potential for a species to contribute to community diversity in different domains is determined by a number of biological factors. Clearly, protozoa that are able to adapt to widely diverse environments, and do not multiply too rapidly compared to other members of a community, are likely to contribute substantially to species diversity in widely varying sectors of the biosphere. Consider, for example, the water column of a lake. A flagellate that can invade widely different regions of an environment (eg, from the surface throughout the water column into the surface layers of the sediment) is likely to contribute to species diversity in communities within each of these locales. By proliferating in balance with other species, the diversity of species is increased by its presence. This is especially the case if the flagellate also has a specific growth rate that is close to that of other members in the community, thus contributing to equivalent population densities among the species. As a first approximation, those protozoa with moderate to low specific growth coefficients, facultative heterotrophic nutrition, and tolerance for widely different values of oxygen concentration and E_h are likely to contribute to species diversity within communities of widely diverse environments. By contrast, a rapidly dividing, strictly autotrophic organism is restricted to environments of particular light intensity and quality. Due to their rapid growth, these species may dominate the communities in their specialized habitats. Hence, such a species would not contribute to species diversity among widely different communities. To place these concepts in clearer perspective, the flagellates presented in Table 3.5 have been categorized by mode of nutrition and listed within each category in rank order of increasing μ. In general, those near the top of the list would be expected to have potential as contributors to species diversity among widely different habitats, whereas those near the bottom are likely to dominate a particular habitat. This admittedly simplified classification omits other characteristics of importance that must be considered in making a judgement about the probable contribution of a species to community diversity. For example, the kind of prey or amount of organic nutrients required by a facultative heterotroph may limit its growth in a particular environment, even though it normally has a moderate doubling rate comparable to other species within the habitat. Consider, for example, a species that has a moderately comparable growth rate to other species growing in its absence, but produces a metabolic product that stimulates the growth of one or more of the species. The presence of the stimulatory species may indirectly decrease diversity by stimulating excessive growth of one or more species benefiting from the auxotrophic effect. In general, as a community of organisms matures, species diversity and compositional stability tend to increase. This is a result of more efficient use of resources, reduction in competition by filling of previously unoccupied niches, and balanced growth of the member species in the community.

Succession and Stability

Following initial colonization of a habitat by protozoa, a progression of species assemblages usually occurs, forming what is known as a succession. In general, the pioneers are those species most able to utilize the existing resources. In an illuminated, oligosaprobic environment (with sufficient mineral nutrients), autotrophic species are most likely to be pioneers; whereas in a polysaprobic environment, predators on bacteria and/or fungi are most likely to initiate a protozoan succession. After the initial pioneers have become sufficiently established, predators or more competitive species may appear, forming a second stage to the succession and by their biological activity (grazing on primary producers or competing for primary carbon sources) decreasing the population density of the pioneer species. Consequently, a new metastable assemblage of protozoa is established at each of the stages in the succession. Within each of these progressive stages, an equilibrium can develop between production and consumption, leading to a stable community. At equilibrium, the consumers graze the primary producers at a rate that is balanced in relation to the growth of the producers, establishing a steady flow of carbon and energy through the network of trophic interactions. Clearly, if the consumers overgraze the primary producers, thus reducing their own food supply below a critical level, the stability of the community can be upset, leading to attrition in population density of both producers and consumers. If there are sufficient energy resources and a favorable range of physico-chemical parameters, this may lead to a cyclical pattern of growth and attrition in the community. Each wave of increase in producers is followed by a concomitant rise in consumers to a peak level, followed by overgrazing of the producers leading to their decline and a consequent decrease in the number of consumers.

During a classical example of succession, a sequence of communities develops over time, each stage becoming more stable and increasingly richer in species diversity as the succession proceeds forward in time. In fairly stable environments with suitable moisture, appropriate temperature range, and an adequate source of carbon and other nutrients, such a predictable progression is frequently observed. Not all protozoan communities, however, conform to this classical pattern. This may be partially attributed to rapid, often deleterious environmental changes (as occur in soils that dry out or flood periodically), causing encystment or death of the species in the community. Thus, the succession is not always a smooth progression as implied by a classical macrobiotic example. In many cases, succession leads to the displacement of one species by another over time, forming a continuously changing community of organisms rather than a gradual progression of intergrading groups of organisms. In habitats favoring heterotrophic protozoa, as with organic-rich eutrophic environments, bacteria and fungi appear first followed by heterotrophic or facultatively heterotrophic flagellates, sometimes coexisting with amoebae, and culminating usually in a ciliate-dominated community. Woodruff (1912) was among the first to use laboratory simulations to document such a succession in the decomposition of a hay infusion.

This type of succession, including a major role of flagellates, has been described by Sladecek (1972) in four stages: (1) "Ultrasaprobie," an abiotic initial step; (2) "Hypersaprobie," a bacterial and fungal stage; (3) "Metasaprobie," a flagellate dominant stage; and (4) "Isosaprobie," a ciliate dominant stage. It is important to recognize that amoebae often occur in association with flagellates in early stages of a succession, and these amoebae should not be overlooked in making a full account of the community. Four metasaprobic communities of colorless flagellates, leading into the ciliate-dominated isosaprobic stage, have been described by Sladecek (1972) as summarized in Figure 3.3. The first phase is dominated by the colorless volvocidan *Polytoma uvella* and other species of this genus. This is a pioneer association of many small flagellates such as is commonly found in the initial anaerobic stages of sewage treatment processes. The second phase is dominated by diplomonads represented by *Hexamitus* spp (Figure 3.3, 5 to 9) chiefly *H inflatus*. In laboratory cultures, this association

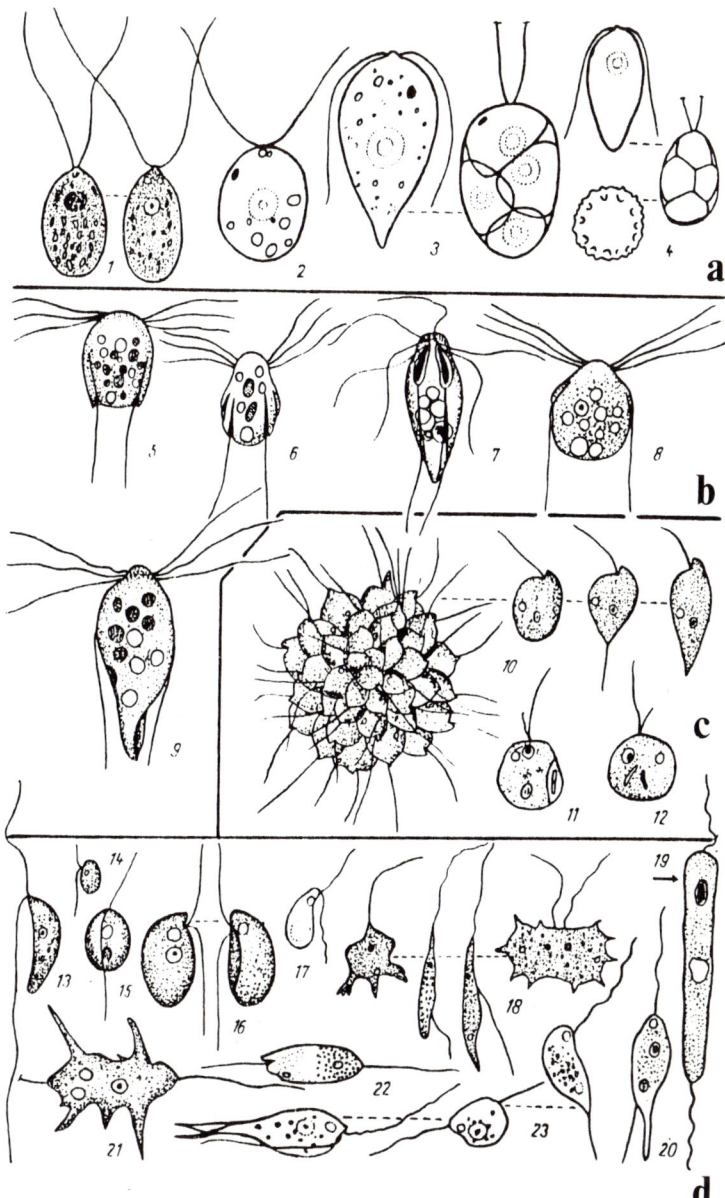

FIGURE 3.3. Colorless flagellate associations that develop in sewage water rich in organics. (a) Association based largely on the presence of *Polytoma* spp (eg, 1. *P uvella* showing a specimen with a large amount of stored starch, left; 2. *P ocellatum*; 3. *P fusiformis*). (b) Association containing largely *Hexamitus* spp (eg, 5. *H inflatus*, 6. *H crassus*, 7. *H fissus*, 8. *H pusillus*, 9. *H fusiformis*). (c) Association based on the presence of *Monas* spp (eg, 10. *Oicomonas socialis*, 11. *Monas vulgaris*, 12. *Monas arhabdomonas*). (d) An association containing Kinetoplastids (eg, 13. *Bodo putrinus*, 14. *Bodo minimus*, 15. *Bodo globosus*, 16. *B edax*, 18. *Cercobodo agilis*, 19. *C grandis*, 23. *C varians*). From Sladecek (1972); with permission.

always follows the *Polytoma* stage. A chrysomonad, *Oicomonas socialis*, dominates the third type of association found mainly in decaying municipal sewage and putrefying sludge. Other less dominant genera include *Bodo*, *Cercobodo*, *Trigonomonas*, and especially *Monas vulgaris* and *M arhabdomonas*. The fourth association is dominated by species of *Bodo* and *Cercobodo* (Figure 3.3, 13 to 23). These kinetoplastids occur in organically polluted environments characterized by extreme saprobic conditions. They occur to a lesser degree also in the first and second associations, and persist into the isosaprobic stage of a succession.

Some of the flagellates in the first type of association are saprozoic and assimilate organics as dissolved compounds or as colloidal aggregates. These organisms utilize metabolic waste products and occur abundantly in aquatic habitats enriched by animal feces and fluid wastes. *Polytoma*, for example, is an acetate flagellate and thrives in low pH, organically-enriched habitats. The species in the second to fourth associations are bactivores, as is characteristic of many small phagotrophic aquatic flagellates. In addition to saprozoic modes of nutrition, they engulf bacteria in food vacuoles. The undulating flagellar motion creates water currents that sweep bacteria toward the oral apparatus or phagocytic site on the membrane of the flagellate. A variety of filtering or trapping mechanisms have evolved to efficiently capture bacteria. Some of these are considered more fully in Chapter 16. As examples, some choanoflagellates use an apical corona of microvilli to snare bacteria propelled toward them by the flagella. *Ochromonas* spp have been observed to use the flagellum to propel bacteria toward a phagocytic site on the plasma membrane. *Bodo* and other small kinetoplastids have a specialized oral groove to collect and ingest bacteria. Bacterivorous flagellates are probably major consumers of bacterial biomass, especially in marine environments, based on laboratory estimates of their feeding rates (eg, Fenchel 1984, 1986a,b; Goldman and Caron 1985). Bacterial numbers in natural seawater usually are fairly stable varying from 10^6 to 5×10^6 per ml in coastal waters and estuaries. Fewer are found in offshore water with a range of 5×10^5 to 10^6 per ml. Fenchel (1986a) has calculated that given a concentration of about 10^6 bacteria per ml, if the flagellate requires approximately two bacteria per day to survive and reproduce, then it must filter 2×10^6 μm^3 of seawater per day to obtain two bacteria. This is equivalent to 10^5 μm^3 per hour! For a small microflagellate, with a volume of about a μm^3, this means that it must clear 2×10^6 times its own body volume each day to obtain sufficient bacteria to reproduce. Estimates of flagellate feeding rates in laboratory and field studies indicate that average rates of predation may be as high as 30 to over 200 bacteria consumed per flagellate per hour (Davis and Sieburth 1984). Hence, during a succession from a hypersaprobic to metasaprobic (flagellate-dominated) stage, considerable attrition in bacterial numbers may occur due to flagellate ingestion.

Normally, however, protozoan predation rarely leads to major decline in numbers of bacterial prey (eg, Alexander 1981). This balanced condition of predation and prey reproduction enhances the community stability. For example, in soil samples inoculated with the bacterium *Rhizobium* at densities greater than 10^8/g, the numbers of indigenous protozoa increased and *Rhizobium* declined concomitantly to about 10^7/g. The population of *Rhizobium* remained at this level despite the presence of protozoa in excess of 10^4/g. There was no decline in number of inoculated bacteria when a protozoan metabolic inhibitor was added to the soil, thus confirming the role of protozoan predation in regulating bacterial biomass. The persistence of predators and prey in a fairly balanced state can be attributed to several factors (Alexander 1981). Predator interactions, especially when they are abundant, decrease predatory efficiency by interfering with feeding. Also, as the prey declines in numbers, a threshold level may be reached where there are too few encounters with the predator to sustain efficient predation. Or, bacterial predators may themselves become prey for other organisms, thus reducing their numbers. In some cases, as the number of prey decline, and feeding is less efficient, the predator may become less active. Thus, its rate of growth is more closely adjusted to the reproductive level of the prey. Finally, a proportion of the prey may be located on surfaces or within inaccessible locations to the predator; thus forming a remnant population that serves as a buffer to maintain a constant supply of prey. Bacteria are particularly likely to grow upon surfaces of suspended matter or inside small spaces within porous organic particles, and thus are protected against protozoan predation.

Community stability can also be enhanced by spatial separation of potentially competitive species. This may occur due to physico-chemical factors that delimit overlap in distribution of species within a rather limited geographical range. For example, in the north Atlantic Ocean, small gyres of water known as warm core rings sometimes separate from the Gulf Stream and penetrate as distinct water masses into the surrounding cold water. These fairly stable ring-shaped water masses possess a distinctive biota and exhibit successional changes over time. However, due to differences in the temperature and hydrodynamics between the peripheral cooler water and the central warmer water, different populations of flagellates occur in the two regions (Figure 3.4). Coccolithophorids (Prymnesiida) are abundant in the peripheral cooler water, while dinoflagellates are most abundant in the central warmer water. Hence, the two species can occupy the ring system with minimum competition owing to their particular temperature and hydrographical requirements for optimum growth. This forcing function of the environment results in territorial separation of populations and increases community stability. The differentiated niches accompanying variations in water quality minimize competition and permit species to flourish within separate domains in a rather restricted environmental space. Hence, two or more species that might otherwise compete so extensively as to lead to competitive exclusion are able to coexist in close but separate locales within a restricted region of the environment. Some of these general principles will be further exemplified as additional physiological-ecological concepts are introduced in subsequent discussions of sarcodinad and ciliate biology.

Summative Perspective

The survival and abundance of protozoan species depend on an appropriate match between physiological requirements of the species and environmental

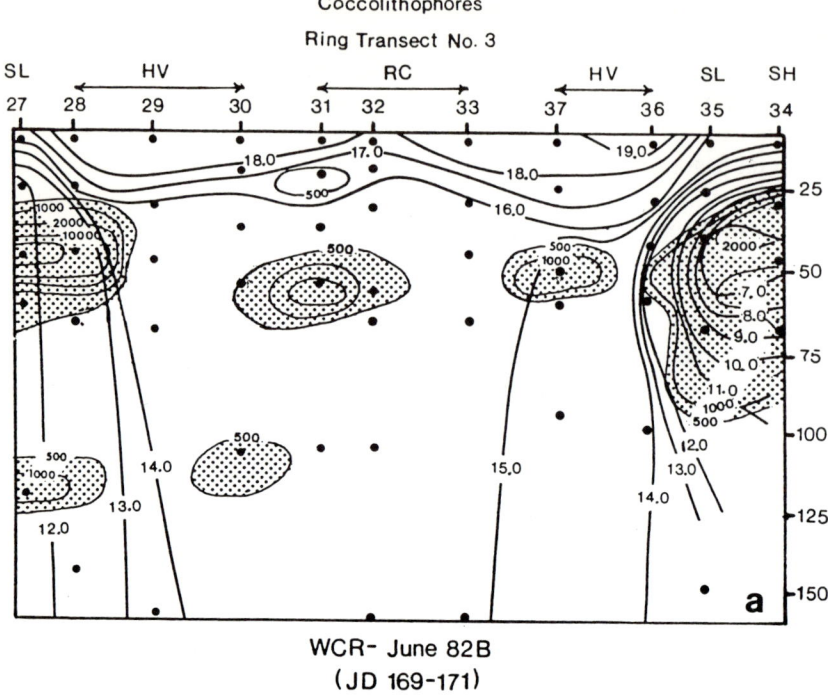

FIGURE 3.4a,b. Segregation of marine flagellates, Coccolithophores (3.4a) and dinoflagellates (3.4b), within different water masses varying in temperature and hydrographic variables producing separate ranges for the species within this small, self-contained marine ecosystem. In Warm Core Ring 82-B in June, dinoflagellate peaks in number are evident in the warm (16–19°C) water in the center of the eddy (b). Coccolithophorids are abundant in the cooler (7–15°C) intrusions of shelf and slope water (a). RC = Ring center, HV = high velocity regions at the ring edge, SL = slope water, SH = shelf water. Courtesy of Professor Patricia Blackwelder, Nova University, Florida.

conditions. Species tolerating a wide variation in environmental conditions, temperature, pH, redox potential, and kind of nutrients (as for some facultative heterotrophs) are likely to be widely distributed. Some occur in a worldwide distribution. Species tolerating particularly harsh environments (eg, low pH) are likely to be the sole species or most dominant species in these locales. Much remains to be learned about the tolerances of many protozoa to variations in environmental parameters. Combined field studies with laboratory culturing experiments can contribute substantially to our understanding of the tolerance limits of protozoa, and potentiating environmental factors that permit their survival in natural habitats. With increasing awareness of possible deleterious effects of modern industrialization and technological by-products on living systems, it is essential to better document the tolerance limits of protozoa to natural and anthropogenic sources of environmental variability. Environmental physical

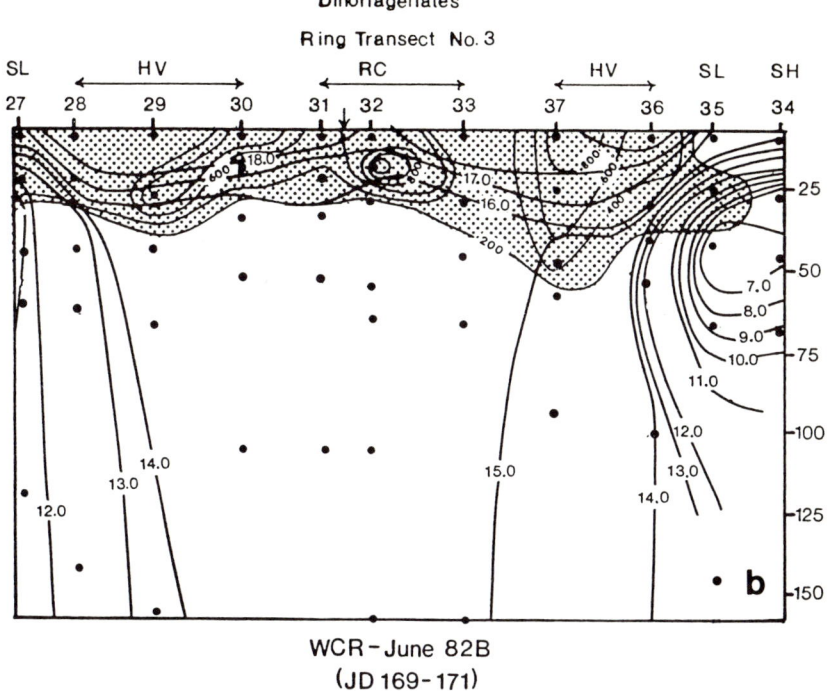

FIGURE 3.4b. *(Continued)*.

and chemical factors influencing life are intricately interconnected, and variations in one or more of the variables can produce complex changes in others, as demonstrated by data in the tables presented in this chapter, and in Figures 3.2 and 3.4. Protozoa are particularly likely to be sensitive to environmental perturbations, as they are intimately associated with their surrounding aqueous environment. Some possess remarkably efficient means for surviving in hostile environments by maintaining a constant intracellular milieu. This is achieved in some cases by specialized surface membranes which are impervious to toxic materials, or which regulate uptake of potentially useful substances that may be toxic in excess quantities. Those capable of forming cysts, as, for example, *Euglena* spp, are able to resist unfavorable conditions for some periods of time by effectively secreting a wall and sealing themselves off from deleterious environmental factors. Some insights into the physiological mechanisms of protozoan survival and possibly evolutionary mechanisms of adaptation can be achieved by studying the complex events associated with protozoan encystment and excystment. The kind of intensity of environmental factors triggering these events and the time course of physiological responses occasioned by these same events constitute an intriguing field of modern cellular physiology. The physiological ecology of protozoa offers an intriguing area of research for those eager to discover the intricate ways living cells have adapted to varied and changing environments.

4
Amoebae and Their Relatives (Phylum: Sarcomastigophora; Subphylum: Sarcodina)

General Morphology

Motility by cytoplasmic streaming is an effective mode of locomotion for cells that creep in an amoeboid fashion or move by extended fingerlike pseudopodia in contact with the substratum. Many of the Sarcodina exhibit some form of cytoplasmic streaming during locomotion, or as a means of gathering food particles and other matter from the environment. Some arenaceous species, for example, collect mineral particles and organize them into rather consistent species-specific shell shapes. Many of these arenaceous species, with rather uniform sized grains in the shell, apparently select particles according to size and composition during the process of shell construction. The particles are transported to the shell and arranged in a specific pattern by rhizopodial activity. Complex patterns of cytoplasmic streaming and movement of the rhizopodia are used to arrange the shell particles. Cytoplasmic streaming also occurs in pseudopodia of planktonic sarcodinads and is used to apprehend and dismember large prey such as copepods. These organisms extrude a halo of pseudopodia, varying among species from a weblike rhizopodial array to fine filamentous filopodia or a corona of stiffened axopodia surrounding the cell. In some planktonic species such as radiolaria and planktonic Foraminifera, a mineralized shell or skeleton is secreted as protection for the soft cytoplasm and as support for the web of peripheral pseudopodia. These extracellular mineralized structures serve a twofold function as a skeleton for attachment and support of the pseudopodia, in addition to protection of the enclosed mass of cytoplasm.

The form of the pseudopodia and the presence or absence of a mineralized shell are significant characteristics used in distinguishing major taxa within the Sarcodina. The naked amoebae represented by the familiar *Amoeba proteus*, and the testate amoebae surrounded by a shell and extending lobate or fingerlike pseudopodia are included in the class Lobosea. Although the pseudopodial morphology may be quite variable among the species in this class, they are typically lobose to more or less filiform and arise from a broader hyaline lobe. The organisms move by a cyclic protoplasmic flow that propels the body either through or over the more or less temporary pseudopodia, often giving the appearance of a forward rolling motion. Sarcodina with filopodia (slender, clear to faintly granular pseudopodia, sometimes branching, but not regularly anastomosing) are in the class Filosea. Some of these species possess only a thin organic test, while others are enclosed in mineralized shells either secreted by the cell or constructed from gathered mineral particles. Sarcodina with anastomosing, finely granular pseudopodial networks are in the class Granuloreticulosea. This group includes the Foraminifera. Some are freshwater naked forms, while many species are marine benthic or planktonic organisms surrounded by mineralized tests. The Sarcodina with stiffened, raylike pseudopodia (actinopodia) are placed in the superclass Actinopodea, subsuming four classes. Class 1, Acantharea, have elaborate skeletons formed of strontium sulfate; class 2, Polycycstinea includes the radiolaria with siliceous skeletons; class 3, Phaeodaria, contain siliceous skeletons with substantial amounts of organic matter; and class 4, Heliozoea, are the so-called "sun animals," with a central mass of cytoplasm and radiating peripheral axopodia. They lack a substantial mineralized skeletal framework, although some secrete siliceous scales. The diverse habitats of the Sarcodina, including terrestrial, aquatic, and marine environments, and their remarkable range of functional morphological adaptations, including small amoeboid forms through elaborate planktonic forms, bear witness to the extent of divergent and elaborate evolution in this group. Each of the major classes will be discussed, with substantial attention to the Lobosea (amoebae), Granuloreticulosea (Foraminifera), Polycystinea (radiolaria), and Heliozoea (heliozoans).

Amoebae in Broad Perspective

This diverse group of protozoa is commonly characterized by the naked amoebae (eg, *Amoeba proteus*) that most biologists have encountered at some point in their biological education. There are, however, varied species including testate forms surrounded by an organic test with or without secreted translucent platelets, or attached or imbedded materials (eg, mineral grains). The testate species vary in type of pseudopodia. Some bear lobose pseudopodia (Class: Lobosea); whereas others have hyaline filose (long, thinly tapering) pseudopodia (class: Filosea). A discussion of the naked amoebae will precede the presentation of testate amoebae. As will become apparent, the use of "naked" for those forms lacking a shell or test is not entirely appropriate, as even the apparently naked forms as viewed with a light microscope often possess some form of organic surface coat, albeit very thin ("sarcolemma" or "glycocalyx"). Some possess loosely organized surface scales of most remarkable geometric design. Comparative features of "naked" and "shelled" lobose amoebae in generalized perspective are shown in Figure 1.1g and h, and Figure 4.1. Both forms commonly possess one or more contractile vacuoles, typically located at the posterior portion of the cell. A thin, sometimes hyaline, layer of cytoplasm known as the ectoplasm or hyaloplasm lies immediately beneath the plasms membrane, and surrounds an inner more granular mass, the endoplasm or granuloplasm.

58   4. Amoebae and Their Relatives

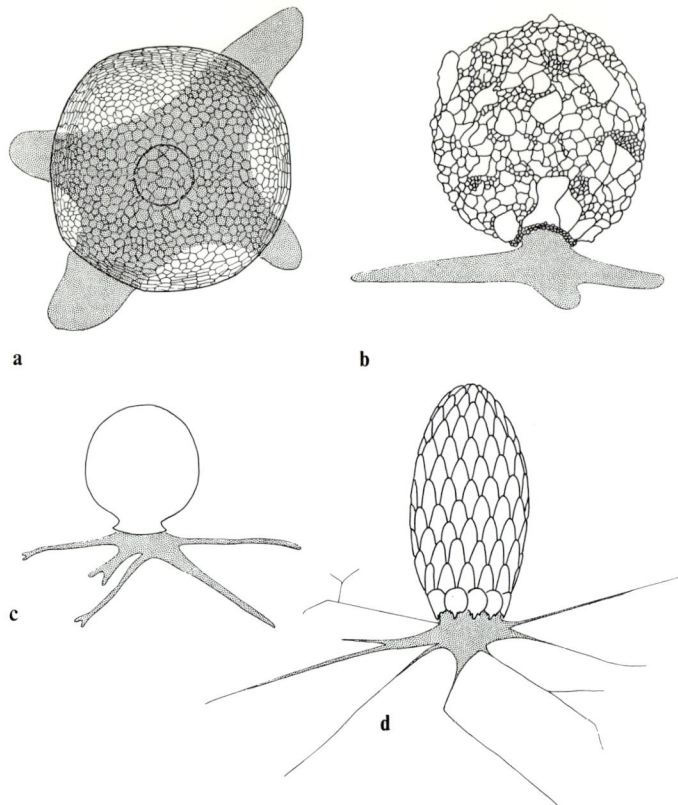

FIGURE 4.1. Representative testate amoebae. (a) *Arcella vulgaris* with a proteinaceous shell protruding about four lobose pseudopodia; (b) *Difflugia gramen* has an agglutinate shell and usually one or two large lobose pseudopodia; (c) *Cryptodifflugia oviformis* has a smooth calcareous shell, the pseudopodia appear lobose or transitional between lobose and filose; (d) *Euglypha rotunda* with a siliceous secreted shell and typical filose pseudopodia. From Ogden and Hedley (1980); with permission.

Naked Amoeba (Class: Lobosea)

The pseudopodia, while varying significantly in size, granularity, and mode of locomotion, are typically fingerlike or blunt lobes projecting from the "anterior" part of the cell. A mass of cytoplasm (uroid) at the posterior end of some naked amoebae appears to be associated with excretion, although food particles may also be attached to the surface and appear to be engulfed in this region. The shape of the uroid, when present, includes (1) morulate forms with knobby surface texture or with the appearance of a mulberry (Figure 1.1h), and (2) villous-bulb type with fine, rather rigid, and short hairlike pseudopodial extensions that may be attached at times to the substratum (adhesive type). The naked amoebae are identified largely by the shape of the body, form of the pseudopodia, and mode of locomotion. When these features are very similar, then taxonomic distinctions are based on the organization of the nucleus, or its mode of mitosis, and the occurrence of cysts. Variations in pseudopodia (Figure 4.2) include: (1) cylindroid, granule-filled pseudopodia varying in size relative to the cell body and generally characterized as lobopodia; (2) hemispheroid, bubblelike projections from the surface of the cell; (3) clear, conical, round-tipped (mammiliform) pseudopodia; (4) fan-shaped, translucent lobes; and (5) broad leading edges, usually without distinct projections except for occasional or temporary bulges. Fine filose projections, subpseudopodia, occur on the surface of some lobopodia. Subpseudopodia vary from fairly lobose to very fine, but their relatively small size compared to the larger mass of the cell clearly distinguish them from lobopodia. The number and arrangement of the pseudopodia permit distinctions among species. Forms with several pseudopodia are polypodial, whereas those with a single broad pseudopodium are monopodial. The latter include tongue-shaped species known as the limax or clavate forms. At least one group (order: Leptomyxida) are highly branched or reticulated, and resemble the slime molds now placed in the class Mycetozoea.

In addition to the organization of the pseudopodia, the form of locomotion is often distinctive. Some amoebae move by a steady forward motion with an apparent rolling quality of the cytoplasm during advance. Others move with intermittent spurts of motion accompanied by a sudden bulging outward of pseudopodia alternating to each side of the cell as the amoeba advances. When disturbed by agitation of the water or otherwise dislodged from the substratum, most amoebae assume a characteristic stellate floating form (Figure 4.2l). The shape of the floating form is also a useful aid in making species determinations (eg, Page 1976a). Flagellated stages, induced, for example, by a reduction in oxygen and dilution of the medium, are found among the amoeboflagellates (Fig. 4.2i). The flagellated, motile stage undoubtedly enhances survival of the amoeba by increasing its capacity to invade new territory. After swimming for a period of time, the flagellated form eventually settles down and resumes an amoeboid habit. The swarmers are either biflagellate or quadriflagellate. The mode of attachment of the flagella, either within a cytostomelike opening or near a beaklike ridge (rostrum), is used to distinguish among species.

Comparative Morphology

The genus *Chaos* is representative of the larger polypodial amoebae (Figure 4.2a), although monopodial forms also occur. It is multinucleate (hundreds of nuclei) and bears a hyaline cap as the tubular pseudopodium elongates. Species grow to sizes of several hundred microns to 2 mm. This genus is closely related to the genus *Amoeba*, with a single nucleus, one dominant pseudopodium during locomotion rather than several, and pseudopodia with hemispherical tips. Although the relatively small-sized *A proteus* (15 to 20 µm) is cited as typical of this genus, other forms such as *A navina* reach sizes of several hundred microns.

60 4. Amoebae and Their Relatives

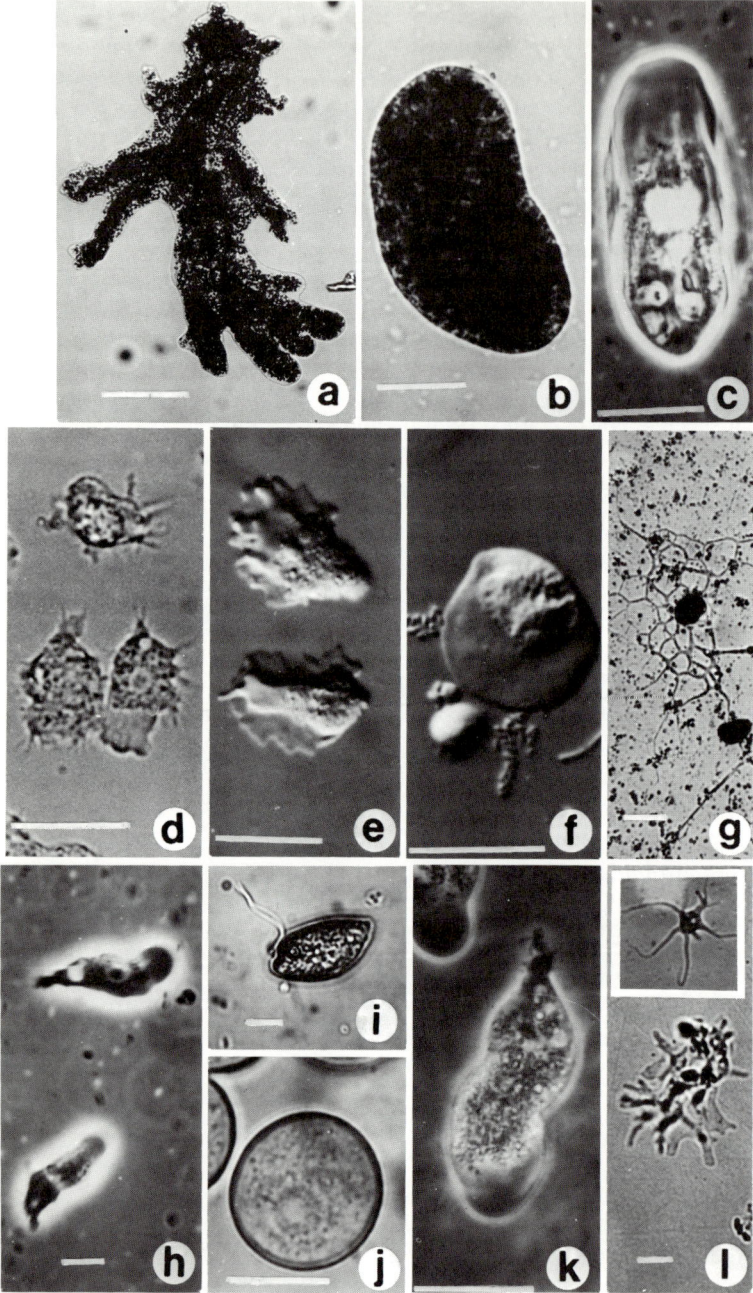

By contrast, the large, multinucleated, *Pelomyxa palustris* (Figure 4.2b), with dimensions of less than 100 to several hundred microns, illustrates a predominantly monopodial form. It moves rather slowly by fountain streaming or hemispherical eruption at the leading edge of the cell. A uroid at the posterior bears short villi. There are no mitochondria. *Pelomyxa* dwells in polysaprobic habitats (low in oxygen) such as the bottom mud of ponds and ditches. Sand grains or other mineral particles occur within the cytoplasm of this limax amoeba. The smaller limax amoeba *Naegleria gruberi* (Figure 4.2h), ca 30 to 40 µm during locomotion, moves by more or less eruptive, hyaline, hemispherical bulges as characteristic of the genus. When sedentary and rounded up, *N gruberi* is 15 to 30 µm in diameter. Temporary flagellated stages (Figure 4.2i) bear two equal flagella, but lack a cytostome. The nucleus is in the anterior region. Cysts are spherical to ovoid and usually smooth, but in some rough strains, the outer wall is separated producing an overall irregular form. *N gruberi* is probably the most common freshwater amoeba with a worldwide distribution. A group of somewhat flattened, fairly symmetrical amoebae, rarely with discrete pseudopodia during locomotion, is exemplified by *Thecamoeba* (Figure 4.2c) and *Platyamoeba* (Figure 4.2f). *Thecamoeba striata*, 30 to 40 µm, exhibits the prominent parallel dorsal ridges or folds characteristic of this genus. A thin pelliclelike surface becomes wrinkled during locomotion and apparently accounts for the formation of the longitudinal folds. No cysts have been reported. The cell is typically ovoidal to elongate with a rather smooth outline. The genus *Platyamoeba* often possesses only slight wrinkles near the sides of the cell, especially when turning. It usually exhibits a substantial posterior mass of granular cytoplasm and a more hyaline peripheral edge, particularly on the advancing front edge of the cell. Conical, not filose, subpseudopodia are produced at the hyaline leading edge of *Paramoeba* (Figure 4.2e), a rather unique saltwater species with a cytoplasmic, DNA-rich inclusion "Nebenkörper" or parasome (sometimes more than one) near the nucleus. The presence of subpseudopodia or dactylopodia, resembling small fingerlike protrusions from the surface of the hyaline perimeter, clearly distinguishes this genus from *Platyamoeba*. Markedly developed but few, blunt, hyaline, nonfurcate dactylopodia are characteristic of the genus *Dactylamoeba* (Figure 4.2l). *Acanthamoeba* (Figure 4.2d), by contrast to most of the forgoing genera, exhibits a marked spiny appearance at the periphery owing to the several or many slender, tapering, flexible, sometimes furcate

◄

FIGURE 4.2. Naked free-living amoebae. (a) *Chao nobile* (scale = 100 µm); (b) *Pelomyxa palustris* (scale = 50 µm); (c) *Thecamoeba striata* (scale = 20 µm); (d) *Acanthamoeba polyphaga* (scale = 20 µm); (e) *Paramoeba pemaquidensis* (scale = 20 µm); (f) *Platyamoeba bursella* (scale = 20 µm); (g) *Corallomyxa chattoni* (scale = 100 µm); (h–i) *Naegleria gruberi*, amoeboid and flagellate stages (scale = 10 µm); (j) cyst of *Naegleria gruberi* (scale = 10 µm); (k) *Acrasis rosea* (scale = 20 µm); (l) *Dactylamoeba bulla* (inset, floating form) (scale = 20 µm). Illustrations courtesy of Dr. F.C. Page, Cambridge, England.

projections (acanthopodia) emerging from a broad hyaline zone. This common inhabitant of soils produces cysts of rather distinctive morphology which are used for species designations. A dense, cytoplasmic-rich endocyst is usually surrounded by a thin, hyaline ectocyst. The shape of the endocyst, its relation to the enclosing ectocyst, and its degree of rippling or wrinkling are significant taxonomic characteristics. Some species are reported to be pathogenic, as discussed in Chapter 9. The rather uniquely reticulate morphology of *Corallomyxa* (Figure 4.2g), found in nearshore marine habitats, exhibits an extreme form that some complex multinucleated amoebae assume.

Slime Molds (Class: Mycetozoea)

The formation of multinucleated, reticulated plasmodia is particularly characteristic of some slime molds that superficially may resemble the larger reticulate amoebae such as *Corallomyxa*, but differ substantially in their life cycle. Slime molds, unlike true amoebae, typically produce sporocarps and release spores that are distributed by air or water. Some species produce sclerotia (crustose or rope-like dessication-resistant resting stages) that persist through unfavorable environments and resume growth as plasmodia when adequate moisture and nutrients are available. Flagellated stages arise from the spores of some species and then transform into amoeboid forms that persist until sporulation, each producing a sporocarp; or by aggregation, fuse to yield a plasmodium. A representative life cycle is presented in Chapter 20. The defining characteristics of the class are as follows. The trophic stage is typically an amoeba, or a uninucleated to multinucleated holozoic (phagocytic) plasmodium. Three subclasses are defined by the morphology of the amoeboid stage and mode of sporulation: (1) a single amoeboid cell (which may or may not be derived from fragmentation of a larger plasmodial form) differentiates into simple, stalked sporocarps containing one to few spores. (2) Aggregation of simple amoebae form multicellular pseudoplasmodia that produce stalked multispored sporocarps. (3) Multispored sporangia (or equivalent structures) are produced from a multinucleate, often sexually derived diploid plasmodium. The commonly studied *Physarum polycephalum*, with a robust, yellow-pigmented plasmodium (eg, Figure 20.2), is a member of the third subclass. The small limax type amoeboid trophic stage of *Acrasis rosea* (Figure 4.2k) illustrates a species of the second subclass. The individual rose-colored amoebae aggregate and form a single or branched stalk with chains of spores. The form of the sporocarp, and when needed, the shape and color of the spores, are significant in species identification. The spore case atop the stalk of the sporocarp can be simply a globose structure, multiple spheroidal cases clustered together, or spongy to filamentous spheroidal or elongate bodies. The form of the fine filaments (capillitial threads) surrounding the spores within the sporocarp is also significant in species determination. More substantial information on the biology and sytematics of the slime molds can be found in Alexopoulos (1963); Gray and Alexopoulos (1968); and Lee et al. (1985a).

Testate Amoebae (Classes: Lobosea and Filosea)

The testate amoebae produce either lobate pseudopodia (Lobosea) or filose pseudopodia (Filosea). Species are distinguished largely by shell shape and composition. The shells of the testate amoebae can be categorized into four broad groups: (1) proteinaceous shells composed either of (a) numerous closely assembled alveoli as building blocks or (b) a homogeneous organic wall; (2) agglutinate shells either formed by (a) gathering sand grains or other mineral particles or (b) use of shell plates obtained by ingesting smaller testates; (3) siliceous shells composed of secreted scales; and (4) calcareous shells usually of rather smooth appearance or composed of shell plates. The cytoplasm either fills the shell as in *Euglypha* (Figure 4.1d) or only partially fills the shell and is attached at places by thin cytoplasmic threads as in *Difflugia* (Figure 4.1b, 4.3d). The morphology of the shell including shape, composition, and geometric plan of the pseudostome (aperture) are important taxonomic criteria. Some major comparative features of shell morphology as exhibited by scanning electron microscopy (Figure 4.3) are discussed.

Lobosea

Shells composed of proteinaceous subunits as exemplified by *Arcella* (Figure 4.3a), with a basically circular profile and centrally located pseudostome, appear either smooth in texture or with a very fine reticulate pattern produced by the close assembly of the protein aveoli. In some species with proteinaceous shells, additional agglutinated particles may be included or there can be spinelike extensions (eg, *Centropyxis*). The shell of *Centropyxis* (Figure 4.3b) is typically somewhat flattened dorso-ventrally, spheroidal at the posterior portion, and tapering toward the apertural region. The aperture is typically invaginated and without a raised rim. *Difflugia* (Figure 4.3d), an arenaceous species, builds a shell that is typically elongate or spheroidal with a terminal aperture. The texture and organization of the gathered mineral grains varies among the species in this arenaceous group. In some cases the particles are highly rugose or sharp-edged, while in other species, the shell is composed of remarkably smooth, flattened particles imposing a tilelike texture to the surface of the shell. *Nebela* spp. (Figure 4.3e) generally exhibit elongate shells, either pyriform or oval in shape, with a neck of varying length terminating in an aperture sometimes with teeth around the opening. The shell may be compressed laterally to varying degrees among species, and in some cases it possesses a lateral ridge. The agglutinated particles are of various types, some being largely ovoid, or irregularly rodlike, and are characteristic of the species. The genus *Lesquereusia* (Figure 4.3g) builds shells that are ovoid or circular in profile, often with a distinct long neck terminating as a nearly circular aperture. In some cases, the shell may be slightly flattened with an unsymmetrically located neck. The shell particles are often characteristically composed of numerous small, smooth, curved, siliceous rods held together by organic cement plaques. The elegantly molded, square, tabular shell units of *Quadrullela* (Figure 4.3h), clearly secreted by the cytoplasm of the cell, are regularly arranged to

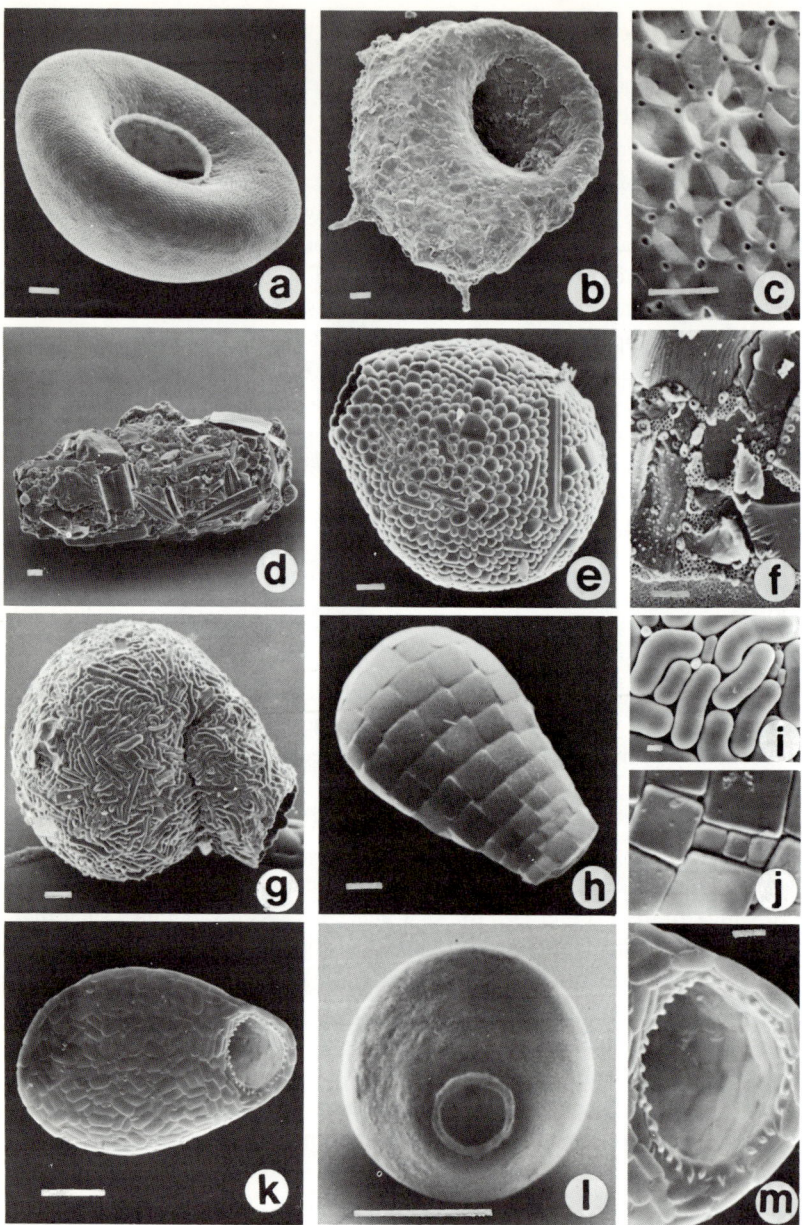

FIGURE 4.3. Scanning electron micrographs of testate amoebae. (a) *Arcella vulgaris*; (b) *Centropyxis hirsuta*; (c) test detail of (a); (d) *Difflugia lacustris*; (e) *Nebela dentistoma*; (f) test detail of (d); (g) *Lesquereusia spiralis*; (h) *Quadrullela symmetrica*; (i) test detail of (g); (j) test detail of (h); (k) *Corythion dubium*; (l) *Cryptodifflugia oviformis*; (m) test detail of (k). Scale bars = 10 μm (a,b,d,e,g,h,k,l); 2 μm (c,f,i,j, and m). Illustrations courtesy of Dr. C.G. Ogden, British Museum of Natural History.

form the ovoid or pyriform shell. The square plates are usually deposited in rows with smaller plates arranged near the aperture. The shell may be compressed laterally, especially in the region of the aperture which is terminal, not located laterally. *Corythion* (Filosea), by contrast, possesses an ovoid shell (Figure 4.3k) composed of several hundred oval siliceous shell-plates that overlap sometimes in a haphazard arrangement so that many are either incompletely or completely covered. The aperture is subterminal (located laterally). Fine filose pseudopodia emerge from the aperture, thus placing this genus in the class, Filosea. About 30 small, oval apertural plates surround the aperture, and each projects a small tooth. The colorless, circular or ovoid shell of *Cryptodifflugia* (Lobosea) (Figure 4.3l) has a smooth surface composed of an outer organic layer covering a thick inner layer of amorphous calcium phosphate. The aperture is terminal, round, and borded by a thin organic collar. The pseudopodia are lobate or finger-like.

Filosea

The testate amoebae with filose pseudopodia produce tests of varying form. The ovate shell of *Euglypha*, with terminal aperture, composed of numerous regularly shaped imbricating scales, is typical of one group that secretes surface scales. *Corythion* (Figure 4.3k), *Trinema*, and related genera, have lateral apertures and ovoid, somewhat unsymmetrically compressed shells. Some produce organic spheroidal to flask-shaped tests with round to elliptical apertures, from which the fine hyaline filopodia emerge; eg, *Gromia*. Others produce arenaceous shells varying from vaselike to elongate (nearly spindle-shaped), usually with a flared rim (eg, *Amphorellopsis*). The testate amoebae with lobose pseudopodia and those with filose pseudopodia often occur in the same habitat, especially aquatic and moist terrestrial environments such as those associated with loose organic-rich soil or among the tufts of moss plants.

Foraminifera (Class: Granulorecticulosea)

Foraminifera are Sarcodina enclosed within an organic or mineralized test with one or more apertures. Rhizopodia with granular cytoplasm emerge from the aperture and form a weblike array. This rather diverse class of organisms also contains rhizopod-bearing organisms not included among the Foraminifera. These include nonshelled, plasmodial organisms with branching and reticulated rhizopodia, and those with a flexible, unmineralized or nonagglutinated, organic, single-chambered test. Foraminifera, by contrast, form substantial organic-agglutinated, or mineralized tests. They are abundant in marine environments. Benthic forms, attached to the substratum or on the surface of floating or attached plants, comprise the majority of the speices. The others are planktonic, floating with the currents and extending a halo of rhizopodia around the calcareous shell enclosing the central mass of cytoplasm. All of the benthic species are spineless. Planktonic foraminifera are either nonspinose with relatively smooth, spiral mutilocular shells, or spinose, with an array of long, sometimes flexible calcite spines radiating out from the shell surface. The spines help to support the mass of weblike and filose pseudopodia that are attached to the spine surfaces and extend beyond their tips.

There are six major groups of living Foraminifera based on the composition and organization of the test and the habitat of the organisms: (1) test thin, membranaceous or proteinaceous without agglutinated material cemented into the wall; (2) test agglutinated, containing mineral grains or other foreign matter held together within an organic or mineralized matrix (Figure 4.5a,b); (3) test wall composed of aggregated needles of magnesium-enriched calcite resembling fired porcelain when viewed by reflected light (Figure 4.4a); (4) bilamellar calcite wall, with an outer and inner layer of calcite separated by an organic sheet, benthic species (eg, Figure 4.4f); (5) bilamellar wall and planktonic habit (Figure 4.5g through i); and (6) test composed of aragonite (a form of calcium carbonate) deposited in a flattened spiral form.

The Foraminifera have a long evolutionary history extending back to the Cambrian, and their shells compose a substantial part of marine sedimentary deposits above the lysocline (the depth in the ocean where calcium carbonate dissolves due to physico-chemical factors associated with depth). Only some representative living species will be discussed here. More substantial information can be found in the *Catalog of Foraminifera*, Cushman Foundation (Ellis and Messina 1940) and in texts (eg, Loeblich and Tappan 1964; Boltovskoy and Wright 1976; Haynes 1981).

Species with a single-chambered naked organic wall, found in brackish and marine waters, typically also gather detrital or mineral matter from the environment and cluster it around the periphery of the text. The genera include *Iridia*, *Myxotheca*, and *Nemogullima*. The test is not as rigid as in the second group which forms firm agglutinated walls. The benthic organisms with cemented agglutinated particles (eg, *Allogromia* sp with a ovoidal test, or *Astrohiza* and associated genera with tubular or branching tests) live in detrital rich environments and protrude branching and reticulated pseudopodia into the surrounding environment. Organic detrital particles, bacteria, and small algae are gathered as food. Agglutinated tests may also be multichambered, consisting of ovoidal chambers arranged in two staggered rows (biserial test) or in a regular or imper-

▶

FIGURE 4.4. Scanning electron micrographs of benthic foraminiferan shells. (a) *Spiroloculina hyalina* (scale = 20 µm); (b) *Quinqueloculina striata* (scale = 200 µm); (c) *Peneroplis planatus* (scale = 200 µm); (d) *Amphisorus hemprichii* (scale = 1 mm); (e) *Spirillina* sp (scale = 200 µm); (f) *Metarotaliella* sp (scale = 10 µm); (g) *Calcarina calcar* (scale = 200 µm); (h) *Elphidium crispum* (scale = 200 µm); (i) *Elphidium crispum* view of edge with closed apertural region (scale = 200 µm); (j) *Amphistegina* sp (scale = 22 µm); (k) *Marginopora ammonoides* (scale = 200 µm); (l) *M. ammonoides* view of edge showing closed apertural region and fine pores along the rim of the shell (scale = 200 µm). Illustrations courtesy of Dr. John J. Lee, City University of New York.

Foraminifera 67

68   4. Amoebae and Their Relatives

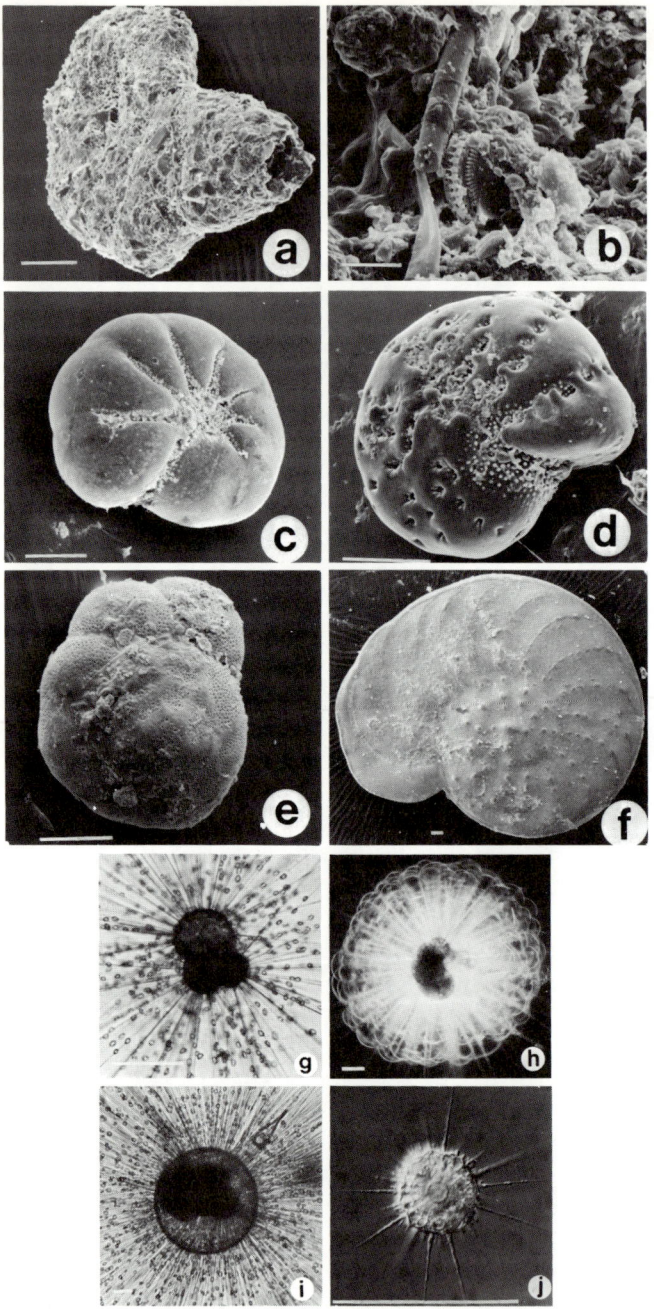

fect spiral. The "porcelaneous" species vary in shell organization. Some as with *Spirolocullina* and *Quinqueloculina* (Figure 4.4a,b) have flattened tests with elongate chambers added on alternating sides. The aperture of each successively added chamber alternates 180° with respect to the previously deposited chamber. Thus, each new chamber is oriented with its aperture at the opposite end of the shell compared to the previous one. Moreover, each new chamber encloses the aperture of the older chamber so that only one aperture is open to the environment. A tooth varying in complexity may occur on the inner rim of the aperture. Other forms in this group include compact spiral species forming disk-shaped shells (eg, *Sorites*), or irregularly fan-shaped shells (eg, *Peneropolis* and *Archaias*). In some cases the shell can be rather massive, with thick walls surrounding the interconnecting, spirally arranged chambers. Hyaline, single lamellar walls sometimes appearing glassy with transmitted light characterize some benthic species living in shallow or deep marine waters. The test varies among species from a conical, uniseriate row of chambers to spiral shells either somewhat flattened (planispiral), or snaillike shells with a more elongated spiral axis (trochospiral). The walls of the chambers may be perforated. A halo of rhizopodia typically radiates outward from the periphery of the shell. The nonspinose planktonic foraminifera with bilamellar walls are typically spiral-shelled, often resembling benthic species. The multichambered shell has one or more apertures, either interiomarginal near the inner part of the leading edge of the last chamber or tending toward extra-umbilical to umbilical. The shell may be rather conical (eg, *Truncatulinoides*) or flattened with a thickened reinforcing peripheral rim (keel) as with *Globorotalia*. Spinose species possess long tapering spines (Figure 4.5, g through i) that may be numerous and round in cross-section, or more sparse and triangular in cross-section (triradiate) as with *Hastigerina pelagica* (Figure 4.5h), a large, thin-walled species reaching sizes of several centimeters, including the spines. Species vary in shell surface texture, varying from smooth to irregularly granular, to finely ridged with hexagonal pore pattern. The position of the aperture (whether lying within the plane of the spiral, or slightly displaced and bridging across two earlier chambers) and the number of apertures varies among species. *Orbulina universa* is exceptional, as it develops first as a spinose spiral stage that subsequently become enclosed within typically one large

◄

FIGURE 4.5. Some comparative views of sarcodinads. Figures (a–f), salt marsh benthic Foraminifera; Figures (g–i), planktonic Foraminifera from open ocean locations, and Figure (j) a heliozoan. (a) *Ammotium salsum*; (b) enlarged view showing cement and mineralized particles within the test wall of this arenaceous species; (c) *Discorbis* sp; (d) *Elphidium incertum/clavatum*; (e) *Ammonia beccarrii*; (f) *Operculina ammonoides*; (g) a juvenile spinose planktonic foraminifer; (h) *Hastigerina pelagica*; (i) *Orbulina universa* mature stage showing the spherical final chamber surrounding an internal spiral stage, tintinnid prey in rhizopodia; (j) *Actinophrys sol*, a freshwater heliozoan. All scales = 100 µm, except (b) = 20 µm. Illustrations (a–f) courtesy of Dr. John J. Lee, City University of New York, and (j) Dr. F. C. Page, Cambridge, England.

spherical peripheral shell supported upon the penetrating spines of the spiral stage, and also bearing its own set of dense surface spines (Figure 4.5i). This outer spherical shell (forming the universa stage) is perforated by numerous small pores, and also by many larger nearly circular apertures through which major strands of rhizopodia emerge. Many of the spinose species harbor algal symbionts in their rhizopodia, which confers a decidedly golden or tawny color to the cytoplasm. Although the peripheral rhizopodia in general form a weblike network with free radiating filopodia, variations in the organization of the rhizopodia and extra-shell cytoplasm occur among the spinose species of planktonic Foraminifera. In some cases, the inner layer of rhizopodia is more densely arranged with a more open network at the periphery (eg, *Globigerinella aequilateralis*). A rather dense radial mass of rhizopodia surrounds the hexagonally-ridged shell of *Globigerinoides sacculifer*. In the last stages of maturation it often produces a saclike chamber. *H pelagica*, by contrast, has a mass of bubblelike alveoli interspersed among the rhizopodia. It bears no symbionts, but occasionally contains numerous, apparently commensal, pyriform-shaped dinoflagellates (*Pyrocystis fusiformis* and *P noctiluca*) in the extrashell cytoplasm.

Radiolaria (Class: Polycystinea)

Radiolaria and planktonic Foraminifera are among the most common larger Sarcodina found in the open ocean. Radiolaria, moreover, are the most significant source of zooplankton siliceous fossils deposited in marine sediments. Their ancestors have been traced to the Paleozoic, with a substantial increase in number of species during the Mesozoic into modern times. Radiolaria (Figure 4.6) are distinguished from other Sarcodina by the presence of axopodia, and an organic wall with pores enclosing the central cell body (intracapsulum). Strands of cytoplasm (fusules) pass through the pores in the capsular wall and continue outward as axopodia. Sometimes a frothy layer of cytoplasm immediately surrounds the capsular wall, and is the site where food particles are digested after being snared by the axopodia. Many species secrete a siliceous skeleton of remarkably complex geometric design (eg, Figure 4.6, N and S2). The pattern of the skeleton is species-specific and is used in identification. Some large

▶

FIGURE 4.6. Some comparative views of radiolaria categorized according to present knowledge of their trophic role (columns). Organisms spanning two columns are assigned to both categories. (C) Colonial radiolarian with large gelatinous sphere (ca 1 cm) containing hundreds of central capsules (CC) interconnected by rhizopodial strands containing algal symbionts (Sy). (S1) Large gelatinous or spongiose skeletal Spumellarida (ca 0.5–1 mm), (S2) small more robust skeletal species (ca 80–200 μm), and (N) Nassellarida with typically conical, ringlike or other elaborate skeletal design (ca 50–150 μm). (P) Phaeodarian with a large geodesic lattice skeleton and dense internal central capsule. From Anderson (1983); with permission.

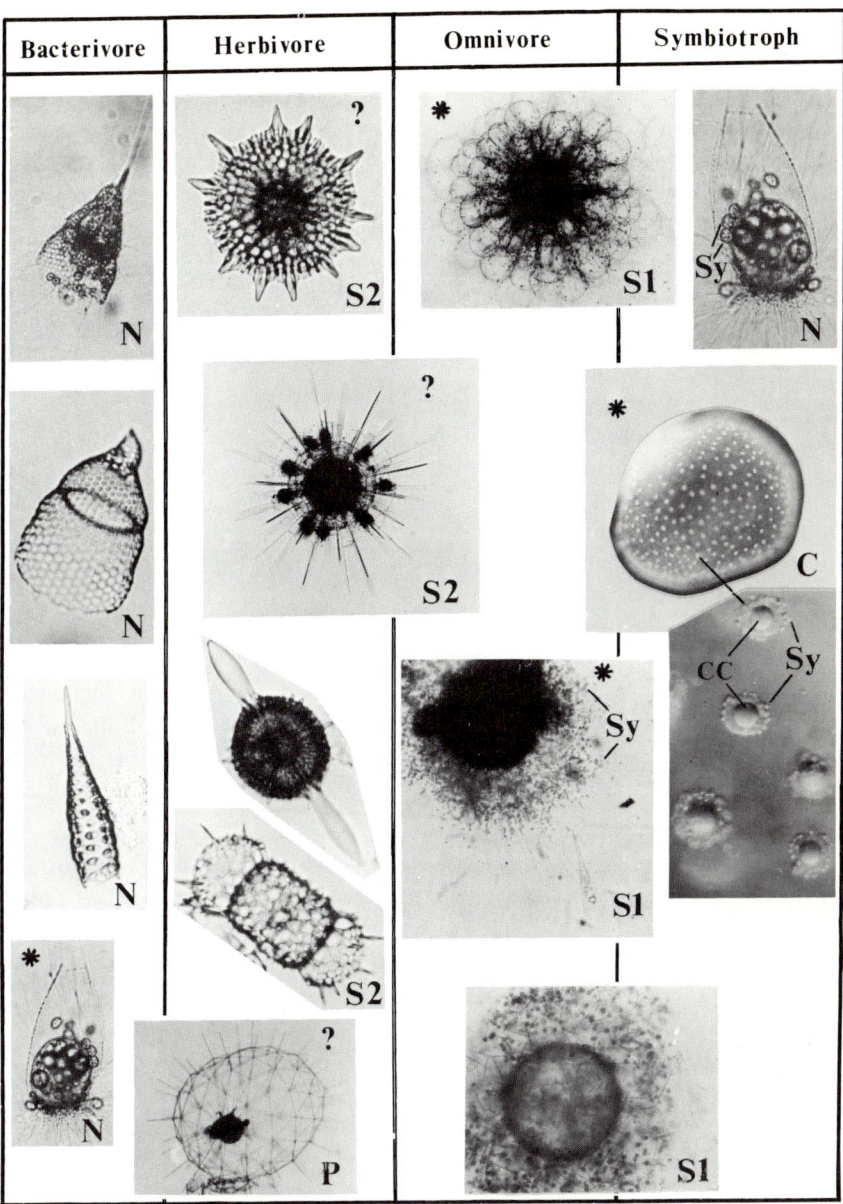

gelatinous species do not have skeletons, and their identification depends solely on cytoplasmic features including size and shape of the central capsule, axopodial arrangement and when present, the organization of vacuolated cytoplasm within and surrounding the central capsule.

There are two major subdivisions within the Polycystinea: (1) Spumellarida, usually with a spherical or spheroidal central capsule uniformly perforated over its entire surface with axopodia radiating outward around the periphery of the cell; and (2) Nassellarida, with an ovate to oblong central capsule bearing a pore field at one pole of the central capsule where relatively massive fusules emerge. The skeleton of the Spumellarida are frequently constructed of concentric porous spheres. Some species secrete elaborate skeletons composed of geodesic lattices, with radially arrange spines or hourglass shaped skeletons with an internal spherical shell at the narrow central part of the skeleton. Other spumellaridan species have diverse skeletons of varying shape, including biconical shells with nested conical frameworks extending symmetrically along an axis to either side of the central capsule. Some species produce spongiose skeletons (spherical, discoidal, or quadrangular) (Figure 4.6, S2). The Nassellarida vary in skeletal type from those with simple spicules near the base of the central capsule, through more complex ring or coronet-shaped structures enclosing the cell, and extending to those with porous helmet-shaped, conical, or elaborate lattice-embellished skeletons (Figure 4.6, N). There are probably several hundred living species, but inasmuch as none have been cultured in the laboratory through successive generations, it is not possible to determine accurately how many of the forms, differing only in small features, are truly different species versus different ecophenotypes. Ecophenotypes are morphological variants of a species induced by environmental factors.

Heliozoans (Class: Heliozoea)

The Heliozoans are granular, axopod-bearing (usually spherical) organisms, without an organic wall separating the central cell mass from the surrounding cytoplasm. In some species, a thin mucous coat lines the cell membrane and is penetrated by the axopodia and fine pseudopodia surrounding the cell. Some species secrete siliceous scales lying tangentially to the surface of the cell, or needlelike radiating spicules surrounding the central cell body. Indeed, *Sticholonche zanclea*, a unique protozoan with oarlike axopodia by which it propels itself through the water, is now assigned to the Heliozoea. It was previously considered to be a radiolarian. The often delicate quality of the central cytoplasmic region, and the sunburstlike arrangement of axopodia of many species (Figure 4.5j) led to the common name of "sun animals" for this group of organisms. Although many are planktonic, others are attached to the substratum by a cytoplasmic stalk. In *Clathrulina elegans* and *Hedriocystis reticulata*, a porous organic capsule surrounds the cell body, with axopodia extending from the large pores.

Actinophrys (Figure 4.5j) is a commonly observed floating aquatic organism. A single, centrally located nucleus is surrounded by a lacunar ectoplasm.

Granule-studded axopodia, extending from the nuclear surface, are thicker at the base than at the tips. Several contractile vacuoles occur in the peripheral cytoplasm. *Actinosphaerium eichhorni* is a naked, spherical heliozoan with vacuolated ectoplasm surrounding a granular endoplasm containing small vesicles and several contractile vacuoles. There are many nuclei, each 5 to 10 µm in diameter at the periphery of the endoplasm. Each of the many axopodia extends from the surface of a nucleus. *Echinosphaerium nucleolfilum* resembles *A eichhorni*, but the axopodia are thinner and longer (up to 500 µm) and not all of its nuclei have attached axopodia. There are several hundred nuclei, each 4 to 8 µm in diameter. Curved, spindle-shaped spicules (15 to 25 µm long) adhere to the gelatinous coat surrounding the cell and enclosing the axopodia of *Raphidiophrys pallida*. *Rabdiophrys annulifera* is a spherical organism with a thin surface coat to which are attached many radiating needlelike spicules of two lengths, 8 to 12 µm and 3 to 4 µm, with expanded bases and (often) swollen tips. The nucleus is eccentrically situated. Other members of the group have long radiating spicules, and in one species, *Raphidiocystis lemani*, there are also small goblet-sized spicules of several sizes (ca 1/10 body diameter) surrounding the cell. Algal symbionts are abundant in the ectoplasm of *Heterophrys myriopoda*, enclosing a distinct, granular endoplasm with a central granule surrounded by bases of the axopodia extending peripherally through the ectoplasm.

Summative Perspective

The organisms included in the broad category of amoebae and their relatives are classified in the subphylum Sarcodina. This is a diverse group of organisms, which form pseudopodia varying from lobopodia or simple hemispherical projections to reticulopodia and stiffened axopodia. Their taxonomy has been clarified substantially by application of transmission and scanning electron microscopy, especially for the naked amoebae (eg, Page 1976a,b, 1978, 1980) and the testate amoebae (Ogden and Hedley 1980). The so-called naked amoebae, however, are seldom truly naked, as some possess an organic coat or surface layer of scales. The slime molds, forming large plasmodial masses in some genera, are now included among the Sarcodina although previously they were thought to be fungi by botanists and amoebae by protozoologists. The Foraminifera and radiolaria are probably more phylogenetically advanced (or at least more specialized in their cytoplasmic organization), and secrete remarkably complex mineralized shells or skeletons. The Heliozoea, bearing axopodia but lacking a distinct capsular wall as found in radiolaria, are widely divergent organisms including sessile, stalked species and free-floating forms spanning freshwater and marine environments. Some species secrete siliceous scales (eg, Patterson and Dürrschmidt 1986) or produce external perforated shells that make them, along with the Foraminifera and radiolaria, interesting subjects for the study of biomineralization and its possible significance in evolutionary biology.

5
Some Comparative Perspectives on the Ecology and Habitats of Sarcodina

General Ecological Concepts

The attached sarcodines, living on terrestrial or aquatic surfaces, are especially favorable subjects for the study of ecological niches. The organisms are sufficiently restricted in their movement, and the environmental surfaces are adequately stable to permit periodic observations of the same group of organisms and their response to changes in the environment. Planktonic organisms, carried by currents, are less easily periodically examined due to their constantly changing location. Moreover, the possibility of mixing with other groups of planktonic organisms due to changes in current and turbulence effects must be taken into account when analyzing niches. The concept of niche is widely used by ecologists, but not always with the same meaning. We will use niche to mean the biological response and accommodation of an organism to the range of environmental and biotic factors that determine its distribution within an ecosystem. In broader perspective, the niche of an organism includes its biological activity (eg, diversity of prey, efficiency of food assimilation, tolerance to variations in temperature, salinity, pH, and E_h) in response to environmental constraints and potentials that determine the geographical range and success of an organism in space and time. In practical terms, one can seldom assess all potential parameters that define a species' niche. Typically, several key factors are identified (as for example, temperature, pH, salinity, oxygen abundance, and available carbon for nourishment) and their range for a species is determined. This composite set of ranges is used to operationally define the niche of the species. Clearly, species that are capable of adapting to and utilizing a wide range of environmental resources are likely to have broader niches than those that require a very specialized environment. The historical development and some current thought on niche theory are described more fully by Vandermeer (1972), Lee (1974, 1980a,b), and Austin (1985). Vandermeer divides the concept of niche into three subcategories: (1) fundamental niche; ie, the potential range of habitats that an organism can exploit when no other competing organisms are present including those of the same species; (2) partial niche; ie, the part of the fundamental niche that can be occupied when a species is competing for resources; and (3) realized niche; ie, the niche range occupied when a species is in equilibrium with other organisms occupying the habitat. A partial niche is determined in part by constraints arising either by intraspecific competition or interspecific competition. Thus, when a niche is first occupied by a species, there is likely to be little interspecific competition, but as the individuals multiply, one of the first likely effects is a restriction on the exploitation of the habitat due to intraspecies competition. However, as other species arrive, further restrictions on the exploitation of a niche may be incurred due to interspecies competition. When a niche is fully occupied by competing individuals, it is designated a realized niche. This is the boundary within the environment where a particular species occurs under the constraints of a balanced and stable interaction with other individuals. In general, the success of a species in inhabiting a particular habitat depends on a mixture of abiotic (physico-chemical) and biotic (organismic) factors. These factors can be further delineated as density-dependent and density-independent. Density-dependent effects on a species' capacity to survive in an environment are those factors that are determined by numbers of individuals. For example, competition for resources is a density-dependent phenomenon. As the numbers of individuals increase within a specified environmental space, competition is likely to increase. Density-independent factors are those not directly related to numbers of individuals. The effects of temperature, water velocity, supply of nutrients, and seasonal variations are density-independent. These variables influence organisms independently of the crowding (density) of the individuals.

Niche Concept

It is generally assumed that two or more species cannot occupy the same niche, since competition for resources will eventually lead to one species excluding the rest. This is clearly a density-dependent principle. At very low densities, as may occur early in colonization of a habitat, two or more species may occupy the same niche. But, as the population of each one increases and competition becomes more intense, it is likely that the more competitive or better adapted individuals will succeed.

Since overlap of niche is likely to result in competition and readjustment of populations within a community to maximize utilization of resources, it is desirable to have a means of quantifying the extent of overlap in niche among different species. The problems of translating conceptual definitions of niche into quantitative terms have been discussed extensively in publications by Hurlbert (1978), Petraitis (1979), and Abrams (1980). Some illustrative measures are introduced here. If the amount of available resources (environmental space, nutrients, food supply, suitable physico-chemical factors, etc) are fairly constant within the environment, then the following formula can be used as a measure of niche overlap between two species:

$$C_{xy} = 1 - [1/2 \cdot (\Sigma \, |px_i - py_i|)], \qquad 5.1$$

where $px_i = x_i/X$ and $py_i = y_i/Y$, the proportion of x and y individuals to the total, respectively. The difference in proportion of the two species is summed across the environments from $i = 1$ to $i = N$. If the proportion of individuals is equivalent among all of the sampling sites, then there is maximum overlap and C_{xy} has a value of 1.0. If, however, the individuals of each species are so distributed that there is little co-occurrence in each of the sampling sites, the value of C_{xy} approaches 0.0.

When as in most cases, available resources vary, an alternative measure is recommended by Hurlbert (1978) for niche overlap L:

$$L = (A/XY) \cdot \Sigma (x_i y_i / a_i), \qquad 5.2$$

where A = total sum of niche resource measures, X = sum of individuals in species x, and Y = sum of individuals in species y. x_i = number of individuals of species x at sampling site i, y_i = number of individuals of species y at sampling site i, and a_i = the resource abundance at sampling site i. To apply this formula, it is necessary to quantify environmental resources. Resources may include available substrate area, amount of prey per unit area or volume of water, etc. This measure is potentially more representative of niche relations, since it takes into account the density of co-occurring organisms relative to the amount of resources available to them. Abrams (1980) concludes, however, that the availability of a resource is less significant than the demand that a species places on the environment for a resource. He suggests using a measure of competitive interaction among species or applying the more simple formula of Equation 5.1. As a general rule, the coefficient chosen should be the simplest, but most representative measure that is consistent with the rationale for the research. If a study is undertaken where there is good evidence that the amount of resources available to a species is likely to be critical in expressing niche overlap, then Equation 5.2 may be a useful measure. There are limitations in making cross community comparisons with this formula when resources allocations included in one community have no relevance in another community (Abrams 1980).

Competitive ability of a species co-occurring with other species in a habitat can be assessed by the relative growth coefficient. The growth of the species in the presence of its competitors is assessed in relation to its growth alone, but under the same resource conditions. For the case of a species (labeled a) competing with two other species (1 and 2), the coefficient is given as $(u_1 + u_2)/2u_a$, where u_1 is the growth of species (a) in the presence of competing species (1), and u_2 is the growth of species (a) in the presence of competing species (2), and u_a is the growth of species a in the absence of other species. The lower the value of the coefficient, the less the competitive ability of the reference species (a). Muller (1972) has used a similar coefficient based on feeding rate:

$$\text{I.S.C.} = \frac{F_1 + F_2}{2F_a}, \qquad 5.3$$

where I.S.C. is the interspecific competition coefficient, F_1 is the amount of food consumed in the presence of species 1, F_2 is the amount of food consumed in the presence of species 2, and F_a is the amount of food consumed when the same number of individuals of the reference species are feeding alone. This requires a measure of food consumption. Typically, this can be done by labelling the prey with radioisotopes and determining how much of the radioactive label is consumed per unit time, thus permitting a calculation of how many prey cells were consumed (eg, Muller 1972; Lee 1974; Anderson 1983). Some examples of ecological research using these principles is presented for major groups of organisms among the Sarcodina.

Naked and Testate Amoebae

Niches

The niches of naked amoebae have been studied less than those of other groups in the Sarcondina. However, Bovee and Jahn (1973) and Page (1976a,c) have cited some of the common habitats and predatory behavior of free-living amoebae. Some illustrative examples are given. *Amoeba proteus* and *Amoeba discoides* are usually found in shallow, shaded, clear, slowly moving waters of lakes, ponds, and streams, at 10 to 24°C, near pH 7.5; the two sometimes occurring together. Both are carnivorous, entrapping prey by pseudopodial engulfment at the anterior end. *A proteus* will subsist well on ciliates such as *Tetrahymena* plus a bacterium (*Pseudomonas* sp). However, *Paramecium caudatum* alone as a food source is eventually toxic. *Polychaos dubium* occurs frequently in grassy water-filled ditches with chlorophyllous algal growth. It is largely herbivorous, and in addition requires fungal spores and chlorophyllous flagellates, as well as *Chilomonas paramecium* as prey. *Chaos carolinense* occurs in less open waters, in swampy pools or marshy backwaters of streams below 20°C. It ingests other protozoa and small invertebrates, but can subsist on a diet of only *P caudatum*. *Pelomyxa palustris* frequents the muddy bottoms of stagnant ponds and streams, often under a screen of heavy algal mats. It is a scavenger ingesting nonmotile debris, including sand and other mineral matter along with detritus and tests of other amoebae. It is largely herbivorous, and will grow well when fed only on *Spirogyra* or filamentous cyanobacteria (blue-green algae). Among the larger amoebae, *Trichamoeba* spp (eg, *T villosa*, *T osseosaccus*, and *T myakka*) are also found in debris-cluttered backwaters of streams, ponds, and lakes. They are largely herbivores. The large amoebae *Thecamoeba terricola* and *Thecamoeba papyracea* occur in algal growth, moist earth, mosses, and fresh water. They are omnivores and can engulf large strands of algae by "peristaltic suction." They snare bacteria and small prey within "trap vacuoles" on the ventral side, or enclose larger prey within phagocytic vacuoles formed by pseudopodial engulfment at the anterior end. *Thecamoeba sphaeronucleolus* can grow well on either a *Vanella* sp (a small amoeba) or bacteria. It occurs in wet forest soils or streams, often among mosses. The larger species of *Mayorella* (eg, *M vespertilio*, *M dofleini*, and *M bigemma*) are likely to be found in plant and algal associations of shallow, sunlit, quiet waters. In general, smaller soil and freshwater species such as *Naegleria gruberi*, *Saccamoeba* spp, and *Vahlkampfia* spp can ingest bacteria or very small algae. The *Acanthamoeba* spp (eg, *A castellanii* and *A glebae*) are moist soil and freshwater inhabitants, ingesting bacteria.

The amoebae in marine environments, especially the open ocean, have not been studed as extensively as freshwater species. Page (1983) described 22 genera and 43 species found in marine habitats. Among the genera are *Vahlkampfia*, *Hartmannella*, *Saccamoeba*, *Thecamoeba*, *Platyamoeba*, *Mayorella*, *Vexillifera*, *Acanthamoeba*, and the large reticulate amoeba, *Corallomyxa*. An open ocean scale-bearing amoeba has been found living with ciliates and flagellates among the fronds of the floating filamentous cyanobacterium (blue green alga) *Trichodesmium* sp, where it preys upon bacteria (Anderson 1977b). Oceanic amoebae from three sites in the North Atlantic Ocean (Narragansett Bay, Rhode Island; open ocean water in the Straits of Florida; and along a transect from Rhode Island to Spain) were identified in cultures from water samples by Davis et al (1978). They found an average of 33.6 amoebae/liter in the surface microlayer of the ocean. This is equivalent to 1,413 organisms/liter assuming they live in a 5 μm thick layer and a 70% efficiency in sampling. Subsurface water contained an average population of 1.4 organisms/liter to a depth of 3,090 meters. The genera *Acanthamoeba*, *Clydonella*, and *Platyamoeba* were isolated most frequently from the Florida Straits and transatlantic stations. *Paramoeba*, though among the most frequently encountered species, was not obtained in samples from the Florida Straits. *Clydonella vivax* was the only species cultivated from samples at all three oceanic stations. In all, 13 species were obtained. Among these, only *Acanthamoeba polyphaga* was obtained from samples taken from the aphotic zone (> 2,500 m depth). Species found in several locations were *Acanthamoeba polyphaga*, *Paramoeba aesturina*, *P pemiquidensis*, *Clydonella vivax*, and *Platyamoeba* sp.

Ecology of Naked and Testate Amoebae

The ecology of testate amoebae has been studied extensively, probably owing in part to the ease of identification of their tests. These persist for long periods in the natural environment, thus enabling a thorough documentation of distribution and production by accounting for both living and dead organisms. The abundance and kind of protozoa in relation to soil habitat has shown that in general, testate amoebae are most abundant in soil of slow but steady productivity, in cool to cold climates. By contrast, other protozoa such as flagellates, naked amoebae, and ciliates are prominent in soils of high productivity, characteristic of tropical and warm temperate climates (Stout 1965). Variations in numbers of testacea within a climatic zone, however, is related to the kind of vegetation (Bamforth 1971). Testacea are most abundant in soils under coniferous and mixed forests (10,000 to 24,000/g net weight in litters, and up to 8,000/g in soils). Ciliates ranged from 100 to 1500/g, but the mean was less than 600/g in litters, and less than 300/g in soils. In deciduous forest soils, testacea were less abundant (generally ranging from 400 to 1000/g substrate). In all, there were thirty-four species of testacea.

Species distribution showed no relationship to pH, but a geographic (climatic) trend was found. The genus *Nebela* was rare in warmer soils such as those in pine forests in Louisiana. The common species *Assulina seminulum* and *Corythion dubium* were infrequent in very moist, subtropical silty clay loams in open pastures and under deciduous forests. By contrast, the flattened shelled species *Arcella discoides* and *Microchlamys patella* (common aquatic organisms) were found only in moist soils. The most abundant members in all samples were largely *Assulina seminulum*, *Centropyxis aerophila*, three species of *Euglypha phryganella* sp, and three species of *Trinema*. The dominance of testacea in low productivity soils can be explained in part by their capacity to subsist on minimal biological sources of carbon. Some testacea can survive solely on sterilized forest soil, and others, eg, *Nebela collaris* and *Trinema complanatum* ingest lignin particles, a highly refractory component of wood (Bamforth 1971).

The abundance of soil-dwelling testacea is in large measure a response to the structure of the substrate; in particular, its porosity, moisture content, and constancy. Large and spiny testacea occur in bogs, wet and open textured leaf litter, or wet moss. More rounded and smaller species occur at greater depths in the soil. Vertical distribution of species within the soil reflects this principle. In the porous upper 2 cm, common testacea include the elongate-shelled *Corythion dubium*, *Euglypha* sp, *Assulina muscorum*, *Nebela* sp, *Trinema enchelys*, and *Arcella discoides*. In the next lower 2 cm, *Euglypha strigosa*, *Arcella vulgaris*, *Nebela bigibbosa*, and *Difflugia globulosa* are common. Below 4 cm the round-shelled species *Phryganella nidulus*, *Centropyxis hirsuta*, *Cyclopyxis eurystoma*, and *Plagiopyxis* sp are most abundant. Abundance of the testate amoeba *Nebela tincta* on the moss *Polytrichum formosum* also exhibits a distinct zonal distribution along the axis of the plant (Gnekow 1981). A maximum population density occurs midway up the axis at 3 to 4 cm, where the green leaflets are brown and decomposing. During drought or intermittent drying of the moss, *N tincta* produces a resting stage. It withdraws into its test, and forms a rounded, motionless mass of cytoplasm surrounded by a cyst membrane attached to an organic plug that closes the pseudostome of the test. The plug is composed of organic waste particles, and more or less parallel layers of the same type of siliceous platelets that are used in construction of the shell. During excystment, water is taken up. The cell swells and ruptures the cyst membrane, and at the same time, the plus is disintegrated. The cystic membrane and components of the plug are ingested as the pseudopodia become active and protrude from the pseudostome.

A comprehensive ecological study of naked and testate amoebae (Bovee 1965) at seven stations within the course of a 7.5 mile stream near Gainesville, Florida provided information on abiotic environment factors supporting growth of various species (Table 5.1). In general, both naked and shelled amoebae were most abundant when pH was at or near 7.6. They were less abundant at pH 7.3 to 7.5 or 7.7 to 7.8; and there were relatively still fewer at pH 7.0 to 7.2 and 7.9 to 8.0. Testate amoebae were occasionally found up to pH 8.2, and certain naked species occurred at pH as high as 8.6. Among the larger amoebae, the very large *Chaos*

TABLE 5.1. Occurrence of naked and testate amoebae in a freshwater stream.*

Organism	Temperature Range	Temperature Most Often	pH Range	pH Most Often	Stations Where Found	Stations Most Often	No. of Times Found
Acanthamoeba spp	16.3–28.4	below 20	7.2–8.3	7.7	a–g	a	9
Hartmannellid spp	9.8–28.9	19.6–20.5	5.2–7.9	7.6	a–g	d	24
Amoeba proteus	21.1–23.0	21.1–23.0	7.3–7.4	7.3–7.4	g	g	2
Amoeba discoides	19.7–28.4	above 21	7.7	7.7	a,e	a	3
Polychaos dubia	19.6–21.1	19.6–21.1	7.4	7.4	a,g	a,g	2
Chaos carolinensis	—	20.0	—	7.7	a	a	1
Pelomyxa limicola	19.6–23.3	below 21	7.4–7.7	7.7	a,d,e,g	a	4
Pelomyxa lucens	20.3–27.3	24	7.3–7.8	7.5	b,e,g	b	4
Pelomyxa villosa	20.3–28.4	above 21	7.6–8.6	8.0–8.6	g	g	4
Trichamoeba villosa	17.2–23.3	below 21	7.3	7.3	e,g	e	3
Mayorella cultura	19.3–30.5	below 21	7.1–8.6	7.8	a,c,d–g	a	7
Mayorella dofleini	16.2–28.9	below 21	6.5–7.9	7.3	a,b,e	a	10
Mayorella spp	15.5–28.4	23.0	7.1–8.6	7.5	a–c,e,g	a	20
Vexillifera telma	21.0–25.0	25.0	7.6–8.6	7.6	b,g	b	2
Vexillifera spp	17.1–21.8	above 21	7.1–7.9	7.6	a–c,e,g	a	14
Flabellula mira	20.5–26.2	26.2	7.5–7.6	7.6	c,e	e	2
Flabellula simplex	16.3–20.5	below 21	7.6	7.6	d,e	e	2

Naked and Testate Amoebae

Species					
Thecamoeba striata	15.8–24.3	15.8	7.5–7.6	7.6	e,f	f	2
Thecamoeba lanceolata	19.6–21.6	below 20	7.3–7.8	7.3	a,e,f	e	3
T sphaeronucleolus	16.2–25.0	below 21	7.3–7.6	7.6	b,e,g	e	3
Cochliopodium bilimbosum	17.2–20.5	below 21	7.3–7.5	7.5	e	e	2
Arcella discoides	10.3–28.4	20.9	7.4–8.1	7.8	a,e,f,g	g	10
Arcella vulgaris	15.5–28.4	21.0	7.2–8.1	7.6	a,b,d,e–g	e	24
Difflugia acuminata	21.1–25.0	above 21	7.7–7.9	7.7–7.9	a,g	g	2
Difflugia elegans	16.2–18.7	below 20	—	—	a,g	g	2
Difflugia globulosa	15.5–28.4	below 21	7.4–8.6	7.6	a,b,d,g	g	13
Difflugia lobostoma	28.3	28.3	7.6	7.6	g	g	1
Difflugia oblonga	15.5–25.0	below 20	7.1–8.1	7.8	a,b,e–g	a	15
Difflugia oviformis	20.5–28.4	28.4	7.3–8.1	7.6	a,b,f,g	a	12
Difflugia tuberculata	19.6–28.4	above 28	7.6–8.6	7.6–8.1	a,g	g	4
Difflugia urceolata	19.6–27.5	23.0	7.2–8.1	7.7	a,b,g	a	11
Lesquereusia spp	21.7–25.0	25.0	7.7–8.1	8.1	a	a	2
Centropyxis spp	21.7–24.0	above 21	7.4–7.7	7.4	a,d	d	3
Cyphoderia ampulla	16.2	16.2	7.6	7.6	g	g	1
Trinema lineare	15.8–17.8	17.2	7.1–7.7	7.3	a,d,e,f,g	e	11
Gromia spp	20.3–25.0	25.0	7.6	7.6	b,g	b	2

*Adapted from Bovee (1965).

carolinense was observed only once after growing up in culture at 20°C in water obtained from a drainage ditch near a storm sewer outlet. *Amoeba proteus* was found only twice at one station with water temperature of 21° to 23°C, pH 7.3 to 7.4, with high dissolved oxygen (11.3 to 11.8 ppm). *Mayorella dofleini* was the most often encountered large naked amoeba in the same location as *C carolinense*. The temperature range was 16.2° to 28.9°C, pH 6.5 to 7.9, but most abundant occurrences were at pH 7.6, and below 21°C. Some species were present over wide variations in environmental conditions. For example, *Arcella vulgaris* was present in active form at all but one station, whether sunlit or shaded, and at temperatures from 15.5° to 28.4°C, with pH 7.2 to 8.1. Other species tolerating a broad range of temperatures were *Mayorella dofleini*, *Trichamoeaba* sp (16.2° to 28.3°C), *Arcella vulgaris* (15.5° to 28.4°C), *Arcella discoides* (10.3° to 28.4°C), *Difflugia oblonga* (15.5° to 25.0°C), and *Difflugia globulosa* (15.5° to 28.4°C).

The distribution and numbers of species in eight families of testate amoebae in three environments (1) a shallow basin, (2) littoral zone of a lake, and (3) river were examined by Moraczewski (1965). In the shallow basin, the most common family was Difflugiidae, contributing approximately 44% of the fauna. Centropyxidae, Arcellidae, and Euglyphidae contributed between 12 to 16% of the total fauna, while Cochliopodiidae, Cyphoderiidae, Cryptodiiflugiidae, and Nebeliidae accounted for less than 5% each. In the littoral region of the lake, Arcellidae and Centropyxidae were most dominant, accounting for over 40% and 30% of the fauna respectively. The lesser contributing families were Difflugiidae (ca 22%), Euglyphidae (ca 8%), and Cochliopodiidae (< 5%). The river-inhabiting fauna was characterized by Arcellidae and Difflugiidae (ca 30 to 35% respectively), Centropyxidae (17%), and Cyphoderiidae, Euglyphidae, and Nebelidae (< 10% each). A more detailed analysis of species abundance was made over a two year period at the location of a man-made lake formed by a dam across a river (Moraczewski 1967). Samples of testate amoebae were taken at four locations across a shallow part of the basin. Site 1 was 40 m from the edge of the lake at a steep sandy slope without a dike, where the deepest part of the water was on the average 7 m. This site was the location of an old lake, with dark well-mixed, gelatinous sediments. The water column ws occasionally stratified thermally and aerobically, but not to a great degree. Site 2 was 200 m from the edge of the lake, where the water depth was about 3 m. At this site, once an old prairie, there was a large stand of *Poligonum ambiguum* and the substratum formed a thin layer of mud. Site 3 was located 15 m from the opposite shore, where the water was no more than 0.5 m in depth and was occupied at times by stands of herbaceous and aquatic plants (eg, *Poligonum ambiguum*, *Potamogeton* sp, and reeds). The fourth site was at the edge of the lake along the original riverbed. The depth of the water was up to 5 m. The bottom was covered with sand and a thin layer of mire. Among the species indigenous to the original shallows near the river bed, 62% had invaded the aquatic environment of the reservoir. Well over 50% of the species occurred in the aquatic environment and a substantial number were associated with mosses and sphagnum. In total, sixty-six species and 17 varieties of testaceans were identified in the environment of the lake. A condensation of the faunal list with species densities is presented in Table 5.2. The diversity of the testacean communities in each habitat, based on the Moraczewski's total data, is also reported at the base of each column. The diversity in the first year is greater than in the second for all four sampling sites. This may be explained by annual variations in climate, especially rainfall, or other factors altering the amount of moisture and/or nutrients. In the first year, sampling sites 1 and 3 had the most diverse communities, whereas in the second year, sites 3 and 4 were most diverse. Based on these data, niche overlap (ie, Cxy) is also calculated for some representative pairs of species to illustrate the quantitative expression of this concept. The value for each of the pairs during the first year is: *D globulosa* with *C ecornis* ($Cxy = 0.64$); *D globulosa* with *C constricta* ($Cxy = 0.48$); *D acuminata*

TABLE 5.2. Testate amoebae distribution in a lake.*

Species[†]	Stations with Highest Abundances	
	Year 1	Year 2
Difflugia globulosa	1–4	1–4
Difflugia gramen var *globularis*	1,3,4	2,3
Difflugia acuminata	2,4	1,2,3
Difflugia oblonga var *longicolis*	1–4	1,2
Centropyxis constricta	1,3	1,4
Arcella gibosa	1	1,2
Centropyxis cassis	1,2	3
Difflugia corona	3	2,3
Cyphoderia ampulla	1	1,2,4
Cyclopyxis arcelloides	3	4
Difflugia elegans	1	1,2,4
Difflugia urceolata	3,4	1,3,4
Euglypha loevis	1,2	1,4
Euglypha acanthophora	1,2	3
Euglypha rotundata	1,3	1
Centropyxis discoides	1	—
Arcella vulgaris	1	—
Trinema lineare	1	—
Lesquereusia modesta	—	4

Diversity Indices	Stations							
	Year 1				Year 2			
	1	2	3	4	1	2	3	4
Diversity (H)	4.71	4.12	4.54	4.08	2.96	3.83	3.88	3.57
Max. Possible H	5.43	5.17	5.46	4.81	4.09	4.32	4.46	4.39
Diversity (Δ) Possible range 0–1.0	0.97	0.93	0.96	0.94	0.88	0.95	0.95	0.91

*Adapted from Moraczewski (1967).
[†]Arranged according to approximate decreasing abundance. Stations: (1) Sandy slope at edge of lake; (2) 200 m from edge of lake, water depth = 3 m; (3) Opposite shore, water depth = 0.5 m; and (4) Edge of lake along original river bed. See text for more detailed descriptions.

with *D oblonga* (*Cxy* = 0.78). Among these three sets of comparisons the *D acuminata* and *D oblonga* niche overlap is the greatest. This can be understood intuitively by noting the high correleation of the frequency of occurrence for the two species in each of the four habitats. It is clear that there is a close correlation of occurrence. The other pairs of species, however, do not have such a clear co-occurrence in each of the habitats. As stated earlier, this measure only considers frequency of co-occurrence in a given habitat without regard to possible differences in resource utilization. In the current data set, space is at least fairly constant since the frequencies were presented as per cm^2 of surface in the habitat. However, the equation does not take into account factors such as amount of food, oxygen content of the water, etc.

COMPARATIVE HABITAT DATA

Some interesting comparative data can be made provisionally between Bovee's (1965) and Moraczewski's (1967) data for testate amoebae assuming, among other limitations, that both authors used the same taxonomic criteria. *D globulosa* was found relatively frequently by both researchers. In the Florida natural stream, it occurred in four of the seven sites, while in the Polish reservoir it was consistently abundant in all four sites for both years. It is interesting to note that in the reservoir area it was most abundant in site 4, which was the site of the original river channel. In the Florida stream, it was most abundant in station g, which was fundamentally a rural stream location with only occasional moderate pollution from runoff. *A vulgaris* occurred frequently in the stream samples in six of the seven sites, and most commonly in site e, a location receiving frequent runoff from mesic hammock woods and scattered homesites. In the reservoir, however, it was infrequent, with 0.6 individuals per cm^2, and only at site 1 which was near the outer edge of the reservoir near a steep incline, presumably with some runoff from land, although there were few homesites nearby. *D urceolata* occurred with a relatively high frequency in the Florida stream at three locations, especially site a (an open, rural, fertilized, nonpolluted section). It was moderately abundant in the Polish reservoir at site 3 at about 0.5 m depth with vegetation, and also at site 4 near the edge of the original river. In year 2, it was also abundant at the steep sandy slope on the opposite side of the reservoir. No data on pH or temperature are available for the reservoir samples, so it is not possible to make further ecological comparisons. Moreover, care must be taken not to make extensive extrapolations using data from only two studies at widely different geographical locations. However, for the most productive species of both locations, there is a good correspondence between the gross habitat indicators and the occurrence of a given species.

PRODUCTIVITY

Production studies of testate amoebae are enhanced by use of the empty shells of dead specimens in addition to those alive in making a full account of productivity over time. The complex set of processes involved in assessing populations of

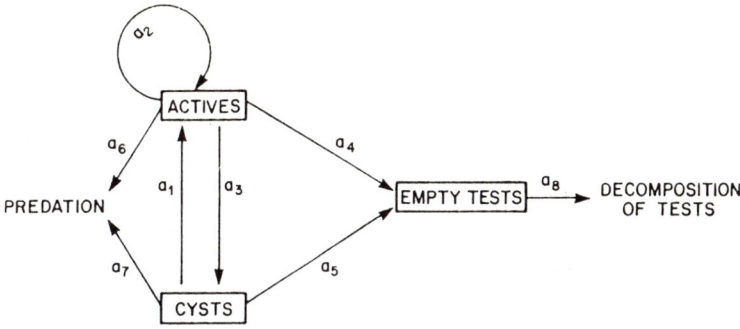

FIGURE 5.1. Environmental and population processes that can effect the turnover of testacean numbers (a_1 = excystment, a_2 = reproduction, a_3 = encystment, a_4 = death of active testate amoebae, a_5 = death of encysted amoebae, a_6 = predation on active testate amoebae, a_7 = predation on encysted individuals, a_8 = decomposition of empty tests). From Lousier (1974); with permission.

shelled amoebae is diagrammed in Figure 5.1 based on research by Lousier (1974). Decomposition of the tests is an important limiting factor in making long-term production studies if sampling intervals exceed the residence time of the empty tests. In general, tests vary in durability in the natural environment. Estimates of endurance vary from 1 to 2 weeks, with significant loss after 4 to 6 weeks. However, shells composed of platelets (eg, *Euglypha*) may be more labile than those formed from sediment particles. With a good estimate of the rate of decay of tests, the total production of individuals within a time period is given by the following equation based on the rate terms shown in Figure 5.1:

$$a_2 = N_t - N_0 + (T_t - T_0 + a_8 + (a_6 + a_7)), \qquad 5.4$$

where a_2 = production numbers, N_t = number of individuals at the end of the time period, N_0 = number of individuals at the beginning of the time period, T_t = number of empty tests at the end of the time period, T_0 = number of empty tests at the beginning of the time period, a_6 and a_7 are respectively loss of active and encysted individuals by predation, and a_8 = tests decayed during time period.

The annual population dynamics and production ecology of testate amoebae in aspen woodland soils were studied by Lousier (1982; 1984a,b,c,d; 1985) in the Rocky Mountains of Alberta, Canada. There were 28 taxa of living Testacea, and 14 were considered constant. Mean annual biomass and total annual production were assessed as 0.72 and 206 g wet weight per m², respectively. For the Centropyxidae, mean annual biomass and total annual production were 0.14 and 41.1 g wet weight per m², respectively. Comparable figures for Euglyphidae were 0.35 and 131 g wet mass per m², respectively, and for Nebelidae 0.03 and 6.4 g wet weight per m², respectively. Some general methods for isolation and analysis of soil-dwelling protozoa and lists of the most commonly encountered species have been presented by Stout et al. (1982). Among the testate amoebae, the shape of the test commonly reflects the moisture regime and structure of the substrate. For example, species with flattened plano-convex tests normally occur in open structures such as loose soil and litter (*Arcella, Microchlamys*), while those with high-vaulted tests normally are associated with forest litters and mosses (*Nebela, Difflugia*).

By keeping account of living specimens and dead individuals, and using sampling periods of two weeks, Schonborn (1977) made an extensive study of testate amoeba production in two rivers and a moss community in Germany. One of the rivers was relatively unpolluted (beta-meso to oligosprobic) and the other was more polluted (alpha-mesosaprobic). The production values he reported are cited as individuals per square meter per year (i) and grams of carbon per square meter per year (g). For the beta-mesosaprobic environment, $i = 24 \times 10^6$ and $g = 1.0$; in the alpha-mesosaprobic river, $i = 3.2 \times 10^6$ and $g = 0.35$. In the moss community $i = 145 \times 10^6$ and $g = 0.11$. The smaller carbon production in the moss compared to the river environment, in spite of larger numbers of individuals, is probably due to the smaller size of the species in the moss community. With respect to the comparative data on rivers, in general, diversity is higher and the population densities greater in the less polluted river. This is consistent with the data of Bovee (1965) and Moraczewski (1967) indicating a general lack of productivity in organic polluted environments.

Benthic Foraminifera

The abundance and diversity of benthic Foraminifera in a wide range of ocean sedimentary environments has received substantial attention. This is probably due to their prominence in benthic communities and the importance of fossil benthic foraminiferal shells in micropaleontological research (eg, Arnold 1974; Lee 1974; Boltovskoy and Wright 1976; Murray 1976).

The assemblages of living benthic Foraminifera found in a variety of geographic locations is summarized in Table 5.3. The number of species and diversity tends to increase with depth (Gibson and Buzas 1973). Some examples are given with depth followed by number of species and diversity, expressed as H, in parentheses: 23 m (19, 2.5), 46 m (44, 2.8), 104 m (47, 3.1), 190 m (61, 3.4). At greater depths, the diversity is variable between 2.2 and 3.0. Substantial numbers of benthic Foraminifera occur in the benthos of major oceans. Bernstein et al (1978) estimated that as many as 120 species and 10,310 shell fragments were found in 10 × 10 cm box core samples of sediment surface in the North Pacific. Their data, and also that of Schafer (1971) (who studied shallow environments), indicate considerable surface patchiness which may be related to the reproductive mode of the Foraminifera or colonization of opportunistic sites where food is abundant.

Lee (1974, 1980a; Lee and Muller 1975), who has contributed substantially to our knowledge of the physiological ecology of benthic Foraminifera, has also

TABLE 5.3. Habitats and representative species of benthic foraminifera.

Habitats	Species
Tidal Marshes	
Texas Coast: Salinity variable with freshwater runoff from land, mean = 30‰ in summer and 20‰ in winter; tidal range small <30 cm to 75 cm. Temperature range ca. 2–26°C. Marsh floral zonal with *Spartina alterniflora* in lower periodically flooded parts, *S patens* at a slightly higher level and *Salicornia* on highest parts. Vegetation is scattered and some parts are pools and flats of sand and mud.	*Ammotium salsum*, *Miliammina fusca*, and locally, *Arenoparrella mexicana*, *Tiphotrocha comprimata*, *Trochammina inflata*, and *Ammonia beccarii*.
Lagoons	
A. Long Island Sound: Salinity min. 25‰ in spring, max. 29‰ in autumn. Depth 20–40 m. Temperature, 2°C in winter to 25°C in summer. Sediments, sands, & silts.	*Elphidium clavatum*, *Buccella frgida*, and *Eggerella advena*.
B. Madre, Texas: Salinity 36–47‰ with local extremes of 63‰. Shallow water with vertical mixing. Temperature, 10°–35°C. Sediment largely sand, but grading into silt. Submarine vegetation (*Diplantheria wrightii* and *Ruppia maritima*).	*Ammonia beccarii*, *Triloculinella obliquinoda*, and locally abundant *Quinqueloculina wiesneri*, *Quinqueloculina* spp.
Mississippi Delta	
A. Interdistributary bay: Salinity 1–10‰; depth 0.1–2 m; temperature 0°–38°C.	*Ammotium salsum*, *Elphidium* spp, *Miliammina fusca*, and *Ammonia beccarii*.
B. Fluvial Marine: Salinity 1–32‰; depth 1.6–10 m; temperature 8°–29°C.	*Ammotium salsum*, *Elphidium gunteri*, and *Palmerinella gardenislandensis*.

TABLE 5.3. *(Continued)*.

Habitats	Species
C. Deltaic Marine: Salinity 34–36‰; depth 10–40 m; temperature 19–30°C.	*Nonionella opima, Bolivina lowmani, Buliminella cf. B basendorfensis,* and *Ammonia beccarii.*
D. Sound: Salinity 18–36‰; depth 1–10 m; temperature 15–30°C.	*Ammotium salsum, Elphidium gunteri,* and *Ammonia beccarii.*
E. Open Shelf: Salinity > 35‰; depth 0–120 m; temperature 17–31°C.	*Ammotium salsum, Ammonia beccarii,* and *Nonionella opima.*
Oceanic Bay	
Baltic Sea: hyposaline (ca 32‰), depth 40–80 m and stratified with greater temperature fluctuations (annual range 2°–16°C) in surface water than at depth.	*Elphidium incertum* (present throughout), *E excavatum* at depth and *Ammotium cassis* in the transitional depths.
Shelf Seas	
Vineyard Sound: Salinity 32‰; channel max. depth 35 m; temperature 16°–16.6°C. Sediment coarse sand and gravel with local sand waves.	*Eggerella advena, Rosalina* spp, *Trochammina ochracea*; species attached to rocks: *Cyclogyra involvens, Patellina corrugata, Rosalina* spp and *Spirillina vivipara.*

Based on data from Murray (1976).

documented the patchiness of benthic Foraminifera in salt marshes near Long Island, New York. For example, hundreds of thousands of individuals can be found in one location, and less than 0.5 m away few, if any, may be found. The density of benthic Foraminifera can reach 10^3 per cm^3. The concentration of individuals (number/cm^3) of other major Protista epiphytic on the large thallus-forming alga *Enteromorpha* include: diatoms, 10^3 to 10^4; euglenoids and colorless flagellates, 10 to 10^2 each; chlorophytes, 10 to 10^3; amoebae, ca 10 to 10^2; and ciliates, 10 to 10^2. Among the benthic Foraminifera, during a two year study, *Ammonia beccarii* was twice as abundant as any other species and comprised 40% of the total foraminiferal population. The remaining samples contained the following dominant species: *Protelphidium tisburyensis*, *Allogromia laticollaris*, *Elphidium incertum*, *Quinqueloculina seminulum*, and *Trochammina inflata*.

Among the several factors contributing to differences in distribution of benthic Foraminifera across space and time, climactic events, seasonal changes in water turbidity, flow, and temperature need to be considered. Some of these are discussed conceptually by Lee (1974, 1980). The quality of the substratum is significant in determining distribution. Certain species prefer to attach to fine grain sediments, while others inhabit surfaces composed of larger sized particles. Others colonize the organic rich surfaces of macrophytic algae and submerged leaves of plants.

Species found largely in the sediment included *Trochammina inflata*, *Quinqueloculina seminulum*, and *Ammotium salsum*. *Elphidium incertum* was abundant in both fine grain sediments and in epiphytic communties. The epiphytic species also included *Elphidium advenum*, *E gunteri*, *E galvestonense*, and *Quinqueloculina lata*. *Rosalina leei* was present during two years of sampling, but was not found in the third year. Changes in annual rainfall and the quantity of lower temperature water running off from tributaries may account for some of the seasonal and annual differences. Variations in abundance and kind of macrophytic algae and submerged grasses also influence foraminiferan abundance and diversity. Lee (1974) reports statistical data showing that *P tisburyensis* was found most often on *Enteromorpha*, while species of *Quinqueloculina* were less likely to be so distributed. *A beccarii* and *Elphidium* spp apparently had no substrate preference. Large standing crops of benthic Foraminifera were found in the sublittoral epiphytic communities of the "marsh grass" *Zostera marina*, and large algae including *Enteromorpha intestinalis*, *Ulva lactuca*, *Polysiphonia* spp and *Ceramium*. The species diversity index was low (0.581), but the standing crop was high for communities on decaying *Enteromorpha* compared to a diversity of 0.94 on young green patches. Lower diversity on the decaying algal masses may reflect greater productivity of a few well-adapted species that are able to exploit the bacterial-rich, eutrophic environment within these moribund masses. In general, the principle that high productivity is inversely related to diversity is demonstrated by this special case of the salt marsh community. By contrast to the moribund algal masses, young green patches of algae are likely to have diverse food sources and provide open habitats where a wide variety of pioneering species can invade with less competitive exclusion. The low productivity of individual species coupled with ample resources sometimes yields high species diversity. Likewise, environments subjected to turbulence or other high energy conditions, causing moderate attrition of organisms on the surface, may provide open surfaces where additional pioneering species can invade periodically. This constant recruitment may lead to low biomass, but high diversity.

Indices of diversity for other communities, categorized according to the plant surface where the foraminiferans grew, were *Zostera* (0.82), *Polysiphonia* (0.86), *Ulva* (0.77), and *Codium* (0.196). Studies on natural growth rate (r = organisms produced per day) and density-dependent competition were reported for three species: *Allogromia laticollaris* (r = 2.533), *Spiroloculina hyalina* (r = 1.472), and *Rosalina leei* (r = 0.272). Crowding affected the feeding rate and reproduction of *A laticollaris* more adversely than for *S hyalina* or *R leei*. The feeding of *S hyalina* was not influenced by crowding although the reproductive rate *was* influenced (Muller 1972). Interspecific competition as measured by the relative decrease in feeding during crowding with another species compared to feeding alone (I.S.C., Equation 5.3) was determined as follows: *S hyalina* (I.S.C. = ca 0.25), *A laticollaris* (I.S.C. = ca 0.65), *A beccarii* and *R. leei* (I.S.C. = 0.75). Some environmental variables influencing survival of some benthic and planktonic Foraminifera are given in Table 5.4.

Planktonic Foraminifera and Radiolaria

Planktonic Foraminifera are exclusively open ocean Sarcodina that dwell in unpolluted water, and are therefore fastidious and difficult to culture in the laboratory. Hence, much of our knowledge of their ecology is based on field studies of natural populations in plankton communities, and on using short-term cultures of those species that are amenable to growth in the laboratory (eg, Bé 1977; Bé et al 1977; Anderson et al 1979; Anderson 1984). Samples of species taken at stations in the major oceans of the world at high to low latitudes show that many nonspinose species dwell in cool water either in the surface water of high latitudes and subtropical latitudes during winter, or at great depths in the ocean at lower latitudes. Spinose species, generally, are warm water, surface-dwelling species, and many possess dinoflagellate symbionts that provide photosynthetic products to their host. Five major faunal provinces have been identified based on the kind of planktonic Foraminifera that dwell there, and also on the average water temperature (Bé 1977). In general, the number and diversity of planktonic foraminiferal species increases in samples as one progresses from northern to more equatorial latitudes. This is reflected in the number of marker species (in parentheses) for each of the following five provinces: (1) Arctic (five species); (2) subarctic (eight species) provinces in the polar regions; (3) transition zones between cold and warm water regions (18 species); (4) subtropical provinces north of the equatorial belt subject to rather large fluctuations in temperature during an annual climatic cycle (20 species); and (5) tropical provinces occurring near the Equator and typified by rather constant warm water tempera-

TABLE 5.4. Some environmental variables influencing survival of representative benthic and planktonic foraminifera.*

Species	Temperature Tolerance I °C	Salinity Tolerance I ‰	Diet
Benthic			
Allogromia laticollaris	14–32 (R)	20–45	Schizothrix, calcicola
Ammonia beccarri	20–32 (R)	13–56	Amphora sp, Nitzschia sp, & algae
Bolivina doniezi	18–22 (R)		Nitzschia angularis
Calcituba polymorpha	–	–	Chyrosophyceans, diatoms
Discorbis sp	< 18 (R)		Nitzschia sp, Navicula spp
Elphidium articulatum	0–20	–	–
Elphidium crispum	–	19.5–48	Diatoms
Elphidium excavatum	7–25	–	–
Miliammina fusca	10–25	–	–
Patellina corrugata	18–25 (R)	–	–
Rosalini leei and spp	15–25	18–47	Amphora sp, Nitzschia, and Dunaliella sp
Rotaliella heterocaryotica	15–24	24–37	Dunaliella praemolecta
Spiroloculina hyalina	14–24 (R)	17–44	Phaeodactylum sp and Amphora sp + algae
Planktonic			
Globorotalia truncatulinoides (N)	4–27	–	Flagellates and diatoms
Globigerinella aequilateralis (S)	12–30	–	Copepods, algae, & small metazoa
Hastigerina pelagica (S)	15–29	–	Copepods, small metazoa
Orbulina universa (S)	10–30	–	Copepods, algae

*Based on data from Bé and Tolderlund (1971); Lee (1974); Boltovskoy and Wright (1976); Anderson et al. (1979); and Lee (1980a). (R) = reproductive range, (N) = nonspinose, and (S) = spinose.

ture (24 species). The nonspinose species are generally herbivorous or consume colorless flagellates, while the spinose species are omnivorous, with the exception of *Hastigerina pelagica*, which is apparently totally carnivorous. The seasonal abundance of planktonic Foraminifera is correlated with phytoplankton productivity. In midlatitude provinces as in the northern Sargasso Sea, a maximum abundance of planktonic Foraminifera occurs between December and March, but in the subarctic waters maximum numbers occur between May and October. The variation in peak numbers of organisms also increases progressively from low to high latitudes. In the northern Sargasso Sea, the difference between the maximum and minimum number of planktonic Foraminifera during the annual cycle is several-fold, whereas in the subarctic water, the difference is ten times higher. Seasonal successions occur regularly in the north Sargasso Sea as the water temperature increases from winter to summer of each year. Nonspinose species (eg, *Globorotalia truncatulinoides* and *G hirsuta*) flourish in winter between December and April. They are gradually replaced by a summer fauna (eg, *Globigerinoides sacculifer*, *Orbulina universa*, and *H pelagica*) that appears most abundantly between June and September. In subpolar and polar regions, where temperature cycles are less pronounced, there is little or no evidence of a species succession.

Radiolaria are also open ocean dwelling Sarcodina that often co-occur with planktonic Foraminifera. There are distinctive species that mark oceanic climatic regimes (Casey 1971; Anderson 1983; Swanberg and Bjørklund, 1986). High latitude species tend to be smaller Nassellarida and Spumellarida, while in the equatorial surface water, large gelatinous solitary and colonial species are common. One also finds certain smaller nassellarian and spumellarian species characteristic of warmer environments. However, at great depths beneath the warm surface water, communities of cold water radiolaria occur, and are believed to be continuous in distribution with those dwelling in the cold surface water of the high latitude regions. Thus, the polar species form a continuous community from the surface water at high latitudes into the deeper cold water underlying the equatorial belt. Since radiolaria and planktonic Foraminifera dwell in the same water mass, competition for nutritional resources would be expected to be high unless there were some differentiation in feeding behavior to reduce niche overlap (eg, Anderson 1983). In general, from limited studies on feeding behavior, it appears that some larger solitary radiolaria consume algal prey more heavily than do spinose planktonic Foraminifera and therefore do not compete for zooplankton prey as vigorously as do the planktonic foraminiferal species. The smaller radiolarian species may also reduce competition with planktonic Foraminifera by consuming largely bacteria, and pigmented and colorless microflagellates. Experimental studies using C-14 labeled bicarbonate have shown that at least some of the larger solitary radiolaria and the colonial radiolaria can acquire sufficient carbon nutrients from their photosynthetic algal symbionts to maintain themselves in the absence of phytoplankton and zooplankton prey. This adaptation may further reduce their need to compete with other microplankton for food, and permits the radiolaria to successfully exploit oligotrophic ocean regimes where primary production is not especially high. They occur abundantly, for example, in the surface water of the Sargasso Sea gyre. There is also good evidence that variations in prey specialization occur among the radiolaria (eg, Swanberg et al. 1986). Analyses of natural prey in food vacuoles of some large spongiose skeletal radiolaria compared to a gelatinous solitary species *Physematium muelleri* indicate that *P muelleri* had captured more silicoflagellates, Acantharia, and mollusc larvae. By comparison, the spongiose skeletal species captured more copepod prey.

The density of radiolaria in equatorial waters reaches levels as high as 10,000 to 16,000/m^3. However, the density varies with depth depending on latitude and hydrographic conditions. In the Pacific, for example, Petrushevskya (1971) found that although radiolaria occurred throughout the water column from the surface to 5,000 m, most were in the upper horizons at depths in a range from 0 to 25 m, to 100 to 150 m. The density of radiolaria in these upper strata was 5,000 to 15,000/m^3, and the temperature was 23 to 26°C with a salinity of 34 to 36‰.

At deeper horizons, 100 to 500 m, the density of radiolaria was about $1,000/m^3$. At depths greater than 500 m, the number was not higher than 100 to 500 individuals/m^3. Two major assemblages of radiolaria were found at this site: (1) a surface-dwelling group (0 to 100 m) at a temperature of 20 to 28°C and a salinity of 33.9 to 35.9‰, and (2) deeper dwelling forms (75 to 100 m), temperature 7 to 19°C and salinities of 34.2 to 36‰. This area of the Pacific Ocean is highly productive, and other sites have far fewer radiolaria. Densities are in the range of 10s per m^3. Further examples of abundance and geographic distributional data can be found in Anderson (1983) or Casey (1977). As in the case of planktonic Foraminifera, radiolarian faunal assemblages not only vary with depth, but also according to latitude, or in relation to varying hydrographic regimes. Thus, a thorough knowledge of their ecology is of substantial significance in understanding modern and ancient ocean environments.

Summative Perspective

The amoebae and their relatives occur in widely diverse habitats (terrestrial, aquatic, and aerial on the surface of mosses) and serve diverse ecological roles within protistan communities. Although it is possible to map their variation in space and time in correlation with changing environmental variables such as water quality, temperature, texture of the substratum, and availability of light and nutrients, much remains to be discovered about the complex environmental factors that determine their abundance and diversity. Experimental studies combining laboratory experiments with field-based observations of natural communities may help to clarify the physiological and environmental variables that interact to determine the range and productivity of sarcodines. Communities of testate amoebae, benthic Foraminifera, or other attached forms are more easily traced in space and time compared to planktonic forms. The changing structure of the water mass and its confluence with other water masses makes field observations difficult over extended periods of time. This is especially the case for marine planktonic species that may inhabit patches a mile or more in surface diameter. A more thorough understanding of planktonic protozoan ecology can be achieved by combining data from carefully designed laboratory studies of species autecology with field observations of community ecology at the sampling sites. The free-living "naked" amoebae deserve particular attention, as they have not been studied as intensively as some other species. Perhaps this is due to their delicate quality, which makes it difficult to collect undamaged samples from the natural environment. Their significant role in terrestrial food webs, contribution to nutrient recycling, and widespread occurrence in aquatic environments suggest that innovative and more intensive research is warranted on their physiological ecology.

6
Ciliates (Phylum: Ciliophora)

General Morphology

The capacity to control motility in fine gradations and to swim in complex patterns is a decided advantage for aquatic organisms, especially for those that invade intricate spaces in search of food or pursue motile prey. The ciliates possess a marked variety of motile systems based on ciliary motion. Thus, the organisms are able to move through the environment in search of prey often with remarkable agility and swiftness, or when sedentary, to create a flow of water carrying food to the organism. The arrangement of the hairlike organelles on the cell surface and the degree of coordinated action made possible by close proximity of the cilia varies considerably among groups. In some species, much of the cell surface is covered by rows of cilia (eg, Figure 6.1). These are called holotrichs, as the entire surface is usually rather uniformly ciliated. In other species, specialized groups of cilia are clustered at sites on the surface of the cell, either as somewhat loosely arranged tufts or organized into cirri (Figure 6.2c). A cirrus is a composite group of cilia (few to hundreds of cilia), often tapering toward the distal end and acting in unison. They typically are found in localized regions on the cell surface. When the cirri are largely localized on the ventral and anterior surfaces the ciliates are categorized as hypotrichs, signifying that the cirri are positioned beneath the cell. In other species, cilia are arranged in spirals (eg, in the spirotrichs) or in ribbonlike assemblages near the anterior (sometimes broader end) of the cell. Although many ciliates exhibit rapid and impressive locomotion, others such as the suctorians (eg, Figure 19.5) are sedentary and lack cilia during most of their life cycle. These feeding (trophont) stages are anchored to the substratum by a stalk or other form of holdfast. They snare prey on tentacles projecting from the cell surface. Ciliated stages occur during reproduction, and aid dispersal.

While the cilia are clearly one of the distinctive features of the Ciliophora, the nuclei, rather than the cilia, provide the main distinguishing feature from other protozoa. There are two types of nuclei: (1) the macronucleus (one or more), regulating metabolism; and (2) small micronuclei, mediating sexual recombination and reproduction of the cell. The macronuclei arise by polyploidy from

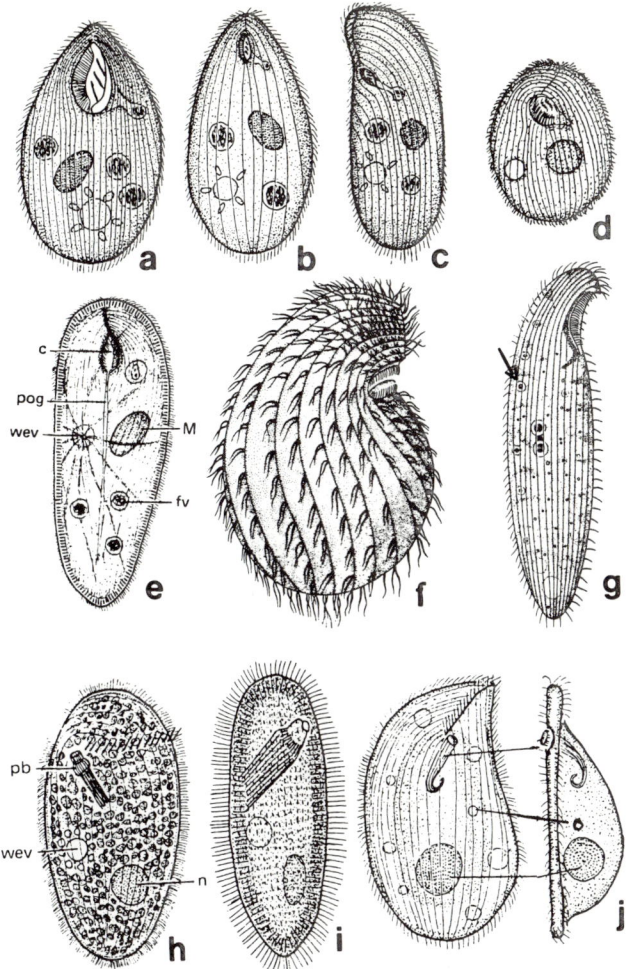

FIGURE 6.1. Comparative morphology of some common ciliates. (a) *Tetrahymena patula*, 80–160 µm; (b) *Tetrahymena pyriformis*, 40–60 µm; (c) *Colpidium campylum*, 50–70 µm; (d) *Glaucoma scintillans*, 45–73 µm; (e) *Frontonia leucas*, 150–600 µm; (f) *Colpoda cucullus*, 100 µm; (g) *Loxodes rostrum*, 125 µm; (h) *Nassula aurea*, 200–250 µm; (i) *Paranassula microstoma*, 80–95 µm; (j) *Chilodonella cucullus*, 130–150 µm. Symbols: c = buccal cavity, pog = post oral groove, wev = contractile vacuole, M = macronucleus, fv = food vacuole, pb = pharyngeal basket, arrow = Müller vesicles, possibly geosensory organelles as described in Chapter 18. Figures (a–e) and (h–j) from Theodore L. Jahn, Eugene C. Bovee, and Frances F. Jahn, *How to Know the Protozoa*, 2d ed. Copyright 1979 Wm. C. Brown Publishers, Dubuque, Iowa. All rights reserved. Reprinted by permission. Remainder from Grell (1973).

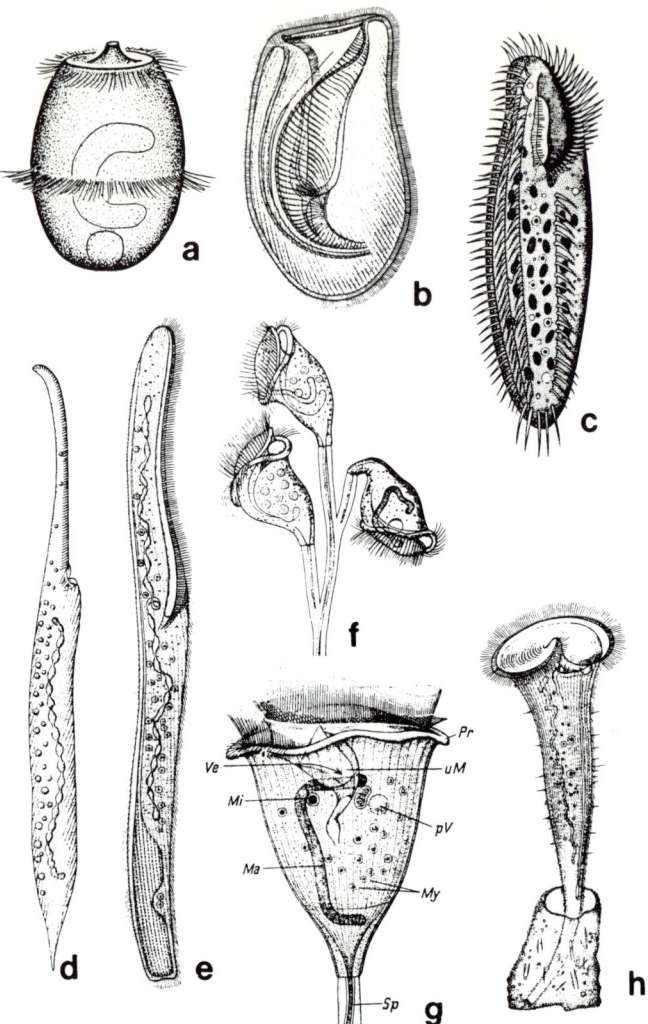

FIGURE 6.2. Comparative morphology of some larger ciliates. (a) *Didinium nasutum*, 80–200 μm; (b) *Bursaria truncatella*, 500–1000 μm; (c) *Keronopsis gracilis*, hypotrich, 100 μm; (d) *Dileptus anser*, 250–500 μm; (e) *Spirostomum ambiguum*, 1–3 mm; (f) *Carchesium polypinum*, 120–125 μm; (g) *Vorticella nebulifera*; (h) *Stentor roeseli*, 600–700 μm. Symbols: Ma = macronucleus, Mi = micronucleus, pi = peristomal margin, ve = vestibulum, uM = undulating membrane, My = myonemes, Sp = spasmoneme contractile sheath. Figure (a) from Theodore L. Jahn, Eugene C. Bovee, and Frances F. Jahn, *How to Know the Protozoa*, 2d ed. Copyright 1979, Wm. C. Brown Publishers, Dubuque, Iowa. All rights reserved. Reprinted by permission. Remainder from Grell (1973).

micronuclei. This nuclear duality is not found in other protozoa, where typically there is a single nucleus. When multiple nuclei occur in these other protozoa, they are all of similar morphology and presumably identical function. One or more contractile vacuoles are typically present in freshwater ciliates to regulate ion balance of the cytoplasm and expel excess water. Some marine species also possess contractile vacuoles, but these are probably not used as a water-expelling organelle, since the tonicity of the surrounding seawater is closer to that of the cytoplasm. Rather, the contractile vacuole probably acts as an ion regulating organelle by expelling excess ions accumulated from the environment. Various extrusomes (ejectile organelles) occur beneath the surface of the cell, including mucocysts that secrete surface substances either to coat the cell to improve its immediate environment, provide material for cyst walls, or aid in the capture of prey. In addition to the commonly observed mucous coat secreted by some ciliates, at least one species *Lepidotrachelophyllum fornicis* (Nichols and Lynn 1984) produces a surface coat of loose organic scales among the cilia. Species of *Coleps* bear somewhat quadrangular, complex calcareous plates surrounding the cylindroid body. Specialized extrusomes with narcotizing ejectile filaments for apprehending and immobilizing prey (toxicysts) occur in some predatory species. Trichocysts, ribbonlike pointed filaments, are ejected by *Paramecium* spp when subjected to noxious stimuli. The functional significance of the response is not clear, as trichocyst ejection seldom prevents predatory attacks, and their release in the presence of noxious chemical stimuli appears to have no ameliorating value. Digestive wastes are released at a specialized area of the cell surface, called a cytopyge, usually in the posterior part of the cell. A small crease or indentation may be present to mark the pore site where the wastes are defecated.

The complex membranous system and associated cilia surrounding the cell, known as the cortex, is analyzed into several levels of organization. In a broad-scale perspective two major regions of ciliature are identified: (1) the somatic region consisting of the cilia covering most of the body surface; largely used for (a) locomotion, (b) attachment to substrate, (c) a protective covering, and (d) sensing the environment; and (2) the oral region where food is accumulated and ingested. The organization of the somatic cilia and the architecture of the oral region are important features in ciliate systematics. Both the somatic and oral regions are underlaid by a characteristic infraciliature within the cytoplasm, consisting of the kinetosome (basal body) of each cilium, and associated bands of cytoplasmic, proteinaceous fibrils. The cilia are organized into a motile unit known as a kinetid (Figure 6.3). Each kinetid consists of three components: (1) one or sometimes two (paired) cilia, (2) the subpellicular kinetosome of each cilium, and (3) a complex array of fibrils situated beneath the cell surface and connected to the kinetosome. Detailed information is presented in Chapter 14. The fibrils typically consist of three types: (1) the kinetodesmal fiber, a cross-striated basal fiber extending from the kinetosome into the cytoplasm, (2) the postciliary microtubular ribbon extending toward the posterior direction, and (3) a transverse fiber arching laterally from the side of the kinetosome. The surface of the cilium is covered by a plasma membrane that is continuous with the plasma

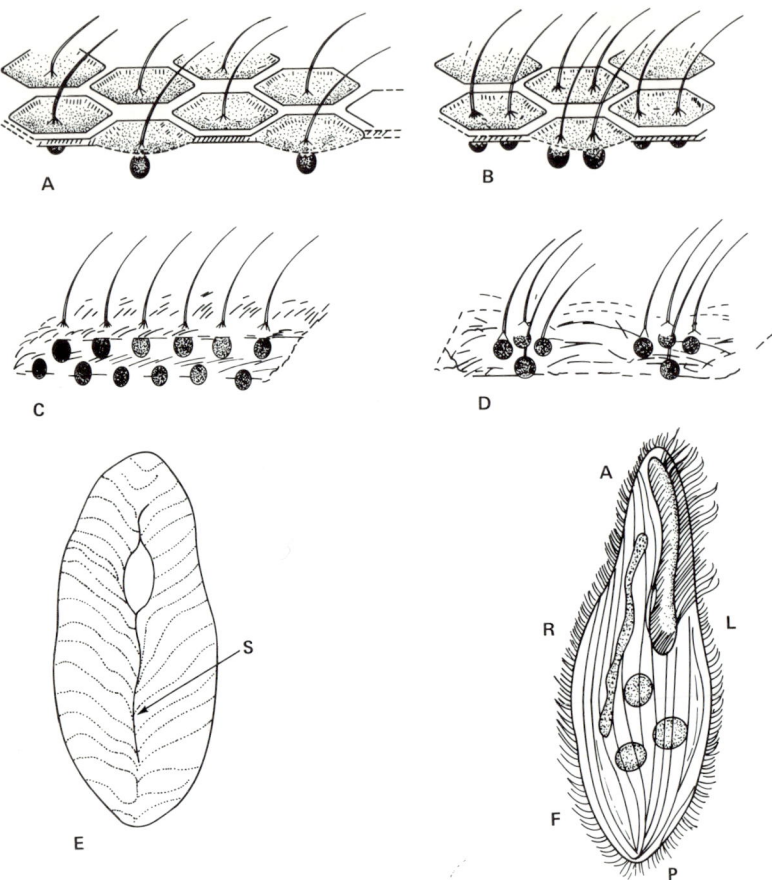

FIGURE 6.3. Somatic microanatomy of ciliates. (a) Monokinetids; (b) dikinetids; (c) polykinetid (haplokinety) with only one row bearing ciliary projections (the other is barren); (d) groups of polykineties; (e) suture line where kineties converge from right and left side of the body; (f) ventral surface of a ciliate showing the anterior (A), posterior (P), and right and left sides (R and L, respectively).

membrane, enclosing the outer surface of the cell. The plasma membrane is often underlaid by a second membrane that forms a set of alveolar spaces surrounding each kinetid (see Figure 14.3). This duplex membranous envelope forms the pellicle of the ciliate, covering the entire cell except at the oral region, where specialized surface membranes usually occur. Comparative fine structural details of the cortex in some representative ciliate species are presented in Chapter 14.

Kinetids are organized into rows called kineties. A kinety is a structural and functional entity consisting of a row of cilia and their overlapping postciliary fibrils. Lateral fibrils connect adjacent kineties. Somatic (body) kinetids may have one, two or more kinetosomes closely spaced, forming a monokinetid, dikinetid, or polykinetid (Figure 6.3a to d), respectively. The arrangement of the kineties and the pattern of occurrence of dikinetids are used in species identification (for example, some taxa have anterior zones with dikentids). In general, the kineties form distinctive patterns of rows of cilia on the cell body, sometimes arranged parallel to the long axis of the cell or in spirals encircling the cell. The line formed by the convergence of encircling rows of kineties, merging from the right and left sides of the body, is known as a suture (Figure 6.3e). Sutures are designated preoral when they extend from the oral region toward the anterior, and postoral when they extend from the oral region toward the posterior. By application of special stains (eg, the protargol or Chatton-Lwoff silver staining procedures, see Lee et al 1985, p. 5) involving the precipitation of silver on the infraciliature, the kineties are visualized as dark lines within the pellicle. A detailed description of ciliate patterns and their significance for systematics is presented by Small and Lynn (1985).

In addition to the somatic ciliature, there are specialized regions for food ingestion. Ciliates ingest food mainly by three ways: (1) phagotrophy (engulfment of food in food vacuoles) through a large unciliated surface area of a cell, probably representing an ancestral mode of nutrition; (2) ingestion through a cytostomal region of the cell aided by specialized cilia, and cytoplasmic fibrils organized within a groove, canal or other special architectural feature of the cell to enhance particulate food engulfment; and (3) absorption and/or pinocytosis across the plasma membrane. Those species with a cytostome often possess specialized kineties to enhance food accumulation and its movement toward the site of food vacuole formation. In some species, a row of cilia borders the right side of the buccal cavity, and often appears as a prominent flickering membranelle projecting from the edge of the buccal cavity. This is the paroral membrane (also known as endoral membrane or undulating membrane). It is a haplokinety characterized by a zigzag arrangement of the ciliary row. Upon closer inspection with proper stains, this kinety is found to be in fact a double kinety, but the inner row is barren; that is, it consists only of the kinetosomes without a ciliary projection, while the outer row is ciliated (Figure 6.3c). The term haplokinety is used for this type of double kinety, since only one half of the double row of kinetosomes is ciliated. Other rows of cilia also occur within the buccal cavity leading into the cytostome where food vacuoles form. These polykineties may consist of four rows of closely spaced cilia, each forming a so-called "peniculus." The number and arrangement of polykineties within the buccal cavity is a significant taxonomic feature of some genera.

The location of surface structures is referenced in relation to the ventral and dorsal surfaces. In general, the ventral surface is the flattened surface bearing the oral cavity when it is present. The dorsal surface is the opposed, sometimes more rounded surface. The anatomical right and left sides of the body are defined in relation to the midline of the cell. As many references are made to the right and left side of the organism, it is important to clearly understand the conventional way of referring to these spatial relations. Imagine that the organism is oriented with the ventral side down; that is, with the oral apparatus pointed forward and downward. Now visualize yourself lying on your stomach parallel to the long axis of the organism, with your head looking toward the front of the organism. Your right arm is toward the right side of the ciliate, and your left arm is toward the left side of the ciliate. That is, the right side of the ciliate is the direction to the right of the oral apparatus when it is oriented downward. If the organism is displayed with the oral apparatus facing upward, then the anatomical right side appears on the left of the figure and the left side appears on the right side of the figure. This can be better understood by looking at your right hand with the palm facing down. Assume that the small finger marks the right side of the hand and the thumb marks the left side. This is directly comparable to the ciliate when it is oriented with the ventral side facing downward. Now rotate your right hand with the palm facing upward. You will see that the small finger (formerly on the right) representing the right hand portion of the body is now facing left, and the thumb representing the left hand portion of the body is facing right. If it is helpful, you may want to draw a small circle on the palm of the hand to represent the oral apparatus of the ciliate, in order to clarify how the anatomical right and left sides are defined. Figure 6.3f illustrates the right and left sides of a generalized ciliate with the oral region facing upward. This is the conventional way of clearly displaying the oral apparatus.

Among several surface features used in systematics, the site of the cytopyge and its prominence, number, and location of contractile vacuoles, the position of the oral apparatus and the number and kind of polykineties when present, and the organization and distribution of somatic kineties are of importance. The organization of the infraciliature (especially the arrangement and orientation of the subpellicular fibrils) are used in class definitions.

The phylogenetic radiation of the ciliates has produced remarkably wide variations in the morphology and mechanisms of feeding organelles, complexity of locomotory systems, size and shape of the cell, and surface structures enclosing the cell. It is not possible to produce a comprehensive taxonomic review of all the ciliates here. Therefore, some representative types will be described to illustrate the range and diversity of forms in this highly successful and elegant group of protozoans as documented more fully in other sources (eg, Corliss 1979; Curds 1982; Curds et al 1983; Small and Lynn 1985). An overview of the major classes described here is presented in Table 6.1.

Class: Nassophorea

One of the most commonly studied ciliates in general biology is *Paramecium*, or one of its relatives in the class Nassophorea, characterized in general by numerous rows of monokinetids or dikinetids (seldom polykinetids) covering the cell surface (Figure 1.1d). The oral cavity is not surrounded by specialized kineties markedly different from those of the somatic regions. *Paramecium* and its close

TABLE 6.1. Major morphological features of some classes of ciliates.

Class	Morphological Features
Karyorelictea	Vermiform, flattened and elongated. Some genera have one barren surface lacking cilia. Extremely contractile, kinetids with overlapping postciliary ribbons or postciliary fibrils. Two to many macronuclei. Includes *Loxodes*, *Tracheloraphis*, *Remanella*.
Spirotrichea	Some genera with dikinetids or somatic polykinetids with overlapping postciliary ribbons. The serial oral polykinetids lead (usually clockwise) as a spiral into the oral cavity. Conspicuous right and/or left preoral ciliature commonly present. Some are loricate, attached to substratum or not. Includes *Blepharisma*, *Stentor*, *Fabrea*, *Spirostomum*, and *Stylonychia*.
Prostomatea	Body with monokinetids, but oral ciliature dikinetids, and tangential to the perimeter of the oral area in some species. Somatic postciliary ribbons slightly convergent and with forward directed kinetodesmal fibril. Cytostome apical to subapical and typically with rhabdose (tubular cytopharyngeal apparatus). Includes *Coleps*.
Litostomatea	Body morphology elongated to subcylindrical, somatic ciliature monokinetids with tangential transverse ribbon, slightly convergent postciliary ribbon and laterally directed kinetodesmal fibril. Oral cilia simple, not polykinetids. Includes *Lacrymaria*, *Didinium*, and *Balantidium*.
Phyllopharyngea	Ciliated stage mainly with somatic monokinetids, sessile stage can be stalked and with tentacles as in suctoria. The oral region of ciliated forms contains radially arranged microtubular ribbons (phyllae), which in the cyrtophorids are enclosed by large, rod-shaped nematodesmata (rods of hexagonally packed microtubules). Includes *Chilodonella*, *Acineta*, and other suctorians.
Nassophorea	Body ciliature variable, monokinetids, dikinetids, or polykinetids. Polykinetids form cirri in several groups. Body alveoli are well developed and a prominent cyrtos (pharyngeal basket) occurs in some groups with at least nematodesmata in the oral apparatus. In the Peniculida (as in *Paramecium*) the oral cavity bears rows of cilia. Includes *Nassula*, *Microthorax*, *Pseudomicrothorax*, *Paramecium*, *Frontonia*, and hypotrichs; eg, *Euplotes*.
Oligohymenophorea	Body cilia monokinetids with distinct overlapping kinetodesmal fibrils, divergent postciliary ribbons, and in most, radial transverse ribbons. Oral apparatus when present differentiated from somatic ciliature. Distinct paroral dikinetid and 1 to many polykinetids. Oral ciliature and apparatus, within ventral oral cavity or deeper infundibulum, may extend onto and surround peristome. Includes *Tetrahymena*, *Glaucoma*, and *Vorticella*.
Colpodea	Body ciliature dikinetids, each complete with cilium. Kinetids have 1 transverse microtubular ribbon and 1 postciliary microtubule associated with the anterior basal body. Posterior kinetosome with 1 transverse ribbon, 1 postciliary ribbon, and 1 kinetodesmal fibril. Resting cysts are common in freshwater and terrestrial species. Includes *Colpoda* and *Bursaria*.

Details of class descriptions can be found in Corliss (1979) and Small and Lynn (1985). Descriptions of kinetid structure and ciliary ribbons and fibrils are presented more fully in Chapter 14.

relatives are placed in the order Peniculida. The peniculine ciliates typically exhibit three peniculi on the left wall of the buccal cavity. They are arranged parallel to the long axis of the cavity and spiral gradually toward the cytostome (eg, Figure 14.11). The right paroral membrane (dikinetid), though detectable, is not prominent. There are two main groups of paramecia based on cell shape, the aurelia type with rather elongate cigar-shaped body, as found in *P aurelia*, and the bursaria type with a profile resembling the sole of a shoe characteristic of *P bursaria*. The structure of the feeding apparatus consists of a scooplike oral groove leading into a conical buccal cavity that connects to a bulbous cytopharynx by a narrow neck. There is one large centrally located macronucleus and one or more micronuclei nearby, sometimes situated in a small depression of the macronuclear membrane. There are typically two contractile vacuoles in paramecia, one situated anteriorly and the other posteriorly.

Nassula, also a nassophorean about 200 µm long, has a decidedly different oral apparatus. The cytostome consists of an oral pore with a cone of rodlike nematodesmata extending into the cytoplasm (Figure 6.1h). The nematodesmata (Figure 6.1h, pb) are proteinaceous rods acting as supporting and reinforcing structures to aid food ingestion. They define the region where food particles enter the cytoplasm and help direct the particles toward the base of the cone where food vacuoles are formed. There are no polykineties within the oral apparatus, as it consists entirely of the intermeshing, conically arranged nematodesmata known collectively as a cyrtos. The cyrtos is most clearly visualized when the organism is oriented on its side, thus permitting a clear perspective of the cone projecting into the cytoplasm. The nassophoreans are ciliates of medium to large size (some > 100 µm long) and are usually densely ciliated. In some species, external oral polykinetids are situated at the left rear of the cytostome and extend laterally onto the dorsum, forming a strip of cilia known as a frange. Members of this group are distinguished on the basis of the position of the cyrtos; ie, whether anterior, lateral (to one side of the body), or posterior. The number and arrangement of polykineties when present and the shape of the body are also of taxonomic significance. In contrast to *Nassula*, *Pseudomicrothorax* is about one half as large, and has more widely spaced kineties running the length of the cell. There are cross-striations between adjacent kineties. The cyrtos is in the forward one third of body. The small sized *Microthorax*, ca 20 µm, has few kineties, widely spaced on the cell body, and the cyrtos is located in the posterior part of the body within a ventral pellicular fold.

The hypotrichs (Figure 6.2c) are a prominent group of ciliates commonly found in aquatic habitats. They are nassophoreans with ovoid to ovorectangular bodies, usually somewhat flattened on the ventral side and rounded on the dorsal side. Dikinetids occur sparsely on the dorsum, and large cirri (closely spaced polykinetids) occur in regular patterns on the ventrum. The characteristic sporadic "crawling" or "running" locomotion associated with spurts of cirral activity on the ventral surface is characteristic of some members of this group. The number and position of cirri (caudal at the posterior, marginal at sides of body, medial near midline of body, or cephalic at anterior) are significant in species identification. There are two major groups, those with a necklike constriction at the anterior separating a distinct cephalic portion from the rest of the body, and those without the cephalic constriction. A prominent polykinety occurs along the left border of the oral cavity in some species.

Class: Oligohymenophorea

Tetrahymena (Figure 6.1a) is a widely studied ciliate grown in laboratory culture for biochemical and physiological research. It is a member of the class Oligohymenophorea. This group, in contrast to the Nassophorea, often possesses a distinct oral ciliated apparatus, clearly distinguished from the numerous kineties covering the body surface. There is a fairly prominent paroral membrane formed by a dikinetid at the edge of the buccal cavity. This "flickering membrane," composed of rows of cilia, aids in the collection of food particles that are swept into the oral cavity. An additional one to many polykinetids line the oral cavity, or extend into the deeper vortex-shaped infundibulum (eg, *Vorticella*, Figure 6.2g) and extend outward onto or surrounding the peristome. The peristome is the perimeter of the oral opening.

Tetrahymena (meaning four membranes) has in addition to the paroral membrane, three polykineties, sometimes referred to as oral membranelles, within the oral cavity. In general, the presence of four oral membranelles is characteristic of the suborder Tetrahymenina (including *Tetrahymena* and related genera). Discriminations among some small members of these genera may be difficult owing to similarities in body shape and ciliature. Four commonly encountered and easily confused species (Figure 6.1a through c) include *Tetrahymena patula* (80 to 160 μm), *Tetrahymena pyriformis* (40 to 60 μm), *Colpidium campylum* (50 to 70 μm), and *Glaucoma scintillans* (45 to 73 μm). These genera have longitudinal arrays of kineties and a prominent paroral membrane on the right edge of the oral cavity, with three internal membranelles to the left. *Tetrahymena* tends to be more narrowly pyriform, but the most convenient method of determining differences is probably counting the number of kineties extending posteriorly from the edge of the oral cavity. Both *T patula* with five and *T pyriformis* with two have rather widely spaced kineties at the base of the oral opening. There is one in *Colpidium*. Moreover, in *Glaucoma* the axis of the mouth is oriented at approximately 45° to the long axis of the body; while in *T patula*, the mouth is large and oriented parallel to the long axis of the cell body. The cells can be visualized more readily by killing and staining them with Lugol's iodine solution.

The peritrichs are a group of Oligohymenophorea with a prominent peristome typically at the broad anterior end of the cell body as in *Vorticella*. The peristome is surrounded by a ciliary fringe (hence the name peritrich or peripheral hairs) formed by a single file of dikinetids. An inner whorl of polykinetids originates from within the oral cavity (infundibulum) and spirals outward onto the anterior surface of the cell (Figure 6.2f,g). As with *Vorticella*, some are sessile and attached to the substratum by a contractile stalk. The cell body at the tip of the stalk is globose or bell-shaped and typically exhibits a rather prominent, elongate, curved or C-shaped macronucleus. Food vacuoles are usually clearly visible at the base of the infundibulum. Colonial stalked peritrichs include *Carchesium* and *Zoothamnium*, with a treelike array of individual vorticellalike zooids on branching stalks. In *Carchesium*, each zooid has a separate contractile stalk, whereas in *Zoothamnium* the stalks contract in unison. The genus *Ophrydium* is a colonial form, with a group of zooids attached to a common base and enclosed within a gelatinous moundlike matrix. Each of the zooids is enclosed within a pocket in the gelatin, surrounded by a sheath inhabited by green algal symbionts.

Some species of peritrichs secrete a flasklike lorica, with or without an external stalk for attachment to the substratum. Others are free-swimming, as is *Astylozoon* with a prominent C-shaped macronucleus and a fringe of cilia at the broad anterior end. It resembles a *Vorticella* without a stalk. The anterior cilia, directed forward during swimming, provide locomotion. Other genera (eg, *Opisthonecta*) have a telotrochal band (circumferal arrangement) of cilia at the posterior end of the cell.

The astomates are a group of Oligohymenophorea inhabiting the guts of annelids. They lack a mouth, hence the name Astomatia (without a mouth). They are typically large ciliates (ca 100 μm or larger) that bear rows of monokinetids similar to other Oligohymenophorea. Some have hook or suckerlike attachment sites for anchoring to the host tissue. Their shape varies from uniformly ciliated ovoid cells to more complex elongate forms with segmented body regions. An interesting group of epibionts living on the surface of crustacea are the apostomates (eg, Bradbury 1966). These ciliates have a rosette of kineties surrounding the oral region and usually three polykineties in the oral groove. The somatic cilia are monokinetids.

Class: Colpodea

The class Colpodea is characterized by somatic oral ciliature. *Colpoda cucullus* (40 to 110 μm, Figure 6.1f), commonly observed in freshwater cultures from (or prepared from) soil samples has a reniform body with widely spaced dikineties arranged in a somewhat sigmoidal pattern. There are two nearly equal-sized oral polykinetids within the indented mouth clearly visible in side view and slightly anterior to the middle of the cell. *Bursaria truncatella* (Figure 6.2b) is a large (several hundred μm) carnivorous member of this class with a large apical mouth and lateral slit leading to a massive oral cavity (pouch) lined with polykinetids on the right and left walls. Prey, including other ciliates, are engulfed within the pouchlike oral cavity. The macronucleus is cylindrical and elongate, curved around the buccal cavity. One sessile genus in this class, *Mycterothrix* (eg, *M tuamotuensis*), secretes a gelatinous cuplike envelope surrounding the posterior part of the cell. The anterior end with elongate dikinetids projects from the envelope. A somewhat lateral oral opening with two nearly equal-sized polykinetids leads into a narrow groove, where food is directed into a cytopharynx.

Class: Karyorelictea

Dikinetids are also characteristic of the class Karyorelictea, which otherwise is rather diverse, containing typically long, vermiform, flattened organisms. Sometimes one surface lacks cilia. An extremely contractile body with numerous macronuclei is common. Representative genera include *Trachelocerca*, with an elongate anterior portion and apical oral aperture surrounded by cilia that are longer than the body cilia. Somatic dikinetids cover the entire body. By contrast, *Tracheloraphis* (see Figure 7.4), which also has an elongate anterior portion, lacks cilia on a ventral strip of the body about one eighth to one half of the cortical perimeter. *Loxodes* (Figure 6.1g) is less elongate and has a round lateral oral area behind an anterior rostrum.

Class: Spirotrichea

The class Spirotrichea is characterized by a diverse group of organisms with somatic dikinetids, and a typically spirally arranged oral ciliature consisting of polykinetids leading by a vortical pattern into the oral cavity, either from a broad anterior end as with *Stentor* (Figure 6.2h) or along the anterior and left margins of the body (Figure 1.1e). Members of this group are either naked or enclosed in an organic lorica, sometimes with embedded mineral grains. The familiar *Stentor*, with a contractile trumpet-shaped body and longitudinal rows of dikinetids, is representative of sessile forms. When sufficiently disturbed, however, it will dislodge and swim by action of the oral polykinetids. The attractive *Blepharisma* (Figure 1.1e), often with pink to purple red pigmentation, is a sluggishly motile, large, pyriform to ellipsoidal ciliate (ca 300 to 400 μm) with a prominent paroral flickering membrane lining the right side of an elongate oral aperture. The name *Blepharisma* is derived from the Greek word for eyelash, signifying the action of the paroral membrane. A series of adoral membranelles line the opposite (left) side of the oral aperture and form a shallow spiral onto the rim of the cytostome. The macronucleus is elongate, extending at least three quarters the length of the body. *Fabrea*, with a rather teardrop shaped body (ca 150 μm), exhibits a prominent left oral band of polykinetids that extend down the anterio-lateral edge of the cell body and spiral into the oral cavity. The arrangement of the polykinetids enhances the capture and ingestion of larger prey, including small flagellates and other protista.

A large and elegant group of loricate marine spirotrich ciliates known as the tintinnids possess a conical or bell-shaped body with poorly developed somatic cilia on most of the posterior part of the cell. The broad anterior is surrounded by prominent oral polykinetids which are used to locomote and feed. A remarkable, delicate and sometimes ornate, vase-shaped or tubular lorica, either sculptured or made of agglomerated grains, surrounds the body. If the organism is subjected to threatening or noxious stimuli, it may shed the lorica and swim away; subsequently a new lorica is secreted. Figure 4.5i shows a tintinnid prey.

Class: Protostomatea

The class Protostomatea includes a diverse group of ciliates with somatic monokinetids and an apical to subapical cytostome surrounded by dikinetids. A shallow, precytostomal cavity is present in some species, while others such as *Coleps* have a large nondepressed apical oral aperture at the anterior end of the rather barrel-shaped body. The cell is covered by calcareous plates secreted by the cell and arranged to form four body segments, one anterior, one posterior, and two medial. Sparse cilia emerge from pores in the sutures of the plates. Spines project from the surface of the anterior and posterior plates.

In contrast, the elongate ciliate *Dileptus* (Figure 6.2d) represents a group of the Protostomatia with an eccentric apical cytostome at the base of a ciliated proboscis. The remainder of the body is urn-shaped, with a markedly tapered posterior. *Lacrymaria*, with a long, flexible, contractile proboscis bearing a terminal oral apparatus offers an interesting contrast to *Dileptus*. When the proboscis is extended, this ciliate is ca 300 to 400 µm in length. The tip of the proboscis forms an apical bulb with elongate ciliature used in locomotion and feeding. There is no permanent cytostome, but upon contact with a food particle the anterior end forms an invagination, whereby the food is engulfed into a food vacuole. Both *Dileptus* and *Lacrymaria* have complex cytoplasmic supporting rods in the oral region to give strength and maintain the form of the oral region.

Didinium nasutum (ca 100 µm) bears two girdle bands of cilia, one near the anterior pole and the other at the midline of the body (Figure 6.2a). A pronounced apical cone is used in the capture and ingestion of other ciliates. Upon contact with suitable prey, the apical cone becomes inverted to engulf the prey organism.

Class: Phyllopharyngea

The suctorians, with a sessile tentacle-bearing trophont stage (Figure 19.5) are remarkably different from most ciliates. They are clearly ciliates based on the motile reproductive stage. These dissemules, formed by cytoplasmic budding within a saclike pouch of the sedentary parent trophont, possess cilia (monokinetids). The arrangement of the monokinetids and the fine structure of their infraciliature indicates that they belong to the large class Phyllopharyngea. This class includes a wide variety of forms, spanning genera with sessile and often bizarre vaselike body shapes to free-swimming, more typical ciliates with monokinetids on much of the body surface. In many, the oral apparatus is underlaid by radially arranged microtubular ribbons called phyllae. In some species, the phyllae are further reinforced by nematodesmata, forming a pharyngeal basket resembling, but not identical to, those of *Nassula* and its relatives. A large number of the species are motile or possess the capacity to adhere for some period of time to the substratum. *Chilodonella cucullus* (C 130 to 150 µm, Figure 6.1j) is a motile, representative member of the class with a flattened ventral surface and rounded dorsal side. The oral apparatus on the ventral side has a distinctive cytoplasmic pharyngeal basket formed by the phyllae and nematodesmata. The nucleus is typically posterior and there are several contractile vacuoles. The chonotrichs (chonotrichia) are a sedentary or sessile group of Phyllopharyngea, with body cilia only on the walls of the perioral funnel (the cone-shaped region projecting from the anterior part of the cell). The base of the tapered body (tomite) may be modified as a type of holdfast (podite) to permit attachment to the substratum. These organisms are ectosymbionts on crustacean appendages. The tendency toward an increasingly sessile habit reaches a certain climax in the suctoria, where elaborate tentacled forms with stalks of varying length and thickness form the trophont stage. The shape of the body, which can be spheroidal (eg, *Podophrya*, ca 10 to 30 µm) or inverted pyramidal (eg, *Tokophrya*, ca 100 to 200 µm), and the form of the tentacles (whether with blunt ends, knobs or pointed tips) determine the genera. The macronucleus varies in shape from spheroidal in some species to cylindrical and elongate, or with several small side branches. The ciliated reproductive stage is characteristic of the species. The dissemule of *Tokophrya*, for example, has several median bands of cilia surrounding the cell, and a caudal tuft. This stage, upon settling to the substratum, becomes attached to the caudal cilia, forms tentacles, loses all cilia, and forms a stalk. By contrast, the swarmer of *Acineta* has bands of cilia only on one surface of the oblate spheroidal cell, and the swarmer of *Tokophrya* has bands partially spiralling around the cell.

Summary of Ciliate Morphology

The presence of cilia covering the cell or restricted to certain portions of the organism is a major defining characteristic of this group, and hence the phylum name Ciliophora. However, a major cytological distinction is the presence of two kinds of nuclei, macronucleus and micronucleus. The adaptive radiation of this group during evolution has produced some remarkably elegant and diverse species. Some are sessile (eg, suctorians or *Stentor*) and capture food by tentacles that penetrate and draw in prey cytoplasm, or by complex water-propelling membranelles that carry particle-laden streams of water into the buccal cavity where food vacuoles are formed. Although most ciliates are "naked," some produce mineralized lorica (tintinnids) or secrete organic scales (eg, *Lepidotrachelophyllum*). The arrangement of the cilia on the body surface and in the region of the oral apparatus, the presence of specialized feeding structures, and the organization of the subpellicular bands of microtubules are important criteria used in creating taxonomic categories. Some of these major features are summarized in Table 6.1.

7
Perspectives on Ciliate Ecology

Basic Concepts

The growth of populations and ciliate community structure have received considerable research attention. This may be attributed in part to the relatively large size of some species, thus permitting rapid and convenient identification, and the significant role of ciliates as consumers of primary production in ecosystems (Curds and Vandyke 1966; Fenchel 1968a, 1969; Taylor and Berger 1976; Muller and Lee 1977; Finlay 1981). Substantial attention has been given to their taxonomy (eg, Corliss 1979) and their role in water purification (eg, Curds 1973), further enhancing the perceived practical significance of understanding the role of ciliates in aquatic systems.

The growth of populations can be represented mathematically by the logistic equation (7.1), which has been widely applied in studies of population dynamics, although it is often only an approximation to the course of population increase (especially in natural environments). The number of individuals (N) in the population at a given time (t) is expressed as follows:

$$N_t = \frac{K}{(1 + C e^{-rt})}, \qquad 7.1$$

where K is the upper bound of the function and represents the maximum size of the population, C is a constant expressing the initial state of the population, e is the base of the natural logarithm (ca 2.71828), and r is the rate of natural increase of the population. Conceptually, r is the rate of increase due to reproduction minus the loss due to mortality. However, in asexually reproducing populations with little or no mortality the reproductive rate can be used for r. In the case of protozoa producing by binary fission, the intrinsic growth rate (μ in Equation 3.3, the natural logarithm of the increase in number per unit time) is often used (eg, Fenchel 1968b). A plot of the logistic function is shown as Figure 7.1. The initial population density N_o at time 0 gradually increases with time and approaches a maximum (K). This is denoted by a dashed line representing the upper bound of the function, with coordinate K on the ordinate. The curve is initially concave upwards, but inflects at the value of $N = K/2$ with $t = \ln C/r$. The

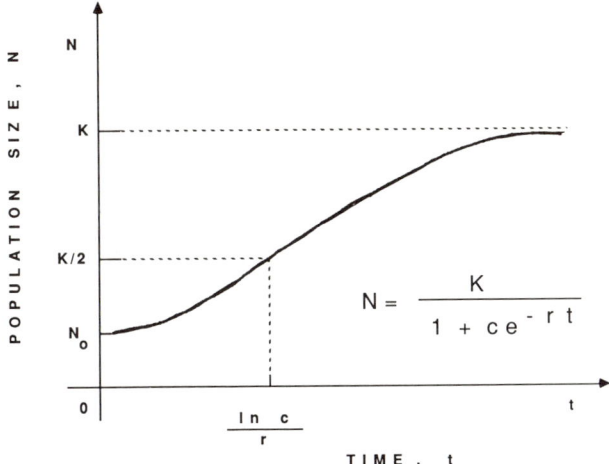

FIGURE 7.1. A logistic growth curve showing population increase (N) with time (t). $K =$ population maximum number, $r =$ natural growth coefficient and c is a constant characteristic of the species. The population reaches one half maximum value at a time equivalent to $(\ln c)/r$.

maximum value of the growth curve (K), and the intrinsic rate of natural increase (r) will vary for different populations of organisms depending on their genetic capacity and the potential of the environment to support growth. Thus, a given species may have an inherent (genetic) capacity to grow at a certain rate, expressed as a maximum population increase per unit time, but if the physical environment is not favorable, or if competition or predatory pressures occur, then the rate of increase will be less than this potential. Likewise, the achievement of maximum population density (K) may vary according to the amount of resources available in an environment. Hence, the values to be used for r and K must be determined for each species relative to carefully defined environmental conditions. Even then, the equation may not exactly approximate the trace of population increase, owing to physiological and environmental factors not appropriately modeled by the logistic equation. Some of the variations in form of the equation have been discussed critically by Wangersky (1978), who also includes modifications to account for predator-prey interactions in communities and to better approximate growth of populations in the natural environment.

The growth characteristics of a species have been described conceptually in terms of the logistic equation with reference to r and K variables. Populations of organisms that grow rapidly, use resources less efficiently, and rarely achieve maximum potential population density are termed "r-selected" species. Those populations that grow more slowly, utilize resources conservatively, and typically mature to their maximum potential (that is, they actually come to the equilibrium or maximum growth level (K)) are termed "K-selected" species. In some cases, although not invariantly (Boyce 1984), the *r*-selected species tend to be smaller and sometimes more rapidly moving, whereas *K*-selected species are larger and less rapidly moving. The use of *r* and *K* in these categorizations refer respectively to the tendency of a species to adapt by maximizing growth rate (*r*) or by maximizing stable, high density populations (*K*). In general, *r*-selected species can exploit environments where resources are sporadically available. Growth is rapid, and resources are often depleted before the population achieves its maximum development. Among protozoa, this is characteristic of some opportunistic feeders that depend on unpredictable but sometimes substantial patchy sources of nutrition in the environment. The resource patch is invaded and rapidly consumed, resulting in nutrient depletion. Unless the individuals are capable of encysting or producing other forms of resting stages, massive death or dispersal must occur and only a small residual population may persist using whatever resources remain. The population never reaches a level where density dependent regulation becomes a major factor in determining population size. By contrast, *K*-selected species tend to utilize resources more effectively toward building a stable biomass and thus often grow more slowly. However, they eventually achieve a population level where the balance between available resources and competition for their utilization yield an equilibrium population density. The merits and limitations of the *r*-*K* selection model have been critically discussed by several researchers (eg, Dawkins 1982; Boyce 1984), and adequate attention should be given to the underlying assumptions of the model before making generalized applications in protozoological research. It is a useful construct for examining the growth characteristics of individual populations in model ecosystems in the laboratory, and as an organizing model to help to explain some observations of field-collected data on protozoan population dynamics. Boyce (1984) argues, however, that although the model may have legitimate use in understanding the life history of organisms, it must be interpreted strictly as it was originally formulated, as a model of density-dependent natural selection.

Population growth of a species in the natural environment is seldom isolated from the effects of other biota, and thus interspecies competition becomes an important factor in analyzing the organization and development of communities. The concept of community structure, although used in a variety of ways by different authors, can be defined generally as the categories and number of species interacting with one another in a habitat, and the mechanisms they have evolved to partition resources and modulate competition to permit a quasistable association within the limiting conditions of the abiotic environment. The species categories as used here can be based on the trophic role of the species, as, for example, primary producers, secondary producers, consumers, etc. Interaction includes predator-prey relationships, auxotrophic and inhibitory effects, nutrient exchange by soluble products released into the environment, and other activities that may alter the environment to either make it more or less hospitable for other species. A complete analysis of community structure, therefore, requires a substantial effort, especially when there are many species. In addition to carefully monitoring all relevant abiotic factors (eg, temperature, salinity or ionic composition, pH, E_h, nutrient concentrations, illumination, and other physical parameters of water velocity, turbulence, etc), the numbers of individuals in each species, how they interact with each other and respond to changes in the abiotic environment should all be determined. This is diagrammed in Figure 7.2.

The complexities of the abiotic and biotic factors influencing the growth and interaction of species often make it difficult to construct exact models of community structure. For example, the mathematical representation of abiotic factors influencing primary production should include significant variations in temperature, nutrients, light intensity and quality, and their temporal patterns. Sometimes the exact data cannot be obtained and approximations must be used. In some cases, it is preferable to attempt to predict changes in communities based on mathematical models of organismic-environmental interactions. In marine environments, the gross primary production rate (P_r) has been estimated based on light intensity (I), amount of chlorophyll per unit mass of carbon (Chl/C), and efficiency factors in light absorption and conversion into chemical energy (Kiefer and Mitchell 1983).

$$P_r = \varphi + a \, (Chl/C) \, I, \qquad 7.2$$

where φ is the quantum efficiency (ca 0.06) expressed as gram-atoms of carbon per unit of light intensity (Einstein), a = an absorption coefficient (17 m²/g Chl a), Chl = chlorophyll concentration in the sample of water, and I = light intensity in Einsteins/m²/day. The total gross carbon production in one day in units of carbon is obtained by multiplying P_r by C. These data were compiled using the diatom *Thalassiosira weissflogii*. Variation in primary production of diatoms as a function of fluctuations in light intensity have been modeled using time series analyses (eg, Neale and Marra 1985), and provides the kind of basic information needed to construct a more complete analysis of community response to changing

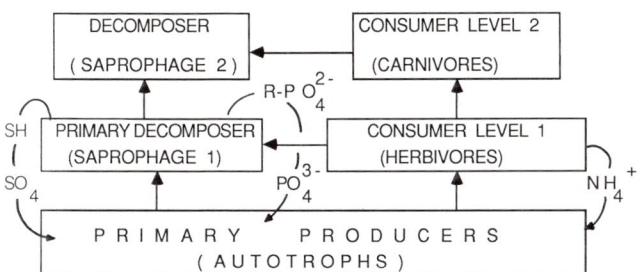

FIGURE 7.2. An hierarchical model of community trophic interactions. Primary producers (primarily autotrophs) fix inorganic carbon and are preyed upon by herbivores that in turn are consumed by carnivores. Omnivores consume both autotrophs and heterotrophs. Primary decomposers (eg, bacteria) also known as saprophages assimilate organic matter from autotrophs or other organisms within the community, and serve a vital role in converting organic bound phosphate, nitrate, and sulfur into inorganic forms suitable for uptake by primary producers. Carbon flow is indicated by the direction of the arrows.

abiotic factors. Competition among biota for environmental resources has been analyzed quantitatively as early as 1934 by Gause, using two species of *Paramecium*, and more recently Leslie (1957) demonstrated statistically that these equations representing growth of two competing species, with some simplifying assumptions, do predict actual growth during competition in culture. These equations are based on modification of the logistic equation to include limiting effects due to interspecific competition. An example is given as Equation 7.3.

$$N_1^{-1} = K_1^{-1} + K_2^{-1} C^{-1} e^{-kt} + C'_1 e^{-r_1 t}, \qquad 7.3$$

where K_1 and K_2 are respectively the maximum growth of species 1 and 2, and $k = r_1 - r_2$. The values of r_1 and r_2 are respectively the rates of natural increase (μ) for species 1 and 2 as defined in Equation 7.1. By inspection of Equation 7.3, it is clear that as the value of r_2 is increasingly greater than r_1, the term e^{-kt} will become increasingly large. This is due to the fact that k will be negative, and when substituted into the expression will produce a positive-valued, large exponential coefficient. Since the dependent variable (N_1) is expressed as the inverse, as the right-hand side of the equation increases N_1 will decrease. Hence, large growth of the competitor species will tend to depress the growth of species 1.

Population Growth and Interspecies Competition

Population Growth

The reproductive potential of some marine benthic ciliates was assessed by Fenchel (1968b), who determined the division rate as a function of body volume of eight species: (1) *Uronema marina*, (2) *Philasteridae* sp, (3) *Aspidisca angulata*, (4) *Euplotes vannus*, (5) *Litonotus lamella*, (6) *Diophrys scutum*, (7) *Keronopsis rubra*, and (8) *Condylostoma patulum*. He obtained the following relationship for a temperature of 20°C:

$$r = 105.9 \times v^{-0.44}, \qquad 7.4$$

where r = divisions per day and v = volume in cubic microns. The range in growth rate was r = 6.65 to 0.36 for *Uronema marina*, the smallest ciliate, and *Condylostoma patulum*, the largest ciliate in the series. The growth response to temperature varied among the species studied, but in general all had a maximal growth rate above 20°C, which was above the mean temperature of the natural environment. The smaller species *Uronema marina*, *Aspidisca angulata*, *Euplotes vannus* and *Litonotus lamella* were able to grow at temperatures as low as 4 to 6°C. Larger species, *Lacrymaria marina*, *Keronopsis rubra*, and *Condylostoma patulum* grew at temperatures as low as 12°C. In general, it appears that the larger the protozoan, the lower the rate of growth. This relationship has also been found in metazoa.

Some intensive studies of population growth in monoxenic or defined cultures have been done for individual species of ciliates. The growth of the small ciliate

Colpidium campylum in monoxenic (single food organism) batch culture was monitored by Taylor and Berger (1976). They examined variation in growth rate and its implications for ecological studies in the natural environment. As with all laboratory simulations, one must be careful not to overgeneralize the findings to the natural environment without carefully considering the limitations of the artificial environments. Nonetheless, some fundamental principles of protozoan growth are often elucidated by carefully described laboratory investigations. Under constant culture conditions, variation in population growth rate among cultures ranged from 0.097 divisions per hr to 0.174 divisions per hr. Some of this variability was attributed to differences in growth rates of exponential versus stationary cells in the culture. When cultures were inoculated with early logarithmic growth phase cells, variation in growth rate was decreased. During the course of population growth, cells entering late logarithmic phase continue to divide although the biomass of each cell declines. Thus, the number of individuals in a starving population increases. This may have survival advantage in the natural environment by ensuring that a large number of progeny can be disseminated, increasing the probability that some will encounter favorable growth conditions and thus ensure continuity of the species. When starving cells are inoculated into fresh medium, however, they increase in size substantially before dividing (eg, increasing from 5,000 to 50,000 μm^3). Since ingestion is known to be interrupted during division and larger cells are able to ingest more prey, the delayed division and enhanced size of *C campylum* may have survival value by permitting starving cells to make maximum utilization of resource patches in the environment before division commences. Similar responses have been found in other ciliates, including *Didinium nasutum* (Salt 1975), *Perspira ovum* (Dewey and Kidder 1940), and *Uronema* sp (Hamilton and Preslan 1969), although the timing of changes in sizes of cells varied among the species during the growth cycle.

Maximum growth of *C campylum*, fed a bacterial diet, is attained at a concentration of 1×10^6 bacteria per protozoan (Laybourn and Stewart 1975), but is temperature dependent. Growth expressed as the amount of protoplasmic volume produced during a 24 hr period divided by the number of cells initially present at the beginning of the 24 hr period was related to temperature of the medium. The average increase in volume per initial cell was 10 to 15×10^4 μm^3 at 10°C, 20 to 25 $\mu m^3 \times 10^4$ at 15°C, and 40 to 45 $\mu m^3 \times 10^4$ at 20°C. An increase in temperature of 5°C within the range examined produced a 40 to 50% increase in cytoplasmic volume. Thus, each ten degree rise in temperature produced a 3.4-fold increase in volume, or a $Q_{10} = 3.40$.

The growth of five ciliates fed a monoxenic diet, consisting of a single species or bacteria strain offered to the ciliate as food, was studied by Curds and Vandyke (1966). Nineteen different bacterial strains (15 species) were offered individually as food. The maximum growth expressed as divisions per day and the bacterial food producing that growth presented in parentheses for each ciliate species is as follows: (1) *Paramecium caudatum*, 2.45 doublings/day (*Bacillus subtilis*); (2) *Histriculus vorax*, 3.90 doublings/day (*Pseudomonas fluorescens*); (3) *Vorticella microstoma*, 2.75 doublings/day (*Streptococcus faecalis*) and 2.55 doublings/day (*Bacillus cereus*); (4) *Opercularia coarctata*, 2.34 doublings/day (*Klebsiella aerogenes*); and (5) *Epistylis plicatilis*, 1.46 doublings/day (*Escherichia coli*). The latter three species of ciliates showed unfavorable or toxic responses to 12 or more of the bacterial strains offered as food, and thus were more restrictive in the range of these bacteria they could accept as food than the first two species.

Trophic Behavior and Population Growth

The trophic behavior (prey preference), population abundance, biomass, species composition, and size distribution of planktonic ciliates were analyzed along a gradient of 20 lakes ranging from oligotrophic (low in nutrients) to eutrophic (high in nutrients and bacterial production) by Beaver and Crisman (1982). The abundance and biomass of ciliates were positively related to trophic state of the lakes, increasing as the lakes became more eutrophic. Oligotrophic lakes were dominated principally by small ciliates (40 to 50 µm) in the Oligotrichida (eg, *Strombidium*, *Tintinnidium*, and *Tintinnopsis*), while eutrophic lakes were codominated by smaller ciliates (20 to 30 µm). The latter belonged to the Scuticociliatida (eg, *Uronema* and *Cyclidium*), Oligotrichida (eg, *Strobilidium*), and Haptorida (eg, *Dileptus*, *Didinium*, and *Spathidium*). The trend toward a decrease in size with increasing trophic state of the lakes is probably related to the presence of larger algae (diatoms and flagellates) in oligotrophic water preyed upon by larger ciliates, contrasted to the abundance of bacteria and microflagellates in eutrophic water utilized more efficiently by smaller ciliates. Variations in size and species composition of the planktonic marine ciliate *Tintinnida* were studied during a one year period in nearshore water of the New York Bight (Gold and Morales 1975). In general, ubiquitous and recurring genera included *Stenosemella* and *Tintinnopsis*. Seasonal species included the hyaline loricate species, *Favella arcuata* and *F ehrenbergii*, occurring mainly in late summer and early autumn; *Helicostomella subulata*, present in midsummer through mid winter (but absent late December through April); and *Metacylis angulata* and *M annulifera*, present in late summer. The size of the lorica of a given species varied inversely with water temperature, being smaller in summer or autumn and larger in winter. For example, the sizes of the loricae in *Tintinnopsis levigata* during autumn occurred largely in the range of 50 to 60 µm, whereas in winter the range was more varied and the size was larger (75 to 100 µm). The abundance of tintinnids was highly variable, their distribution was typically unpredictable, and many of the factors determining their distribution in the natural environment remain uncertain.

Factors regulating population numbers of ciliates have been investigated in laboratory and field studies, including food limiting factors, the role of rate of growth, and capacity to adjust to environmental fluctuations. Luckinbill and Fenton (1978) examined the effects of varying population density and fluctuations in amount of food on two protozoan species in laboratory culture. *Colpidium campylum*, a small rapidly reproducing organism (larger r), was compared to *Paramecium primaurelia*, a larger, slower dividing organism. A third intermediate-sized species, *Paramecium tetraurelia*, was also used in some experiments as a contrast to the two extreme species. When logarithmically growing cultures of each species are experimentally altered by changing population density (increasing it in one culture and decreasing it in the other) under carefully controlled sampling methods, and the cultures are allowed to continue growing, all cultures tend to converge with time toward a mean population density (Figure 7.3). That is, the more dense population declined and the less dense population increased, converging at a mean value approximately equivalent to the mean of the unaltered culture. However, the more rapidly dividing species (eg, *C campylum*) converged toward the mean of its control culture more rapidly than the other two slower growing species, but *C campylum* fluctuated more erratically around the mean value during subsequent growth than the slower growing species. Thus, as might be expected, these data suggest that species with large r tend to adjust more rapidly to altered environments by more promptly achieving an equilibrium density than do species with smaller r. The population density equilibrium,

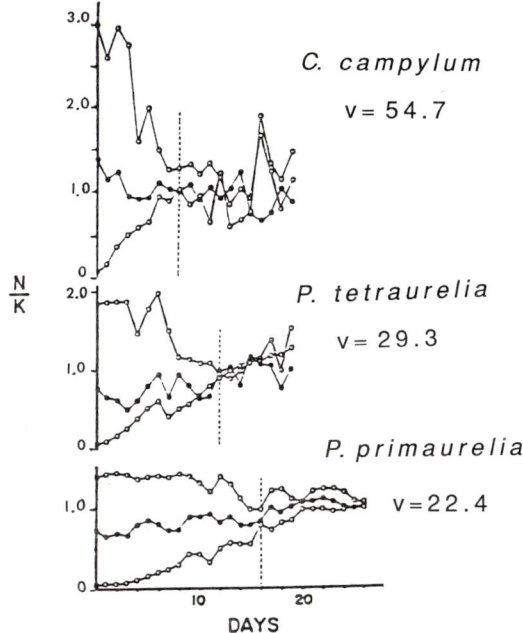

FIGURE 7.3. Stability of three ciliate populations as a function of time after being concentrated (upper trace) or diluted (lower trace). The fast growing *Colpidium campylum* adjusts more rapidly toward a stable mean population level (vertical dashed line), but exhibits greater variability around the mean (less stability) than the slower growing species *Paramecium tetraurelia* and *Paramecium primaurelia*. Although they require a longer time to reach a stable mean population level, their variance is less ($V = 29.3$ and 22.4, respectively). From Luckinbill and Fenton (1978); with permission.

however, is more stable for slower growing species. This is reflected in the variance of the population numbers around the mean, as recorded in Figure 7.3. The variance of population numbers in *C campylum* was 54.7, compared to *P primaurelia* with 22.4. Since field conditions would almost certainly be less constant than those in the laboratory, Luckinbill and Fenton also examined the ability of the two extreme species to adjust to variations in food supply, as bacterial prey density became progressively more variable over time. Thus, *C campylum* and *P primaurelia* were grown in cultures where the bacterial density was varied at progressively shorter intervals, forcing the ciliates to adjust to food levels that were progressively more unpredictable. The results showed that *C campylum* population size adjusted more rapidly to changing food densities when the frequency of change was low, but its rapid growth rate caused severe uncontrolled fluctuations in numbers when the frequency of change was high. This resulted in extinction when the ciliates were forced to respond to wide fluctuations in food at weekly intervals. By contrast, the slower growing *P primaurelia* was able to adjust with greater stability to the rapid fluctuations and maintain a more constant equilibrium population level. The extinction of the *C campylum* is, apparently, due to wide overshoots in adjustment to variations in food supply as documented in the initial experiment. This leads to overproduction when bacterial numbers are high, and a consequent catastrophic decline due to lack of food resources when bacterial numbers are low. The slower growing *P primaurelia*, being larger and consuming more bacteria per individual, are able to build reserves and adjust more regularly to changing environments, and hence are capable of subsisting in environments when food resources vary relatively rapidly. Also, it was found that these differential population effects among species varying in r were related more to the frequency of variations in food supply than to the magnitude of the changes in food abundance. Thus, changes from very abundant amounts of food to very low amounts were less perturbing than rapid oscillations between high and low amounts of food, independent of the magnitude of the fluctuation in food abundance.

Fluctuations in density of protozoa in mixed populations containing bacteria have been noted in laboratory and field studies (eg, Woodruff 1912; Bamforth 1958). The origin of these fluctuations is uncertain, but may be due to a combination of climatic cycles, variations in nutrient supply, and changes in bacterial and protozoan cell composition, size and physiology during population growth and maturation. Jensen and Ball (1970) examined the effects of variations in nutrient supply on bacterial and protozoan population fluctuations in laboratory culture. They found that fluctuations in the protozoan population was significantly linked to the patterns of nutrient (sucrose) replenishment in the medium. Large fluctuations in population density of bacteria and protozoa were produced when sterile nutrient solution was added periodically at long intervals (weekly). These fluctuations were substantially reduced in magnitude and were no longer cyclic when the nutrient solution was added at shorter intervals (daily). The decrease in fluctuation magnitude was interpreted to be the result of a longer generation time of the protozoa when nutrients were added daily. This rather straightforward study indicates that among other forcing functions in the environment, variations in nutrient supply, including recurrent patchiness of nutrients, may account for some of the major fluctuations observed in mixed populations of protozoa, especially when nutrient replenishment recurs at long time intervals relative to the intrinsic population growth rates. One of the interesting corollary findings is that when the sterile nutrient medium was provided on a daily basis as 0.5 mg/l of culture, compared to a weekly aliquot of 3.5 mg/l culture, the density of bacteria and protozoa was less. The protozoa exhibited a rather steady decline in numbers over 18 days from several hundred per cc to less than one hundred per cc. The cause of this decline was not examined, but it may be due to enhanced competition among species or perhaps accumulation of bacterial metabolites deleterious to the protozoa. No analysis of the protozoan species composition was reported, which limits interpretation of why the stability of the mixed population declined.

Population Stability and Community Diversity

An analysis of the stability of a ciliate community in relation to the diversity of its component species was made by Hairston et al. (1968). The laboratory simulations of natural communities consisted of a bacterial food source, *Aerobacter aerogenes*, and two unidentified bacilliform species isolated from the natural environment. The bactivorous ciliates used were two varieties of *Paramecium aurelia* and an unidentified variety of *P caudatum*. Two ciliate predators, *Woodruffia metabolica* and *Didinium nasutum*, served as higher trophic level members of the community. Diversity was operationally defined as the number of different species added to the culture at the outset of an experiment, and stability was defined as the ability to withstand change. Stability was indicated by a relatively constant number of individuals among the several species in the community, or by a slow change in their abundance over time. The growth rate of the bactivorous ciliates was inversely related to size, as established by Fenchel (1968b) and Finlay (1977), and was highest when the ciliates were fed a combination of two species of bacteria as opposed to offering one or three. In total, the basic growth data showed that the three species of *Paramecia* were quite different in their utilization of bacteria as food. Communities with two trophic levels (bacteria and bactivorous ciliates) exhibited classical competitive exclusion. In every case, one of the ciliates increased in abundance while the other, less competitive ones, declined. However, the decline and loss of ciliate species (an indication of lack of stability) could be minimized by increasing the diversity of bacteria introduced at the outset of the culture. Thus, stability was enhanced by greater diversity of prey at the lowest trophic level. However, increasing the diversity of the higher trophic level (ie, using three as compared to two species of *Paramecium*) did not produce a statistically significant increase in stability. Indeed, over 14 days, species persistence was greater in a two species rather than three species community. Addition of a ciliate predator always led to complete elimination of the prey *Paramecium* and death of the predator. This was true whether one or two predators were added. *Paramecium* persisted longer in the presence of *Woodruffia* than with *Didinium*, and *Woodruffia* persisted significantly longer when alone than it did in competition with *Didinium*. In overall view, this study with a simple laboratory simulation of ciliate community structure indicates that stability is enhanced by increasing diversity of species at the lowest trophic level, but not at higher trophic levels. The generalizability of the findings, however, is clearly limited, since no attempt was made to use ciliate predators of varying growth rate, nor to provide more complex environments where refuge locations could help protect the prey from complete annihilation. Nonetheless, it provides some empirical evidence to show that one cannot make sweeping generalizations regarding the following: that community stability necessarily increases with increasing diversity over all trophic levels, and that competitive interactions within limited environments can lead to destabilization of the community.

In a complementary study, Luckinbill (1979) further examined the effects of diversity on stability of a three-tiered trophic community. However, in his cultures, the predator was *Didinium nasutum* and the prey was the ciliate *Colpidium campylum*. This small fast-growing ciliate is not preyed to extinction when grown alone with *D nasutum*, apparently because *Didinium* is not as effective in capturing it as larger prey such as *Paramecium*. Thus, it is possible to examine the effects of increasing prey diversity on the stability of the system. In general, for this system as in the preceding study, increasing the diversity of prey at an intermediate trophic level by adding one or two species of *Paramecium* led to destabilization and extinction of the prey and predator. Likewise, enriching the culture by adding additional nutrients also caused collapse of the community. These effects were attributed to the enhanced vigor of the predator when additional prey is available, resulting in overgrazing on both *Colpidium* and *Paramecium*. Thus, in this simple system, a balanced growth of predator and prey can be maintained if the diversity of prey is less and the dominant prey is consumed only moderately by the predator. The decline in the community when *Paramecium* is added may also be attributed to the added competition of the *Paramecium* for bacterial prey, thus decreasing the available producers at the base of the food chain. Additional research varying the size, growth rate, and feeding efficiency of the bactivorous ciliates cohabiting with *C campylum* could help elucidate the role of competition of second trophic level members in maintaining ciliate community stability.

Competition

Competitive structure (patterns of competition) among ciliates was studied by Vandermeer (1969), who examined the growth rate of three species of *Paramecium* and one species of *Blepharisma* when cultured alone, or when grown together as pairs, or when combinations of four species were used. The growth of the individual organisms was adequately predicted by the logistic equation (Equation 7.1), and competitive exclusion was demonstrated consistent with the equations of Gause (Leslie 1957) as exemplified by Equation 7.3. When all four ciliates were cultured together, however, no interaction effects were found above those accounted for by the logistic equations and pair-wise competition. Thus, the data indicate that higher order interactions above pair-wise competition produce little or no effect on the dynamics of this artificial community under rather carefully controlled abiotic conditions. Further studies designed to test for higher order interactions are warranted while varying abiotic factors such as temperature, nutrient supply, salinity, or other stress-inducing variables that may have contextual effects on the higher order interactions.

The comparative competitive capacity of two species in the *Paramecium aurelia* complex (Pa2 and Pa5), when subjected to variations in temperature, was investigated by Gill (1972). The two strains exhibited clearly different growth responses to variations in temperature over a range from 15° to 25°C. Pa5 was broadly tolerant of temperature with respect to biomass, maintaining a relatively steady population density (K), but was extremely sensitive with respect to productivity as indicated by depressed growth rate (r) at lower temperatures. The results for Pa2 were converse, and showed a marked decrease in K with increasing temperature, but only a moderately linear decline in r with decreasing temperature. Thus, in both species, the two niche dimensions r and K were inversely related, but the temperature response of the two strains was clearly different. When the two species were grown together under three temperature regimes (15°, 20°, and 25°C), Pa5 grew for a short time and then declined abruptly and approached extinction. Its competitive capacity declined appreciably as temperature dropped. Pa2 grew well and approached its K value at all three temperatures. In general, the data show that the species with least perturbation in r was most successful, as is consistent with the data of Hairston et al (1968) showing that the smaller faster growing species tend to dominate. This is also in agreement with the results of Jensen and Ball (1970), which indicate a tendency for faster growing species to dominate when food supply is large but sporadic. However, at 25°C, where Pa5 has slightly higher r, and where Pa2 shows substantial decline in K, theoretically one would expect to see a reversal in the competitive dominance with Pa5 increasing in abundance. The absence of this effect suggests that competitive ability overall is not consistently related to either K or r in these simple laboratory environments. A finer analysis of the data indicated that the suppression of Pa5 by Pa2 may have been more a result of interference of Pa2 with the growth of other species. This can be attributed to killer endosymbiotic bacteria harbored by this strain. The killer endosymbionts produce substances that are deleterious to other nonkiller-containing ciliates.

Further studies of the capacity of Pa2 and Pa5 to invade and simultaneously exploit the same natural environment (Gill and Hairston 1972) indicated that Pa2 was much more adaptable to varied temperature and other abiotic factors of the natural seep environment than was Pa5. This may arise in part from the reproductive strategies of species Pa2 and its lesser sensitivity in growth rate to variations in environmental temperature.

In general, it appears that as might be expected, protozoan competitive strategies are complex and vary with the demands of the environment, and in relation to their genetic potential to adapt to diverse abiotic and biotic variables of their habitat. On the whole, increasing evidence points toward the efficacy of rapid growth and swift exploitation of available resources (*r*-strategy) when environmental resources are patchy, occurring sporadically with low frequency, but abundantly. On the other hand, the ability to grow more slowly and accumulate reserves resulting in a more stable population carrying capacity (persistence or *K*-strategy) is more appropriate in environments where resources are less variable. An inferential test of this principle was made by Taylor (1978), who examined the relative abundance in pond water samples of two kinds of ciliates: (1) those that grew at rates more rapidly than would be predicted based on their size (using Fenchel's function, Equation 7.4), and (2) those ciliates that grew slower than predicted based on their size. The species in category 1 represent "*r*-strategists" and those in category 2, "persistence strategists." Samples were taken from relatively stable sites in the pond. The assumption of the study was that in this rather stable environment, unpredictable patches of food, such as dead animal remains, would be exploited by *r*-strategist species that bloomed up and then dispersed upon food depletion. Hence, their numbers would be low in randomly collected samples. But, those species that tended to maximize their stability by a persistence strategy should have been more consistently present in samples across wide areas of the pond. The data supported this conclusion, and suggest, as found by Legner (1973), that sporadic and massive nutrient enrichments of ciliate cultures favor initial bursts of dominance by *r*-strategists (eg, *C campylum*). On the other hand, more steady levels of nutrient supply, or rapid fluctuations of high abundance nutrients, favor dominance by persistence-strategists (eg, *Halteria grandinella*, *Cyclidium glaucoma*, and *Uronema turbo*). Moreover, in cases where both *C campylum* and *C glaucoma* were found, a bloom of *C campylum* preceded *C glaucoma* in a succession. These data indicate that community composition and the structural relationships among the member species are determined in part by complex interactions between patterns of nutrient supply, variations in climatic factors, life cycle strategies, and the opportunity for migration to and colonization of microhabitats, especially when resource patchiness is common. Moreover, generalizations about persistence of species in relation to food abundance and frequency of variation must be tempered by a clear understanding of interspecies variability in physiological responses to these changing environmental variables. Interspecies variations due to encystment strategies versus prolonged survival at minimal growth rates, or other variations in reproductive strategies and mechanisms for dispersal, must be taken into account in predicting changes in population density.

Structure of Natural Communities

Freshwater Communities

The distribution of freshwater ciliates in relation to environmental factors is rather predictably related to temperature, pH, and nutrient abundance. The physico-chemical and food correlates of the distribution of some freshwater ciliates in a variety of standing and flowing water habitats was examined by Noland (1925). Some comparative data from his study and other sources are presented in Table 7.1. He particularly noted that the ciliates collected from 18 locations could be grouped according to mean temperature of their habitat. Warm water species included *Halteria grandinella*, *Coleps hirtus*, *Caenomorpha medusula*, *Loxodes rostrum*, *Oxytricha ferruginea*, *Metopus sigmoides*, *Spirostomum ambiguum*, *Spirostomum teres*, and *Urocentrum turbo*. A low temperature group included *Trochilia palustris*, *Chilodon cucullus*, *Lionotus anser*, *Oxytricha pellionella*, *Ophrydium versatile*, and *Vorticella campanula*.

In general, bactivorous species, or those preying on bactivores, were more tolerant of lower oxygen levels than those preying on algae or other protista. As expected, the environments rich in bacteria are also likely to be low in oxygen and high in reducing and acidic compounds. Among the species found in these environments were: *Microthorax sulcatus*, *Caenomorpha medusula*, *Mesodinium pulex*, *Glaucoma scintillans*, *Loxodes rostrum*, *Urocentrum turbo*, *Dileptus gigas*, *Spirostomum teres*, and *Colpidium colpoda*. These were particularly tolerant of high carbon dioxide concentrations. Some of these ciliates (eg, *Loxodes*, *Spirostomum*) are abundant in the sediments of lakes. Their distribution within the sediments and overlying water column, however, is seasonal (Goulder 1972, 1974; Finlay 1980, 1981, 1982). In lakes subject to summer stratification, the upward migration of the reducing horizon in the sediments eventually forces even these anoxic-tolerant species up into the hypolimnion near the anoxic-oxic boundary, where less anoxic conditions prevail. Other more oxygen-dependent species either encyst or migrate to more favorable environments. This pattern is reversed when winter destratification and convectional cycling increase oxygen concentrations at depth in the water column. Some of anoxic tolerant species also occur in locations where nitrite is high in addition to low oxygen partial pressures, largely attributed to bacterial metabolism. Some of these anoxic-tolerant species are physiologically adapted to reducing environments. Many lack typical eukaryotic mitochondria (Fenchel et al 1977) and appear to contain organelles resembling hydrogenosomes. Moreover, at least *Loxodes* may have nitrate reductase activity that permits it to compete with bacteria for nitrate as an oxidizing source (Finlay and Fenchel 1986).

Further studies on community structure in natural environments (Taylor 1979; Landis 1982) indicate that although bactivorous ciliates occur in clumped (patchy) distributions and a clear discrimination can be made between those living in aerobic surface water as opposed as benthic species, ciliate distribution appears to be random. Thus, no clear structural differentiation seems to exist apart from opportunistic exploitation of microhabitats. Landis (1982) specifically examined the distribution of *Paramecium bursaria* and *P aurelia* in the littoral zone of two small Indiana ponds. Both species were found in randomly distributed patches. *P bursaria* was found at the water surface and in the mud–water interface, and the population structure indicated an outbreeding strategy with diffuse population distribution. *P aurelia* occur only in the mud–water interface and are inbreeders, largely sharing a restricted gene pool. They occur in

TABLE 7.1. Environmental variables associated with habitats of some freshwater ciliates.*

Species Name	Food	Temperature	Oxygen (cc per liter)	Oxygen (% saturation)	Free Carbon Dioxide (cc per liter)	Free Carbon Dioxide (cc per liter)	pH
Amphileptus claparedei	P	15.3	5.0	82.8	4.3	63.6	7.6
Aspidisca costata	BDA	13.1	5.6	74.5	2.2	65.5	7.7
Blepharisma lateritia	ABD	10.7	3.4	40.8	3.9	73.0	7.6
Chilodon cucullus	T	7.6	5.7	67.6	4.2	61.4	7.5
Chilodon uncinatus	B	10.7	5.7	60.8	5.2	69.3	7.7
Cinetochilum margaritaceum	BDa	8.7	5.6	66.9	3.4	63.9	7.6
Coleps hirtus[†]	SPa	14.3	5.7	81.1	1.5	55.7	7.9
Colpidium colpoda	B	11.9	5.3	68.7	6.0	72.5	7.3
Cyclidium glaucoma	B	9.4	4.2	50.6	10.1	67.4	7.2
Dileptus gigas	P	9.8	3.3	37.0	7.2	80.6	7.4
Euplotes patella	ABd	12.8	3.7	48.1	4.4	74.0	7.4
Frontonia acuminata	A	11.1	5.1	65.1	3.5	61.7	7.7
Glaucoma scintillans	B	9.8	2.2	27.4	11.1	87.2	7.2
Halteria grandinella[†]	Ba	16.1	5.6	79.7	3.7	48.4	7.8
Lacrymaria olor	P	5.1	6.0	66.0	4.7	49.5	7.4
Lionotus anser	P	7.5	6.3	73.5	2.8	64.0	7.7
Lionotus fasciola	P	3.9	5.8	61.5	4.7	72.8	7.5
Loxodes rostrum[†]	Bda	15.1	3.1	42.7	14.7	53.4	6.9
Mesodinium pulex	Bp	7.9	3.3	38.1	12.2	54.8	7.2
Metopus sigmoides[†]	B	12.2	1.2	15.2	19.0	62.9	6.9

Species	Diet				
Microthorax pusillus	Bda	11.9	2.3	29.1	48.1	6.9
Microthorax sulcatus	Bda	4.5	0.7	8.5	61.0	6.6
Oxytricha ferruginea[†]	Ab	15.6	6.9	99.4	55.1	8.2
Oxytricha pellionella	Ab	7.8	6.8	80.7	60.3	7.7
Paramecium bursaria	ABd	9.5	7.3	88.9	44.3	7.7
Paramecium caudatum	Bda	11.3	7.0	64.4	71.1	7.6
Pleuronema chrysalis	BAd	11.9	6.3	82.8	57.9	8.1
Spirostomum ambiguum[†]	BDa	20.3	5.5	92.6	40.0	7.9
Spirostomum teres[†]	BDa	16.0	4.3	60.1	70.2	7.4
Stentor igneus	APB	3.5	7.5	79.4	60.9	7.8
Stentor roeseli	ABP	8.6	5.4	64.9	67.3	7.5
Trachelophyllum pusillum	Bd	9.8	4.3	51.7	65.5	7.5
Urocentrum lurbo[†]	Bd	16.4	5.0	75.1	49.6	7.7
Uroleptus piscis	Ab	13.0	5.8	59.0	62.6	7.7
Vorticella campanula	Bda	5.7	6.3	69.4	65.1	7.6
Vorticella nebulifera	BAd	9.3	3.7	45.0	77.3	7.5

*From Noland (1925).
[†]Indicate high temperature adopted species.
Legend: A = algae; B = bacteria; D = nonliving detritus; P = protozoa; T = diatoms; S = scavenger. Small letters indicate that the item so marked makes up a minor part of the food of the species.

patches and probably exploit occasional nutrient abundances. Similar nutrient-regulated abundances of *P aurelia* were found by Hairston (1967) in a woodland seep. The broader distribution and lack of inbreeding strategy in *P bursaria* are probably a result of the greater autonomy of this algal-symbiont bearing species. Individuals are able to exploit broader habitats as long as illumination is sufficient, and thus are not as likely to be restricted to fortuitous patchy sources of food. Outbreeding (requiring conjugation between individuals of two different mating types) and the existence of a plurality of mating types in *P bursaria* ensures that among the few encounters that occur, the two individuals will likely be compatible conjugants. A striking contrast exists for *P aurelia*, where conjugation between individuals of distant populations can lead to 98% mortality in the F_2 generation. Hence, this is consistent with its tendency to invade and colonize localized resource patches where inbreeding is more likely to occur.

The number of ciliate species in various feeding categories and their trophic interactions were studied in a eutrophic stream by Small (1973). In this environment, some distinct co-associations ordered around trophic interactions were detected, and the abundance and diversity of species in general was directly related to the degree of eutrophication. The number of species in each feeding category was: algivore, 22; bactivore, 87; carnivore, 56; and omnivore, 8, in a total of 109 genera. Four transient "microbiocoenoses" were described: (1) free-swimming microphagous ciliates at moderate temperatures were associated with specific bacterial substrates and supported a number of predatory ciliates; (2) sessile bactivorous ciliates (peritrichs) feeding on nonattached bacteria, but highly selectively (as, for example, *Carchesium*, which predominately preyed on *Aerobacter aerogenes*, *Alcaligenes faecalis*, *Bacillus* sp, and *Escherichia freundi*); (3) free-swimming carnivorous gymnostomes (eg, *Amphileptus* sp, *Hemiophrys pleurosigma*, and *Litonotus* spp) preying on the peritrichs which in turn are fed upon by suctorians attached to the stalks of the peritrichs; and (4) the stalked bactivorous peritrichs and attached suctorians preyed upon by invertebrates (a snail (*Physa* sp), larval chironomids (midges), and various species of rotifers).

Marine Communities

Ciliates in marine environments are cosmopolitan in distribution and inhabit most of the remarkably diverse habitats available in the benthic and planktonic realms (Borror 1980). The marine environment offers widely varying habitats for ciliates, ranging from the organic-rich sediments of the deep ocean and porous sands of beaches to the remarkably large surface area of floating filamentous or dendritic clumps of algae and particulate matter suspended in the water column. Borror (1980) has summarized current knowledge of marine ciliate distribution with reference to five habitats: (1) interstices of particulate matter (sediments and beach sands), (2) filamentous and planar habitats (mats of finely branched filamentous algae in tidal marshes and *aufwuchs* on flat marine surfaces), (3) deep sea and antarctic cold environments, (4) planktonic habitats favoring free-swimming individuals, and (5) symbiotic associations. Ciliates in sediments and fine sands exhibit structural modifications complementary to the environment. Those in well sorted sands of 0.1 to 0.2 mm mean grain size are very narrow, with posterior elongation, cephalization (a bulbous anterior end, especially as in hypotrichs), fragility, thigmotaxis, and some peculiarities of nuclear configuration. Ciliates exposed to strong wave action on beaches may have a tough cortex, oblong or circular shape, and an abundant intracytoplasmic fibrillar network. Distribution and abundance of ciliate species are related to sediment composition and particle size, temperature and pH tolerance, redox potential, wave action, and oxygen concentrations.

Ciliates inhabiting algal mats and similar surfaces exhibit a broader array of adaptations including variations in locomotory and feeding mechanisms, specialized modes of attachment, and presence of loricae in some species. A variety of specialized sessile forms include *Lycnophora*, a heterotrich, and *Stichotricha*, a hypotrich with elongate body and pointed anterior, surrounded by a lorica. Various peritrichs, suctoria, and foliculinids are also common surface-dwellers. Co-existing congeneric species within the varied microhabitats of the surfaces exhibit specific adaptations. Larger, slower-growing species have fewer prey types than smaller, faster-growing congeners. Differences in salinity and oxygen tolerances are commonly observed among species within the same marsh, owing to the wide variations in microhabitats created by the tidal cycle and substratum variations. For example, in a tidal marsh in the northeastern United States, Borror (1980) found that although there tended to be a cosmopolitan distribution of ciliates, species inhabiting the upper marsh, where ponds surrounded by *Spartina* and other flora occur, were characterized by *Strombidium purpureum*, *Coleps tesselatus*, *Euplotes quinquicarinatus*, and *Dysteria marina*. The lower marsh, formed by numerous narrow tidal channels 1 m deep and dominated by a vegetative cover of *Spartina*, was inhabited characteristically by *Strombidium kahli*, *Uronema filificum*, *Euplotes bisulcatus*, and *Dysteria monostyla*. Muller and Lee (1977) found that the salt marsh hypotrich *Euplotes vannus* grew well over a broad range of temperatures and salinities, but that it required higher densities of food ($> 1 \times 10^4$ cells/ml) for rapid reproduction than do other herbivores such as *Allogromia laticollaris* and nematodes. If, however, food was abundant, the algae was consumed at a rate faster than it could reproduce, leading to collapse of the community. However, with lower initial algal cell concentrations of about 1×10^5 cells/ml, a balance in growth developed between the hypotrich and the algae. In overall view, the data of Muller and Lee suggest that *E vannus* is best adapted to being a migrating initial colonizer of fresh algal blooms, and further supports the general concept that ciliates with large intrinsic growth rates are probably best adapted to opportunistic, exploitative survival strategies.

Ciliates inhabiting the benthos are not particularly different structurally from those found in shallower locations. But, as with those in Antarctic habitats, they have specialized physiological adaptations to survive at low temperatures. By contrast, a considerable diversity of structural adaptations occur in planktonic species, including the loricate tintinnida, holotrichs, nonloricate oligoitrichs, and a hypotrich (*Gastrocirrhus*). The red-pigmented, algal-symbiont-bearing holotrich, *Mesodinium rubrum*, sometimes achieves such large densities as to color the water red. A variety of commensal and symbiotic associations are also formed by ciliates with other animals, including those dwelling in the flagellated chambers of sponges, and the gastrovascular cavity of cnidaria and ctenophores, and surface dwellers on the gills of fishes and epidermis of whales.

Although many species of marine ciliates are distributed worldwide, there is growing evidence that geographic isolation, while not contributing to marked structural changes within species, may have led to physiological and genetic segregation (Borror 1980). Thus, local populations may have become partially or completely genetically isolated during evolution. Inbreeding within these small groups could enhance survival in particular local microhabitats through sexual reproduction within a more narrow gene pool adapted for that environment.

A substantial analysis of ciliates in the microbenthos of high latitude shallow marine environments on islands near Denmark and Sweden was made by Fenchel (1969). He examined community structure and species composition in relation to physico-chemical environmental factors including salinity, temperature, and redox potential. For convenient reference to classifying data, he identified three major microfaunal communities: (1) a "Sublittoral sand microbiocenosis," (2) an "Estuarine sand microbiocenosis," and (3) a shallow-water community, "Sulphuretum," dominated by the sulfur-cycle and rich in sulfur-reducing bacteria.

Sublittoral Sand Microbiocenosis

In general, the sublittor sand microbiocenoses exhibit three major zones along a depth gradient in the sand. The oxidized zone (E_h = ca 400 to 200 mv) in the upper strata, a reduced zone above the sulfide rich layer (E_h = 200 to 0 mv), and a sulfide zone where reducing conditions (E_h = < 0) are also accompanied by high sulfide concentrations. The oxidized zone is characterized by herbivorous ciliates feeding on diatoms and phytoflagellates of varying size and their predators. The representative genera include *Blepharisma*, *Strombidium*, *Frontonia* (*arenaria*), and *Tracheloraphis*, also found in the reduced zone. Some common herbivorous and omnivorous species in the reduced zone are shown in Figure 7.4. The carnivorous predators in this reduced zone (Figure 7.5) include *Loxophyllum*, *Dileptus*, and *Lacrymaria* spp. Although *Lacrymaria* occurs in all zones, only *L marina* is characteristic of the oxidized zone. Bacterivorous ciliates include small hymenostomes, *Aspidisca*, *Pleuronema*, *Euplotes*, and *Metopus*. Histophagus scavengers exploiting opportunistic masses of animal tissue include *Coleps*, *Ophryoglena*, *Paraspathidium*, and *Plagiopogon*. In the reducing zone, members of the family Loxodidae are dominate, eg, *Remanella*, and members of other families including the genera *Trachelophoris*, *Homalozoon*, and *Mesodinium*. Small hymenostomes and other bacterivorous ciliates are also present. When a sulfide layer develops, ciliates consuming sulfur bacteria appear in the vicinity of the redox discontinuity layer, eg, *Plagiopyla*, *Metopus*, and *Sonderia*.

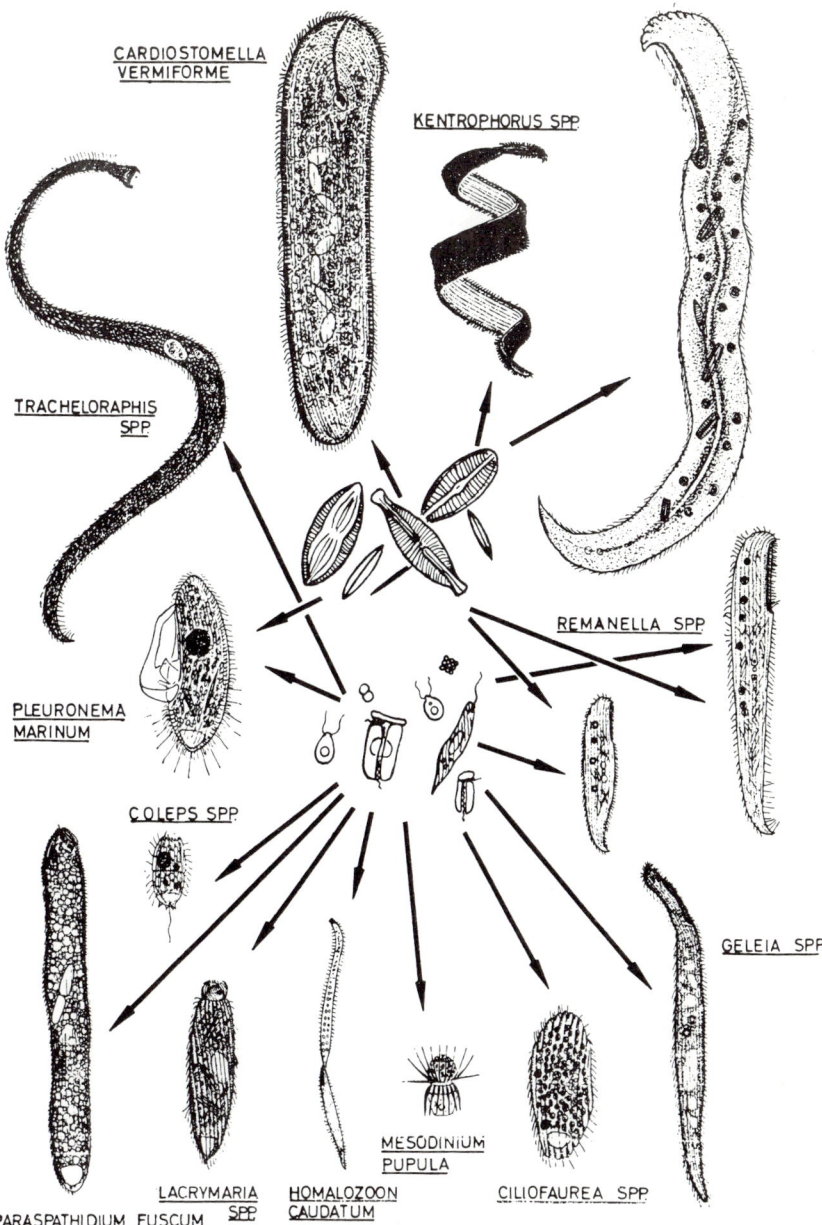

FIGURE 7.4. Some important herbivorous ciliates and their prey found in the reduced zone of sublittoral sands of marine environments. From Fenchel (1969); with permission.

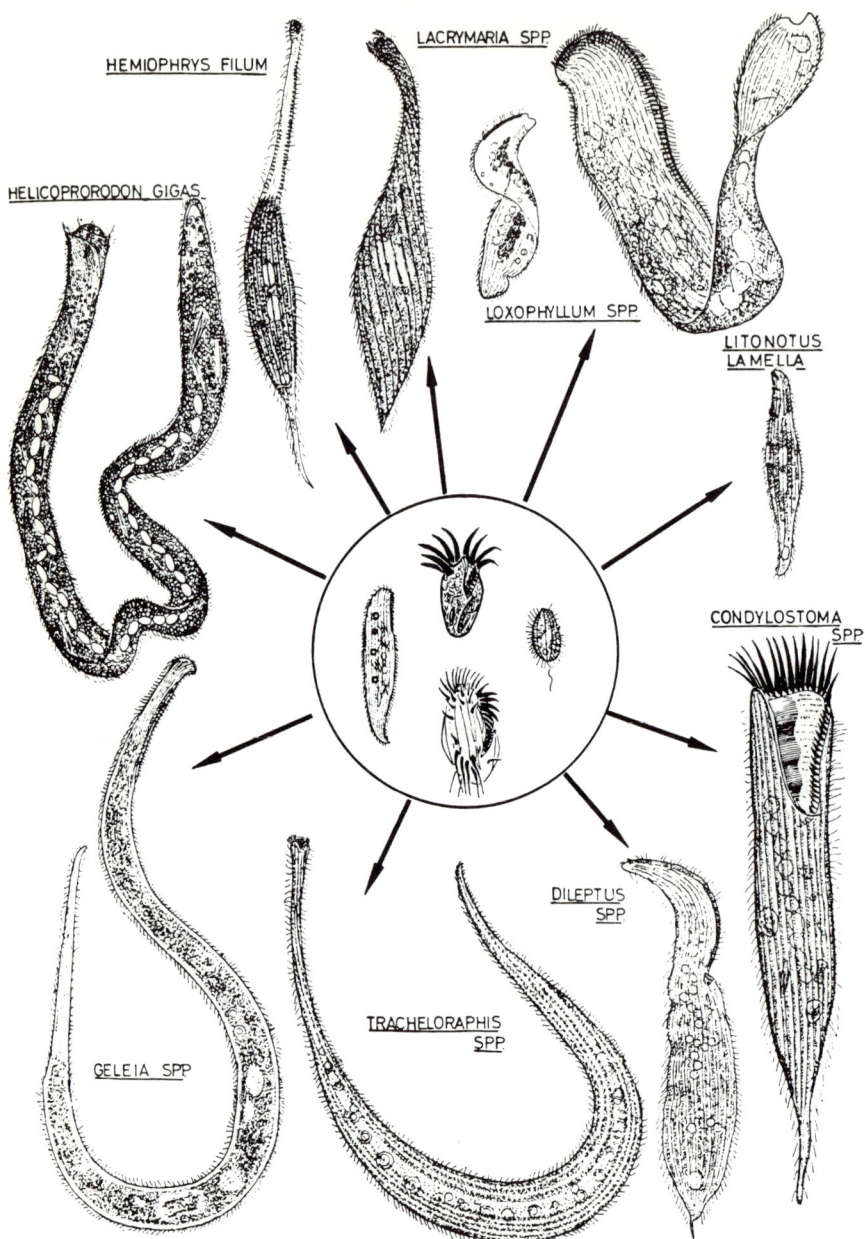

FIGURE 7.5. Some important carnivorous ciliates and their prey found in sublittoral sands of marine environments. From Fenchel (1969); with permission.

ESTUARINE SAND MICROBIOCENOSIS

The presence of greater amounts of organic matter in the estuarine sand produces a stronger reducing environment than in the sublittoral environment and a shallower zonation along a redox gradient. Thus, the major zones are more compressed and nearer to the surface of the sand. The fauna, similar to the species of the sublittoral sand microbiocenoses, is distributed in much the same zonal pattern within the shallower strata. The stronger reducing conditions favor growth of photoautotrophic bacteria, including abundance of filamentous cyanobacteria. A layer of purple sulfur bacteria occurs below the algal mat. A felt of filamentous, white sulfur bacteria, especially *Beggiatoa*, appears to provide mechanical support and food for the microfauna. Among the major ciliates are diatom predators: *Tracheloraphis kahli* (as many as 586 individuals/cm^2), *Frontonia marina* (also consuming cyanobacteria), *Blepharisma clarissimum*, and *Strombidium* spp. Purple and white sulfur bacteria are also consumed by *Blepharisima salinarum*, *Sonderia* spp, and *Plagiopyla frontata*. Predators at the surface include *Loxophyllum*, *Lacrymaria*, and *Tracheloraphis*. In increasingly reduced zones, *Metopus*, *Sonderia*, and "sulfide-resistant forms" *Caenomorpha levandri*, *Saprodinium halophila*, *Parablepharisma chlamydopherium*, and *Mesodinium* sp are common.

SULPHURETUM

The strong reducing environment characteristic of the benthos and sometimes overlying water of the sulphuretum, produced largely by the sulfate-reducing activity of *Desulphovibrio*, favors the development of sulfide-resistant ciliates normally found at lower strata in other benthic environments. Sulfate-reducing bacteria are often sufficiently abundant to form a skim on the surface of the water. These are sometimes mixed with autotrophic filamentous cyanobacteria and photosynthesizing protista yielding oxygen during the day, but also contributing to high oxygen demands during the night. This can lead to large fluctuations in surface water redox potential and in the amount of free sulfide. For example, in one location Fenchel (1969) reports that the free water above the sediments varied in H$_2$S concentrations from 0 to about 20 mg/l, and in redox potential from +329 to −190 mV. Reducing conditions are generally prevalent during the night. In the summer months when the sulphuretum is most productive, 45 species of ciliates were found among the sulfate-reducing bacteria. These included *Prorodon*, *Lacrymaria*, *Tracheloraphis*, *Loxophyllum*, *Sonderia*, *Fontonia*, *Blepharisma*, *Metopus*, *Euplotes*, and *Uronychia*.

Summative Comment

The similarity in co-associations among the ciliates and the close distributional pattern following the redox gradient in both of the sand biocenoses, though more compressed in the latter case, clearly illustrate the close coupling of ciliate community structure with environmental composition. These data illustrate the following principle. Although many ciliates are opportunistic, and in rather uniform oxic environments are distributed in apparently random patches, when marked physico-chemical gradients occur, accompanied by strong biotic conditioning variables, predictable community structures correlated with the environmental gradients can be found. This is augmented by observations of ciliate community distributions in eutrophic streams showing a close correlation between degree of eutrophy and ciliate abundance and community composition. As with most organisms, tolerance to marked and wide changes in physico-chemical environmental factors varies among species. Some peritrichs, for example, are remarkably resistant to wide variations in pH, light intensity, temperature, and ion composition (eg, Finley and McLaughlin 1962). These studies, combined with some of the foregoing ones on sarcodinad and flagellate distribution and ecology, further demonstrate the remarkable adaptive plasticity of many protozoa to widely diverse environments. This physiological tolerance, exhibited by many species, supports diffuse and widespread occurrences in habitats only marginally suitable to their survival, and permits rapid exploitation and colonization of fortuitous microenvironments that more perfectly match their optimal habitat. Further differentiation between rapidly reproducing species (r-strategists) and those that develop stable long term populations (persistence or K-strategists) provide additional refinement in niche exploitation. Variations in the kind of prey consumed, intensity in feeding, and time delay in responding to increments in nutrient resources also permit finer niche differentiation among species resulting in richly diverse, stable communities of protozoa, sometimes within patchy microhabitats of remarkably small dimension.

8
General Perspectives on Protozoan Ecology

Protozoan Communities

In the preceding chapters, some comparative data on ecology were presented for each of the major groups of protozoa. An integrative overview of major principles is included here as a summative perspective. Protozoa serve a significant role in the maintenance of energy flow and conservation of nutrients within many ecosystems. Their rapid rate of growth, relatively efficient use of resources compared to larger forms of life, and their role as prey for other organisms makes them a significant link in the food webs of aquatic and terrestrial environments. Moreover, the capacity of many species to encyst or otherwise form resting stages, combined with the foregoing physiological characteristics of protozoa, makes them significant conservators of carbon and mineral nutrients essential to the stability and productivity of plant and animal communities.

The hierarchical organization of biological communities based on photosynthetic primary producers supporting a network of consumers (Figures 7.2 and 8.1) has been extended and revised in several forms (eg, Porter et al 1985). A block diagram (Figure 7.2) of community hierarchical structure has been presented in Chapter 7, and a network model for planktonic communities is presented here (Figure 8.1). According to this perspective of biological communities, life is organized into trophic levels beginning with photoautotrophs utilizing solar energy to fix carbon into biological compounds. Based on recent discoveries in the deep sea of chemoautotrophic bacteria, utilizing sulfur compounds expelled from fissures in the ocean bottom, the concept must be broadened to also include primary producers that are chemoautotrophic and fix carbon in the absence of illumination. Also, some carbon dioxide can be incorporated by *Euglena* spp and other protozoa directly into intermediates of metabolism (eg, Peak and Peak 1981), which represents another, though less substantial source of new carbon production in trophic networks of organisms. Heterotrophs, comprising the second major group of organisms in this model, obtain their nutrition by utilizing products of autotrophs, either by preying upon them as food (first order consumers) or by utilizing their metabolic byproducts or decayed remains as sources of nutrition (eg, first order saprophages). First

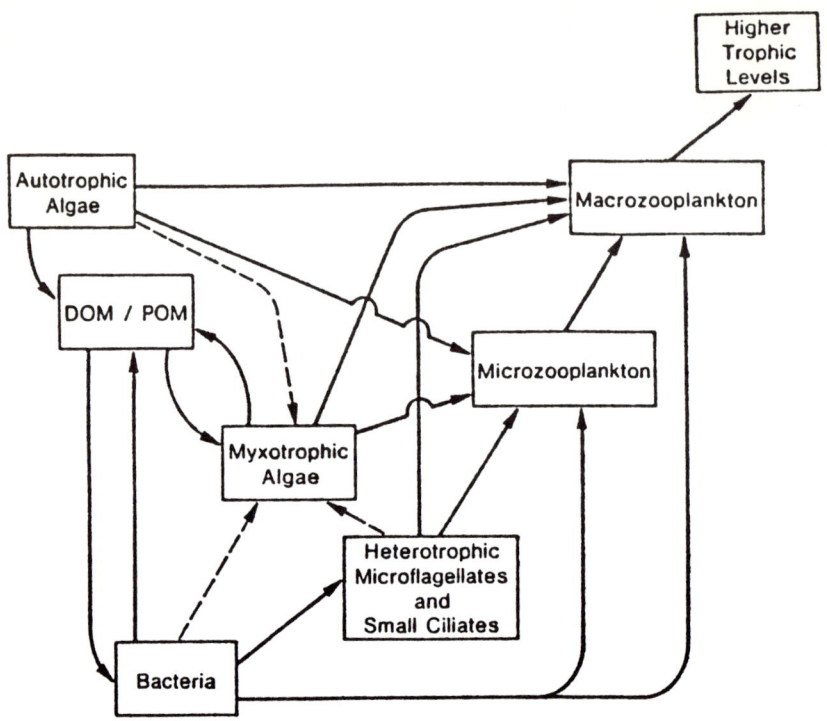

FIGURE 8.1. Interactions in planktonic food webs showing paths of carbon flow. All groups eventually contribute to the pool of dissolved and particulate organic matter (DOM/POM). From Porter et al (1985); with permission.

order saprophages also consume waste products and dead remains of first order consumers. Among these organisms are the bacteria and detrital consumers in an ecosystem. The bacteria have been identified as a separate group in this category in order to clearly delineate their role in mineral cycles and to more fully establish their critical contribution as a food source of many protozoa at higher levels in the trophic structure. The arrows in the diagram connecting these major groups indicate the flow of carbon and energy between the various groups. The head of the arrow points toward the consumer. Within the hierarchy, further trophic relations are indicated by the transfer of carbon and energy to higher levels. Thus, second order consumers can ingest first order consumers, as, for example, a carnivorous ciliate (second order consumer) that preys upon an herbivorous ciliate (first order consumer). Second order consumers (eg, bactivorous ciliates) may also prey upon first order saprophages (bacterial). This pattern of interaction is repeated at higher levels within the hierarchy, where typically larger, and sometimes less efficient consumers and saprophages utilize the products of organisms lower in the trophic web. Since no form of life is 100% efficient in the transformation of energy, losses occur during respiration and due to motion against friction, and heat is generated.

Interactions with the abiotic environment include the flow of mineral nutrients, and their remobilization through cycles linking the biotic activity of living things with the chemical dynamics of the environment. Nitrogen and phosphorus, essential components of proteins and energy conserving molecules in cells, are linked with environmental and biotic processes that regenerate compounds suitable for reuse by living systems. Nitrogen is consumed by autotrophs typically as nitrate, and also ammonia. Nitrate is reduced to ammonium during metabolism and is released into the environment, where it is reoxidized to nitrate. Some bacteria and perhaps also at least one ciliate (*Loxodes* sp), as described in the preceding chapter, can utilize nitrate as an oxidizing source to produce nitrite. This is especially significant in anoxic environments where organisms respiring oxygen would be excluded. The ammonium and nitrite excreted by these heterotrophs is recycled through the metabolic activity of autotrophs. Some heterotrophs are capable of assimilating reduced nitrogen compounds, especially ammonia. Some flagellates have been documented to be particularly significant as ammonia sinks. Phosphate bound into organic compounds is released by degradative processes in the natural environment, including lysis by exogenous enzymes released by bacteria and protozoa, thus regenerating the free phosphate in a form assimilable by cells.

Sulfur-reducing bacteria utilize sulfate as an oxidant in their metabolism. Sulfate is produced as a metabolic product by other organisms, and also by chemical oxidation of reduced sulfur compounds. The oxidized sulfur is reduced to sulfide (eg, H_2S) in a process that is comparable to the reduction of oxygen to water during respiration by other organisms. The sulfur regenerative cycle linking bacterial products and the oxidizing (oxygenated) environment is shown for convenience only for the bacteria in the first order saprophage category. The activity of sulfur-reducing bacteria provides a pool of carbon as a source of nourishment for other organisms, including protozoa, but the strong reducing environments they create also limits the species that can survive under these harsh conditions.

The flow of carbon through a trophic network, and its biosynthesis into large organic molecules, is among the most significant sources of biomass increase in ecosystems. The source of new carbon production is fixation of carbon dioxide by autotrophs. Although CO_2 is seldom limiting in terrestrial environments, it may be limiting in some aquatic environments. Oxidized carbon compounds may be chemically isolated as precipitates or other nonmetabobilizable forms. Regeneration of oxidized carbon by respiration and fermentation of living organisms provides a link between higher trophic organisms and primary producers. Protozoa are especially efficient in carbon utilization compared to metazoa (Fenchel 1974; Stout 1980). A metazoan of comparable size to a protozoan metabolizes at a rate 8.3 times greater, but its intrinsic growth rate is only twice that of the protozoan. Hence metazoa, relative to protozoa, conserve less carbon as biomass and more rapidly recycle it as CO_2. The wide diversity and large populations of protozoa, including facultative autotrophs that reincorporate oxidized carbon into organic molecules, provide an efficient "carbon buffer" within biological communities. The protozoa sustain efficient fixation of carbon into biological molecules, enhances its rapid reutilization through saprophagic activity, and mediate its transfer to higher trophic levels.

Stability and Productivity of Communities

In general, the stability and productivity of biological communities depends less on the total mineral abundance in the biosphere than on the efficiency and rate of recycling (Stout 1980). Bacteria and protozoa serve as significant links in this process of mobilizing and converting organic and inorganic metabolic resources for use by higher trophic level organisms. The efficiency of utilization of prey by protozoa in the biosynthesis of biomass has been investigated by several researchers (eg, Laybourn and Stuart 1975; Curds and Bazin 1977; see also Stout 1980) and some representative data are presented in Table 8.1. In general, energy conversion efficiency can be as high as 50 to 70% (Calow 1977). Phosphate regeneration by bacteria also appears to be stimulated in the presence of protozoa. It has been well established that bacterial growth is stimulated in the presence of protozoa, but it is not fully determined whether this is due to utilization of metabolic products from the protozoa, an effect of grazing by the protozoa that selectively removes less rapidly growing bacteria, or a combination of several factors of these kinds. Barsdate et al. (1974) reported that release of phosphate from *Carex* plant litter expressed as µg P/bacterium/hr $\times$ 10^{-7} was 0.025 when bacteria were grown alone, but increased to 1.09 in the presence of protozoa. Bacterial phosphate assimilation (expressed as µg P/bacterium/hr $\times$ 10^{-7}) increased from 16.7 to 191.3 when protozoa were added. Studies in terrestrial environments also indicate a stimulatory role for protozoa increasing bacterial nitrogen fixation (Cutler and Bal 1926) and increased ammonification (Meiklejohn 1932).

TABLE 8.1. Growth yield for some representative protozoa.

Protozoan	Prey	Yield
Acanthamoeba sp	*Saccharomyces cerevisiae*	0.37
Colpoda steinii	*Escherichia coli*	0.78
Entodinium caudatum	*Escherichia coli*	0.50
Tetrahymena pyriformis	*Klebsiella aerogenes*	0.50
Tetrahymena pyriformis	*Klebsiella aerogenes*	0.73
Stentor coeruleus	*Tetrahymena* sp	0.72

Data from Curds and Bazin (1977) and Laybourn (1976). Yield is expressed as biomass in weight units gained per biomass weight of prey consumed.

Although protozoa may account for only a small part of the total biomass in an ecological system, their "catalytic effect" on other organisms and their contribution to mineral recycling and carbon flow indicate a far more significant role in maintaining a balanced ecosystem than inferred from their population densities. In some environments, including fine sediments, protozoan biomass is of the same order or larger than other microfauna. In Fenchel's studies of Scandinavian waters, ciliates numbered 10^6 to $10^7/m^2$, constituting up to 2.3 g wet weight. Coarser sands, however, contained fewer ciliates and the metazoa were more dominant. Large numbers of dinoflagellates (10^5 to $10^7/m^2$) and fewer Euglenida ($< 5 \times 10^5/m^2$) were commonly found in these marine environments. Naked amoebae apparently played even a lesser role based on numbers, seldom exceeding $10^5/m^2$. However, numbers alone do not accurately reflect the significant role that many protozoan groups play in conditioning the environment or serving as food links in the complex web of trophic interactions that occur in many natural environments. There is good evidence that the tendency of protozoa to overgraze prey to extinction in laboratory cultures does not occur so commonly in the natural environment. Natural ecosystems with larger volumes and refuges for the prey support a more moderate growth of protozoa and a more steady predator-prey interaction. In general, when the numerical response of a predator protozoan is equivalent to or longer than the rate of natural increase of the prey, a stable oscillating predator-prey relationship can be established. Under these conditions, the carrying capacity of the prey (maximum density) regulates the stability of predator-prey dynamics. When the prey has a low density, the growth of the predator is limited and hence it cannot overgraze the prey. Sparse populations of prey, and those that are motile, moreover, place additional energy demands on the predator. The predator must continuously move about in search of food, and realizes fewer successful encounters than with more abundant and nonmotile prey. Consequently, there is once again less likelihood that predator overproduction will occur. High prey densities produce large oscillations in predator and prey densities characterized by large peaks and low minima. These wide oscillations can lead to a precipitous decline and extinction of the prey when overproduction of the predator coincides with a pronounced minimum in prey density. The presence of prey refuges, coupled with a moderate intrinsic rate of growth, helps to ensure a steady but low level of prey abundance and hence a more stable predator-prey interaction. These observations reinforce the need to interpret results from laboratory simulations with caution, and the wisdom of cross-checking results from laboratory simulations with evidence obtained from the natural environment. In particular, the vastness of many natural habitats compared to laboratory culture vessels permits a more diversified set of habitats and allows migration from one site to another as resources become limited. The spatial dispersion of protozoa and their colonization of opportunistic resource-rich sites in large-scale natural ecosystems is probably one of the factors contributing to a nearly constant but low density of many naturally occurring protozoan populations.

Protozoan Distribution and Colonization Rate

Distribution

Cairns and associates (eg, Cairns and Yongue 1977) have examined the distribution of protozoa in a variety of habitats, and determined the rate of colonization of artificial substrates anchored in aquatic environments. It is becoming increasingly clear that protozoa are cosmopolitan in distribution. Their co-occurrence in a given location is largely a result of favorable localized physico-chemical factors, sufficient food, and opportunistic arrival of cohort species that inhabit and exploit the locale. A capacity to rapidly encyst and excyst, coupled with extended longevity of encysted stages, provides further potential for opportunistic exploitation of occasional favorable habitats. The diversity of protozoa, and the similar roles served by particular species across phyla, suggest that the composition of localized communities of protozoa may be quite variable from one place to another. The various trophic and environmental conditioning roles may be served by one collection of species in locale A, but by a different collection of species in locale B. Hence, a predictable and simplified model of protozoan community structure of the type constructed for higher organisms in a given ecosystem may not be possible with protozoa. Rather, particular functional roles may need to be identified and the kind of species fulfilling these roles categorized. Within this conceptual framework, a well-ordered system of interactions among species forming the community can be constructed, as has been done for metazoan communities. Thus, protozoan assemblages are not random collections of individuals whimsically forced together by unpredictable environmental events, but an integrated set of species fulfilling functionally predictable roles within a community structure (Yongue et al 1973).

The distribution of protozoa in the natural environment in space and time is a continuum, and one seldom finds a distinct collection of organisms within a well-defined geographic locale. For example, in a study of species distribution in a river (Cairns and Yongue 1977) the amount of species overlap from one sampling station to another was appreciable, ranging from 25 to 40% across four stations, but less constant from one sampling time to another (ca 18 to 21%). Rapid changes in weather and concomitant fluctuations in protozoan populations contribute to a less predictable gradient in time. A protozoan community, therefore, is a transient entity and consists of a collection of species in a given locale at a given time. The numbers of individuals, the dynamics of their interaction, and the particular roles they serve are potentiated or limited by the physico-chemical factors of the environment. Nonetheless, some commonly occurring species appear to be consistently found together. This is a logical expectation based on the premise that protozoan species are cosmopolitan and that some of them have similar environmental requirements. Cairns (1964) and Cairns and Yongue (1977) report that out of a possible 190 pairs of species found most commonly in sampling sites, 44 were statistically different than chance combinations. Based on an examination of the environmental ranges for these species, Cairns concluded that: (1) pairs or larger groups of co-occurring species always had virtually identical ranges of environmental conditions in their natural habitats, (2) the species also tolerated rather broad ranges of environmental conditions, and (3) in spite of the foregoing tendencies for co-association, identical ranges of tolerance to environmental conditions *does not ensure* that species will be associated together more often than by chance. The latter conclusion was based on the observation that only 44 of the 190 associations were statistically significant.

Colonization

In studies of the colonization rate of small masses of artificial sponge, Cairns et al (1969) found that although different assemblages of colonizing species were found at different locales, the process of colonization was remarkably identical. In the early stages of invasion of a new substrate, the growth of populations greatly exceeded the rate of loss. As the number of species increased with time, however, the rates of increase and extinction began to equal one another and the population densities approached an equilibrium level. A similar exponential rise toward a plateau in numbers was found by Maguire (1977), who investigated the rate of colonization and species composition in beakers placed in various locations distant from a natural body of water. This negatively accelerated exponential increase (rising to a plateau) as a function of time is consistent with island colonization models as proposed by McArthur and Wilson (1963, 1967).

The number of different species inhabiting an artificial substrate at near steady state levels is exponentially related to the volume of the substrate (Cairns and Ruthven 1970). The mean number of species in one experiment is given by $N = 0.64 + 3.39 \log x$, where x is substrate volume in mm^3.

Although these data may seem to be only of academic interest, or only relevant to artificial substrates, the diversity/volume relationship may be significant in understanding community dynamics during exploitation of patches of litter or other small-volume sites in the natural environment. It is not known to what extent sessile or larger species are trapped in the sponge during sampling, and thus are underrepresented in the data. As with any method, one must be aware of the limitations. In general, however, the technique has yielded significant information on population distributions and dynamics of colonization.

The number of species initially colonizing a substrate is limited, and during the first several days to weeks only the population density of established species increased. Thereafter, there was a marked increase in the number of different species detected within the substrate. This delayed increase may be due to a conditioning of the substrate by the initial arrivals, which made it possible for additional species to successfully invade and inhabit the spaces within the plastic sponge (Cairns et al 1969). The nature of this conditioning process is not known. It may include alterations in pH (Yongue and Cairns 1971), enhancement of the nutrient properties of the medium through increased carbon production and release of organics, or the establishment of a richer bacterial flora serving as food following biological conditioning of the substrate. However, as the substrate becomes more fully colonized and diversity begins to reach a limit, further new arrivals appear to be excluded. Exclusion may occur because the new arrivals cannot compete with the robust populations already established. This hypothesis was strengthened by the observation that periodic removal of part of the existing populations by squeezing the sponge substrate permitted invasion by additional species in greater numbers than in undisturbed control substrates (Cairns et al 1971). During the course of community maturation, the rapid increase in species diversity following the initial colonization gradually decreased to a rather steady, but oscillating level around a mean typically of 30 to 35 species. The oscillations were within the range of ±6 species in 75% of the cases (Cairns and Yongue 1977). When the artificial substrates are disturbed, as for instance by sampling them, this balance is temporarily upset, but is restored by a succession of species reinvading the substrate in a predictable pattern resembling the initial colonization events.

Succession and Community Structure

Laboratory and Experimental Studies

In laboratory cultures of protozoa initiated with natural organic matter (eg, dried hay) or organic nutrients (peptone and carbohydrates) in sterile pond water, a regular succession of species is observed (eg, Woodruff 1912; Bretthauer 1980). In hay infusions, a rich bacterial flora appears within 48 hr and occasions the onset of massive protozoan growth comprising saprozoic, herbivorous, carnivorous, and omnivorous forms. Flagellates (monads) are among the first organisms to appear in quantity at about the second day and are accompanied in varying numbers by naked amoebae. The flagellates reach maximum densities of several hundred to several thousand per cc at periods from 4 to 20 days depending on the organic enrichment of the medium, but thereafter enter a decline. Ciliates appear in succession, beginning typically with *Colpoda* sp, various Hypotrichida, sometimes accompanied or succeeded by hymenostomes (eg, *Paramecium* sp), and eventually culminating in the appearance of sessile forms (eg, *Vorticella* sp). On occasion, carnivores such as *Didinium* may occur in sufficient numbers to suppress the hymenostome populations. The *Colpoda*, though among the first protozoa to appear, sometimes enter an abrupt decline after peaking at densities of hundreds to thousands per cc. In one culture of Woodruff's, for example, *Colpoda* reached an average maximum number of 4500 per cc on the fifteenth day after initially appearing on the seventh day. But by the twenty-second day, there was a rapid decline to about 40 per cc. In some cultures, however, *Colpoda* persists as the dominant species. The hypotrichs, in general, were most adjusted to dwelling on the surface of the infusions *after* the flagellates and *Colpoda* had subsided in density. *Vorticella*, representative of sessile forms, typically may appear as early as the third to seventh day after inception of the infusion, but usually does not reach maximum growth until about the twentieth day or later. At peak abundance, it reaches densities of several hundred per cc. On the whole, *Vorticella* typically attains abundance after the hymenostomes. Amoebae were detected with low frequency throughout the development of most successions, but when abundant, they peak late in the succession at about the twentieth to fiftieth day. Variations within this general successional pattern are to be expected, depending on the initial composition of encysted forms, the kind of organic enrichment used, bacterial biomass and species composition, and physico-chemical variables including pH, E_h, and available oxygen.

In illuminated and aerated aquaria containing water enriched with a peptone nutrient source, a succession of flagellates was observed after seeding the system with natural water systems (Bretthauer 1980). Among the first monads to appear in the plankton phase were *Bodo*, and *Cercomastix* followed in about 10 days by additional species of *Bodo* and *Hexamitus*, which were closely succeeded by *Euglena* and *Cryptomonas*. The early appearing monads growing upon surfaces (periphyton) included the phytoflagellate *Chromoulina*, and zooflagellates *Cercomastix*, *Cercobodo*, *Rhynchomonas*, *Pleuromonas*, *Bodo*, *Oicomonas*, and *Monas*. Later, at about the tenth day, *Euglena* and *Cryptomonas* became more dominant as most of the preceding species receded in numbers. The changes in physico-chemical parameters during the 20 day period of the experiment showed a rather steady O_2 saturated level. Ammonium and bicarbonate concentrations reached maximum levels at about the second day and remained elevated until the fourteenth day. Coincident with the ammonia and ammonium maximum, the first flagellate populations declined, followed by an increase in *Euglena* and *Cryptomonas* populations. At this point, nitrate concentrations abruptly increased. In general, flagellates preceded the appearance of ciliates as reported in the earlier studies by Woodruff (1912), and in systems arranged to simulate an upstream to downstream flow. Ciliates appeared later in the downstream locations compared to the upstream (Bretthauer 1980). While the flagellates remained fairly constant in abundance, the ciliates reached a peak at the fifth to tenth day and then declined. In the more downstream vessels, however, densities increased once again, maintaining a steady but lower level than the peak at the tenth day.

In water purification systems where organically rich media and initially polysaprobic conditions prevail, a succession typically begins with a high bacterial population preyed upon by heterotrophic flagellates, some bactivorous ciliates, and amoebae. In general, the flagellates precede the ciliates and become less abundant as more efficient bactivorous and predatory ciliates dominate the system. Organic nitrogen is liberated as ammonia, and the pH is typically low with high carbon dioxide levels produced by heterotrophic respiratory activity. Although some diatoms and cyanobacteria may be present, this stage is dominated by heterotrophic metabolic processes and has been designated the "Heterotrophic stage" by Bick (1973). As the bacterial populations decline and autotrophic organisms (diatoms, euglenoids, and filamentous algae) become more dominant, carbon dioxide is consumed, pH tends to rise, and ammonium is present in increasing amounts as ammonia. The presence of nitrate and nitrites favor autotrophy, and although some of the more delicate ciliates are killed by the elevated ammonia levels, some of the herbivorous, more resistant forms become dominant. In general, the system passes from a polysaprobic or alpha–mesosaprobic stage into a beta-saprobic and oligosaprobic phase (Bick 1973; Bretthauer 1980). This is designated the "Autotrophic stage" by Bick. During a succession in a cellulose-based organic medium, Bick (1973) found that early dominant flagellates included *Ochromonas* and *Chlamydomonas*. This is consistent with findings of Bretthauer (1980), who reported that during the anaerobic phase of purification, *Ochromonas sociabilis* and also *Chrysomonas ovalis* are important components of the surface mat, where oxygen supply is greatest. In several studies, eg, those of Bick (1973), the subsequent rise in the ciliate population, including *Uronema* (which preyed on *Ochromonas* and other predatory ciliates), resulted in increasing dominance of ciliates. This produced an early peak in *Colpidium* and *Cyclidium*, which attenuated by the tenth to fifteenth day. Thereafter the major ciliates were *Halteria, Litonotus, Stylonichia, Microthorax, Chilodonella,* and *Paramecium*. These free-swimming species were succeeded by sessile forms, including *Vorticella*. The major abiotic factor correlated with this succession is the gradual shift from an acidic, reducing environment, increasingly enriched in inorganic nitrogen toward an oxidizing environment. In this phase, autotrophs utilize inorganic nutrients, and produce oxygen and consume carbon dioxide, thus elevating the pH. The general increase in autotrophs and favorable physico-chemical environmental factors favor the development of a population of herbivorous ciliates.

In further studies of environmental factors regulating species abundance and stability, Taub (1971) used microcosms consisting of connected continuous flow vessels simulating stream flow. She examined the effect of light intensity, nutrient limitation, and dilution rate on community development. The cultures contained the single alga *Chlamydomonas reinhardtii* (an autotroph), the bacteria *Pseudomonas fluorescens*, and *Escherichia coli* (saprophage), and *Tetrahymena vorax* (consumer). Nitrate was the limiting nutrient in the system. At high dilution rates of 1.0 to 1.4 changes of volume per day, algal populations and *Pseudomonas* were low in the upstream community. The ciliates and *E coli* were not able to reproduce at a rate sufficient to compensate for the flow-through and became extinct. At lower dilution rates (0.57 and 0.73) all of the organisms were able to maintain steady state densities. When light intensity was reduced to a level where the algae were just able to maintain themselves at a sparse steady state at the lower dilution rates, the protozoan population was able to maintain a growth rate of 0.5, but not 0.75. At a dilution rate of 0.73, they were washed out of the system. The growth of communities in the downstream flasks were generally more dense and less perturbed by changes in environmental conditions. This can be attributed to the upstream flask serving as a buffer. When productivity is low in the upstream flask, some residual nutrients are available to flow into the second flask. Rapid dilution rates that tend to diminish growth in the upstream flask also increase net input of nutrients into the downstream flask, thus increasing productivity and numbers of individuals and increasing community stability.

Increased community stability with larger numbers of individuals and more diverse taxa (higher diversity) has also been demonstrated during colonization of beakers of water in the natural environment (Maguire 1977). The coefficient of variation for number of taxa (larger variation representing less stability), and variability in population densities, were decreased when many different taxa grew in the beakers. Thus, as the beakers became more fully colonized and the number of different protozoan taxa increased, the variations in population density and fluctuations in kinds of taxa in the beakers became increasingly lower. Hence, the communities under these conditions were less easily perturbed. These data from complex protozoan communities appear to support the general principle that community stability tends to increase with increasing diversity of species. This is in contrast to the findings of Hairston et al (1968), who found that in some cases increasing diversity in laboratory cultures of bacteria and protozoa did not yield greater stability. The differences in range of species in Maguire's vessels and the more complex interactions between abiotic and biotic variables that can occur in the natural environment may explain the differences in results between these two studies.

Natural Community Structure

The structure of natural protozoan communities was examined as early as 1917 by Hausman, who described five "Environmental Types." Environmental type I (marsh pools), with warm water and abundant decaying vegetable matter emitting a "Swampy odor," contained 21 dominant species including, among others, amoeboid protozoa (*Amoeba limax*, *Amoeba proteus*, *Arcella vulagaris*, *Difflugia acuminata*); flagellates (*Oikomonas* sp, *Peridinium cinctum*); and ciliates (*Coleps* sp, *Carchesium*, *Stylonychia*, *Stentor*, and *Vorticella* species). Environmental type II (clear cold waters lacking plant growth) was dominated by flagellates (including *Astasia*) and some lesser numbers of ciliates (*Colpidium*, *Paramecium*). Environmental type III (clear, flowing water with abundant plant life) was dominated by species of *Amoeba*, ciliates (including *Chilodon*, *Colpidium*, and *Colpoda*), flagellates (*Monas*, *Chilomonas*), and testate amoebae (*Difflugia*). Environmental type IV (clear, small pools with abundant decomposing organic sediment) were found typically in rock depressions containing leaf litter and filamentous algal growth with average temperature about 60°F. The major genera represented were *Chilodon*, *Coleps*, *Paramecium*, *Monas*, and *Difflugia*. Lesser numbers of genera such as *Colpoda*, *Euglena*, *Pleuronema*, and *Stylonichia* were also found. Environmental type V (warm water pools with abundant algae) were dominated by flagellates (eg, *Chlamydomonas*, *Euglena*, *Monas*, *Synura*, and *Peridinium*). Lesser numbers of the ciliates *Colpoda campyla*, *Frontonia*, and *Vorticella* were present. Occasional individuals of the heliozoan *Actinophrys sol*, and the testate amoeba *Arcella vulgaris* and *A mitrata* were observed.

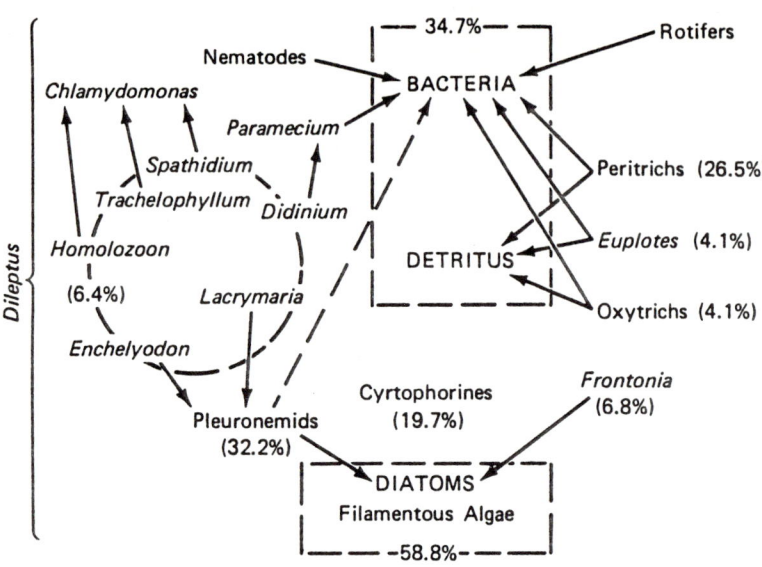

FIGURE 8.2. A blue-green (cyanobacterial)-based association found in an Antarctic pond. Percentages are based on the proportion of each group found at sampling sites. Species enclosed within dashed lines are in three major groups: (1) general mixed feeders on detritus and bacteria (34.7%), (2) herbivores (58.8%), and (3) carnivores (6.4%). The percentage reported by each species name is based on the proportion of individuals within each of the major groups. From Dillon and Bierle (1980). Reproduced by permission of the U.S. Department of Energy, Office of Scientific and Technical Information.

The classification of protozoa into two types of communities based on food webs (Figure 8.2) was proposed by Picken (1937). The first type (a blue-green association) consisted of bactivores (eg, the ciliates *Colpidium, Glaucoma, Vorticella*, and flagellates *Peranema*) and a predator, *Lionotus*, which preyed on *Colpidium* and *Glaucoma*. Several facultative bactivores and herbivores included the ciliates *Euplotes, Chilodon, Frontonia*, and the amoeba *A proteus*, which along with *Peranema*, also preys on small flagellates. *Stentor*, an omnivore, and its predator *Dileptus* formed a higher trophic level in this community structure. The second type (a fungal association) was richer in composition, and bacteria formed the major food source. This community included the carnivorous ciliates *Dileptus, Stentor,* and *Lionotus*; herbivorous naked and testate amoebae *Amoeba proteus, Euglypha,* and *Difflugia*; and diatom feeding ciliates *Frontonia* and *Chilodon*. Bactivores including the ciliates *Glaucoma, Colpidium, Colpoda,* and *Spirostomum*; and saprophytic flagellates, colorless euglenoids, and protomonads were also common. The pigmented flagellates *Synura* and green euglenoids were major autotrophs among the monads.

Picken made a significant observation that the filamentous strands of either algae or fungi served a major role in the integrity of these communities. They provided a physical network of surfaces upon which many of the protozoa remained attached. The distribution and community organization of the protozoa were dictated by the topology of the environment. Few species, except those capable of gyrational motion when free-swimming, were able to migrate to other portions of the environment without following the filament network.

The microbiocenoses in an Antarctic pond studied by Dillon and Bierle (1980) conformed closely to the blue-green algal community structure (Figure 8.2) as proposed by Picken (1937). No protozoan successions were observed during the course of the study and most ciliate species peaked at about the same time. The proportion of ciliates in order of dominance was pleuronemids > peritrichs > cyrtophorines > *Frontonia* > oxytrichs = *Euplotes*. Similar cyanobacterial-fungal associations were reported by Cowling and Smith (1986) in fellfields of the maritime Antarctic (Signy and Orkney Islands), where a protozoan fauna of at least 24 species was found. In the unvegetated stony fellfield fines, however, only a single colorless flagellate occurred, *Heteromita globosa* Stein (5 to 8 µm). A maximum density of $1.4 \times 10^4/cm^2$ and a standing crop biomass of 2.5 µg/cm² were found during the summer sampling period. Population size was significantly correlated with moisture content of the fines ($r = 0.76$). The flagellate grew well in culture at 5°C and subsisted on an indigenous strain of *Pseudomonas* sp as its sole nutrient source.

Modern perspectives on freshwater and marine community structures (eg, Steele 1974; Porter et al 1985; Pratt and Cairns 1985) have considerably expanded our understanding of the complex interactions among protozoa and algae in conserving carbon and remineralizing nutrients essential to growth of higher organisms in aquatic ecosystems. A diagrammatic summary of major trophic interactions (Figure 8.1) indicates the significant contribution of autotrophic algae to the supply of dissolved and particulate organic matter (DOM and POM) in aquatic systems. Bacteria (consuming as well as contributing to the DOM and POM) are grazed by heterotrophic microflagellates and small ciliates, and to a largely undertermined degree by myxotrophic monads (phagocytic and osmotrophic pigmented monads) which are closely linked as consumers and contributors to the pools of DOM and POM. They are also a food source for the heterotrophic microflagellates and ciliates. All of these groups are consumed in varying degrees by microzooplankton which contribute carbon to organisms in higher trophic levels. Although classical theory, based on analyses of larger organisms, suggests that every additional link in a food chain decreases the efficiency of energy transfer, the presence of bacteria and bacterial grazers within the complex food web of aquatic ecosystems appears to increase the efficiency of the total system. This is accomplished by the protozoa recovering carbon that would be lost as algal exudates, fecal matter, organics lost during "sloppy" predation by larger organisms, and refractory organic matter such as plant wall residues (Porter et al 1985).

The ecological categorization of algae and protozoa is based upon their size and trophic role in aquatic ecosystems (eg, Sieburth 1979). Three groups are commonly recognized. (1) Picoplankton (0.2 to 2.0 µm) include the bacteria and microflagellates, autotrophic and heterotrophic, that either utilize energy of sunlight to fix carbon or subsist on the dissolved carbon exuded by other organisms. These are among the most abundant and often most highly productive smaller members of an aquatic community. (2) Nanoplankton (2 to 20 µm), which encompass pigmented flagellates and diatoms that dominate the plankton in terms of numbers and primary production. The colorless flagellates and facultative autotrophs (eg, *Ochromonas*, *Cryptomonas*, and colorless euglenoids) are included here. More broadly, members of the kinetoplastida (*Bodo*), Choanoflagellida (choanoflagellates), Dinoflagellida, and bicoecids (flagellates with complex loricas) are also included here. Many of the heterotrophic monads are bactivores and are probably less dependent on direct uptake of DOM by osmotrophy. The bacteria, when abundant, can more favorably compete for the dissolved organic nutrients, leaving less for the flagellates. Thus, the flagellate nanoplankters are compelled to feed on the bacteria to obtain carbon and nutrient resources. Hence, the nanoplankters are significant links in transferring organic carbon from picoplankton to higher trophic levels. (3) Microplankton (20 to 200 µm) include ciliates, tintinnids, amoeboid protozoa, larger diatoms and dinoflagellates, and small metazoa (eg, copepod nauplii). The composition of each of these size classes and the species filling each of the roles is not fully determined for many aquatic environments, but it is generally known that seasonal variations in numbers and kinds of protozoa produce successions of populations in many aquatic communities.

Freshwater lakes and ponds exhibit a typical annual cycle of succession correlated with the temperature and quality of the water. During summer stratification, benthic dwelling anoxic tolerant species (eg, *Loxodes* and others) are forced to migrate higher into the water column by means of the upward movement of the redox discontinuity bringing increased reducing substances into the water overlying the sediments. Surface water becomes dominated by herbivorous obligate aerobic species. As the water column becomes less stratified in late autumn and during winter, mixing resuspends the nutrients from deeper within the water column and increases oxygen content at depth, thus permitting wider distribution of protozoa throughout the water column, especially among those protozoans tolerant of lower temperatures. Superimposed on this major seasonal pattern is a fairly reproducible cycle of annual planktonic successions. Bamforth (1958) reported that in a small artificial pond, plankton abundance and species composition were correlated with physico-chemical properties of the water during the annual cycle. The flagellates *Glenodinium* and *Pyramimonas* and also diatoms were abundant in spring. Euglenoids appeared in summer and *Cryptomonas* occurred abundantly in early summer and autumn. Chrysomonads and *Chlamydomonas* were among the dominant monads found in winter. The amount of dissolved phosphate appeared to control some species densities. Chrysomonads were not found when the phosphate concentration exceeded 0.02 ppm, whereas cryptomonads and diatoms increased at this concentration. Euglenoids were abundant at concentrations several times this level and especially responded to supply of organic nutrients. Within these major controlling nutrient patterns, nitrogen and amount of vitamins and other growth factors seemed to have an intermediate effect. In general, the moderating role of nitrate is characteristic of aquatic regimes where nitrogen is not limiting. If, however, available nitrogen is below a critical level, it is usually the most significant inorganic nutrient to determine biomass. Increases in sunlight and temperature were favorable for diatoms, but unfavorable for the dinoflagellate *Glenodinium* and the loricate euglenoid *Trachelomonas hispida*. The disappearance of *Glenodinium* during summer suggests that it is intolerant of increases in ammonium and phosphate. It otherwise has a fairly wide tolerance for large temperature variations. This is indicated by its presence, though in small numbers, throughout the year. Too few sarcodina were found to enumerate them, but heliozoan species (eg, *Actinophrys sol* and *Heterophrys glabrescens*) occurred mainly during the cool periods of the year. The majority of ciliates appeared in three population peaks. (1) A January appearance of *Urotricha*, consuming the abundant chrysomonads. (2) Spring-appearing hypotrichs coinciding with the decline in *Chlamydomonas*, and succeeded by the gymnostome *Phascolodon vorticella*, which increased upon the decline of the *Gonium* population. Gymnostomes appeared to prey on green algae, as they were filled with grass-green inclusions. (3) Summer ciliates dominated mainly by the genus *Holophrya* occurred after the rise in ammonia in late May and exhibited wide fluctuations in numbers. They usually followed the abundance of phosphate and turbidity, suggesting that bacteria were their major food supply.

An analysis of seasonal succession in the euphotic zone (illuminated, autotrophic region) of Lake Tanganyika (Africa) also exhibited three phases (Hecky and Kling 1981). An early annual bloom of chloromonads with low total biomass (< 100 mg/m^3) occurred in February to April. This was followed by higher densities of chrysomonads, diatoms, or both in midsummer, with intermediate biomass (100 to 200 mg/m^3). The highest productivity occurred in September through November. Throughout this period, the plankton was dominated by *Strombidium* containing green algal cells in its food vacuoles. Green algae were most abundant during this period. Cyanobacteria, cryptomonads (especially *Rhodomonas minuta*), and chrysophyceans were of secondary abundance. The dinoflagellate *Glenodinium* was generally sparse. At one sampling location, however, it occurred abundantly in April and May. The single euglenoid observed, *Trachelomonas volvocina*, was never abundant. In total, 16 species of heterotrophic protozoa were found, and among the pigmented monads, 13 chrysomonads, 5 cryptomonads, 9 dinoflagellates, and 1 euglenoid (*Trachelomonas*) were collected.

Comparative studies of protozoan populations in several freshwater and saline lakes in Africa (Curds et al 1986; Finlay et al 1987) indicate that contrary to some prior assumptions, protozoa are among the most abundant secondary producers. Protozoan diversity was greatest in the littoral zones of the lakes and in the submerged vegetation surrounding the papyrus swamps. Gymnamoebae were found mainly in the littoral zones but absent in other zones, and these were detected only by making cultures on agar plates streaked with suitable bacterial prey.

Approximately 51 to 79% of the protozoa were bactivores in the freshwater lakes, while almost all of the protozoa in the papyrus swamp were bactivores, including some large ciliates (eg, *Spirostomum minus*, *Frontonia leucas*, and *Paramecium caudatum*). Also present were *Chilodonella cucullus*, *Urocentrum turbo*, *Halteria grandinella*, and *Vorticella longifilum*. The oxygen content of the bacterial enriched water was 0 to 3.7 mg/l. Phytoflagellates included *Chlamydomonas* and some euglenoids. When diatoms were plentiful in the freshwater lakes, phytoflagellates were in general less abundant. Filamentous cyanobacteria were also common. The saline lakes Nakuru and Simbi represented two diverse ecosystems. Nakuru is a shallow, bacterial-rich lake frequented by millions of flamingoes. Their waste deposits enrich the water and produce bacterial numbers of 2.7×10^8/ml, which is an order of magnitude larger than in Lake Simbi. Consequently, the number of flagellates and suspension-feeding ciliates were much larger in Lake Nakuru, and the protozoan population more diverse, with at least 20 different heterotrophic species. Several larger species, eg, *Frontonia*, feeding on larger algae and cyanobacterial filaments, and a carnivore, *Spathidium*, were characteristic of Lake Nakuru. In Lake Simbi, a few *Fontonia* and *Trachelophyllum* were seen and only a single specimen of *Spirostomum teres* was obtained. A relatively large population of the ciliate *Holophrya*, feeding on rotifers, was present in Lake Simbi, and remains of its prey, including the jaws, were observed in food vacuoles. On the whole, the conspicuous abundance of ciliates in Lake Nakuru indicate that they serve a substantial role in the trophic network of the lake, and it is estimated that their consumption rate is in the range of 3 to 14 g/m^3 per day. This estimate, based on a limited set of samples, requires additional investigation over longer periods of time to determine the annual production in these remarkable tropical saline lakes.

Studies of annual production cycles in an oligotrophic lake in central Russia (Petrova and Smirnova 1974) showed two major peaks of productivity, largely dominated by a limited number of protozoan species including the holotrichs *Coleps hirtus*, *Didinium nasutum*, *Stokesia vernalis*, and *Lionotus fasciola*; spirotrichs *Metopus* sp, *Stentor polymorpha*, *Strombidium viride*, and *Spirostomum teres*, and peritrichs *Vorticella campanula* and *V nebulifera*. *L fasciola* appeared in the lake at the beginning of July and persisted until September, but peaked during early August. The planktonic peritrichs were also abundant in August, but declined sharply in September along with *L fasciola*. A second peak of major production appeared at this time. However, it was dominated by numerous small holotrichs and a heavy growth of *S viride*. *Coleps* also appeared during this time and bottom sediments were occupied by *Vorticella*. *L fasciola* is oxyphilic and disappeared as soon as abundant amounts of carbon dioxide built up in the water, whereas *Caenomorpha* and *Metopus* are sulfide tolerant and were common through much of the year. The authors also estimated the amount of production per unit biomass (P/B) for the major occurring species by examining the rate of increase when cultured in bottles suspended in the lake for 20 to 48 hr. A range of 0.02 to 1.30 was found among the species studied. The daily mean P/B for the small holotrichous ciliates was highest (0.67), as would be expected based on their tendency to have high reproductive rates. Among the remaining massively growing species the P/B was 0.44 for *S viride*, 0.34 for *L fasciola*, and 0.37 for *Vorticella*.

The gymnamoebae, though clearly significant in their role as links in the food web between smaller metazoans and primary producers, are often overlooked in broad studies of aquatic succession. This may be due in part to their fragility during collection, or the perception by some researchers that they are ephemeral and not as abundant, and hence perhaps not as significant as other protozoa. Specific studies on the seasonal abundance and succession among amoebae (eg, O'Dell 1979; Kyle and Noblet 1985, 1987) indicate at present that physicochemical variables typically correlated with broader protozoan community successions probably do not account for changes in their abundance. This is particularly true of benthic dwelling species. O'Dell (1979) found that during a one year study of bottom-dwelling gymnamoebae in a Nebraska lake, *Acanthamoeba polyphaga* was the most frequent amoeba and exhibited marked changes in numbers during the sampling period. This pattern, however, was not correlated with water temperature, bacterial counts, or nitrate and phosphate concentrations. The amoebae may be responding to other factors such as available organic sources associated with sediment resuspension or edaphic sources of organic compounds. Other less frequent genera, or those with low constant densities, included *Hartmanella*, *Vahlkampfia*, *Naegleria*, *Paratetramitus*, and *Echinamoeba*. The seasonal abundance of small thermotolerant free-living amoebae, however, appears to be related to water temperature in warm monomictic lakes. Kyle and Noblet (1986) reported that population densities peaked in the summer. Littoral genera were represented by *Acanthamoeba* and *Naegleria* in August, *Hartmanella* in July, and *Vahlkampfia* in May. Profundal populations exhibited dramatic seasonal variations, apparently in response to seasonal chemical changes in the hypolimnion. *Acanthamoeba* again was the most abundant in late summer, representing as much as 82% of the free-living amoebae in the profundal sediment. It is interesting to note that Kyle and Noblet found *Hartmanella* to be relatively abundant in the warm-water pond, but it was not found abundantly in the Nebraska lake. *Hartmanella* and *Vahlkampfia* populations peaked in April to early May, but declined with onset of summer stratification. Both *Acanthamoeba* and *Naegleria* continued to appear during this period, suggesting that they are more resistant to the harsh anoxic and reducing environment of the hypolimnion. The extent to which they are active versus encysted, however, needs additional clarification. Similar seasonal variations were found in the surface water, but *Naegleria* was the most abundant genus. In a warm water lake in the Piedmont region of South Carolina (Kyle and Noblet 1987), free-living amoebae within the water column were most abundant in a persistent detrital layer, however, few amoebae were isolated from a massive (1.5 mm thick) layer of *Oscillatoria*. The abundance of amoebae in the nueston at the very surface of the water in other ponds was considerably greater than in the underlying water, suggesting micro-habitat differences between the surface film and bulk water. Whether this is largely a phenomenon of surface attachment of the amoebae or differences in physico-chemical and biotic factors remains to be determined.

Microspatial heterogeneity in aquatic environments has been documented frequently in protozoan research. Taylor and Berger (1980) examined the patchiness of protozoa in a small pond along five transects on the surface and one from its bottom. They found a mean patch size of about 1.5 to 2 cm and an interpatch distance of 3 to 4 cm. The heterogeneity in abundance of protozoa within temporary and ephemeral patches appeared to be related to behavioral aggregation. This patchiness probably resulted from several factors, including opportunity to invade patchy food sources, reproductive strategies, and dispersal mechanisms. Further evidence for variations in community structure among microhabitats was obtained by Rossman (1959), who examined three sites near a limestone sinkhole in Florida during the winter season. The sites included (1) a spring flowing from a hillside (temperature 69°F), (2) water trickling from a liverwort patch, and (3) surface water at the edge of the pool (67°F, pH slightly basic). He found 43 forms of protozoa, all totaled (16 mastigophorans, 8 sarcodinans, and 19 ciliophorans). The only species common to more than one site was *Centropyxis constricta*. The genus *Monas* was present at all three sites, but represented by different species. *Entosiphon* and *Bodo minima* were the most abundant and occurred largely in the spring water. The flagellate *Astasia longa* and the ciliate *Cyclidium* were very abundant in the liverwort community. The flagellate *Oikomonas* and ciliate *Chilodonella caudata* predominated in the pool sample along with the flagellate *Pleuromonas*. Ciliates of lesser abundance included *Metopus* and *Holosticha*.

Many of the protozoa occurring in aquatic habitats are also found in soils and litters, especially if moisture is present in sufficiently constant amounts (Bamforth 1985a,b). Excess amounts of water, however, resulting in flooding of pore spaces and producing anoxic conditions severely limit the diversity of the protozoan community to those species adapted to low oxygen and high reducing conditions. Terrestrial environments subjected to frequent drying can support only those species that form dessication-resistant cysts. Soil with wide pore spaces ($>$ 10 µm) between larger aggregates of soil particles allows rapid transport of water, nutrients, and diffusion of gases. They are colonized by aerobic organisms. These habitats are subjected to rapid fluctuations such as drying due to water drainage, and the species found here must be able to survive by intermittent spurts of growth and quiescence. Smaller pore spaces (2 to 10 µm) hold more water by capillary action, and provide a nearly constant, though anaerobic environment. Small gymnamoebae and flagellates are particularly abundant in moisture-laden soil with pore sizes of 2 µm or larger. Ciliates such as *Colpoda* are typically found in soil with pore sizes of 30 µm or larger, and depend on films of water approximately 40 to 50 µm thick to swim from one location to another. In the surface litter and soil with larger pore spaces, a wide range of aquatic to edaphic species of protozoa can be found including testacea, gymnamoebae, and ciliates, some having invaded by migration from bryophyte vegetation or temporary ponds. In the rhizosphere, surrounding roots of plants where pore space is small, the predominate forms are gymnamoebae and small flagellates. Ciliates are represented by only two species each of *Colpoda* and *Cyrtolophosis*, and testate amoebae are rare (Bamforth 1985b). As in aquatic environments, there is increasing evidence that terrestrial protozoa serve a major role in the mobilization and remineralization of nutrients essential to the maintenance of higher organisms in the terrestrial community. They serve as an important link in food chains between bacteria and small metazoa (eg, rotifers and nematodes) and also increase bacterial productivity and mineralizing activity associated with their grazing pressure. Further modern perspectives on the ecological significance of protozoa in widely different environments can be found in current reviews (eg, Bamforth 1985a,b; Fenchel 1986b; Foissner 1987).

Summative Comment

The cosmopolitan distribution of protozoa and their intimate association with their environments make them particularly interesting subjects for ecological research. Their rapid growth compared to many higher organisms, and their often rapid response to environmental changes provide useful characteristics for efficient analyses of organismic-environmental interactions. The use of artificial substrates of varying composition and surface area provide useful tools for studying protozoan productivity, colonization rate, and variations in response to changing environmental variables. Further research is needed to determine what kinds of substrates are most useful for assessing sessile as well as free-swimming protozoa. Innovative research may also lead to novel techniques for creating artificial substrate systems that can be recovered from the natural environment with intact protozoan communities for refined analyses of protozoan community dynamics. Since protozoa respond rapidly to changing environments, they can be useful in assessing effects of pollution and environmental stress in natural environments. Some of the experimental methods reported here can be adapted to examining the effects of wastes and pollution on protozoan ecology toward more sensitive measures of the impact of modern society on the natural environment. Moreover, the important role of protozoa in maintaining stable community structures through enhanced primary productivity, carbon flow through the food webs, and nutrient recycling justify increased efforts to understand what factors potentiate the development of stable and productive communities of these remarkable, ubiquitous organisms.

9
Human Parasitic Protozoa

The Concept of Protozoan Parasitism

The clear dependence of many protozoa on exogenous sources of nutrition and their small size have led some to invade and inhabit the cells, tissues, or cavities of animals, where a rather constant environment and rich source of nutriments occur. Some of these parasitic species, during the long course of their evolution within animal systems, have developed an obligate relationship with the host and acquired specialized cellular structures and functions to efficiently utilize the unique environment they inhabit. But, this high level of specialization also has a cost. Many of the parasitic species have become so closely adapted to their parasitic existence that they are incapable of existing outside the host except as resistant stages. Digenetic species with alternating generations in two different hosts (inhabiting two different genera, hence digenetic) often exhibit complex structural and functional changes accompanying the different phases of existence. An example is the malarial parasite that has one form during invasion of the mammalian host and different forms during part of its growth in the insect host. Some have become adapted to invading specialized cells of animals such as erythrocytes (eg, malarial parasites), and the wandering cells of the immune system (eg, leishmanial parasites), where the restricted volume and accommodation to intracytoplasmic physiology have favored the evolution of a very small size. Some of these species also exhibit very simplified intracellular organelles. Those inhabiting host spaces with little oxygen (anaerobic environments) often lack mitochondria. Some of these intracellular species have also evolved specialized, although not unique to them, cellular organelles to support oxidative metabolism and generate ATP using chemical reactions quite different from those found in the mitochondria of free-living species. The discovery of the hydrogenosome (Müller 1980), an organelle of anaerobic eukaryotic cells mediating oxidative metabolism and generating free hydrogen gas, has been a major contribution in understanding the complex physiology of some anaerobic species. Prior to this discovery, these species seemed paradoxically to have no identifiable mitochondrial-like structure to support oxidative metabolism. The significance of this organelle in the broader scope of protozoan physiology and its importance to other protista is considered more fully in Chapter 12.

Each major phylum of protozoa has at least one representative species that has developed a parasitic habit, and some of them are among the most dangerous and deadly of human pathogens. The scourge of many nations is not the familiar diseases of western societies, but rather protozoan diseases that threaten large segments of the population. Protozoan diseases take a heavy toll in death and suffering, especially among the young. Malaria, for example, annually claims millions of childrens' lives or subjects them to chronic and recurrent episodes of debilitating pain or malaise. Our ever increasing awareness of this pandemic problem, and the magnitude of the human cost in physical and emotional suffering, places a heavy burden of obligation on the scientific community to consider lasting and effective ways of controlling if not eliminating these maladies.

Some representative human parasites from each of the major phyla are discussed. Their morphology and life cycles, as representative of their ecological adaptations to a parasitic habit, are compared to those of some nonhuman parasites. Although major emphasis is given here to human parasites as a means of focusing the discussion, the significance of nonhuman parasites should not be overlooked. Some are major disease agents in domesticated and feral animals that provide major sources of food for humans, or are an important component of natural ecosystems. With increasing human utilization of land resources and concomitant pressures of wildlife to occupy overcrowded space, once minimally significant animal parasites may exact a heavy tool in stressed animal populations. Hence, a thorough understanding of their biology and methods of control may be of increasing importance. Scientific investigation of these protozoan parasites in animals and those in plants (some invading solitary algal cells) can provide insights into the variety and range of adaptations protozoa have developed to stabilize their unique intraorganismic mode of life. Research with these nonhuman parasites, as with some free-living forms, may provide essential biological knowledge applicable to biomedical research leading to control of the human parasitic species. In addition to parasitic species within the major phyla previously discussed in the chapters on free-living protozoa (ie, Sarcomastigophora and Ciliophora), some phyla are almost exclusively parasitic or commensal species. These include the Apicomplexa, exemplified by the malaria parasites: Microspora, containing intracellular sporelike parasites that infect cells by discharge of a penetrating filament, and Myxozoa, parasites of tissue and organ cavities of annelids and poikilothermic vertebrates. The Myxozoa produce spores by fission of a multicellular spore-producing body. Among the latter three phyla, the Apicomplexa are probably of major significance in human pathology, and greatest emphasis will be placed on this group. To facilitate meaningful comparative analyses, the parasitic genera will not be discussed strictly within a sequence corresponding to the arrangement of phyla in modern classification systems. Although clear reference will be made to taxonomic designations, the genera will be presented in an order that permits meaningful comparisons based on morphology and parasitic habit. Among the Sarcomastigophora, for example, some parasitic amoebae will be discussed first, followed by parasitic flagellates (ie, *Trichonympha* and members of the Kinetoplastida). Following the discussion of the Apicomplexa, a brief treatment of other groups will be made.

Amoebae

The majority of the known species of amoebae (Lobosea) are free-living and contribute significantly to the stability and vigor of microbial communities. Their benign contributions to the maintenance of microscale ecosystems, supporting the broader order of the natural ecosystems of importance in human affairs, should not be overlooked when considering the few cases of species known to be human pathogens.

Endoparasitic Species

Among the several species documented to be pathogenic, and dwelling largely within human or warm-blooded animals, the intestinal-dwelling *Entamoeba histolytica* (Figure 9.1) is one of the best known (Baker 1973; Griffin 1978). This organism sometimes lives as a commensal without harm to the host. However, in other cases it becomes invasive and pathogenic. It is transmitted from human to human by contamination of food with feces containing the encysted stage. There are three stages in the life cycles of *E histolytica*: (1) trophozoite, an active, feeding stage exhibiting ameboid motion, (2) precyst stage, and (3) cyst. The precyst stage is a transitional phase passing from the active trophozoite to the inactive cyst. The trophozoite is a vegetative stage without fixed shape and moves by means of fairly characteristic single, blunt, clear, hyaloplasmic pseudopodia (lobopodia). The pseudopodia are typically rather abruptly extended. Trophozoites range in size from 15 to 50 μm in "diameter," depending on the strain and physiological state. Locomotive forms can be observed in fresly prepared slides from smears of diarrheic or dysenteric stools of patients or carriers. The slide must be kept warm (near body temperature) to observe locomotion as the trophozoite rounds up and becomes sedentary upon cooling. Trophozoites have vesicu-

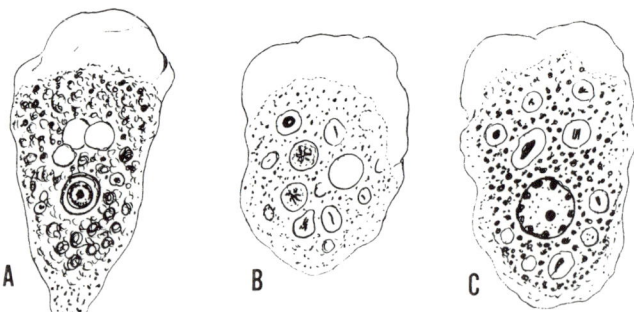

FIGURE 9.1. *Entamoeba* species. (A) *Entamoeba histolytica* (15–40 μm long); (B) *Entamoeba gingivalis* (12–40 μm long); (C) *Entamoeba coli* (20–50 μm long). *E histolytica* and *E coli* dwell in the intestinal tract of humans. *E gingivalis* is found in the mouth of humans, apes, and monkeys.

lar nuclei with a distinct endosome (small, densely staining central body) perhaps composed of condensed DNA, and peripheral dark basophilic granular masses of nucleolar material forming a dense ring next to the nuclear envelope (Figure 9.1). This nuclear figure is quite distinctive in trophozoites, but is not a reliable characteristic during transition into cyst stages as the nucleus is often transformed. Active trophozoites from infected stool are densely granular, and there is not a distinct separation of central granuloplasm and peripheral hyaloplasm. There are no mitochondria in this anaerobe, which dwells within the oxygen-poor environment of the gut. Glycogen particles are of two kinds: alpha configuration, with a typical rosette cluster of glycogen granules, and beta configuration, consisting of individual glycogen grains. During active feeding, food vacuoles become charged with tissue particles and erythrocytes or other organic matter, including intestinal bacteria. It is important not to confuse the genus *Entamoeba*, inhabiting mammalian intestines, with the similar sounding genus *Endamoeba*, inhabiting the guts of roaches and termites. The latter has a temporary uroid, and clear, glossy hyaloplasm with 60 or more nuclei in the granuloplasm. The cyst, moreover, is spherical and 20 to 50 µm with a thick membrane.

The precyst stage of *E histolytica* is a transition to the cyst stage, when the trophozoite ceases moving, rounds up, discards or digests remaining food, shrinks in volume, and begins to form a thin cyst wall. The cyst stage (typically 10 to 20 µm) is inactive, round or sometimes oval, and surrounded by a fully developed thin wall ca 0.5 µm thick. It is resistant to some environmental conditions such as passage through the low pH environment of the stomach, but is destroyed by temperature above 50°C, sunlight, dessication, or extended periods of time in chlorinated water.

During pathogenesis, *E histolytica* multiplies in the intestinal wall and lumen. The details of how infection is initiated are not clear, but it appears that the amoeba penetrates between the epithelial cells of the mucosa and forms lesions at the interface between the mucosa and the muscularis. Lateral invasion results in a localized lesion with a small opening on the mucosa, and flasklike expansion beneath the surface layer where invasive growth is most active. Severity of the infection varies from a simple carrier state without symptoms (the usual case), to invasion of the bowel wall without systemic spread, or penetration of the bowel wall and systemic infection. Mild to severe cases of the disease are accompanied by recurrent diarrhea, flatulence, and malaise. The intestinal mucosa may exhibit only small lesions in mild, localized infections. If the submucosa is seriously invaded, the mucosa becomes undermined, resulting in an "elevated" and "roiled" appearance. If the pathogen invades the underlying lymphatic tissue of the gut and penetrates the portal vein, hepatic and systemic infections are possible, resulting eventually in severe amoebic hepatitis and death. Complete penetration of the intestinal wall can produce severe hemorrhages and peritonitis. When liver invasion occurs, the right lobe is the most likely site of initial infection.

There are probably at least two groups of *E histolytica*, the normal strain cultivated at 37°C and the "Laredo-type" strains that grow at room temperature. In addition to *E histolytica*, there are two other species often observed as commensals in the human intestinal tract, *Entamoeba hartmanni* and *Entamoeba coli*. The trophozoite of *E hartmanni* is 5 to 15 μm in diameter, and like *E histolytica* may possess food vacuoles containing bacteria. There is a single nucleus. *E hartmanni* seldom if ever ingests erythrocytes, and there is no evidence that it invades host tissues. The cyst (5 to 9 μm) is smaller than those of *E histolytica*, but like the latter has four nuclei. *E coli* resembles *E histolytica* in nuclear morphology which, in stained preparations, also contains a peripheral ring of dense basophilic granules, but the endosome is eccentric. The cell moves sluggishly, is about 20 to 30 μm in diameter, and extends and withdraws short, blunt, broad pseudopodia. The cysts, unlike those of *E histolytica*, contain only two nuclei. *Entamoeba moshkovskii*, a free-living amoeba found in sewage, is morphologically similar to *E histolytica* both in trophozoite and cyst stages, and is mainly separated by antigenic analyses. *Entamoeba invadens* is a nonpathogen in the intestine of turtles and a pathogen in lizards and snakes. It resembles *E histolytica* in having a central endosome and peripheral ring of basophilic granules in the nucleus, but the nuclear structure persists into the cyst stage, where four nuclei are found. *Endolimax nana* is a nonpathogenic amoeba inhabiting the lumen of the upper portion of the human colon. The trophozoite (6 to 18 μm diam.) has a prominent nuclear endosome, but lacks peripheral basophilic grains. The trophozoite moves by periodic extension of monopods. The cyst are usually ovoid, and are 5 to 12 μm in diameter. Young cysts contain one to two nuclei, but mature cysts typically contain four.

A common commensal in human colons is *Iodamoeba butschlii*. Trophozoites are 6 to 20 μm in diameter, and stained cells exhibit a large, somewhat spherical endosome surrounded by achromatic granules. Motility is generally sluggish, but progressive, mediated by extension of broad, hyaline pseudopodia. Cysts are very irregular in shape (ovoid, ellipsoid, triangular, pyriform, or round) and 6 to 15 μm in diameter.

Free-living Potential Pathogens

Among the free-living amoebae, several have been implicated as potential pathogens and two genera in particular, *Naegleria* and *Acanthamoeba*, have been clearly linked to animal and human disease. *Naegleria fowleri* is the cause of the serious and usually fatal disease amoebic meningoencephalitis (eg, Griffin 1978). *Acanthamoeba* has also been reported to be a disease agent in animals and humans, although the incidences are not numerous. The pathogenic, free-living amoebae have a rather simple life cycle similar to other free-living forms. The active stage is a trophozoite feeding on bacteria and other organic sources in the natural environment, and invading human tissue in the pathogenic state. A resistant cyst stage alternates with the trophozoite stage, thus allowing the organism to persist during unfavorable environmental conditions. In *Naegleria*, at least, there is a flagellated stage enabling more efficient dispersal. The flagellate stage can rapidly transform into an amoeboid stage, and both are potentially infective.

Naegleria fowleri (ca 10 μm length) has smooth contours and a limax shape. It moves by eruptive bulges from side to side, usually at the anterior. At peak motility, it can move one to two body lengths in a minute. There is no prominent bulbous uroid. *N fowleri* inhabits warm water in the natural environment or in heated pools and baths. Infection of the nasal mucosa occurs by trophozoites, or by inhalation of water containing swarmers or cysts which subsequently revert to an invasive trophozoite stage. The trophozoite invades the olfactory nerve endings in the mucosa and establishes a localized infection of the olfactory nerve tract. The amoeba follows the afferent nerve tract to the central nervous system. Major infection of the subarachnoid space and perivascular tissue of the brain leads within approximately 1 week to severe meningoencephalitis, which is almost always fatal. Since the amoeba is a eukaryote, reistant to antibiotics, and there is no known effective amoebicide, there is little hope for victims of the infection. The symptoms are similar to those of bacterial meningitis, but spinal tap assays contain no bacteria. The histopathology reveals small, rounded cells in the central nervous system that are clearly identifiable as amoebae owing to the round nucleolus in the center of the nucleus.

There has been at least one report of possible human infection with *Naegleria gruberi*, but the accuracy of the identification has been challenged, with the strong likelihood that the amoeba could have been an *Acanthamoeba* (eg, Griffin 1978, p. 528). Given the severity of amoebic meningoencephalitis, it is probably wise to take due precaution when working with both *N gruberi* and *N fowleri*. Care should be taken to avoid contact with culture fluids and to prevent aerosol emission that could carry amoeba-laden droplets into the nose or other potentially infective sites.

Environmental factors favoring the growth of *N fowleri* include an abundant bacterial floral in conjunction with elevated temperatures due to sunlight or thermal spring outflow, and artificial heating as in heated pools or from industrial discharges. In most cases of infection, the contacted water had been chlorinated. Although it is not known why chlorination fails to suppress *N fowleri* and indeed may enhance its growth, one hypothesis is that the chlorine destroys less resistant strains of protozoa and other bactivores, thus reducing competition with *N fowleri*.

Acanthamoeba infections have been reported in a number of cases of human pathologies in a variety of disease patterns, including fatal cerebral granuloma, central nervous system infections, gastritis, and intestinal infections (Griffin 1978). Cultures of suspected pathogens have rarely been obtained, and most of the evidence is based on histopathological examinations. Based on the sizes of the amoeba identified in tissue sections, the putative infective *Acanthamoeba* spp appear to fall into three groups. There is a small form about the size of *Naegleria fowleri*, an intermediate size *Acanthamoeba culbertsoni*, and a group larger than *A culbertsoni*. As described in Chapter 4, *Acanthamoeba* is characterized by several to many fine, spiculelike subpseudopodia extending often irregularly from a broad hyaline pseudopodial surface. The subpseudopodia may be distributed over a considerable part of the periphery of the anterior and lateral sides of the cell, and occasionally from the posterior. The cysts are polyhedral or convex, with a polygonal or stellate endocyst. As the exact identification of many species of amoebae depends on several characteristics, including behavior during locomotion, cyst structure, and sometimes the form of the organism during floating, great care must be exercised in identification of species in a killed and fixed state. Thus, there may be some uncertainty in the exact identification of amoebae in histological preparations. Nonetheless, it is clear that some free-living amoebae are capable of invading human tissue, and reasonable precautions should be taken in working with enriched cultures of amoebae to ensure that no accidental infections occur.

Comparative Perspectives

The large variety of free-living amoebae and the wide variation in commensals living within the intestine of humans and other animals raises the intriguing question of what causes some strains to become pathogenic. Even in the case of *E histolytica*, clearly implicated in severe amoebopathologies, the environmental and physiological circumstances occasioning the onset of infection are not fully understood. In some instances, impaired immune response or other debilitating physiological states of the individual could be identified as antecedent to disease onset. But, in other cases it is not clear why some people are passive carriers of potentially pathogenic amoebae (eg, *E histolytica*), while others are particularly prone to infection. Nor is it clear why a commensal form of the amoeba becomes invasive. Current techniques of using antigens to identify species and strains of organisms may apply to pathogenic amoebae. This technique may help to clarify whether different strains of the amoebae are responsible for varying degrees of pathogenicity, and to better identify individual organisms isolated from infected tissue. Although some strains of pathogenic amoebae (especially *E histolytica*) are inhabitants of the intestine of nonhuman mammals, there is no evidence that these serve as reservoirs for human infection. The possible zoonoses, however, have not been fully explored, and additional future research could profitably be directed toward understanding the degree of host specificity among intestinal dwelling amoebae and the extent to which cross-infections are possible among widely different mammalian species. A zoonosis is a disease involving animal and human hosts. The animal host can be a reservoir for the disease organism, which is transmitted by a vector (carrier) such as an insect that bites animals and humans. Alternatively, transmission between different hosts can occur through contaminated food or water as occurs with *E histolytica*. Much can be done to clarify how pathogens adapt physiologically to different host environments, and to determine what physiological factors of the pathogen and host limit the range of organisms that can be inhabited or infected. The intriguing field of physiological ecology of potentially pathogenic amoebae is an area that could be profitably explored, combining culturing techniques with pathological studies. Some current efforts in this domain are summarized in subsequent chapters on physiology and nutrition.

Flagellates

There are three groups of major human pathogenic flagellates in three orders: (1) trichomonads (Trichomonadida), (2) giardias (Diplomonadida), and (3) leishmanias and trypanosomes (Kinetoplastida). These parasitic protozoa inflict diseases ranging from diarrhea produced by the intestinal dwelling *Giardia*, to serious urogenital tract infections (trichomonads), and various central nervous system (trypanosomes) and visceral or cutaneous diseases (leishmanias). Most trichomonad and giardia diseases are monogenetic, transmitted from one individual to another of the same genus. Some of the most serious and life-threatening human leishmanial diseases are digenetic with zoonoses involving life stages alternating between animal and human hosts. The monogenetic pathogens will be discussed before the digenetic ones, and the latter can be compared more readily to the malarial parasites that also have alternations between two different hosts.

Trichomonads

Three species have been clearly identified as invading humans: (1) *Trichomonas vaginalis* (Figure 9.2a), inhabiting the urogenital tract; (2) *Trichomonas tenax* (Figure 9.2b), found in the oral cavity, and (3) *Pentatrichomonas hominis* (Figure 9.2c), dwelling in the large intestine (the usual trichomonad location). The human colon may also be the site of one additional trichomonad parasite,

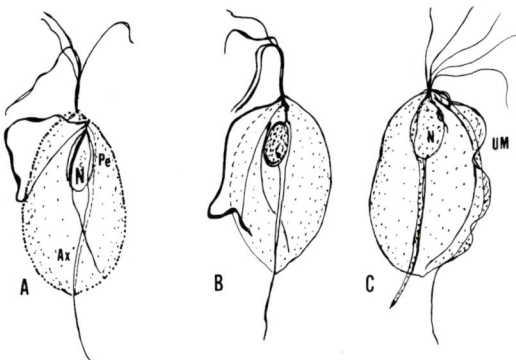

FIGURE 9.2. *Trichomonas* species. (A) *Trichomonas vaginalis* (10 μm); (B) *Trichomonas tenax* (10 μm); (C) *Pentatrichomonas hominis* (10 μm). Compare with Figure 2.3b. Ax = axostyle, N = nucleus, Pe = pelta, and UM = undulating membrane. Based on Lee (1985) and Honigberg and King (1964).

Trichomitus fecalis, isolated repeatedly from the feces of one individual by Cleveland (1928). Based on current knowledge, only *T vaginalis* contains confirmed pathogenic strains. An interesting and comprehensive review of the medically significant trichomonads has been presented by Honigberg (1978a).

Trichomonas vaginalis has a worldwide distribution and causes a serious human veneral disease. It can be isolated from urine or vaginal smears of infected individuals. It is a small flagellate (ca 9.5×7 µm). Normal dividing cells always have four anterior flagella originating in an anterior flagellar complex. In some individuals, the flagella appear to be arranged in pairs of two. The undulating membrane and underlying costa (Figure 9.2a) originate somwhat posterior and dorsal to the site of the anterior flagella. A spatulate axostylar capitulum closely applied to the nucleus connects anteriorly to a cresent-shaped pelta. The trunk of the axostyle is relatively slender and projects for some distance from the posterior end of the cell. The numerous intracytoplasmic granules are undoutedly hydrogenosomes mediating oxidative metabolism in these organisms (Chapter 12). The cell shape is variable, depending on the environment. Those grown in nonliving culture media have a more uniform shape, ovoid or pyriform, than those observed in vaginal secretions or urine. Tonicity of the medium clearly influences shape; in hypertonic saline the trichomonads become spindle-shaped, while in hypotonic solutions they round up and are vacuolated. Shape is also related to growth conditions. Rapidly dividing populations have smaller more rounded cells, and the flagella tend to be more unequal and their undulating membranes and costae are longer in relation to body length (Honigberg 1978a). In general, more actively dividing cells are believed to be more invasive, causing more serious infections which result in inflammation and often bleeding of the urogenital tract. The disease is more commonly recognized in women, as it frequently causes a purulent, frothy greenish-yellow discharge. The potential for serious secondary consequences of infection, including possible cancer of the cervix uteri, appears to be greater in women patients. Although some medical reports apparently treat trichomonad diseases as only a minor ailment, or consider trichomonads as commensals, Honigberg (1978a,b) has presented persuasive biomedical evidence showing the severity of the disease and the necessity to take prompt and appropriate ameliorative action.

The *Trichomonas* spp inhabiting humans and considered to be nonpathogenic are not treated as extensively here as is *T vaginalis*. *Trichomonas tenax* (Figure 9.2b) obtained from scrapings around the gums and gum recesses are somewhat smaller than *T vaginalis* (7.1×4.7 µm), with an oval to ellipsoidal shape. The flagella are more or less unequal in length, and can be discerned as two groups each with unequal flagella. The undulating membrane, shorter than the body, originates posterior and dorsal to the anterior flagella. The free margin consists of the recurrent flagellum and the "accessory filament" external to the flagellum. *Pentatrichomonas hominis* is typically ellipsoidal to pyriform (6 to 14 µm long). Four anterior flagella resembling those of other trichomonads arise from a large kinetosomal complex. A fifth prominent trailing flagellum (6 to 13 µm in length) originates from a kinetosome located ventrally to the base of the bundle of the anterior flagella. The undulating membrane and the costa (both of equal length to the body) arise from the same kinetosomal complex as that of the anterior flagella. A well-developed crescent-shaped pelta is connected to the capitulum. *Trichomitus fecalis* (ca 9.5 × 5.7 µm) bears three very long anterior flagella (18 to 27 µm length). The undulating membrane consists of the attached segment of the recurrent flagellum, as is typical of trichomonads and the "accessory filament." The tip of the recurrent flagellum protrudes for about one third its total length from the end of the undulating membrane. A prominent peltar-axostylar complex extends the length of the cell and projects as a prominent extension of the cell at the posterior end.

The trichomonads, though typically found in the intestinal tract of a wide variety of organisms including humans and wood-eating insects, have clearly diversified to inhabit and infect different body cavities in mammalian systems. This diversification of human pathogenic strains appears to be substantial. There is good evidence, for example, that intestinal species do not cause infections of the urogenital tract (Honigberg 1978a). The physiological ecological basis for this specificity is worthy of additional attention, as well as studies on the extent to which trichomonads in general can be induced to adapt to different environments by gradual changes in parameters of their most favorable environment.

Giardias (Diplomonadida)

This diarrheic disorder of the intestinal tract is caused by invasion of *Giardia lamblia*, usually acquired by drinking contaminated water. The disease results from the flagellate parasite's habit of adhering to the epithelial cells of the intestinal mucosa. Attachment is by cuplike structures on the ventral surface of the diplomonad, with a characteristic double cell organization as described in Chapter 2 (Figure 2.1k). The irritation resulting from attachment of the parasites on the cell surface apparently causes fluid loss resulting in diarrhea and intestinal discomfort. The discharge from the bowel is often malodorous. There is no evidence that the *Giardia* invade the bowel tissue. Hence, they are apparently more like a commensal that overpopulates and induces changes in the physiology of the water balance of the host, thus causing diarrhea.

Leishmanias (Kinetoplastida)

Some of the most debilitating diseases of tropical countries in the New and Old Worlds are caused by leishmanial infections (leishmaniasis). There are three major forms of the disease, caused by parasitic invasion of macrophages by the leishmanial flagellate: (1) cutaneous leishmaniasis, also known as oriental sore; (2) mucocutaneous leishmaniasis, mainly an infection of the respiratory tract (espundia); and (3) visceral leishmaniasis, a disease of the intestinal tract. There are similarities in the mode of infection among all three diseases. However, differences in molecular biology of the kinetoplast DNA, clinical symptoms, and the geographic range of the parasite are used in discriminating species within this

160   9. Human Parasitic Protozoa

genus. A general morphological description of *Leishmania* and its infective cycle is presented before describing each of the diseases in greater detail. Leishmanias are kinetoplast-containing flagellates that alternate between two very different forms during their life cycle, involving a nonmotile, intracellular phase in mammals and a motile phase within the gut of bloodsucking sand flies (Baker 1973). The intracellular form infecting mammalian macrophages is a small, nonmotile, spheroidal cell (ca 2 to 5 µm). The flagellum is reduced to a stubby projection within a flagellar pouch (Figure 9.3). This is the amastigote or micromastigote stage, as it is currently known. It produces the diagnostic "Leishman-Donovan" or "L.D." bodies, named for the initial discoverers, observed in blood smears of

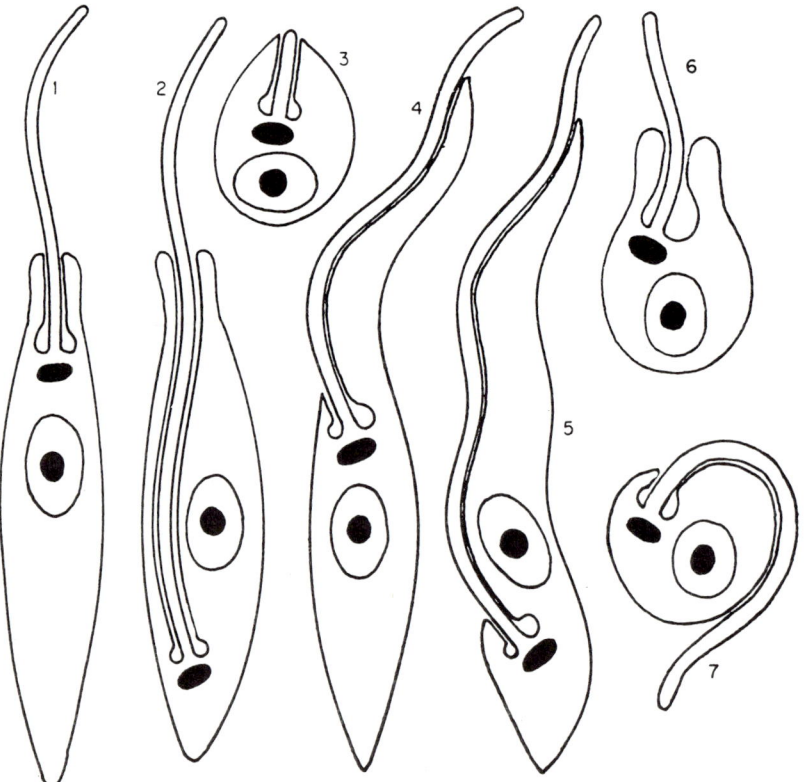

FIGURE 9.3. Developmental stages of trypanosomatids. (1) Promastigote with flagella emergent from anterior flagellar reservoir with nearby kinetoplast; (2) opisthomastigote with deeply recessed flagellar reservoir and flagellum arising below the level of the nucleus; (3) amastigote (micromastigote) with flagellum reduced and nonemergent; (4) epimastigote with flagellum emergent above the level of the nucleus; (5) trypomastigote exhibiting displacement of the flagellar root toward the tip of the cell wall below the level of the nucleus; (6) choanomastigote; and (7) sphaeromastigote. From Vickerman (1976); with permission.

infected patients. The micromastigote multiplies repeatedly by binary fission, eventually destroying the host macrophage. Released micromastigotes infect other macrophages when engulfed by the phagocytic cell, but they remarkably avoid destruction and completely invade and infect the cell. Successive cycles of release and infections of macrophages result in proliferation of the parasite within the infected tissue, causing local necrosis in mild cases, or massive degeneration in severe diseases. When a micromastigote is ingested by the insect intermediate host (the sand fly, a dipteran bloodsucking insect), it differentiates within the insect gut to form a fully flagellated form (Figure 9.3), a promastigote (ca 20 µm long). The process as observed in vitro at 26° begins with elongation of the micromastigote. A flagellar vacuole, lying near the invaginated flagellar reservoir, increases in size. At 20 hr, the flagellum begins to protrude and by 24 hr is fully grown. The promastigote body is fully formed and the flagellar vacuole becomes contracted. This motile extracellular form undergoes repeated division within the gut of the insect, thus increasing in number. Unlike the mammalian infective stage, it is not restricted to intracellular growth. The promastigotes multiply within the gut, sometimes attaching to the epithelium by their flagella. They migrate forward toward the foregut, where they become established in such large numbers that they eventually block the cavity of the proventriculus, pharynx, and proboscis. At this stage, about 10 days after ingesting the parasite, when the insect attempts to feed again its efforts to pump saliva down the blocked proboscis and to ingest blood result in dislodgement of the promastigotes and injection into the host. The promastigotes are ingested by macrophages, lose their flagella, and progressively form micromastigotes, thus initiating the macrophage infectious phase of the cycle.

The leishmanial diseases vary in severity and duration ranging from the mild, self-limiting cutaneous forms to the fatal mucocutaneous and visceral forms (eg, Zuckerman and Lainson 1978). The systematics of the species causing each disease is complex, and likely to change as modern molecular biological techniques elucidate the fundamental genetic affinities of the various disease agents. Each of the three major diseases is described briefly. Cutaneous leishmaniasis is usually a mild and self-limiting disease. A dry lesion at the site of infection gradually becomes ulcerated, resulting in restricted areas of necrosis or in severe cases spreading to other surfaces of the body. In the Old World, a typical causative agent is *Leishmania tropica*. Mucocutaneous leishmaniasis (espundia) is initiated at the site of cutaneous infection by the insect vector, but via metastasis can spread massively, resulting in disfigurement and often resulting in fatal tissue damage localized especially in the nasal and buccal cavities. *Leishmania braziliensis* is a widely recognized causative agent of the disease in South America. Visceral leishmaniasis (kala-azar), caused by *Leishmania donovani* or *Leishmania chagasi* and other species, is characterized by fever, anemia, reduced white cell count, wasting, splenomegaly (enlarged spleen), and serious imbalance of serum proteins. It results in a very high death rate in untreated cases. Effective treatment of the leishmanial diseases includes administration of drugs containing antimony, some aromatic diamidines (with varying success), plant alkaloid derivatives, or antibiotics. The latter, including Amphotericin B (Fungizone), have been reported to be effective against cutaneous leishmaniasis.

The transmission mechanism of the various forms of leishmaniasis and the nonhuman reservoir for the parasites vary considerably according to geographical region. Zoonoses are characteristic of some forms of cutaneous leishmaniasis that are rarely if ever transmitted directly from human to human. Examples include rural oriental sore due to *L major*, with a reservoir in burrowing rodents of the Middle East, and southern Russian and African desert regions. Wild rodents of the New World tropical rain forest harbor reservoirs of *L braziliensis* and related strains. Zoonotic diseases are most prevalent when humans enter areas where the animal host and vectors are abundant, and under conditions where stress or reduced resistance may make them more susceptible. In the case of visceral leishmaniasis due to *L infantum* and *L chagasi*, the reservoir hosts are dogs and foxes. Not all cases of dog reservoirs, however, are effective in transmitting the disease. The domestic dog, for example, has been reported to be a natural host for *L tropica*, but there is no evidence that it can serve as a link in human cutaneous leishmaniasis. In zoonotic forms of the disease, transmission from human to human by the sandfly vector is unlikely, owing either to the low probability that the vector will feed at an infected lesion, or because the parasite does not circulate in the peripheral blood stream. Hence, the animal reservoir becomes a significant link in the zoonotic sequence of animal to fly to human. In other cases, as with kala-azar, the parasite *L donovani* occurs in the peripheral blood stream, and the incidence of vector transmission by the fly directly from human to human is high, thus forming an anthroponosis. Clearly, in all cases it appears there is little likelihood that transmission can occur by human contact alone; thus, the vector is an important and also vulnerable link in the spread of leishmanial diseases. Effective disease control often involves elimination of the insect vector.

Trypanosomes (Kinetoplastida)

COMPARATIVE MORPHOLOGY AND LIFE STAGES

The trypanosomes are closely related to leishmanias, and in recognition of their morphological similarities, have been classified together in a group known as hemoflagellates. Some comparative features of trypanosomal and leishmanial infective cycles are presented in Figure 9.5. The trypanosomes possess a kinetoplast, and the anteriorly located flagellum can either be reduced as in micromastigote stages of *Leishmania*, or emerge as an anteriorly directed, fully formed undulipodium. In some stages, the flagellum is directed posteriorly, attached to the surface of the cell, and forms an undulating membrane. The genus *Trypanosoma* (Figure 9.4) includes parasites of the blood and other tissues of vertebrates that spend at least part of their life cycle in the gut of bloodsucking invertebrates. During the complex phases of their life cycle, trypanosomes can assume one of several different stages (Figure 9.3). The form of flagellum,

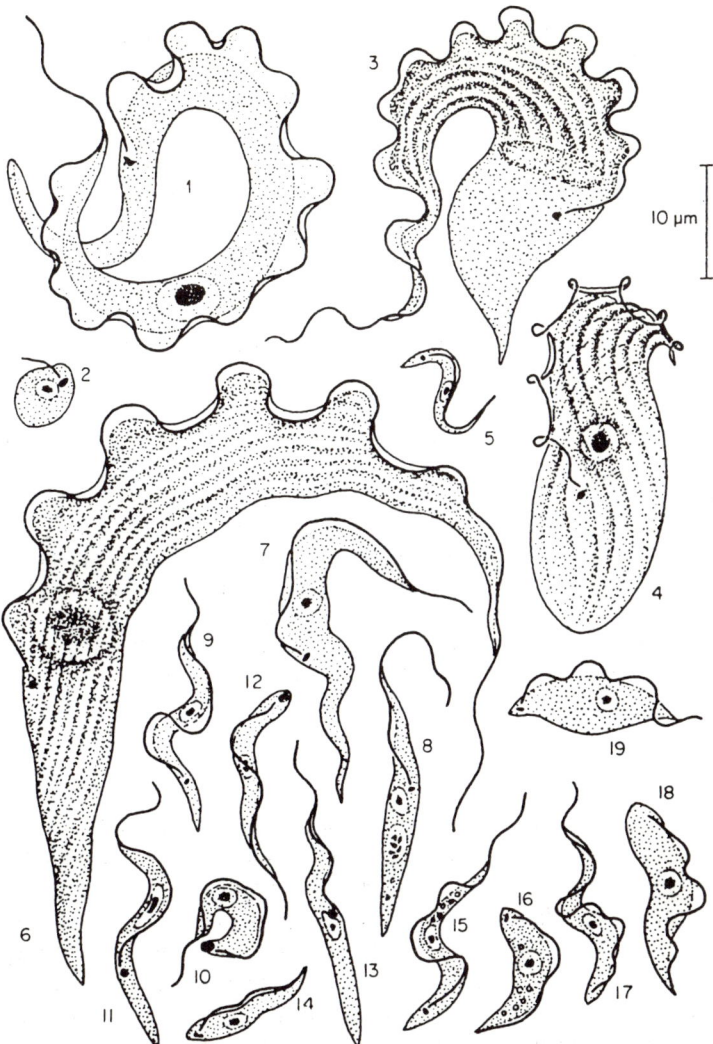

FIGURE 9.4. Some representative genera of the family Trypanosomatidae. (1–2) *Trypanosoma raiae*; (1) trypomastigote from elasmobranch blood; (2) sphaeromastigote from gut of leech; (3) *T mega*, trypomastigote from blood of frog; (4–5) *T rotatorium*; (4) trypomastigote from frog blood; (5) trypomastigote from stomach of leech; (6) *T grayi* from crocodile blood; (7–8) *T (Megatrypanum) cyclops*; (7) from blood of monkey *Macacus* sp; (8) from culture showing pigment vacuole; (9) *T (Herpetosoma) rangeli* blood trypomastigote; (10–11) *T (Schizotrypanum) dionisii* of bats; (12–13) *T (Duttonella) vivax*; (12) blood trypomastigote; (13) epimastigote from blood-sucking insect *Glossina*; (14) *T (Nannomonas) congolense* blood form; (15–19) range of slender and stumpy forms of blood stages of *Trypanosoma* spp. From Vickerman (1976); with permission.

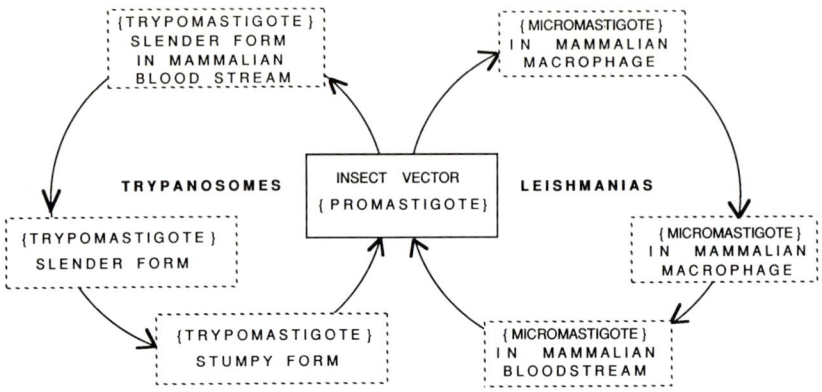

FIGURE 9.5. Comparative life cycle stages in digenetic disease-producing trypanosomes (left) and leishmanias (right) showing the promastigote stage in the insect vector and the blood stream stages in the human host. The trypomastigote in the blood stream phase of *Trypanosoma* spp divides repeatedly and multiplies; some trypanomastigotes become stumpy stages that are ingested by the insect vector where they multiply in the gut. The micromastigote stage of *Leishmania* spp invades macrophages, where it grows and divides, initiating repeated invasion and infection of macrophages. Some of the released micromastigotes may be ingested by a feeding insect vector, where they transform into a promastigote in preparation for the next infective cycle in a human.

whether anteriorly directed or posteriorly deflected and associated with an undulating membrane, and the position of the kinetoplast characterize the different stages. An amastigote, as with *Leishmania* micromastigotes, bears a stubby, reduced anterior flagellum but retains a kinetoplast near the basal bodies of the flagellum. The promastigote has a fully formed, anteriorly directed flagellum arising from an anterior invaginated reservoir. The kinetoplast is anterior near the base of the flagellum. Opisthomastigotes possess a deeply invaginated flagellar passage and the nucleus is decidedly lateralized. The kinetoplast at the base of the flagellar apparatus is anteriorly located. In the epimastigote, the flagellum emerges anterior to the nucleus, while in the trypomastigote it emerges posterior to the nucleus at one side of the cell, forming an undulating membrane by adhesion to the cell surface. The choanomastigote stage bears a collarlike flagellar reservoir, while the sphaeromastigote (as the name implies) is spheroidal with a reflexed tapering flagellum. Not all forms occur in a given genus, and some are characteristic of developmental stages in animals as parasites or their commensals. The terms appear sufficiently in the literature that one should be aware of them.

The genus *Trypanosoma* contains parasites of the blood and sometimes other tissues of vertebrates, with an alternating generation usually in the gut of blood-sucking invertebrates. They exist as trypomastigotes in at least part of their life cycle in both hosts. Two major groups of trypanosomes are distinguished based on the mode of transmission and differences in the physiology of the parasite. The Salivaria are transmitted in the saliva of the vector. Stercoraria are deposited in feces of the vector and introduced into the mammalian host by ingestion or by rubbing the parasite into the epidermis at the site of the vector bite. Transfer also occurs by rubbing the eyes and transferring the parasites to the conjunctiva.

SLEEPING SICKNESS

The sleeping sickness disease of Africa is one of the more familiar and deadly Salivarian-type trypanosomiases. It is caused by *Trypanosoma gambiense* and *Trypanosoma rhodesiense*. Sleeping sickness caused by *T rhodesiense* is an anthropozoonosis. Wild game is an animal reservoir, and the disease is localized in East Africa. The vector is the dipteran (two-winged) tsetse fly *Glossina* of the moristans group. *T gambiense* occurs in West and Central Africa. It is entirely an anthroponosis, transmitted by the *Glossina palpalis* group of tsetse flies from human to human. Mortality rates due to both forms of the disease are close to 100%.

The morphology and life cycle of the parasites are closely related to that of *Trypanosoma* (*T*) *brucei*, infecting wild and domesticated animals. The major characteristic of each group and the diseases transmitted by them are summarized in Table 9.1. The trypanosomes of the three groups all occur in the mammalian host as a trypomastigote (ca 30 µm long). They are transmitted by the dipteran vector as metatrypanosomes that are short and fairly slender (ca 15 µm × 2.5 µm). While in the subcutaneous tissue, they are converted to the so-called blood forms with undulating membranes including long, slender trypomastigotes (20 to 40 µm) and stumpy forms (15 to 25 µm × 3.5 µm) present in blood and tissue fluids. Some of the stumpy forms have the nucleus near the posterior end,

TABLE 9.1. Comparative data for trypanosome disease organisms.

Attribute	*T brucet*	*T gambiense*	*T rhodesiense*	*T cruzi*
Disease	Nagana in cattle	Chronic human sleeping sickness	Acute human sleeping sickness	Chagas' disease with invasion of cardiac muscle
Geographical location	Broad regions of Africa, except North Africa	West Africa, West & North Central Africa	East Rwanda, Africa, SE Africa & Ethiopia	South & Central America
Hosts	Wild & domestic animals	Humans	Humans, wild & domestic animals	Humans & small rodents (eg, wood rat)
Vector (insect)	*Glossina moristans* group of Tsetse dipterans	*G palpalis* group of Tsetse dipterans	*G moristans* group of Tsetse dipterans	*Triatoma* spp "Kissing bugs" sylvatic insects
Transmission cycle	Game, Tsetse, & riverine animals	Human, riverine Tsetse, human	Animals or humans, Tsetse, to animals or humans	Humans or rodents, hematophagous insect, to humans or rodents

Compiled from Baker (1973) and de Raadt and Seed (1977).

as is characteristic of the species. These are called the "posteronuclear" form. The flagellum of the slender forms is long and fully formed (up to 6 μm), and the position of the kinetoplast is subterminal. By contrast, the flagellum of the stumpy forms is usually less than 1 μm in length, or absent. The kinetoplast is near the posterior end. During the course of infection, intermediate forms between slender and stumpy can occur. During the initial rising parasitemia, slender forms predominate, followed by the appearance of stumpy forms. These may be a response to the attack by the immune system of the host, and are probably a form acquired by the vector after ingesting a blood meal. They have mitochondria and cytoplasmic features resembling the form in the insect gut. The wide variability in forms of the parasite while in the mammalian host is probably a reflection of the cyclical character of the disease, which passes through several periods of varying intensity. This is also related to the insidious, but biologically remarkable capacity of the flagellates to regularly change their surface-coat antigens, thus eluding the immune response of the host. When ingested by the dipteran vector, both slender and stumpy forms pass through the proboscis into the digestive tract and reach midgut, where apparently only the stumpy forms multiply and transform into slender trypomastigotes. However, these insect midgut trypomastigotes have the kinetoplast halfway between the posterior end and the nucleus, instead of in a terminal or subterminal position as occurs in the blood forms. The curves of the flagellum during undulation stay closer to the body than in the blood form, rendering its shape considerably less tortuous. After development in the midgut of the fly, the trypomastigotes undergo a complicated passage through the gut, finally emerging at an anterior location by penetrating a semisolid, "gelatinous" part of the cardial wall. They migrate toward the hypopharynx in the proboscis and then retrogradely into the salivary glands. When the tsetse fly feeds again, the trypomastigotes may be inoculated into the mammalian host with the saliva. Transmission can also occur if the vector is crushed while biting the host, thus forcing infective forms from the gut through the proboscis into the injection site. The contrasting physiological demands placed on the parasite when alternating between the mammalian blood stream and the anaerobic environment of the insect gut are clearly diverse, and must require very different morphological and functional responses by the parasite. Some of these changes have been mentioned for shape and behavior. Cytological differences also are notable, including a change in the appearance and activity of mitochondria. In the stumpy forms, a more efficient use of oxidative respiration occurs, as would be expected in the low oxygen environment of the vector. The slender forms, largely found in the blood stream, apparently have no functional mitochondria and utilize an alternative, less efficient mode of metabolism requiring higher levels of oxygen, as occurs within the circulatory system.

The symptoms and pathology of sleeping sickness are the same whether the disease is caused by *T rhodesiense* or *T gambiense*. Two to 3 days after subcutaneous introduction of the trypanosome by fly bite, a rather inocuous seeming local inflammation occurs, accompanied by itching, erythema (reddening), swelling, pain, and local heat. At about the sixth day, a typical trypanosomal chancre begins. This is characterized by a circumspect rubbery induration, adherent to the skin and extending underneath and around the visible swelling. Subsequently, between the fifth and twelfth day after infection, trypanosomes become detectable in the blood and a cyclical sequence of parasitemias is begun. The cyclical bursts of parasitemia last about 2 to 3 days and occur at irregular intervals from 2 to 10 days. The number of parasites in the blood can reach levels as high as 10^5 per ml in *T rhodesiense* infections, but there are consistently fewer from the start in *T gambiense*. Fever and nonspecific symptoms such as headaches, pains in the joints and limbs, and general malaise accompany the illness. A second phase of the disease is more debilitating, and involves invasion and degeneration of the central nervous system and cardiac tissue resulting in lethargy or coma. If untreated, the duration of the disease is between a few weeks and 9 months for *T rhodesiense*, or from months to several years for *T gambiense* infections.

CHAGAS' DISEASE

Chagas' disease, a Stercorarian-type infection, is a serious health threat to inhabitants of Central and South America. This trypanosomal infection caused by *Trypanosoma cruzi* is transmitted by reduviid insects (eg, *Rhodnius prolixus* and others). The disease is most common among inhabitants of squalid conditions, especially in mud-walled huts where the vectors accumulate. The parasite develops in the human host as a large micromastigote (amastigote) C 4 μm in diameter in lympho-macrophage cells, and muscle tissue including heart muscle. Consequently, patients with chronic Chagas' disease may die of heart failure. It is not a sleeping sickness disease as found in the African forms of trypanosomiases.

During the course of the disease in the vertebrate, the amastigote forms elongate, protrude a flagellum, and undergo metamorphosis, developing through promastigote and epimastigote stages into trypomastigote forms. These are rather slender flagellated cells (ca 20 μm long) with an acutely pointed posterior end, large kinetoplast located anterior to the nucleus, and a free flagellum. The other flagellum is recurved and forms an undulating membrane. In blood smears these trypomastigotes appear as small C- or S-shaped bodies.

Development of the trypanosome within the insect vector is largely within the gut, which is characteristic of Stercoraria. After ingesting a blood meal, the trypomastigotes accumulate in the midgut and rapidly transform into epismastigotes by forward movement of the kinetoplast and flagellar apparatus. These divide repeatedly by binary fission and gradually migrate toward the hindgut, where metacyclic trypomastigotes develop. This is the infective stage being passed out in the insect vector's feces. The insect vector typically feeds at night. During or shortly after the blood meal, it excretes feces. Disturbed, sleeping victims may crush the vector or scratch the bite, thus either forcing metacyclic trypomastigotes from the gut into the body tissue through the proboscis or introducing them into the puncture wound by rubbing the skin. Chagas' disease can be very debilitating, causing massive disruption of infected tissue and resulting in malfunction of the intestinal tract and/or cardiac arrest.

Animal Parasites and Commensals

A variety of vertebrate hosts are invaded by members of the trypanosome group, resulting in benign or disease states. Among these, the infection of rats worldwide by *T lewisi* is of historical significance, as it was first described by Lewis in 1878 as a hemoflagellate. It is a nonpathogenic parasite in rats, and all evidence suggests that it is totally intravascular. It lives and multiplies in the peripheral blood system, and perhaps also in small blood vessels of the organs. The vector is a flea, that as with reduviid bugs spreading Chagas' disease, deposits metacyclic trypomastigotes in feces on the mammal's skin. Scratching or licking by the host causes infection. The blood-dwelling trypomastigote first forms what is known as a broad trypomastigote, which transforms into an epimastigote. These reproductive forms undergo unequal multiple fission, and by incomplete separation at first produce a rosette of connected daughter epimatigotes that subsequently are set free and initiate further reproductive growth. Epimastigotes may undergo several reproductive cycles before maturing into trypomastigotes. A complex sequence of events accompanies development within the flea after ingesting a blood meal. the trypomastigotes eventually enter epithelial cells of the gut, where they become "pear-shaped" and produce 8 to 10 daughter kinetoplasts and nuclei by multiple fission. Daughter flagella appear near the new kinetoplasts and the cell becomes spheroidal. Cytoplasmic multiple fission produces separate trypomastigotes, which burst forth from the epithelial cell into the lumen to initiate infection at other sites and continue the reproductive cycle. Ultimately, the epimastigotes migrate to the rectum, where they undergo repeated division, some becoming transformed into small, club-shaped trypomastigotes. These are the metacyclic trypomastigotes that are infective to rats when released from fleas.

Trypanosomes found in nonmammalian hosts include parasites of fishes, amphibia, reptiles, and birds (eg, Mansfield 1978). These are usually larger than mammalian parasites (ca 50 to 100 µm long). The trypanosomes of aquatic vertebrates are transmitted by leeches, and the infective forms develop within the salivary glands (as occurs with the Salivaria parasites causing sleeping sickness in humans). Terrestrial parasitic trypanosomes are transmitted by bloodsucking insects or occasionally mites, and the metacyclic trypomastigotes develops in the hindgut and are deposited in feces. *Trypanosoma rotatorium*, a parasite of the frog, is interesting as its morphology is a normal trypomastigote in the tadpole, but after the latter matures to the adult frog stage, the parasite becomes enlarged and "leaflike" (Figure 9.4, 4). These interesting variations on a common developmental cyclical theme among parasitic trypanosomes living within diverse hosts, and the remarkable biological transformations accompanying their alternation between insect and vertebrate hosts, raises intriguing questions as to how this complex process emerged during evolution. We may also inquire as to what genetic characteristics of the flagellate group, and those destined to become trypanosomes in particular, potentiated this remarkable duality in physiological ecology of the parasites (Figure 9.5). The elucidation of these and other fundamental questions about the physiology of these unique protozoa may contribute substantially to our understanding of some very fundamental principles underlying the remarkable evolutionary radiation of life from primitive forms, and ultimately also to eliminate some of humanity's most devastating diseases.

Malarias: *Plasmodium* (Apicomplexa)

The malarial parasite, once thought to be on the way to elimination through combined efforts to destroy the insect vector (mosquitoes) and pharmacological control of the parasite, is beginning to recur with alarming vigor. This reversal is attributed in part to less effective control of the insect vector by insecticides, and some evidence of genetic resistance by the parasite to some of the previously most potent antimalarial agents. There are several excellent reviews of the pathology and biology of malarial parasites, varying from fundamental treatments to comprehensive critiques (eg, Baker 1973; Rieckmann and Silverman 1977; Phillips 1983). Malaria is fundamentally a blood-inhabiting parasite, at least during fulmination of the disease, and alternates between human and dipteran insect (mosquito) hosts. There are several forms of the disease varying in intensity, lethality, and residivism (chronic recurrence). *Plasmodium falciparum* is one of the most debilitating, not only destroying red blood cells, but also causing malignant malaria by accumulation of infected red blood cells within small vessels, probably due to their being less plastic and deformable than uninfected cells. Severe tissue damage of the brain, kidneys, and other vital organs can result from impaired blood flow. *P falciparum* occurs worldwide in tropical, subtropical, and warmer temperate regions. It is characterized by periodic chills and fevers at 48 hr intervals (tertian malaria), as also is the case for the less deadly forms, ie, *Plasmodium vivax*, with similar geographical range, and *Plasmodium ovale*, largely in tropical Africa. *Plasmodium malariae* produces chills and fever with 72 hr cycles (quartan malaria). It is worldwide, but scattered mainly in tropical and subtropical regions.

Malarial Infective Cycle

The life cycle, alternating between human and mosquito host, is complex, and a brief overview of the various stages is presented (Figure 9.6). The infective form of the parasite is a motile spindle- to scythe- (haltere) shaped cell (ca 1 μm × 11 μm) called a sporozoite. It moves with a sinusoidal motion produced by intracytoplasmic contractile tubules (microtubules). The sporozoite contains the characteristic "apical complex" of the class Apicomplexa; ie, a terminal apical

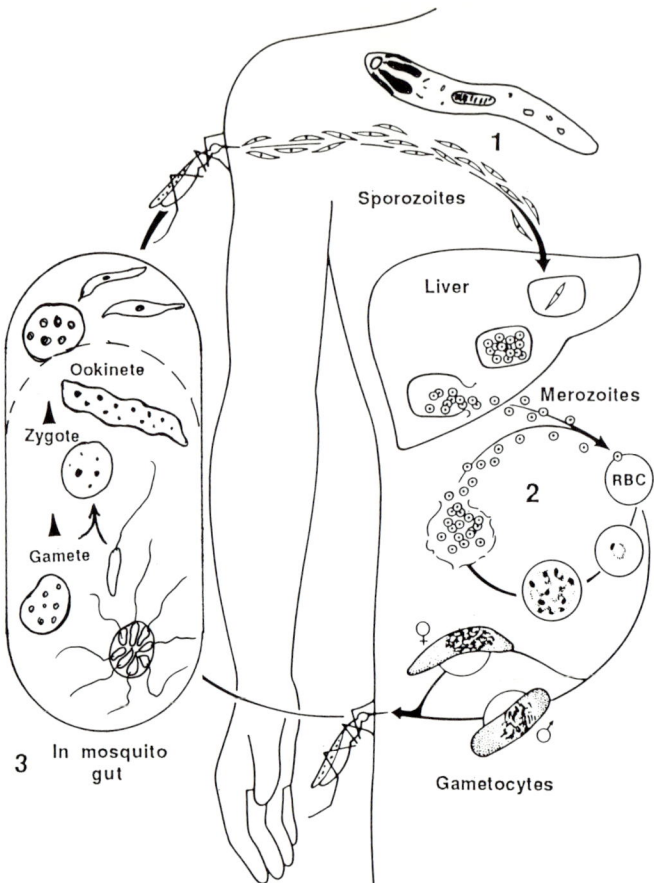

FIGURE 9.6. Life cycle and infective stages of the malarial parasite. (1) Sporozoites injected by the feeding mosquito contain a typical apical complex consisting of an apical ring, dense rhoptries, and smaller micronemes. The sporozoites invade the liver, where they are engulfed in phagocytotic cells and are transferred to parenchymal cells where merozoites form. (2) The merozoites divide and are released from the liver cells and carried by the blood stream, where they invade erythrocytes, initiating the blood stream parasitemia and major debilitating phase of the disease. Cyclical release of the merozoites from the RBCs causes the periodic chills and fever associated with the disease. Some of the merozoites differentiate into gametocytes which are ingested by a feeding mosquito. (3) The gametocytes mature in the gut of the mosquito, forming gametes that fuse to yield a zygote. The zygote matures into a motile ookinete that penetrates the wall of the gut (dashed line) and forms a cyst stage that eventually yields numerous sporozoites which migrate to the salivary gland of the mosquito, completing the digenetic life cycle. From Miller et al (1986). Reproduced by permission of the American Association for the Advancement of Science. Original illustration copyright 1986 by A.A.A.S.

ring, densely staining elongate rhoptries, and small vesicular micronemes. The apical complex apparently aids penetration into host cells. When the sporozoites are released into the blood stream by a feeding mosquito, they are rapidly cleared from the blood by the liver (usually within minutes), where the sporozoites invade hepatic parenchymal cells and commence what is known as the preerythrocytic or exoerythrocytic phase of development. Within the liver cell, the sporozoite transforms into a multiplying stage known as a schizont, appearing rather rounded and enclosed within a vacuole (parasitophorous vacuole). The nucleus undergoes repeated division and becomes dispersed within the as yet undivided cytoplasm. During maturation, the cytoplasm is segregated into pseudocytomeres, with nuclei lining the periphery. Buds form from the surface, each including one nucleus and apical complex (apical ring, rhoptries, and micronemes). Each bud is eventually released as a mature merozoite (Figure 9.6), at this stage bearing a nonfunctional cytostome (ingestion region). The total period of this development is about 8 days. The pear-shaped merozoites (C 1.5 µm in length) must survive passage from the liver through the blood until they attach to and penetrate an erythrocyte. The merozoite must attach by the apical end to invade the cell. The process, aided by the release of material from the rhoptries, results in the engulfment of the merozoite into a vacuole within the erythrocyte. The merozoite rapidly transforms into the feeding or trophozoite stage with a functional cytostome. Hemoglobin is engulfed by phagocystosis through inward budding of the vacuolar membrane and pinching off of vesicles containing erythrocyte cytoplasm. Metabolism of the erythrocyte hemoglobin is incomplete, resulting in the accumulation of residual pigment called haemozoin. During early stages of development, the space between the trophozoite and the vacuolar membrane is distinct, resulting in the characteristic ring figure observed within erythyrocytes in blood smears of infected patients. During maturation, the trophozoite enlarges, fills most of the vacuole, and produces 8 to 24 nuclei by multiple fission. As with hepatic schizonts, merozoites are produced by a process of budding, filling the erythrocyte and eventually released by its rupture. These merozoites attack other erythrocytes, thus exponentially increasing the number of infected cells. The free merozoites, released almost simultaneously from all infected erythrocytes at regular 2 to 3 day intervals, depending on the parasite, are the cause of the cyclical clinical symptoms of chills and fever accompanying the disease. The asexual reproductive cycle can be repeated many times, with increasing destruction of the red blood cells and consequent organ trauma. A sexual stage occurs in only a proportion of the merozoites and results in the production of male (microgametocyte) and female (macrogametocyte) reproductive cells. The enlarged size of the gametocytes and their differentiated cytoplasm make them clearly distinguishable from merozoites. The exact trigger for gametogenesis is unknown, but among the factors implicated are changes in blood plasma constituents, loss of CO_2, pH changes, or concentration of bicarbonate ions. The gametocytes are not capable of completing the sexual phase while in the blood of the host, and must be ingested by a mosquito to form micro- and macrogametes. The diel production of merozoites and gametocytes occurs late in the day, thus increasing the probability that mature gametocytes will be ingested by the female mosquito, which typically feeds in late evening or during the night.

The gametocytes ingested by the mosquito rapidly mature within the gut. The microgametocyte forms small uninucleated swarmers, simply called flagella, by repeated division. These swim rapidly through the blood meal and upon encountering a macrogamete penetrate and fertilize it, forming a zygote. During the next 12 to 18 hr the zygote forms an oval leaf-shaped ookinete (7 to 18 μm in length and 2.5 μm in diameter). The ookinete has all of the penetrating apparatus of the merozoite including apical ring, rhoptries, and micronemes. A nucleus, mitochondria, and endoplasmic reticulum are also present. Within 24 hr, the ookinete penetrates intracellularly through the wall of the mosquito's midgut and comes to rest extracellularly at the interface between the basement cell membrane and the basal lamina of the midgut wall. The parasite rounds up and during the ensuing 10 to 12 days transforms into a large oocyst (10 to over 50 μm diameter). Nutrients for this process are probably absorbed from the surrounding fluid of the host tissue. Repeated nuclear division fills the oocyst cytoplasm, forming a syncytium of apparently haploid nuclei. The point of meiosis is not known, but it is thought to be during the zygote stage, shortly after fertilization. This enlarged multinucleated stage is known as the sporoblastoid body. Eventually it produces peripheral buds which each receive one nucleus. These become elongated into developing sporozoites, forming a rosette around the mother cell. Subsequently, the sporozoites are released into the haemocoel of the insect, where they migrate to and penetrate the salivary gland. They can remain in this state for some time until the mosquito once again feeds, whereupon some of the sporozoites are injected with insect saliva, thus completing the digenetic cycle. The apparently complex set of stages and transformations accompanying the alternation of phases in insect and human host may seem at first perplexing to those encountering it for the first time. It is indeed a remarkable sequence of ordered transformations, producing cells uniquely adapted to each of the specialized sites where they must survive or grow. The complexity of the process, however, is also an aid to biomedical researchers hoping to find a way to completely eradicate the disease. The many different stages, each often with some particular metabolic requirements, offer additional opportunities to find a blocking agent that will destroy one of the stages and thus break the continuity of the life cycle.

Comparison of Hemoflagellates and Malarial Parasites

The digenetic life cycles of hemoflagellates and malarial parasites have some interesting similarities, related particularly to the changes in metabolism and form of the parasites as they alternate between mammalian and insect hosts. The overall similarity in progression of the various stages is punctuated, however, by the specialized forms of the parasite and their physiological responses to the unique environments occasioned by the kind of vector and the location in its body where the parasite resides. The similarities in functional role of the micromastigote (hemoflagellates) and the merozoite (malarial parasites) becomes apparent by inspection of Figures 9.5 and 9.6. Similarly, within broad perspective, analogous roles can be suggested for the metacyclic trypomastigote and the sporozoite.

Other Parasites of Medical and Economic Importance

Within the limitations of scope and space available here, not all of the significant parasitic diseases can be represented. The Microspora (also known as Microsporida), for example, constitute an important group of parasites influencing the economic well-being of society. Many infect commercially important food organisms, and others are of medical significance. This group has been concisely and clearly reviewed by Canning (1978) and general information on their biology and basic pathology is presented by Weiser (1985). In general, these are unique, small protozoa with peculiar structures not analogous to other protozoa. These organisms (typically 1.5 to 4 µm) have scanty cytoplasm, and in at least one stage possess a terminal apparatus (polar filament) that is used to penetrate the host cell membrane and inject the infective stage into the host cytoplasm. The filament is propelled through a tube and is the product of a specialized form of Golgi apparatus, the posterosome. Commercially important diseases include those that decimate fish populations. Historically, they are important as the cause of pébrine (or silkworm disease) caused by *Nosema bombycis*, which threatened the silk industry in Europe during the nineteenth century.

Ciliates are also known commensals or parasites in a number of vertebrates including fish and humans (eg, Baker 1973). Some are serious pathogens in lower vertebrates and humans. The large, ovoid ciliate *Balantidium coli* (60 to 70 × 40 to 60 µm) inhabits the intestine (caecum and colon) of man, apes, monkeys, and pigs. The latter is the most common and probably the natural host. The cell is covered with cilia arranged in longitudinal, slightly spiral rows. At the anterior end, there is a deep groove or vestibulum, lined with slightly longer cilia, that leads to the cytostome. There are two contractile vacuoles and a macronucleus and a micronucleus. *B coli* causes the disease balantidosis (balantidial dysentery), which can be isolated from the feces of diseased organisms. It appears to be nonpathogenic in pigs, but cysts passed out in feces can infect humans if consumed in contaminated food. The irritation of lesions produced by the ciliates burrowing into and through the mucosa and submucosa produces diarrhea, which may become bloody when blood vessels are eroded. The presence of the ciliates is easily diagnosed by the occurrence of free or encysted forms in the diarrheic discharge.

Summative Perspective

The parasitic protozoa yield interesting examples of the myriad ways these single-celled organisms have adapted to diverse environments. The eradication of these serious parasites has been a long struggle, and much remains to be done in understanding the physiological and ecological basis for parasite infection and spread. Modern molecular biological and biochemical techniques promise to yield insights into the basic functioning of the parasites toward development of vaccines or pharmaceutical agents that can at least arrest human infection if not eradicate the disease agent (eg, Honigberg 1978a; Baker 1973; Philips 1983; and Miller et al 1986). In addition to molecular approaches to understanding the physiology of the organisms, one should not overlook basic research on the ecology and mode of transmission of the parasite. Insights into the physiological adaptation of the parasite to its vector, the behavioral biology of the vector in acquiring and transmitting the parasite, and the environmental factors that may potentiate its transmission are significant areas of study. Laboratory simulations of vector habitats (eg, Grunewald 1976; Rausch and Grunewald 1980) and carefully controlled studies of how the vector is attracted to and infects hosts are useful adjuncts to molecular biological approaches. The wide diversity and remarkable adaptability of protozoan parasites suggests an equally broad scientific approach to understanding their total biology. They are notable for their behavioral plasticity in response to sometimes rapidly changing environments, as occasioned by their transfer from one host to another, or during the process of dispersal of successive stages within a given host. This remarkable, but in the case of parasitic species, perplexing plasticity offers yet additional opportunities to explore the intricate mechanisms of survival and adaptation that have occurred among all life forms. Hopefully this will provide additional biological models toward understanding pathological systems and lead to better control if not eradication of these remaining scourges of humanity.

Section II Functional Microanatomy

10
General Principles of Cell Fine Structure

Introduction

The remarkable range in morphology and magnitude of size among microbial organisms, varying from viruses (ca 20 nm) to large free-living protozoa (ca 2 mm), spans nearly five orders of magnitude (10^5). Some basic knowledge of cellular microstructure pertinent to further discussions of protozoan microanatomy and physiology is presented. A more thorough treatment of the subject can be found in standard textbooks of cell biology or cellular ultrastructure.

Plasma Membrane and Cytoskeletal Structures

The envelope surrounding the cell and enclosing the intracellular organelles is a thin macromolecular membrane ca 10 nm thick. Although the chemical composition, proportion of macromolecular constituents, and their spatial organization varies among the different types of cellular membranes, the basic plan of organization appears to be the same. According to current theory, this exceedingly delicate barrier is a bimolecular sheet consisting of two layers of lipid molecules oriented with their hydrophobic (oily) hydrocarbon portion inward and a polar (charged) portion outward (Figure 10.1A). The close association of the hydrophobic chains laterally, and with opposing molecules across the thickness of the membrane, provides structural integrity for the bilayer. Polar associations of the charged groups with the surrounding aqueous medium stabilizes the oily portion of the membrane in the aqueous environment. Other lipoidal molecules including sterols (eg, cholesterol or its derivatives), smaller molecular weight lipids, and other lipid soluble compounds mediating membrane biochemical reactions are also included within the bilayer. Protein molecules embedded within the lipid bilayer (intrinsic) or attached to the surfaces (extrinsic) form a significant part of the structural and functional integrity of the membrane. The intrinsic proteins sometimes bear a terminal chain of mucoproteins (carbohydrate and protein complexes) that project into the aqueous phase, forming the

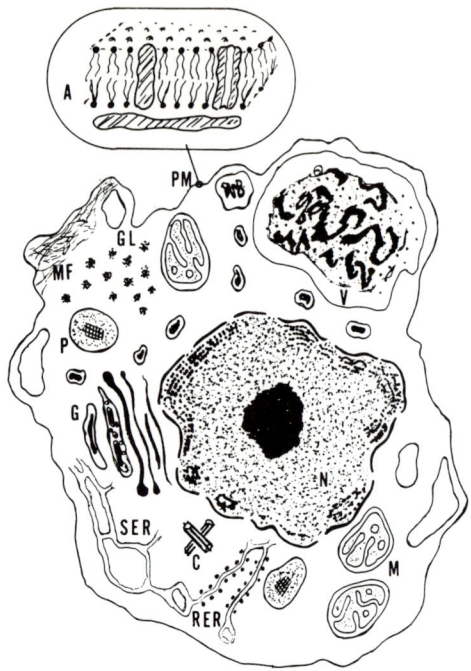

FIGURE 10.1. A schematic diagram of some major features of eukaryotic cells. The nucleus (N) is enclosed by a double membrane envelope, with small pores permitting exchange of molecules between the nucleus and cytoplasm. For example, mRNA (information carrying molecules) can pass from the nucleus into the cytoplasm by way of the pores. Mitochondria (M), surrounded by a double membrane, endoplasmic reticulum, smooth (SER) and rough (RER) bearing ribosomes on the cytoplasmic surfaces, and Golgi apparatus (G), with secretory vesicles distributed in the cytoplasm, are characteristic of major organelles surrounding the nucleus. Peroxisomes (P) with crystalline inclusions, glycogen deposits (GL), and digestive vacuoles (V) are common in the peripheral cytoplasm. A thin web of microfilaments (MF) occurs beneath the plasma membrane (PM) that is composed (A) of a lipid bilayer with internal (intrinsic) proteins and external (extrinsic) proteins forming part of the structural and functional apparatus of the membrane. Centrioles (C) are composed of two groups of cylindrically arranged microtubules and serve as microtubule organizing centers. They are the organizing centers for spindle fibers during mitosis.

extracellular coat of the plasma membrane. These surface molecules serve a variety of functions including increased stability of the membrane, nutrient binding, chemical "identity" for the cell, and cell-to-cell adhesion during processes of cellular aggregation.

Some intrinsic proteins are enzymes, each catalyzing a membrane-bound biochemical reaction essential to the life of the organism. Others are porters, carrying substances across the membrane, either into or out of the cell, thus maintaining a proper concentration of ions and nutrients within the cell and its various compartments (Figure 10.1). Porters attach to substances by substrate-specific chemical bonds and translocate them across the membrane. In some cases, the membrane carriers are cylindrically organized proteins forming a small channel across the bilayer. Substances of the appropriate size and chemical composition diffuse through the channel to the opposite side of the membrane. Other carriers are complex proteins that undergo structural transformations and move the attached substrate from one side of the membrane to the other. This process may involve a protein transformation whereby the protein carrier inverts and thrusts the attached substrate to the opposite side of the membrane. Some of these carriers facilitate transfer of membrane insoluble substances along a diffusion gradient from regions of higher concentration to lower concentration. Others utilize cellular energy to "pump" substances across the membrane from a region of lower concentration to one of higher concentration. Among the substances translocated are nutrients (eg, sugars, amino acids, acetate, and vitamins). Each species has particular porters to translocate substances essential to its normal physiological functions (and in the proper amount to avoid toxic accumulations or excesses). All cells have a low tolerance for sodium; hence, one common translocating intrinsic protein is Na^+/K^+ ATPase. This protein utilizes cellular energy (hydrolysis of ATP) to move sodium from inside the cell to the outside, thus preventing buildup of toxic amounts in the cytoplasm. At each cycle, K^+ ions are translocated into the cell and Na^+ ions are carried out. The intrinsic proteins are arranged in the lipid bilayer in a regular pattern characteristic of the kind of cell or subcellular organelle, but they are also capable of moving laterally within the bilayer, especially demonstrable in the plasma membrane. Hence, this conceptual view of cellular membranes is called the fluid mosaic lipoprotein model.

Cytoskeletal Structures and Undulipodia

The plasma membrane surrounding the outer surface of the cell may be strengthened by additional mucoidlike deposits on its outer surface, and by cytoplasmic proteinaceous fibrillar and microtubular components that give elastic and rigid integrity to the cell. The microfilaments are thin strands of contractile protein (ca 5 nm in diameter) apparently composed of a form of actin and myosin as found in the contractile components of muscle. A web of these fibrils may be especially well-developed immediately beneath the plasma membrane (Figure 10.1, MF), and extends in varying amounts throughout the cellular cytoplasm. The fibrils are especially abundant in regions of cellular motility and contractility. The microtubules appear more stiffened and often form straight, rodlike elements within the cytoplasm. Among other functions, microtubules are cytoskeletal structures providing internal support. They also exhibit gliding motion upon each other, and thus serve as cytoplasmic motile elements. Each microtubule is a hollow protein cylinder (ca 30 nm in diameter) composed of globular protein subunits (tubulin) that are spirally arranged to form the tubule wall.

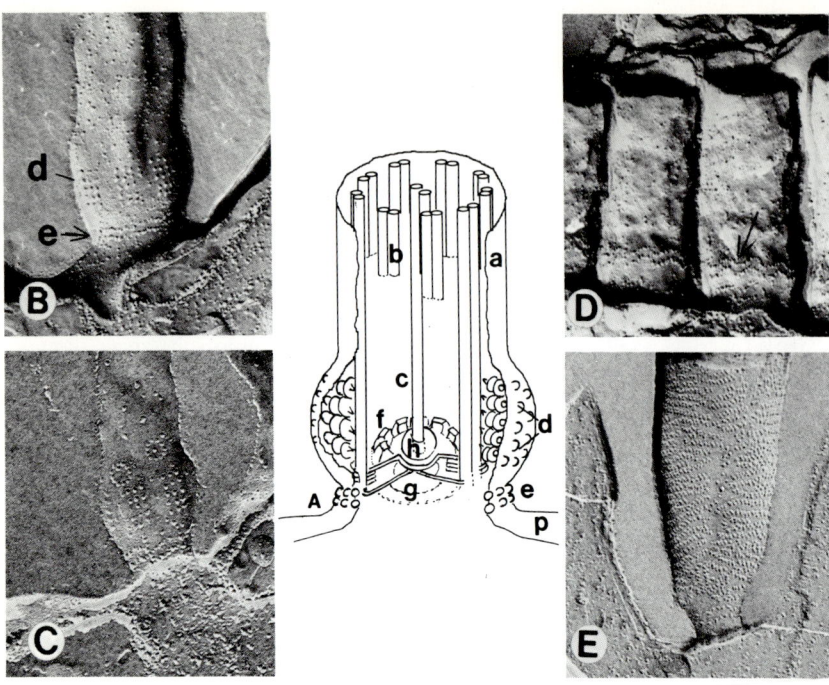

FIGURE 10.2. Ciliary membrane plaque and necklace particles in the proximal segment of a cilium. (A) A diagrammatic, cutaway view of the proximal segment of a cilium showing the ciliary membrane (a), internal circle of nine doublets (b), central pair of tubules (c), one penetrating the basal axosome (h) that is surrounded by a ring of amorphous matter (f). A series of terminal plates (g) occurs at the level of the insertion of the cilium into the plasma membrane (p), which is continuous with the ciliary membrane. Plaque particles (d) occur near the base of the cilium, which is encircled by a ring of necklace particles (e). Based on data from Nanney (1973), Dute and Kung (1978), and Bardele (1980). (B–E) Freeze/fracture images of ciliary particles in (B) *Paramecium caudatum* with double-stranded ciliary necklace (arrow) and ciliary plaques; (C) *Frontonia leucas* with ciliary necklace, ciliary plaques, and ciliary rosettes; (D) *Pseudourostyla weissei* with double-stranded ciliary necklace (arrow); (E) supposed sensory cilium from top kinety of the paralabial organelle from *Ophryoscolex purkinjei*. All freeze/fracture images 60,000 ×. Electron micrographs courtesy of Dr. C. F. Bardele, Tübingen University, adapted from Bardele (1980, 1981).

Parallel arrays of microtubules form the axial rods of the stiffened axopods of some protozoa (eg, radiolaria and heliozoa) and are major structural and functional units in cilia and flagella categorized collectively as undulipodia (Figure 10.2). Macromolecular bead and necklace inclusions within the membrane of the undulipodia may have taxonomic significance (Bardele 1981) and further exemplify the remarkable complexity of these motile organelles. The arrangement of the microtubules in the 9 + 2 pattern within undulipodia is shown in a cross-

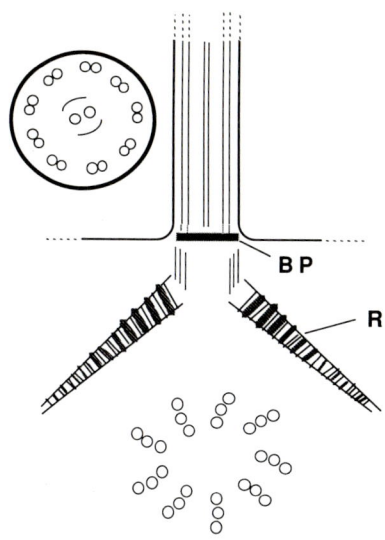

FIGURE 10.3. Schematic drawing of a flagellum showing a cross section of the distal portion containing a peripheral ring of nine doublets and a central pair of microtubules (compare to Figure 10.2), the terminal or basal plate (BP) at the base of the flagellum and basal body with striated rootlets (R) attached. The arrangement of the triplet microtubules in the basal body is shown beneath the rootlets in a cross-sectional view.

sectional view (Figure 10.3). The organization is identical in most cilia and flagella. The two kinds of undulipodia are differentiated only by length (ca 10 µm for a cilium and up to 150 µm for a flagellum). The pair of central tubules is surrounded by nine peripheral doublet tubules. The basal body within the cytoplasm has a spiral arrangement of arms, each containing three microtubules (Figure 10.3). The basal bodies in some species are further augmented by strands of parallel microtubules forming striated fibers anchoring the cilium or flagellum in the cytoplasm. In some cases, the organization of these fibers can account for the pattern of undulating movement observed in flagella. Other configurations of interconnected microtubules provide supporting and form-determining structures in cells of protozoa. For example, groups of microtubules form supportive sheaths within the cytoplasm surrounding the gullet, where the flagella are inserted (in some flagellates including *Euglena*). In the trichomonads, the cytoplasm in the region of flagellar insertion is reinforced by a protective sheath of microtubules that extends posteriorly and partially surrounds the nucleus. A long rodlike extension known as an axostyle continues posteriorly and often produces a spinelike projection of cytoplasm at the posterior end of the cell. A variety of supportive microtubular structures occur beneath the pellicle of many ciliates and under the plasma membrane of other protozoa. The arrangement of ciliary fibers and subpellicular microtubules has been a useful criterion in ciliate taxonomy (eg, Small and Lynn 1985).

Centrioles

Regularly arranged short segments of microtubules also form the intracytoplasmic centrioles in some protista. Two cylinders of microtubules arranged perpendicular to each other form the centriole pair (Figure 10.1, C). During mitosis, the centriole pair duplicates and the daughter centrioles migrate to opposite poles of the dividing nucleus. The centrioles are microtubular organizing centers (MOC) and give rise to the long microtubular spindle fibers that pass through the nuclear region and connect to chromosomes. A gliding activity of parallel microtubules, and their capacity to elongate or contract by regulating the number of tubulin monomers incorporated in the ends of the cylinders, is thought to be intimately involved in the dynamics of chromosome separation during mitosis.

Cytoplasmic Organelles

Endoplasmic Reticulum

A network of intracytoplasmic membranous canals and tubules, called the endoplasmic reticulum, provides a large surface area supporting numerous biochemical reactions. This extensive network of canals, connected to the plasma membrane, also permits communication between the peripheral plasma membrane and intracytoplasmic membrane-bound spaces. Portions of the endoplasmic reticulum may be continuous with the nuclear envelope, which appears as a double membrane in ultrathin sections. Portions of the endoplasmic reticulum and the outer surface of the nuclear envelope bear attached protein aggregates known as ribosomes. These ribonucleoprotein macromolecules are the site of protein synthesis. Endoplasmic reticulum with attached ribosomes is called rough endoplasmic reticulum (RER) and the remaining segments are known as smooth endoplasmic reticulum (SER). The smooth endoplasmic reticulum is the site of lipid synthesis and other metabolic functions.

Golgi Bodies and Lysosomes

The Golgi body, sometimes seen near or in continuity with the endoplasmic reticulum, is a complex array of flattened cisternae, arranged in parallel stacks or sometimes forming a curved array in a horseshoe configuration. It is a secretory organelle and produces free vesicles containing metabolic products that are pinched off at the periphery. Digestive enzymes are contained in some of the vesicles known as primary lysosomes. Primary lysosomes can also arise directly by vesicular budding from the endoplasmic reticulum. Vesicles containing digestive enzymes are also produced from digestive vacuoles by extrusion of residual enzyme enclosed in peripheral buds of the vacuolar membrane. The potentially destructive lytic enzymes are sequestered within the primary lysosome. This protects the cytoplasm from self-destruction and permits them to be directed to food-containing vacuoles where digestion occurs. Particles of food are engulfed within food vacuoles by a process of phagocytosis or pinocytosis. Larger food particles, usually above colloidal size, are enclosed within food vacuoles formed by invagination of the plasma membrane, forming a phagocytic vacuole. Colloidal size particles are carried into the cell by pinocytosis, a process of forming a fine conical invagination of the plasma membrane into the cytoplasm where the suspended food particles are collected in a small food vesicle. Food particles engulfed in phagocytic food vacuoles, or drawn into pinocytotic vesicles, are digested by conversion of the food vacuole to a digestive vacuole. The fusion of a primary lysosome with the food vacuole forms a secondary lysosome or digestive vacuole (Figure 10.1, V). Residual nondigested matter within mature digestive vacuoles (called residual bodies) is either stored in the cell within a very small and contracted vesicle, or as is common with many sarcodines and ciliates, defecated by exocytosis. In the ciliates, the waste vacuoles are defecated at a specialized region of the peripheral membrane known as a cytoproct. Other Golgi-derived vesicles contain either mucopolysaccharides, elaborate scales, or mineral platelets destined to be deposited on the surface of the cell. These substances are released on the surface by fusion of the vesicle with the plasma membrane. The organization and kind of surface structures deposited on the outer membranes of amoebae (eg, Page 1976a,b, 1983) and some flagellates (eg, Hibberd and Leedale 1985) are of taxonomic value.

Other Cytoplasmic Organelles and Inclusions

The cytoplasm surrounding the nucleus, and bounded at its periphery by the plasma membrane, contains other membranous organelles including the mitochondrion (a double-membrane organelle with internal membranous tubules or saccules called cristae), single-membrane-bound organelles including lysosomes, microbodies or peroxisomes (Figure 10.1, P), often with granular enzyme inclusions, and small vesicles containing storage products or other biosynthesized products. The coated vesicles, with an external fuzzy coat of protein, become attached to and glide along the surface of microtubules that apparently guide their movement throughout the cytoplasm. These coated vesicles serve as shuttle organelles moving packets of biosynthetic products within the cytoplasm. Other organelles such as mitochondria and microbodies may also attach to and move along microtubular surfaces. Microbodies function in nitrogen metabolism (eg, degradation of nucleic acids) and in the conversion of fat to carbohydrate (gluconeogenesis) in some protozoan species. The mitochrondria are sites of calcium ion storage, lipid degradation, and aerobic metabolism yielding stored energy in the form of ATP. Mitochondria also contain a strand of circular DNA, and thus are capable of synthesizing some of the proteins required for their activity. The remaining proteins are produced by the cell nucleus. Some protozoa lack mitochondria but possess hydrogenosomes. These are single-membrane-bound organelles, or perhaps with two very thin outer membranes. They provide energy for the cell by forming ATP from ADP, and producing molecular hydrogen (see Chapters 11 and 12).

The hyaloplasm surrounding the membranous organelles contains the soluble products of cell metabolism, free enzymes, carbohydrates including masses of glycogen, amino acids and other macromolecules, and ions that maintain the ionic balance of the cytoplasm. The glycogen granules (20 to 35 nm) are of two kinds: (1) alpha glycogen organized into clusters or rosettes, and (2) beta glycogen consisting of scattered grains. Lipid droplets and other forms of food reserves are commonly observed suspended in the cytoplasm. In rare cases, some protozoa (eg, the amoeba *Pelomyxa*) contain mineral grains (ie, sand, etc) suspended within the cytoplasm. Free ribosomes are sometimes visible within the hyaloplasm. Soluble proteins, suspended within the hyaloplasm, are typically produced by free ribosomes. Ribosomes attached to the endoplasmic reticulum usually synthesize proteins to be packaged into vesicles or distributed within the cisternae of the endoplasmic reticulum. Contractile vacuoles, as described more fully in Chapter 11, are usually situated near the plasma membrane. They are conspicuous by their large size, and by a very thin cytoplasmic layer adjacent to

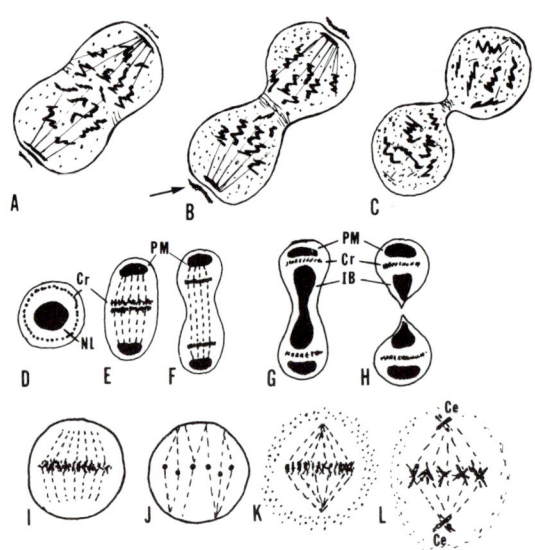

FIGURE 10.4. Illustrations of mitotic figures during nuclear division. (A–C) Elongation and constriction of a nucleus with nonclassic mitosis showing a persistent nuclear envelope, separation of the chromosomes by internal spindle fibers (microtubules), and the dense attachment plaque on the inner side of the nuclear envelope. An external organizing plaque (arrow) occurs in some species. A constriction ring of microfilaments gradually pinches the dividing nucleus into two daughter nuclei by constriction at the midplane. (D–H) Mitotic figures as exhibited by some amoebae. Cr = chromosomes, NL = nucleosome (centrosome), PM = polar mass, and IB = intermediate body. (D–F) is a promitosis with closed nuclear membrane, (I–L) exhibit spindle fiber arrangements in closed mitosis (I–J) and open mitosis (K–L). (L) is a classical form of mitosis with centrioles (metaphase). Based on Anderson (1983, p 160) and Page (1976a).

the plasma membrane, where the contents of the vacuole are expelled during contraction (systole).

Symbiotic associations either with monerans (bacteria) or algae are observed in some protozoa, especially in the sarcodines, including amoebae, Foraminifera, and radiolaria. The host cell engulfs the symbiotic organisms within intracellular vacuoles (Figure 10.4). These vacuoles resemble food vacuoles, but most of the engulfed potentially symbiotic organisms are not digested. The vacuolar membrane is apparently altered to prevent fusion with primary lysosomes. Some species of benthic Foraminifera ingest and disrupt algal cells, releasing the plastids into their cytoplasm. The functional significance and longevity of these "alien" organelles needs further investigation. They may provide photosynthates as nourishment for the cell.

The pigmented flagellates (phytoflagellates) possess plastids. These intracellular organelles are usually surrounded by two or more membranes. The presence of this external membrane, resembling those surrounding symbionts, has led to the theory that these organelles arose during evolution by the gradual reduction in cell mass of an algal symbiont engulfed by the host. Thus, eventually only the pigmented plastid and a whorl of circular DNA remained. It has been hypothesized that the ingested symbiont may have been a pigmented prokaryote resembling modern cyanobacteria. Plastids have internal membranes (thylakoids) containing chlorophyll and other photosynthetic pigments within the lipoprotein matrix of the membrane. Thylakoids may be arranged in stacks forming laminae. The number of thylakoids in each lamina and the number of membranes surrounding the plastid are significant in the taxonomy of the phytoflagellates. In some species, the plastid may be included within an expanded space of the perinuclear cisterna. Thus, the outer membrane of the plastid is continuous with the outer membrane of the nuclear envelope. In other cases, the plastid appears to be surrounded by a portion of the endoplasmic reticulum.

Nucleus

The organization and enclosure of the genetic material (DNA) varies considerably among microbial life forms. In viruses, the nucleic matter (DNA or RNA) is typically a naked strand enclosed within the protective protein coat or capsid. Prokaryotes (bacteria) possess distinct chromosomes, surrounded by cytoplasm, but not protected by a nuclear membrane. The chromosomes are condensed during interphase and hence clearly visible in ultrathin sections viewed with an electron microscope. Mesokaryotes, including many dinoflagellates, have condensed chromosomes, appearing as distinct whorls of DNA, but unlike prokaryotes the nuclear material is surrounded by a membranous nuclear envelope. The nucleus of most eukaryotes, including many of the protozoa, is enclosed by a porous double membranous envelope, and contains nucleoplasm, chromosomes, and dense masses of ribonucleoprotein forming a single nucleolus or scattered nucleolar granules. During interphase, the chromosomes are dispersed into fine strands, but usually condense during mitosis. The pores in the nuclear envelope provide continuity between the nucleoplasm and the surrounding cytoplasm. Substances regulating nuclear function diffuse in. Ribosomes (synthesized in the nucleolar substance), messenger RNA (mRNA) dictating protein synthesis, and other nuclear products diffuse out. Chromosomal material appears as either a fine network of DNA fibrils (euchromatin) or as condensed masses (heterochromatin). During classical eukaryotic mitoses, the nuclear envelope is resorbed at prophase and the separated chromosomes are reenclosed in a nuclear membrane during telophase. But in some protozoa, the membrane remains intact and the chromosomes are separated within the elongated nuclear envelope, which eventually is constricted and pinched apart at the midline (Figure 10.4) to form two daughter nuclei. The constriction ring during anaphase and telophase which pinches the dividing nucleus in two appears to be produced by microfilaments. In the absence of centrioles, the spindle fibers may be attached to the inner surface of the nuclear envelope, or penetrate through the envelope and attach to an external dense plaque located at each pole of the elongating nucleus. Alternatively, as in some dinoflagellates, the nuclear membrane is only partially resorbed, and long spindle fibers originated outside the nucleus pass through channels in the nuclear membrane during mitosis (eg, Dodge 1973). The chromosomes, however, appear to be attached to the inner membrane of the nuclear envelope. In amoebae, two forms of nuclear division have been found (Page 1976b). In promitotic nuclear division, the nuclear membrane remains intact and the centrally located nucleolus elongates and divides into two polar masses. In some species, an elongate interzonal body occurs between the polar masses and associated chromosomes (Figure 10.4). In the second form of nuclear division, mesomitosis, the nucleolus disintegrates, as eventually does the nuclear envelope, giving rise to a typical mitotic spindle apparatus as in metazoan cells, but lacking a centriole. No centrioles have been conclusively observed in amoebae. In some species a centrosphere serving as a polar center has been seen (eg, *Acanthamoeba*, *Cochliopodium*, and *Gocevia*). However, in the absence of centrioles amoebae do not exhibit typical metamitotic nuclear division as found in most metazoans, where a centriole forms the spindle fiber organizing center. It is interesting to note that some euglenoids, including *Euglena*, also do not exhibit a classical mitotic nuclear division. The nucleolus (endosome) and nuclear envelope remain intact as in promitosis of amoebae. The nucleolus, which may be multiple or lobed, elongates and divides along the plane of division axis, and the chromsomes remain condensed during interphase (Walne 1980). The chromatin, however, may be less distinct and more diffuse during mitosis than during interphase. In Volvocida, as represented by *Chlamydomonas* spp (Lembi 1980), the nuclear membrane remains intact during mitosis. In early prophase, the nucleus migrates from the center of the cell to a position near the basal bodies of the flagella, and the nucleolus disperses. Here, the basal bodies replicate to form centrioles that become detached from the flagellar bases. The nucleus, surrounded by the nuclear envelope, elongates and becomes crescent-shaped with the centrioles positioned at the poles of the nucleus (at the tips of the crescent).

Spindle fibers originate from the centrioles and pass through fenestrae in the nuclear envelope. During the remainder of karyokinesis, the nucleus elongates and separates into daughter nuclei, the interzonal spindle disassembles, and new nucleoli are formed.

Among the remarkably large and elegant polymastigote flagellates dwelling in the gut of termites, mitotic division also involves separation of chromosomes within an intact nuclear envelope (Cleveland 1938; Grell 1973). In *Barbulanympha*, for example, the centrioles are always present below the flagellar tuft and appear as rodlike contorted structures with a pointed end surrounded by a spherical centrosome. During phrophase, the clearly prominent spindle fibers adhere to the surface of the nuclear membrane at specialized electron opaque attachment sites (membrane components of the chromatid granules), and penetrate into the nucleoplasm where contact is made with the chromosomes. Most fibers are connected to one another to form the spindle, while others are connected to the spindle granules of the chromatids that are attached to the inner surface of the nuclear envelope. During subsequent stages of mitosis, the nucleus elongates and the chromosomes separate during anaphase, culminating in the constriction and separation of the nuclear membrane to form two daughter nuclei. In the heliozoa (eg, *Actinophrys* sp) centrioles are absent, but the axonemes (bundles of microtubules extending outward into the axopodia) contact the nuclear membrane. Dense "caps" of cytoplasm develop at the surface of the nucleus, corresponding to the opposite poles of the dividing nucleus. Meanwhile, the nucleus begins to elongate and the nucleoplasm becomes transformed into an elongate spindle apparatus. The nucleus, still enclosed within the nuclear envelope, assumes a barrel shape. Separation of the chromosomes and constriction of the nuclear envelope produces two daughter nuclei, rather uniformly surrounded by the remaining polar cap material. The axonemes can, once again, move directly up to the surface of the nuclear envelope.

Some species of protozoa (eg, radiolaria, heliozoa, and some Granuloreticulosea) contain more than one nucleus. All of the nuclei are of similar size and organization. The ciliates are typically dikaryotic with a large somatic nucleus and one or more smaller micronuclei. The organization of the nucleus and its changes during cell division are useful criteria in making separations among closely related taxa that are otherwise difficult to discriminate.

Summative Perspective

The basic organization of the eukaryotic cell is remarkably conservative, and exhibits similar subcellular organelles among a wide variety of unicellular and multicellular organisms. Variations in location, quantity, and specialized function of organelles are important correlates of cellular differentiation or species differences. On the whole, protozoa exhibit the major features of eukaryotic cells, although some organelles are rudimentary or replaced by specialized ones as described in succeeding chapters. All eukaryotes possess a nucleus enclosed within a nuclear envelope and have mitochondria or mitochondrial-like organelles (eg, the kinetoplast of Kinetoplastida) that mediate oxidative phosphorylation, among other functions. Some anaerobic forms lack mitochondria, or possess hydrogenosomes to provide a source of ATP. Phagotrophic cells form food vacuoles that are converted into digestive vacuoles by fusion with primary lysosomes, usually derived from the Golgi apparatus or from the endoplasmic reticulum. The rough ER or free ribosomes are the site of protein synthesis. Cilia and flagella exhibit the same basic internal structure containing a peripheral ring of nine doublet microtubules and a central pair of microtubules. Flagella are longer than cilia and may either be strengthened by cytoplasmic striated rootlets or by bands of microtubules that extend into the cell from the basal body. The basal body (kinetosome) is composed of nine triplet microtubules arranged in a characteristic spiral pattern. Additional reinforcing linkages occur among the triplets at the distal portion of the basal body. Plaque and necklace particles occur in bands around the proximal segment of the cilium and form characteristic patterns that appear to be species specific. Variations in fine structural features, especially those that differentiate protozoa from other eukaryotes, can be significant indicators of structural-functional specializations of protozoa that make them unique along the spectrum of life forms.

11
Protozoan Fine Structure and Functional Microanatomy

A Concept of Protozoan Fine Structure and Function

The diverse morphological features and broad variations in cellular function exhibited by many species of protozoa are explained at the fine structural level by an assemblage of remarkably similar subcelluar structures. Some of the intracytoplasmic inclusions found in protozoa are shown in Figure 11.1. The spatial organization, cytoplasmic distribution, and temporal coordination of the activities of these fundamental organelles accounts in large measure for the diversity of form and function observed at the light microscopic level. Thus, amidst clear evidence of diversity in form and function at the organismic level, there is an aesthetically attractive sense of parsimony and uniformity in organizational features at the fine structural level. There are, indeed, specialized organelles characteristic of many of the taxa. These specialized fine structural features often account for unique physiological functions not found in other groups. Some taxonomic distinctions are based on the presence of these characteristic organelles. For example, the Kinetoplastida are characterized, among other features, by the presence of a mitochondrial-like body possessing a large mass of DNA called a kinetoplast. This organelle is apparently a form of mitochondrion serving as an oxidative metabolic center with production of ATP. No other chondriome is present in the Kinetoplastida. It is interesting to note, however, that the mitochondria of Euglenida and the chrondiome of Kinetoplastida bear discoidal cristae rather than tubular cristae, which, among other characteristics mentioned in Chapter 12, suggest that the two groups may be phylogenetically related (Brugerolle 1985). Microtubular reinforcement of the reservoir region in several flagellates including Euglenida, Kinetoplastida, and Trichomonadida may represent a common structural adaptation to reinforce the cytoplasm in this critical region, where stress is likely to be increased due to flagellar motion.

Endosymbionts, ranging from bacteroids to algae, are observed in some protozoa. These are usually enclosed within vacuoles (eg, Figure 11.1c) that segregate the host and endobiont. This barrier probably provides fine control over exchange of material between the two organisms and also ensures cytoplasmic

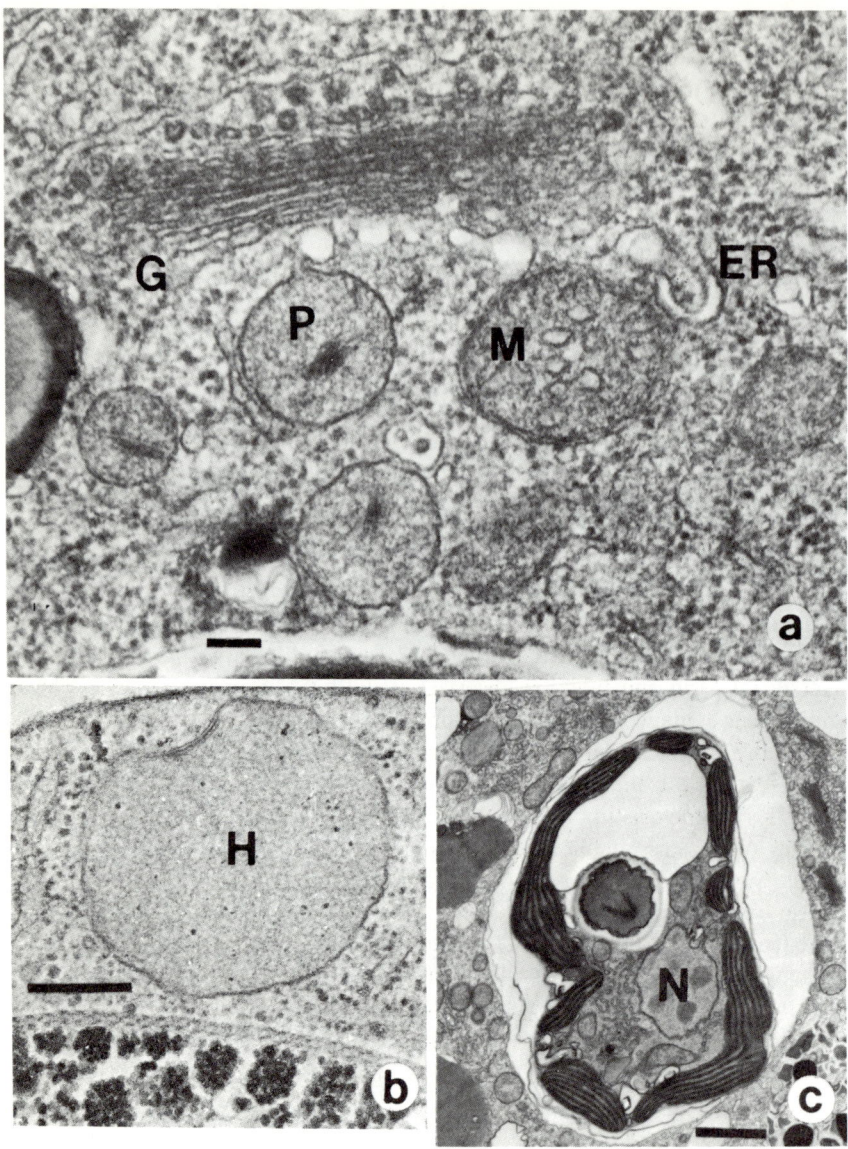

FIGURE 11.1. Transmission electron micrographs of cellular organelles in protozoa. (a) Fine structure of the cytoplasm in a planktonic foraminifer showing the Golgi body (G), mitochondrion (M), endoplasmic reticulum membranous canal system (ER), and peroxisome (P) containing a dense inclusion characteristic of some protozoan peroxisomes. (b) Hydrogenosome (H) contains a finely granular matrix (Courtesy of Ms. Helen Shio of The Rockefeller University, New York). Markers = 0.25 μm. (c) A dinoflagellate symbiont enclosed within a perialgal vacuolar membrane. The mesokaryotic nucleus (N) containing puffy chromosomes is characteristic of dinoflagellates. Marker = 2 μm.

separation of these often genetically diverse organisms. A stable association between two highly diverse organisms living in close proximity requires specialized structures to regulate interaction and provide adequate protection of both participants from adverse physiological influences. The role of endosymbionts in host nutrition and the complex web of physiological interactions between host and symbiont is discussed more fully in the last section of Chapter 16.

Common Features and Specialized Functions of Organelles

Within this spectrum of organellar variation and specialization, however, there are some principles of subcellular organization that are characteristic of a broad range of protozoan cells.

Plasma Membrane

The plasma membrane, though varying in chemical composition and surface decoration, forms an outer limiting boundary of widely diverse protozoa. Variations in thickness, number of peripheral boundary membranes, and the presence of organic or mineralized deposits clearly differentiate among protozoan species (Figure 11.2). However, the apparently ubiquitous occurrence of the basic bilayer membrane design indicates a degree of molecular conservatism and uniformity that suggests this was an early and successful molecular adaptation which has persisted throughout subsequent evolution.

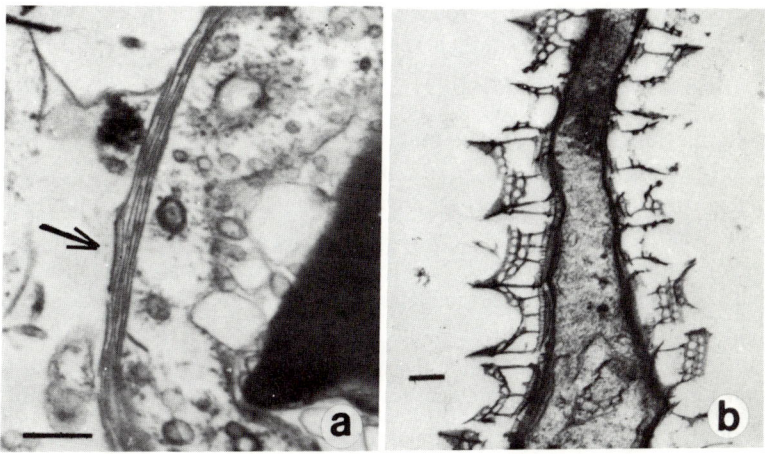

FIGURE 11.2. Surface scales secreted by a coccolith-bearing flagellate showing the imbricated multilayered scales (arrow, panel a), and by an amoeba *Dactylamoeba bulla* (panel b). Markers = 0.2 μm. Illustration (b) courtesy of Dr. F. C. Page, Cambridge, England.

Hydrogenosome and Microbody (Peroxisome)

The hydrogenosome, recently discovered in some protozoa (eg, Trichomonadida), living in reducing environments or lacking mitochondria (eg, Müller 1980, 1985), has helped to explain their survival in environments not inhabited by other protozoa. Moreover, these structural-functional correlates of the hydrogenosome integrate our understanding of some otherwise seemingly disparate physiological data; ie, why some protozoa produce molecular hydrogen as a metabolic product. Comparative fine structure views of a hydrogenosome and microbody are presented in Figure 11.1 (a and b). Both organelles contain a lightly granular matrix. However, the peroxisome contains an internal granular mass (catalase), and microbodies in some protozoa commonly include a nucleoid formed by parallel stacks of membranous inclusions (Anderson and Tuntivate-Choy 1984). The fatty acid degradative enzymes producing active acetate (acetyl-CoA) are sometimes found in a specialized form of the microbody known as a glyoxysome. These organelles also contain enzymes of the glyoxylate cycle, which among other functions, utilize active acetate to produce intermediates in the tricarboxylic acid cycle (eg, oxaloacetate). In many eukaryotic cells, however, the fatty acid degradative enzymes (beta keto-oxidizing system) are found in the mitochondrial matrix.

Mitochondria

Protozoan mitochondria (Figure 11.1, M) usually have tubular cristae (internal membranous projections) where oxidative phosphorylation occurs, producing ATP from ADP. Oxygen is consumed and is reduced to water by action of the enzyme cytochrome oxidase. Some of the chemical reducing potential for this reaction comes from NADH produced during glycolysis or by reactions in the TCA cycle. Much of the respired carbon dioxide is produced by reactions in the tricarboxylic acid cycle of the mitochondrion. Other sources include decarboxylation of carbohydrates during glycolysis (breakdown of glucose), which occurs outside the mitochondrion. The close proximity of mitochondria and peroxisomes (glyoxysomes) in the cytoplasm of some protozoa probably facilitates exchange of compounds between these two organelles. There are common intermediates in the glyoxylate cycle and tricarboxylic acid cycle, as explained more fully in Chapter 15. In the absence of a true mitochondrion, the functions of the chondriome may occur on intracellular membranes (as in many Monera) or in specialized chondriomes such as the kinetoplasts of the Kinetoplastida (eg, Figures 12.15 and 12.16). In some flagellates, the mitochondrion is a single large chondriome that forms a reticulated or branched structure extending through large parts of the cytoplasm.

Contractile Vacuoles

Contractile vacuoles (also known as water expulsion vacuoles) are characteristic protozoan organelles observed in a wide variety of free-living species. They are particularly evident in freshwater species and consist of a central expulsion vacuole surrounded by collecting tubules. The regular swelling (diastole) and contraction (systole) of the central vacuole is readily observed in flagellates, ciliates, and amoebae, and especially among some heliozoans. The prominent contractile vacuole(s) of *Actinophrys*, located at the periphery of the cell, is readily observed during its contractile cycle (Figure 11.3). Considerable quantities of water can be

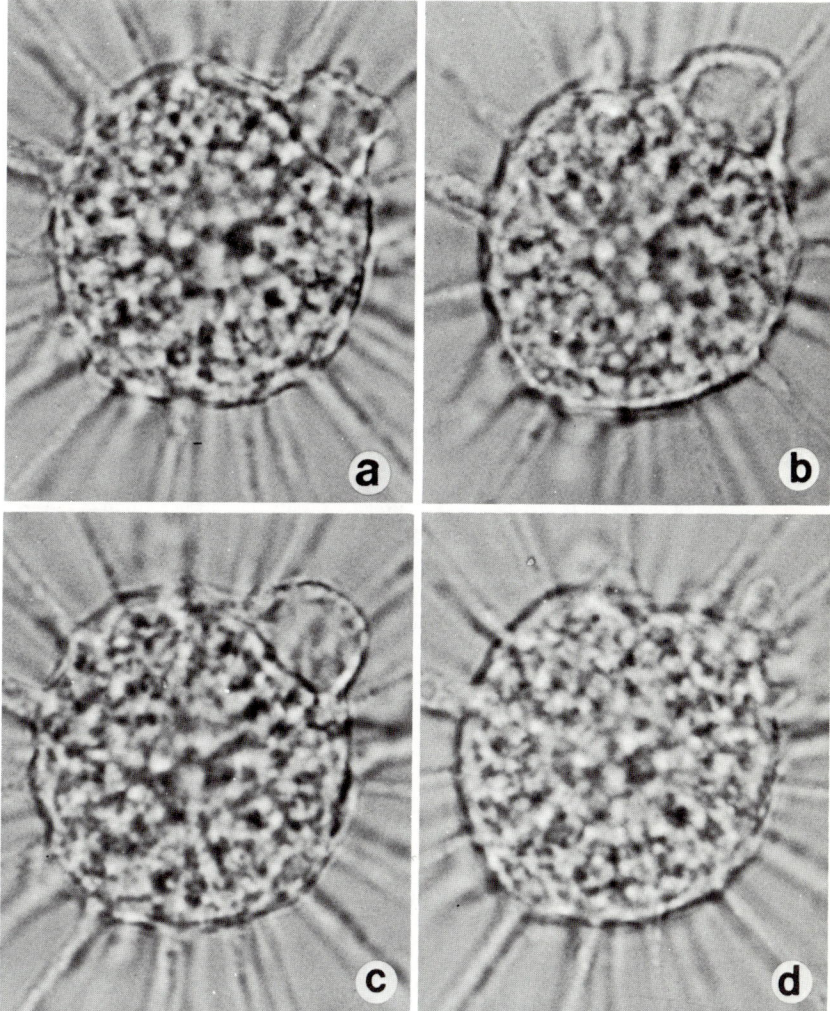

FIGURE 11.3. Light micrographs of a contractile vacuole cycle in a freshwater heliozoan showing gradual emergence of the fluid-filled vacuole (a), its expansion during diastole (b–c), and collapse of the cytoplasmic membrane after systole (d) when the fluid is expelled.

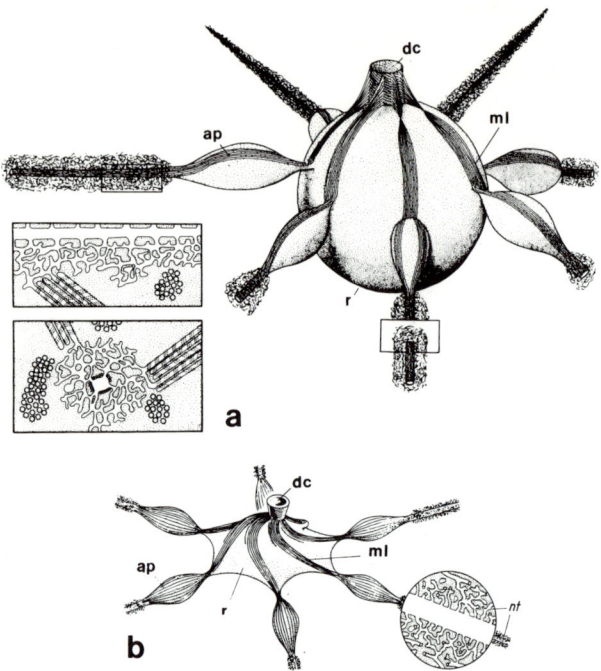

FIGURE 11.4. Contractile vacuole of *Paramecium* during diastole (a) and systole (b). The discharge canal (dc) is reinforced by bands of microtubules (ml) which extend radially onto the surface of the fluid reservoir (r) and spread laterally along the ampullae (ap) of the feeder canals. Feeder canals collect fluid from the cytoplasm by way of decorated parallel tubules (inset, a) that connect to the network of feeder canals. During diastole, when the reservoir is expanding, the feeder canals are contracted, forcing fluid through the ampullae to the reservoir. During systole, when fluid is expelled, the feeder canals and nephridial tubules (nt) expand in preparation for the next filling stage of the contractile cycle. The region surrounding the collecting canals is sometimes referred to as nephridioplasm. Adapted from Rankin and Davenport (1981) and Jurand and Selman (1969).

expelled by the pumping action, and thus eliminate excess water accumulated by osmosis from freshwater environments. The mechanisms of contractile vacuole action and maintenance of internal osmotic balance are discussed more fully in Chapter 17. Although the general microanatomy of the contractile vacuole is remarkably similar among diverse taxa ranging from flagellates through the ciliates, fine details of the organization of the central vacuole and its contributory canals vary among species (eg, Figure 11.4a and b). Variations include specialized arrangements of microtubules to reinforce the connection of the contractile vacuole to the pore, differences in the mode of expulsion either through a differentiated pore site in the cell surface or by opening of a simple closing membrane, and differences in the organization of collecting tubules and the kind of surface structures or surrounding organelles present in the cytoplasm. The position and constancy of location of the contractile vacuole(s) varies among major groups. In many ciliates, the contractile vacuoles have a specific position and are of taxonomic importance. Two vacuoles may be present, one anterior and the other posterior at a particular location relative to the midline of the body. In some amoebae, the contractile vacuole is rather consistently located in the posterior part of the locomoting cell. In many flagellates, contractile vacuoles appear near the base of the flagella either on the outer surface of the cell, or as with some euglenoids and kinetoplastids, emptying into the reservoir of the flagellar pouch.

Cilia and Flagella

The internal organization of the flagellum and cilia is so remarkably similar in design among most groups that it is tempting to conclude this was a very early adaptation which has been conserved during the otherwise highly diverse evolutionary history of the protozoa. Some variations in fine structural detail of the flagellum, including a pattern of only one central tubule rather than two, illustrates that even within an otherwise conservative fine structural feature, some adaptive radiation has occurred during the long and diverse course of protozoan evolution.

Summative Perspective

The specialized fine structural features of protozoa, including variations in oxidative organelles (mitochondria, hydrogenosomes, kinetoplasts, etc), and some of the unique and often elaborate subcellular patterns of organization cited above, reaffirms an observation made earlier. Namely, the protozoa are a highly evolved group of organisms with remarkably specialized and diversified cellular structures and functions. Although they undoubtedly arose from a primitive stock of unicells, during the long history of their evolution over geological time some clear advances in complexity of structure and function have occurred. Hence, modern species of many protozoa are remarkably sophisticated though small, single-celled organisms. The distribution and spatial coordination of the several fundamental organelles of life (including mitochondria, Golgi bodies, endoplasmic reticulum, cytoskeletal structures, and nucleus or nuclei) can account in part for the variations in macroscopic form and function observed among species. Some of the microanatomical features of the major groups of protozoa are presented here in general perspective, and in subsequent chapters, individual groups receive more specialized treatment. Where possible, the functional significance of the fine structural organization will be explored, but we must clearly recognize that for many species our knowledge of the details of functional microanatomy is limited. This is especially true of our understanding of the functional meaning of major patterns of spatial and temporal distribution of subcellular organelles. The static images of cell structure obtained in ultrathin sections viewed with the electron microscope do not accurately reflect the remarkably dynamic patterns of change that accompany cellular activity. The interpretation of these static vignettes of cytoplasmic organization in relation to varied metabolic states of the cell, and their organization into a temporally-ordered series correlated with metabolic changes, may permit a more realistic interpretation of the functional significance of variations in organelle form and pattern. This requires time series studies (usually with cultures of cells in synchronous cell cycle) to correlate fine structural changes with varying metabolic states and growth stages of the cell. The use of combined electron microscopy and biochemical analysis of cellular organelles has contributed much to our understanding of the correlation of structure and function at the subcellular level.

The functional microanatomy of each of the major groups: (1) flagellates, (2) amoebae and their relatives, and (3) the ciliates is discussed in Chapters 12 through 14 as a conceptual orientation to the physiological topics presented in Chapters 15 through 18. The focus is intentionally broad and conceptual. Details of protozoan fine structure can be found in specialized treatments including the earlier works of Pitelka (1963) and modern perspectives on fine structure of specific groups (eg, Dodge 1973; Anderson 1983; Page 1983; Patterson 1977, 1985; Patterson and Dürrschmidt 1986; Lee et al 1985a; Wichterman 1986).

12
Comparative Microanatomy of Flagellates

Flagellate Microstructure

The flagellates are especially diverse in their fine structural features, encompassing the plastid-bearing phytoflagellates and their colorless relatives, and the non-pigmented zooflagellates. The zooflagellates include free-living and parasitic forms, often with unique structural adaptations complementing their specialized environments. The flagellum, though a distinctive characteristic of the group, varies considerably in fine features among species. The internal organization is usually the typical 9 + 2 arrangement of microtubules, but occasional variations occur in the number of central microtubules, and in the presence or absence of a paraflagellar rod near the periphery. The structural significance of surface hairs, scales, and other decorations is not fully understood. It seems clear, however, that these surface structures must influence the hydrodynamics of flagellar activity. Some of the major features characterizing the flagellates are presented with particular reference to the organization of the flagella, distribution of subcellular organelles, and structure of the cell surface.

Flagellum and Haptonema

The two major peripheral projections commonly observed in flagellates are the flagellum and the haptonema. The flagella are locomotory organelles, while the haptonema, observed typically in the Prymnesiomonadida (Hibberd 1980), is an extensible fingerlike body usually emanating near the point of attachment of the flagella. Its fine structure has been characterized carefully in at least two genera, namely *Chrysochromulina* and *Prymnesium*. The function of the haptonema is uncertain, but it may have a tactile or sensory capacity. It is usually coiled close to the cell body, but can be thrust outward, giving the impression of another flagellum. Its internal organization is presented following a discussion of flagellar structure and function.

Flagellar Surface Structure

The flagellum, as observed with a light microscope, may appear as a rather structureless, undulating organelle. When examined with an electron microscope, however, it clearly exhibits marked structural details. The internal organization of flagella (9 + 2 microtubule arrangement) has been presented in Chapter 10. The surface details, however, are quite variable among different species. Four types of flagella are recognized based on the presence of bristlelike projections called mastigonemes: (1) simple flagella without decoration; (2) acronematic flagella with a fine distal projection of varying length; (3) stichonematic flagella with a single file of mastigonemes along one side; and (4) pantonematic flagella with files of mastigonemes along two sides (Figure 12.1). The term acronematic has been used in different ways in the literature to mean flagella with either a thin cytoplasmic extension produced by the protrusion of one of the central tubules, or the presence of one or more fine, hairlike extensions. To avoid this confusion, Manton (1965) recommends the term be abandoned. However, it still appears in the literature and requires careful interpretation, based on context, to discriminate between the two meanings. Mastigonemes vary in length and fine structural organization. Tubular-type mastigonemes appear rigid, have a hollow center, and are probably composed of protein and a carbohydrate. Fine, hairlike mastigonemes appear more flexible and produce a hirsute coat. The mastigonemes of some flagellates (eg, *Ochromonas*) are anchored to globular connections (basal granules) within the membrane of the flagellum. The mastigonemes may also be decorated with fine hairlike projections, either bristling from the tip or radiating from the sides (Figure 12.1). Some flagellates (eg, among the Euglenida) have stichonematic flagella with a single row of mastigonemes (2 to 3 µm) and a fine surface coat of fibrous material, or in some cases, the mastigonemes spiral around the axoneme of the flagellum. The tip is often ornamented with a tuft of slightly stiffer hairs. Walne (1980) summarizes current understanding of the organization of the mastigonemes in *Euglena* as follows. There are about 30,000 fine, nontubular mastigonemes attached to the surface of the flagellum. The mastigonemes are of two types: a short form (ca 1.5 µm) forming a peripheral coat, and longer ones (ca 3 µm) attached to the paraflagellar rod. In the euglenoid

---

FIGURE 12.1. (a–d) Flagellar configurations, (e–f) scanning electron micrographs of dinoflagellate thecae. (a) Pantoneme flagellum of *Synura* sp showing lateral mastigonemes and short haptonema near base of long flagellum. (b) Stichoneme flagellum of the euglenoid *Trachelomonas*, with mastigonemes arising from one side at complex attachment sites on the flagellum and extending to the tip, where substantial numbers of the mastigonemes occur. (c) Acroneme flagellum of *Chlamydomonas* with a single short apical projection. (d) Transverse flagellum lying in the transverse groove or girdle of a thecate dinoflagellate, *Peridinium cinctum*, showing the typical ribbonlike organization of the flagellum. (e) Thecal organization of the armored dinoflagellate *Gonyaulax grindleyi* and (f) *Gonyaulax polyedra*. Scales = 2 µm. Courtesy of Dr. John D. Dodge, University of London, Surrey, England.

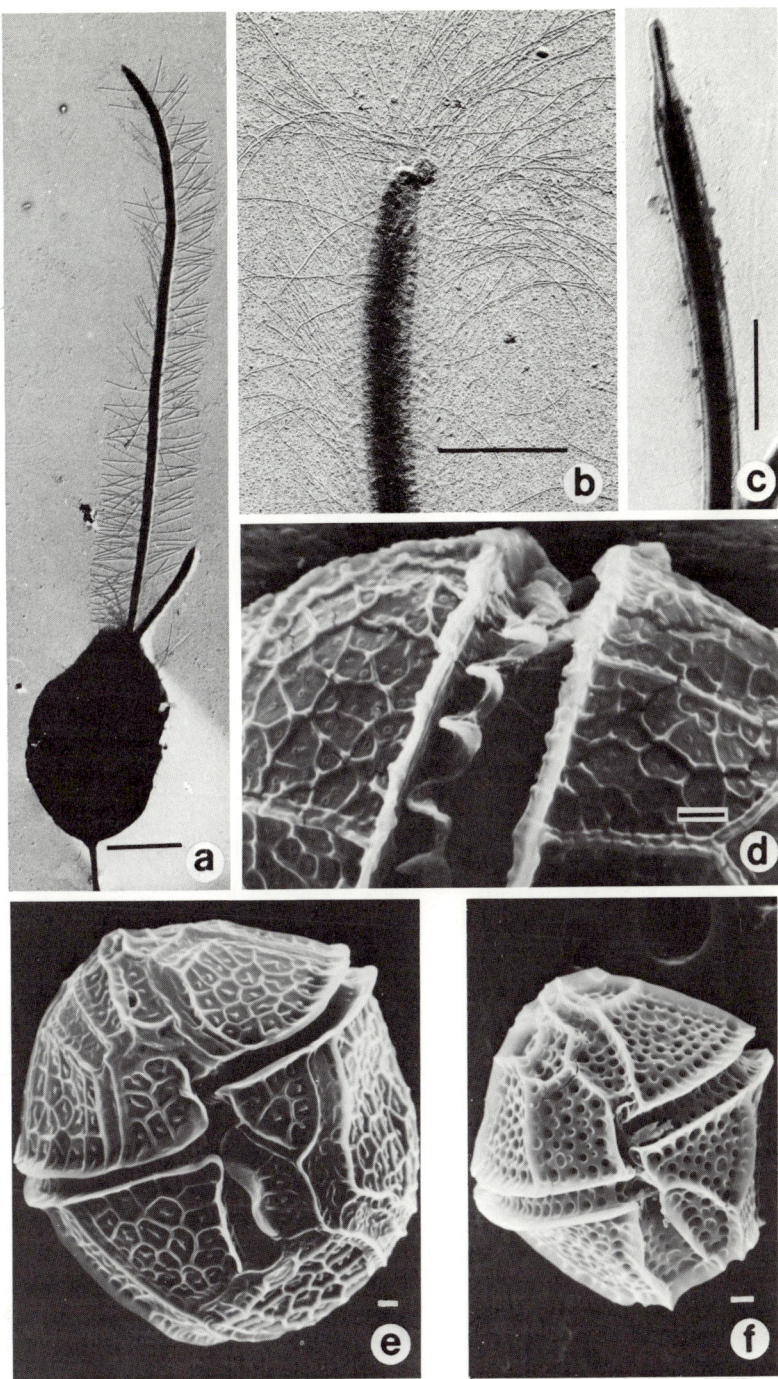

Peranema, the anterior flagellum bears long mastigoneme filaments that appear to arise from shorter tufts and are arranged helically along the flagellum. Two lateral complexes of shorter mastigonemes appear to form a network covering the flagellar membrane surface.

Among some Prasinomonadida and Chrysomonadida, the flagella are covered by a layer of fine scales in addition to mastigonemes. The function of these surface ornamentations is not clear. It appears that the mastigonemes especially may be important in the hydrodynamics of the flagellum. When the flagellate has more than one flagellum, the active flagellum is typically the one that bears mastigonemes, if at all. The broad, bristling appearance of the mastigonemes, at least as viewed in dead cells, suggests that they may increase the surface area of the flagellum and hence improve its power in propelling fluids. Little is known about how the mastigonemes are deployed in living cells during the undulations of the flagellum, or whether the flagellate has the capacity to alter the orientation of the mastigonemes or their spatial arrangement relative to one another. These alterations could have profound effects on the hydrodynamics of the flagella and the speed, pattern of motion, and progress of the flagellate through the aqueous medium. In some phagotrophic Chrysomonads (eg, *Ochromonas*), one flagellum is more active and pantoneme, while the other is simple. The pantoneme flagellum, with its large array of mastigonemes, could aid capture of prey particles and help propel them toward the surface of the cell where ingestion occurs. Little is known about the possible tactile or chemosensory role that these projections may serve.

Paraflagellar Rods

The internal structure of the flagellum is augmented in some species by a paraflagellar rod lying parallel to the microtubules (axoneme), as shown for a trypanosome (Kinetoplastida) (Figure 12.2). A paraflagellar rod has also been observed in the kinetoplastids *Bodo* and *Cryptobia*. The term rod may be misleading, as the internal structure, although variable among species, usually reveals a spongiose or complex cross-linked array of fibers and laminae arranged in a regular geometric pattern. The elaborate pattern of cross-linked fibrils suggests that this flagellar structure may be more than merely a strengthening device. A paraflagellar rod, apparently of somewhat different organization, extends the entire length of the flagellum in some Euglenida, as has been documented for *Euglena*, *Astasia*, *Phacus*, *Rhabdomonas*, *Entosiphon*, and

FIGURE 12.2. Flagellar organization of the trypanosomatid *Herpetomonas megaseliae*. (a) Emergence of the flagellum from the flagellar pocket showing the paraxial structure as a ridge (arrow) on the flagellum (scale = 1 μm). (b) Ultrathin section of the flagellum exhibiting the paraxial structure (arrow), forming a complex network of fibrils as shown in greater detail by scanning electron microscopy of a freeze-dried preparation in cross section (c). Scales = 0.1 μm. Adapted from Farina et al (1986); courtesy of the authors.

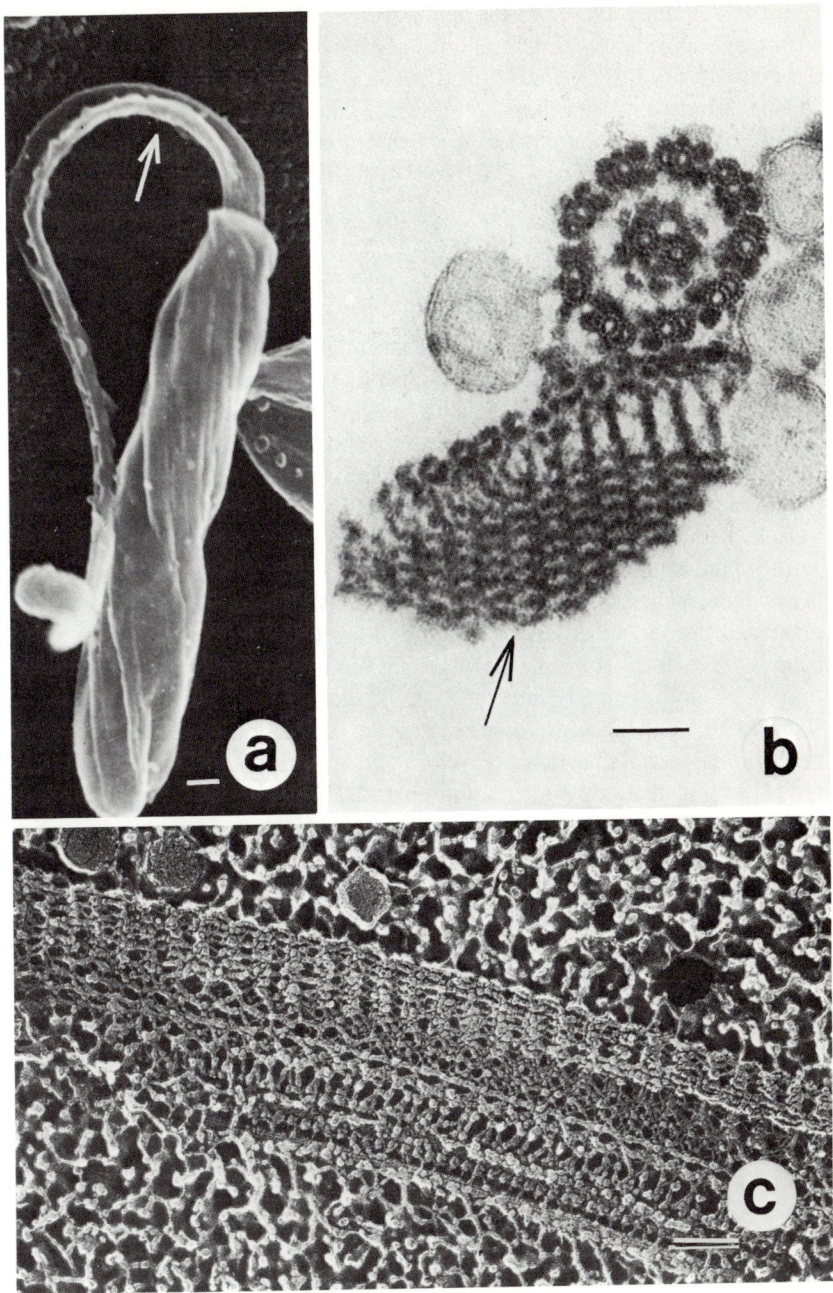

Peranema (Pitelka 1963). In the dinoflagellate *Oxyrrhis*, the paraflagellar rod within one flagellum is composed of four or five fine helically arranged strands (ca 10 to 15 nm thick), cross-linked by short cross-pieces. In other dinoflagellates, the transverse flagellum, lying within the cingulum of the theca (Figure 12.1d), contains an internal narrow striated strand with regular transverse banding and a major periodicity of 66 nm (Dodge 1973).

Basal Body (Kinetosome)

The kinetosome, or flagellar basal body, is the portion of the flagellum anchored within the cytoplasm. It is a cylindrical structure ca 200 nm in diameter and 400 nm long. As shown in Figure 10.3, it consists of a cartwheellike arrangement of nine microtubule triplets. The double central tubules in the distal part of the flagellum end at the dense plate situated approximately at the level where the flagellum is attached to the surface of the cell. Each peripheral triplet of microtubules is attached to the central tubule by a fine radial fibril. The pattern of the basal triplets varies depending on the species. Within a given species, the organization of the triplets changes with depth in the cytoplasm. The most distal portion is usually arranged in a star pattern, which becomes progressively more spirally organized deeper within the cytoplasm. Flagellar roots consisting of striated bands of fine fibrils (rhizoplasts) or groups of microtubules, varying in organization and length, extend from the base of the kinetosome into the cytoplasm (Figures 10.3 and 12.3). In some species, the tapering strands of fibers appear to reach the surface of the nuclear envelope and may be attached to it, but this is not typical of flagellates. Among some phytoflagellates, the proximal end of the rhizoplast is anchored close to the surface of a plastid. The zooflagellates also possess striated flagellar rootlets, either as broad bands or fine striplike projections, and in other cases are augmented by lateral bands of 15 nm diameter tubular structures as in some intestinal Hypermastigida (Pitelka 1963). In some prasinomonads, in which both microtubular and striated rootlets occur, a long

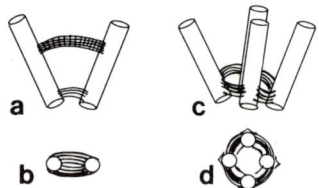

FIGURE 12.3. Comparative illustrations of fibrillar bands linking the basal bodies (kinetosomes) of *Chlamydomonas* (a–b) with two flagella linked by fibrils distally (wide band) and proximally (narrow band), and *Carteria* (c–d) with four flagella linked laterally by four bands of fibrils. The differences in organization of the basal fibrils may explain why *Chlamydomonas* flagella move with a "breaststroke" motion, while those of *Carteria* exhibit an undulating motion as described in the text.

striated rootlet (rhizoplast) projects from the basal granules deep into the cytoplasm and makes associations with the nucleus and chloroplast (Norris 1980), thus perhaps increasing its anchorage. In other species, the rhizoplast ends in the vicinity of a microbody or extends along the surface of the nucleus, located next to the plasmalemma. The massive rhizoplast of *Platymonas*, attached to the plasmalemma, may represent the most advanced type among the prasinomonads. Some evidence for the functional significance of rhizoplast structure has been reported for solitary volvocidans (Lembi 1980). In *Chlamydomonas*, the two flagella are inserted at an acute angle to one another and exhibit a breaststroke waving motion. The striated fibrils are attached between the basal bodies, arranged in a V-shape (Figure 12.3a). There is a distal, more robust, striated fiber bundle and a proximal, less massive bundle attached at the sides of the base of the kinetosomes. Similar striated fibers are observed in the colorless relative *Polytomella*. There are four flagella arranged in pairs and cross-linked by striated fibrils. However, in the quadriflagellate *Carteria*, which swims with a distinctive helical pattern, the striated fibers are connected between adjacent basal bodies, forming a square pattern (Figure 12.3b). The striated fibers are not connected diagonally between bases in a homologous pattern to the basal bodies of *Chlamydomonas*. This difference in striated fibril arrangement may partially account for the difference in flagellar stroke pattern and swimming behavior of *Chlamydomonas* and *Carteria*. In *Polytoma* and *Chlamydomonas*, microtubular fibers have also been observed radiating from the basal bodies in a cruciate pattern, forming four bands that diverge into the peripheral cytoplasm (Lembi 1980). These may help to anchor the flagellar bases. The organization of the microtubular bundles, including the number of constituent microtubules in each band, is of potential taxonomic significance. Chrysomonadida possess both striated fibrillar rootlets and microtubular rootlets. The striated rootlets project deeply into the cytoplasm and branch outward over the nuclear envelope. Microtubular rootlets also radiate from the basal granules and spread under the plasma membrane in the peripheral cytoplasm (Pienar 1980).

In the Prymnesiomonadida, possessing two flagella inserted at an acute angle to one another, only microtubular rootlets have been observed (Hibberd 1980). The bands of microtubules radiate from the flagellar bases, and course around the membrane lining the flagellar depression before projecting into the cytoplasm, where contact is made with the surface of the chloroplasts. The difference in function between striated and microtubular rootlets has not been conclusively determined, but it appears that the striated rootlets may be more elastic and perhaps contractile, while the microtubular rootlets provide anchorage and support.

Haptonema

The haptonema of the Prymnesiomonadida (Figure 12.4) resembles a flagellum, but contains a central shaft of seven microtubules arranged in a cylinder (Hibberd 1980). There are no doublets. The fine structure of the proximal transition zone

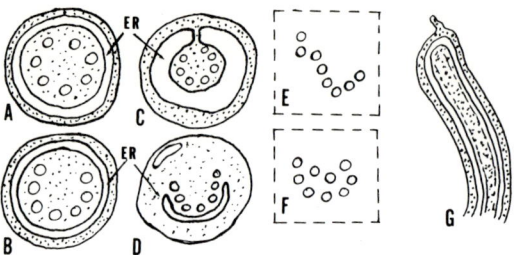

FIGURE 12.4. Diagram of the organization of an haptonema showing a distal portion in cross section (A), with an inner ring of seven microtubules surrounded by a double membrane space formed by the endoplasmic reticulum. At a more proximal level, the microtubules form a c-shaped configuration (B) and near the level of the surface of the cell, the endoplasmic reticular canal is crescent-shaped (C) owing to a thin cytoplasmic bridge connecting the inner core with the surrounding cytoplasm. At the point of entrance into the cell body, the endoplasmic reticular canal is reduced to a narrow crescent-shaped body complementary to the c-shaped configuration of the seven microtubules (D). Deeper within the cytoplasm, the microtubules are augmented by additional ones forming a zigzag configuration of eight (E), and deeper in the cytoplasm is further augmented to form a close packed array of nine microtubules (F). The haptonema in longitudinal section (G) illustrates the inner cytoplasmic core with surrounding ER canal.

is complex, due to a rearrangement of the microtubule core and surrounding membranes. The distal part of the haptonema is fairly straightforward. It is surrounded by a plasma membrane continuous with the membrane covering the cell. The plasma membrane is continuous over the tip of the haptonema and may be smooth, drawn into a tip, or form a spathulate projection. Internally, there is a cylinder of endoplasmic reticulum enclosing the core of seven microtubules. Hence, in cross section there are three concentric membranes: the outer plasma membrane, and the two membranes enclosing the ER cisterna. The space between the plasma membrane and the outer ER membrane is filled by a thin layer of cytoplasm, as in the inner core. Fine cytoplasmic strands, passing through fenestrae in the ER, connect the core cytoplasm with the outer layer of cytoplasm. At the point where the haptonema joins the cell surface, profound changes are observed in the organization of the internal parts. The cylindrical core of microtubules and the surrounding ER become ordered in a crescent shape. Deeper within the cytoplasm, the endoplasmic reticulum is continuous with the ER of the cell and no longer surrounds the microtubule core. The basal segment of the microtubular array is augmented first by an eighth microtubule forming a zigzag array, and then by a ninth forming a hexagonal, close-packed array. This serves as an anchor for the base of the haptonema. In some species, the ER network within the cytoplasm lies in close proximity to the plasma membrane of the cell, and is underlaid by sheets of microtubules that connect to the basal bodies of the flagella.

Cytoplasmic Fine Structure

The organization of the cytoplasmic fine structure and the overall plan of organization of the cell is maintained by the balanced state of dynamic forces among the cytoskeletal and pressure-producing components. These include transitory balanced states among the supportive (microtubular) and tensile (microfilamentous) components of the cytoskeletal system, and their relationships to the osmotic and hydrostatic pressure states of colloidal aggregates and membrane enclosed spaces of cytoplasm (eg, Bereiter-Hahn and Strohmeier 1987; Anderson et al 1987). Some components of the cytoplasm have a distinctive and fairly predictable location within the cell of a given species. This is often the case for the point of flagellar insertion, contractile vacuole position, and plastid location (particularly if it is solitary and large). In many species, the site of the nucleus is fairly constant during a given phase of the cell cycle. In some species, the Golgi body is characteristically located near the nucleus, or near the flagellar base. Some of the dense-staining bodies observed by light microscopists near the flagellar bases, called "parabasal bodies," are in fact Golgi bodies. Not all of the inclusions observed near the flagellar bases are Golgi bodies. Their ultrastructure requires further investigation. Many of the subcellular organelles have no fixed location and are moved about within the cytoplasm according to the metabolic state of the cell. Thus, it is important to realize that illustrations of cytoplasmic fine structure are static, generalized models of a representative state of the organism. For example, the position of the mitochondria, while fairly predictable for some species, is quite variable in others. In some ciliates, peripheral layers of mitochondria are commonly observed near the inner membrane of the pellicle. These may not be fixed at one location, but move inward and outward over time. On an average, however, many would be found near the periphery, where oxygen is more abundant. High resolution light microscopic observation clearly indicates that mitochondria, small plastids when present, vesicles, and other membrane bound organelles are moved along definite paths within the cytoplasm, probably guided by microtubules. Digestive vacuoles move in a fairly predictable path during the digestion process in many protozoa, especially in ciliates. These dynamic processes are discussed more fully in Chapter 16 and illustrated in Figure 16.4.

Phytoflagellates

Phytoflagellate fine structual features vary considerably among major taxonomic groups, especially with respect to the plastid. The chloroplast in pigmented forms, or the "leukoplast" in close relatives without pigments, differ largely with respect to size, shape, number, and the organization of internal membranes. For purposes of orientation, a transmission electron microscopic view of an ultrathin section of *Dunaliella*, a marine representative of the Volvocida related to *Chlamydomonas*, is shown (Figure 12.5a). A diagrammatic representation of the

206   12. Comparative Microanatomy of Flagellates

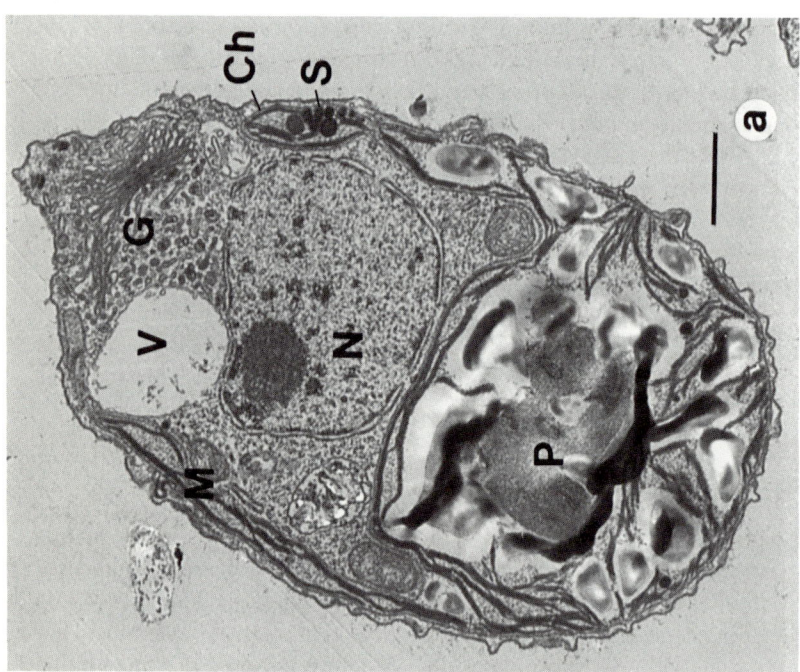

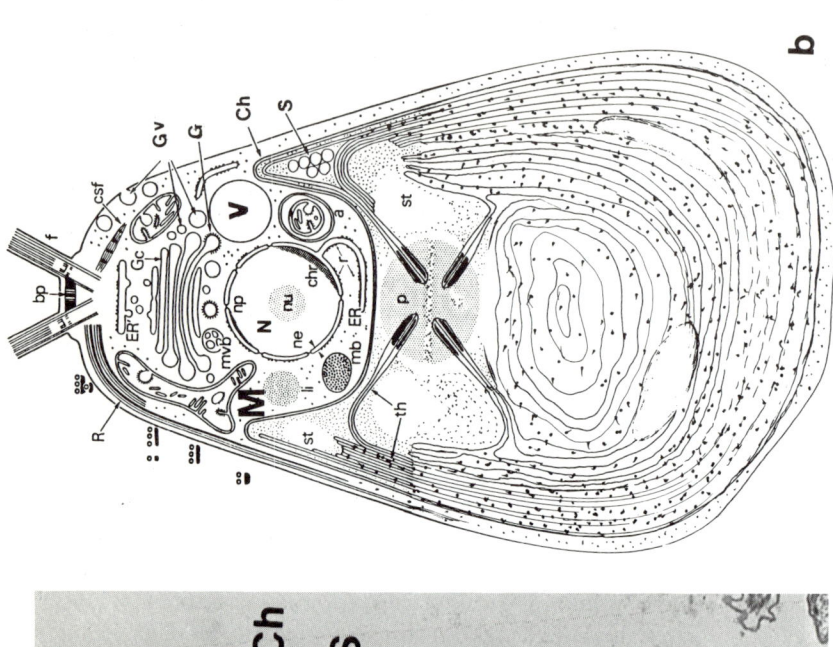

FIGURE 12.5. Fine structure of *Dunaliella* in a transmission electron micrograph (a) and detailed illustration (b) adapted from Eyden (1975). N = nucleus, Gv = Golgi vesicle, G = Golgi body, M = mitochondrion, V = vacuole, and Ch = lobe of chloroplast that surrounds nucleus and expands into a massive segment beneath the nucleus. A prominent pyrenoid (P) surrounded by starch grains is penetrated by lamellae from the plastid. S = stigma bodies. Scale = 1 μm.

major cytoplasmic features is presented in Figure 12.5b. The cytoplasm in longitudinal view is dominated by a single, massive, cup-shaped plastid with a prominent pyrenoid surrounded by starch grains. Individual starch grains also occur between the lamina in other locations in the lobes of the plastid. A large nucleus with fine strands of euchromatin and denser masses of nucleolar material is enclosed within the interior of the cup-shaped space of the plastid. Mitochondria, microbodies, lipid droplets, endoplasmic reticulum, and occasional vacuoles are commonly observed surrounding the nuclear region (Eyden 1975). A pair of Golgi bodies occurs near the flagellar bases at the anterior, more tapered end of the cell. The V-shaped arrangement of the flagellar kinetosomes and the presence of a striped fibrillar band connecting the distal part of these basal segments resembles *Chlamydomonas*. The cytoplasmic organization of *Chlamydomonas* is similar to *Dunaliella*, but the cell is more globose and hence more rounded in profile in ultrathin sections. The Golgi bodies are not always as prominent in the parabasal region, and a distinct wall surrounds the cell. Other members of the order are characterized by the presence of eyespots (ordered arrays of osmiophilic granules) located in the plastid (eg, *Carteria*), and by the organization of the pyrenoid. Variations in the number of internal membranous lamellae and the shape of the surrounding starch grains are distinguishable among species. The structure of starch grains and their abundance are not always reliable taxonomic criteria, since they may vary with changes in the physiological state of the cell. *Polytoma*, a colorless biflagellated relative of *Chlamydomonas*, lacks a well-organized plastid. It has a delicate somewhat cup-shaped membranous leukoplast containing starch grains located at the posterior end of the cell. The leukoplast is surrounded peripherally by the lobes of a branched mitochondrion forming a basketlike calyx around the plastid (Lang 1963). There is no pyrenoid. The nucleus, with a prominent endosome (central nucleolus), lies in the center of the cell surrounded by a granular cytoplasm containing Golgi bodies and endoplasmic reticulum. The contractile vacuole is situated at the anterior of the cell. *Polytomella agilis*, also a colorless quadriflagellated volvocidan, appears not to have a leukoplast corresponding to that of *Polytoma* (Moore et al 1970). The actively growing cell, obtained during logarithmic growth, possesses a central nucleus. The cytoplasmic organelles are arranged in an approximately concentric pattern. The perinuclear Golgi bodies produce vesicles that increase in size with distance from the center of the cell. During their passage from the Golgi body to the periphery of the cell, the smaller vesicles fuse to form larger ones. They contain a carbohydrate storage product. A thin layer of mitochondria occurs beneath the plasma membrane. Endoplasmic reticulum, with associated ribosomes, occurs evenly distributed in the region between the mitochondria and the Golgi. A layer of delicate, membrane-enclosed bodies resembling proplastids lies adjacent to the peripheral ring of mitochondria. These organelles become highly branched, increase in size during late log and early stationary phase of growth, and reach maximum development in late stationary and early precyst stages of development (Moore et al 1970). If these are proplastids, then their elaboration into a branching structure during preencystment stages is of interest, as it may represent a reversion to an ancestral plastid-bearing state. Further investigations are needed to clarify the biochemical composition of these delicate organelles to verify if they are indeed proplastids.

The fine structure of the wall in volvocidan monads, as exemplified by *Chlamydomonas* (Bray et al 1974) and *Carteria* (Lembi and Lang 1965), is a laminated glycoprotein complex, and not a cellulose wall as in plant cells. In *Chlamydomonas reinhardtii*, the wall contains seven layers (Roberts et al 1972). These comprise an inner amorphous layer terminated by a thin osmiophilic layer, three middle layers (central triplet) bounded on the outer side by a thin osmiophilic layer, and a thin outer amorphous layer of variable consistency. The outer two lamina of the central triplet layer contain a highly organized crystalline lattice. The repeating structural unit, as viewed in ultrathin sections, is a parallelogram with sides of 28.5 nm and 23.6 nm and an angle of 80° (Hills et al 1973; Roberts 1974; and summarized by Cox 1980). Similar laminated proteinaceous cell walls have been found in *Chlorogonium*, *Polytoma*, *Carteria*, *Haematococcus*, and *Brachiomonas*, and two other species of *Chlamydomonas*, but the dimensions of the structural units in the crystalline inner layers are different (Cox 1980). During wall synthesis, it appears that the inner amorphous layer is produced first and serves as a template to initiate deposition of the more ordered triplet layer. Major synthesis of wall material occurs at two times in the daily growth cycle, (1) at the end of the light period when the cells reach their maximum size, and (2) in the second half of the dark period when cell division is completed and new walls are being deposited by daughter cells. During this time, the cells devote approximately 15% of their protein synthesizing capacity to production of cell wall proteins (Cox 1980).

▶

FIGURE 12.6. Electron micrographs of green and colorless euglenoid flagellates. (a) Longitudinal thin section of typical cell of the green *Euglena granulata*. White arrow denotes stigma granules adjacent to reservoir (R) containing portion of flagellum and electron-dense paraflagellar body. Note also nucleus (N) with prominent endosome and chromatin material, chloroplasts (Cp) with pyrenoid regions (Py), paramylon caps (black arrow), contractile vacuole (CV), mitochondria (M), Golgi apparatus (G), and numerous cytoplasmic vesicles and vacuoles. Undulating pellicular strips (Ps) around the surface of the cell invaginate at the anterior to form the canal (C), here showing a portion of the emergent flagellum (arrowhead). The canal and reservoir are actually continuous, but due to place of section shown in this micrograph they appear to be separated. Bar = 3 µm. (b) Longitudinal thin section through the colorless phagotroph, *Entosiphon sulcatum*. Ingestion apparatus or siphon (S), seen here as slightly protruded into the anterior flagellar pocket adjacent to the reservoir (R), tapers posteriorly for the length of the cell. Small arrow denotes portions of the two emergent flagella in the reservoir region. One of 12 surface grooves is apparent (large arrow). Note also nucleus (N), mitochondria (M), and numerous cytoplasmic vesicles and vacuoles, including some food vacuoles (Fv), containing ingested bacteria and other particulate matter. Bar = 2 µm. Courtesy of Dr. Patricia L. Walne and J. A. Solomon. Figure a adapted from Walne and Arnott (1967), with permission.

Cytoplasmic Fine Structure 209

The fine structure of euglenoids presents a very different organization, especially with reference to plastid structure and distribution, extracellular products, and the point of flagellar insertion in a depressed region of the plasma membrane known as a reservoir (Figure 12.6). The pellicle is composed of a complex array of interlocking organic strips beneath the plasma membrane, which give a distinctive scalloped appearance to the edge of the cell in ultrathin sections. Each plastid bears a prominent central pyrenoid and starch sheath. The centrally located nucleus contains finely filamentous euchromatin, and is enclosed by a typical eukaryotic double membrane envelope. A swelling near the base of the

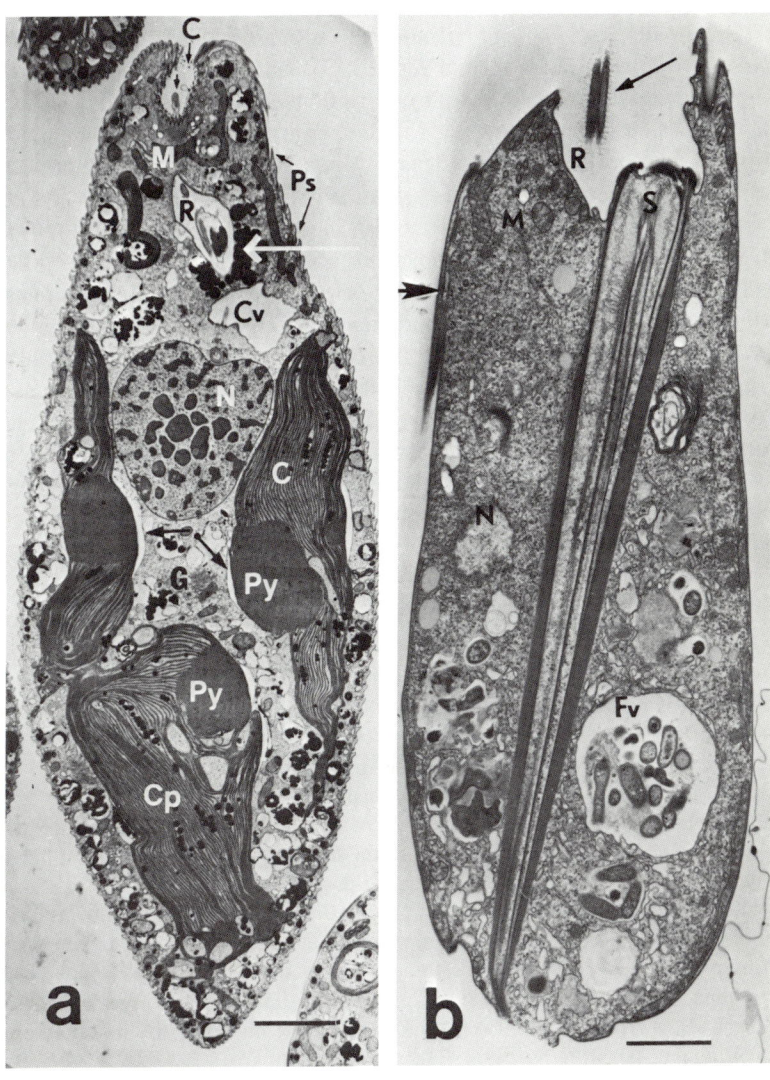

free portion of the flagellum is apparently the light sensing apparatus. Within the cytoplasm, near one side of the reservoir and opposite the flagellar basal swelling, is a mass of osmiophilic granules (stigma) containing carotenoids. This body may help shade the light sensor, but it is not certain that this is essential to the photoreaction of the organism. The orientation of the photoreceptive molecules in the flagellar basal swelling may be more significant in determining phototaxis and responsiveness to light direction (Creutz and Diehn 1976), as described in Chapter 18. There is no evidence of an enclosing membrane (Walne 1980), but the stigma is consistently associated with microtubules. In *Euglena*, *Trachelomonas*, *Phacus*, and *Hyalophacus*, the stigma is encompassed by microtubules and ER. A clear functional relationship has not been found for this association. Mucocysts secreting a mucous substance outside the cell have been observed in some euglenoids; eg, *Euglena splendens* (Hausmann and Mignot 1977), *Peranema trichophorum* (Helenski and Walne 1983), and the marine species *Eutreptiella eupharyngea* (Walne et al 1986). Variations in fine structural detail occur among the species, but in general, the undischarged mucocysts contain a fine amorphous matter of low electron density. Upon discharge, a tubelike projection is formed which in some species has a diamond-shaped pattern, and in *Peranema* and *E splendens* exhibits a few array of filaments projecting from the tip. In *P trichophorum*, moreover, acid phosphatase activity has been localized in the subpellicular mucocysts. This may be the origin of hydrolytic enzymes released into the growth medium (Helenski and Walne 1983). The presence of a mucous layer covering the surface of the cell may improve motility by reducing hydrodynamic resistance at the cell surface. The lorica of *Trachelomonas*, *Ascoglena*, and *Strombomonas* is formed by mineral-impregnated mucilage secreted on the surface of the cell. Fine crystallites containing manganese and silica, but only traces of iron, have been detected within the fibrillar structure of the mucilage in species of *Trachelomonas* (Walne 1980). In general, cytoplasmic fine structure resembles that of other euglenoids. The flagellar canal leads into a deep reservoir

▶

FIGURE 12.7. Feeding behavior of *Peranema trichophorum* engulfing an *Euglena* as prey (A–C). The feeding commences by the outward thrust of the rod organ exposing the cytostome near the flagellar reservoir. Expansion of the cytostome lip permits initial engulfment of the prey (A). Continued contortions (metaboly) of the body of *Peranema* draws the prey deeper within the expanding cytostomal pouch (B). Eventually, the prey is fully engulfed within a food vacuole (C) produced by the expansion and closure of the cytostomal pouch. Fine structural details of the feeding apparatus (D) exhibit the large flagellar reservoir (re) also known as a flagellar pocket. The cytostome (cys) opens laterally at the anterior region of the reservoir. The rod organ consists of microtubules forming two proximal rods, inner rod (IR) and outer rod (OR), that are linked at the tip by a lateral connection. The feeding apparatus is strengthened by fibrillar lamellae, composed of an inner strand (IL) an outer strand (OL), and a cross linking double serrated marginal lamella (DM). Anchoring lamellae (IA and OA) extend around the neck of the flagellar reservoir. Figure (D) from Nesbit (1974), with permission.

extending over halfway into the interior of the cytoplasm. The surface of the cell, however, is covered by a thin membrane and not reinforced by pellicular thickenings (Dodge 1975).

The ultrastructure of pigmented *Euglena*, bleached *Euglena* induced by heat or antibiotic treatment, and permanently colorless relatives (eg, *Astasia longa* and *Khawkinea quartana*) is remarkably similar (Blum et al 1965; Rogers et al 1972;

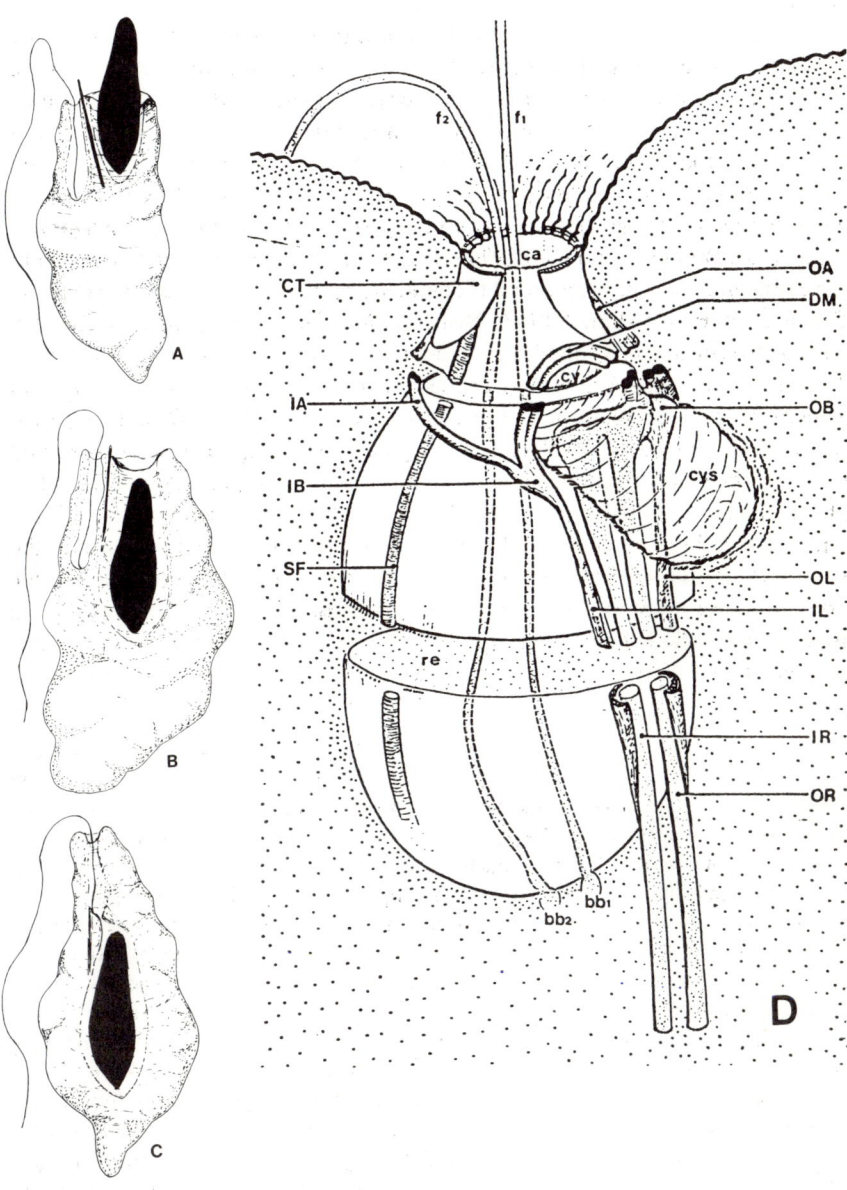

Schuster and Hershenov 1974). Bleached *Euglena* can be distinguished from colorless *Astasia* by the presence of irregular platelike masses of paramylon characteristic of *Euglena* species. Also, dense osmiophilic bodies associated with the paramylon bodies in both bleached and pigmented *Euglena* separate them from *Astasia*. *Khawkinea quartana*, a naturally occurring colorless euglenoid, is identical in fine structure to bleached *Euglena*, and lends additional credence to the hypothesis that *Khawkinea* is phylogenetically derived from *Euglena* by loss of plastids.

The flagellar reservoir is a common feature in many euglenoids, and its fine structure has been of long-standing interest, as some controversy existed in the earlier literature as to whether it served only as a site for flagellar insertion or also as site of food ingestion. It is now clear that the flagellar reservoir per se is not used for food ingestion. The membrane lining the reservoir is strengthened by microtubules and in some species by a thickened fibrillar sheath. This suggests that its main role is protective and supportive. In the colorless phagocytic euglenoids, *Peranema* and *Entosiphon* (Figure 12.6b), there is clear evidence that the cytostomal pouch, where food is engulfed, opens into the anterior part of the flagellar reservoir (Mignot 1963; Nisbet 1974). The feeding apparatus of *Peranema* also consists of a complex rigid device known as a rodorgan (Figure 12.7). This structure, composed of closely linked parallel microtubules, is enclosed in a membranous sac adjacent to the cytostomal sac (Nisbet 1974). It is thrust forward during feeding, displacing the flagella, distending the reservoir lip, and exposing the cytostomal opening. The rod organ is composed of two rods linked at the anterior end by a transverse sheath. Each of the basal rods is enclosed within a membranous canal produced by the ER. Fibrillar laminae attach to the base of each rod and extend anteriorly. The lamina is connected to a swollen node or dense body. At this point, it branches giving rise to two bands: (1) a transverse double lamella supporting the lip of the cytostome, and (2) a lateral lamella encircling the anterior part of the reservoir. The neck or anterior canal

▶

FIGURE 12.8. Comparative perspectives of the pellicular organization in *Euglena oxyuris* (a–b) and *Astasia longa* (c–d). The plasma membrane (arrow in a) covers the organic pellicular strips of *E oxyuris* that interdigitate by overlapping margins subtended by microtubules (M) and the pellicular strip (PS) supporting vertical platelike projections. Endoplasmic reticulum (ER) occurs between the pellicular strip and plate-bearing surfaces. The imbricated pellicular strips can slide upon one another, perhaps by action of the microtubules producing the characteristic euglenoid movements of the species. *Astasia longa* (c–d) possesses rather sigmoidally shaped pellicular strips when viewed in cross section (d). The junction of the overlapping strips is supported by four microtubules (1–4 in c) and knoblike vertical projections that intersect with a groove in the overlying segment of the proximal strip. The knoblike periodic projections (pp in d) arise from a ridge with two of the four microtubules running along each surface (M in d). Euglenoid movement is believed to occur by sliding of the pellicular strips as shown in the right-hand part of (d). Figures (a) and (b) from Suzaki and Williamson (1986a); (b) and (c) from Suzaki and Williamson (1986b); with permission. Scale = 100 nm.

of the reservoir is strengthened by sheaths of dense matter (canal thickenings) attached on the cytoplasmic side of the plasma membrane. The major function of the rodorgan appears to be support of the cytostomal opening and displacement of the flagella and reservoir canal when thrust outward, thus exposing the phagocytic cytostomal opening. The presence of the terminal transverse bar, forming a blunt tip, seems to exclude the possibility that the rodorgan is used to pierce the prey as once presumed based on light microscopy. A facilitatory role,

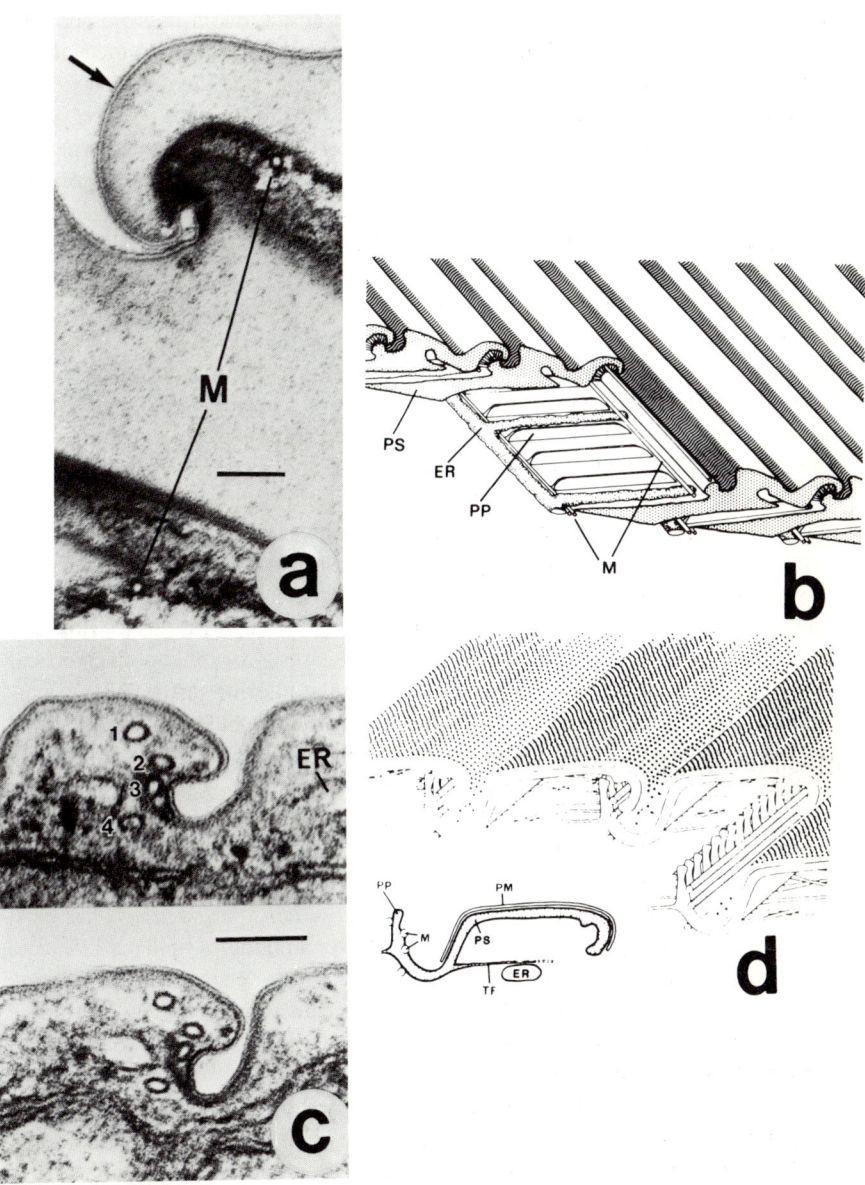

however, in securing and engulfing prey particles cannot be entirely dismissed. The inner wall of the anterior canal, leading into the flagellar reservoir, is further reinforced by the presence of interlocking pellicular strips extending inwardly from the surface of the cell. The pellicular strips curve over the lip of the anterior canal and taper slightly, ending just at the point where the anterior canal joins the reservoir.

Surface Structures

A detailed examination of the surface pellicular structures in *Euglena* spp and *Astasia longa* (Suzaki and Williamson 1986a,b) show structural and mechanical differences in the organization of the pellicular ribbons and their intersection to form the flexible covering of the cell (Figure 12.8). In *Euglena ehrenbergii*, for example, the pellicular strips interlock by an overlapping arrangement whereby a ridgelike rim of the trailing edge of each plate slides into the overlying concave segment of the leading edge of the preceding plate (Figure 12.8a). The shelflike trailing surface of each strip bears vertical platelike projections underlaid by ER and reinforced by microtubules. The ridge–groove intersection between adjacent strips permits sliding motion. This can account for the flexibility of the pellicle during the peculiar twisting and contracting motions of the cell called "metaboly," or euglenoid movements. In *E oxyuris*, the strips are sometimes fused, hindering sliding, and thus accounting for its limited shape changes. *Menoidium bibacillatum* has completely fused strips, and as would be predicted from a sliding strip model, it does not exhibit shape changes (Leedale and Hibberd 1974; Suzaki and Williamson 1986a). In *Astasia longa*, also exhibiting euglenoid movements, a similar overlapping arrangement of pellicular strips was found (Suzaki and Williamson 1986b). The overlapping strips, however, have a different geometry (Figure 12.8b). Each strip in cross section exhibits a somewhat sigmoidal profile. The leading concave surface of each strip overlaps the upward directed trailing edge of the adjacent strip. This trailing edge possesses a row of knoblike vertical projections that engage with a shallow ridge on the interior surface of the overlapping edge of the adjacent strip. A set of four microtubules lies between the intersecting surfaces of the overlapping strips, and Suzaki and Williamson (1986b) hypothesize that the translational motion of the strips generated by these microtubules may account for the shape changes of the cell. Hence, the pellicular strips may serve not only as a protective layer under the plasma membrane, but also as the site of motive force generating euglenoid movements. The chemical composition of the isolated pellicles of *Euglena gracilis* var *bacillaris* has been reported to be largely protein (ca 80%), with lesser amounts of lipid (ca 11%) and carbohydrate (ca 17%) (Barras and Stone 1965). Further chemical identification of protein components of *E gracilis* strain Z indicate 10 polypeptides, and three apparently major pellicle proteins, comprising ca 50% of total polypeptides present (Hofmann and Bouck 1976).

The fine structure of dinoflagellates, especially the armored species with thecal plates (Figure 12.1g, 12.1e,f, and 12.9), has been extensively studied, especially with respect to the origin and organization of thecae (eg, Dodge 1973;

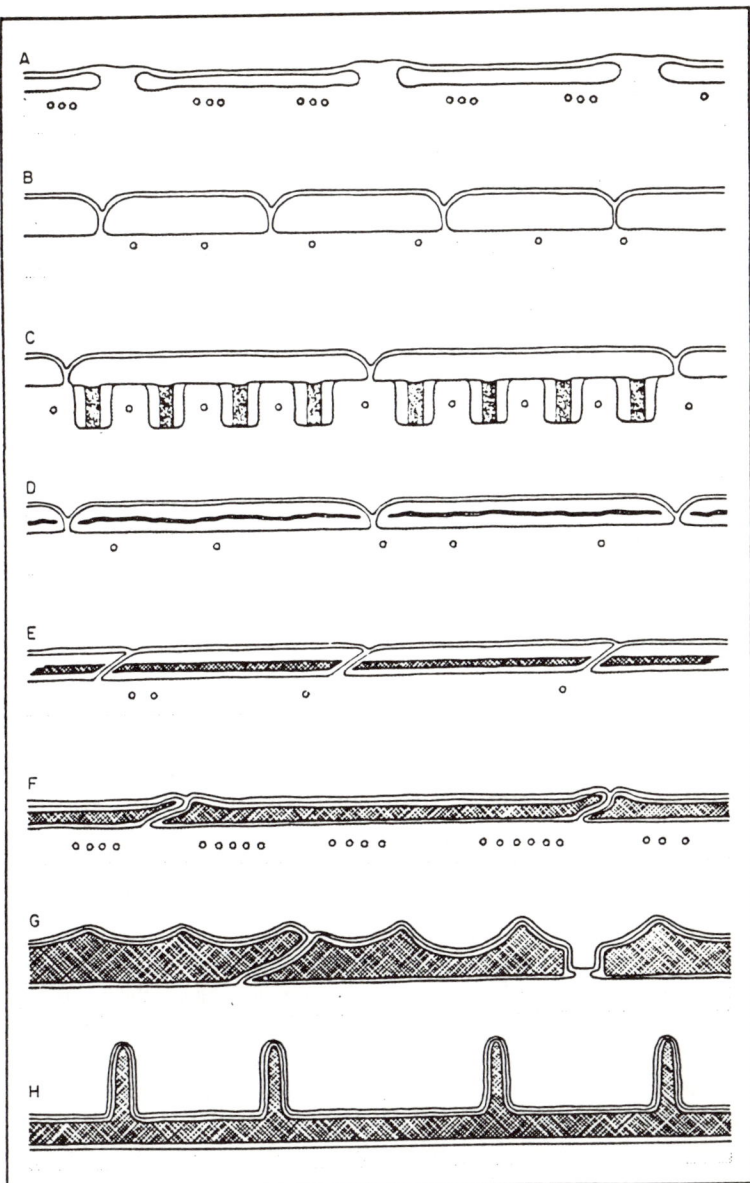

FIGURE 12.9. Diagrams illustrating the structure of various types of dinoflagellate thecae as seen in vertical section. (A) as in *Oxyrrhis*; (B) *Amphidinium*, (C) some species of *Gymnodinium*, (D) *Katodinium*, (E) *Woloszynskia*, (F) *Glenodinium* and *Heterocapsa*, (G) *Ceratium* and some *Peridinium* species, and (H) *Prorocentrum* (= *Exuviaella*). The general trend is from isolated thecal plates with large intervening spaces toward more closely fitted plates that eventually overlap, forming imbricated and interlocking plates, or in the extreme case become fused into a solid thecal wall (H). From Dodge (1973); courtesy of Drs. J. D. Dodge and Richard M. Crawford.

Netzel and Dürr 1984). The thecal plates are deposited within membranous vesicles beneath the plasma membrane at the surface of the cell. Thus, three membranes appear in cross section near the surface of the cell: the two membranes representing the inner and outer surface membrane of the flattened vesicle, and the peripheral plasma membrane. The individual plates vary in organization and arrangement among species, including isolated and separated plates, butted plane plates, interlocking plates that overlap, and complex decorated plates connected by closely interdigitated joints. In some species, the wall is not segmented and forms a continuous, sometimes ornamented outer coat. The internal cytoplasm cannot be simply characterized. It varies substantially among pigmented and colorless forms. In general, the nucleus as reported above is centrally located and contains permanently condensed, often whorled chromosomes (although some species have dispersed chromatin during interphase). The absence of histones permits visualization of the chromatin throughout all stages of the nuclear cycle in most species. Ejectile organelles (trichocysts) are commonly observed beneath the plasma membrane. An extensive membranous canal system ramifying through the cytoplasm forms a unique osmoregulatory organelle called a pusule. This system varies in form, but typically consists of an elaborate network of tubules and dilated cisternae formed by the invagination of the plasma membrane and closely adpressed to a membrane derived from the cell vacuolar reticulum. Thus, this system provides an ideal mechanism for collecting fluids from a large surface area of cytoplasm. The pusule opens to the surface of the cell at a specific site, typically in the base of the flagellar canal. In some species, stigmas consisting of rows of osmiophilic granules within the plastids, or more complex structures with lenses and light sensitive surfaces (ocelli) act as photoreceptors. The mitochondria are of the tubular type, either with densely or only sparsely arranged cristae as in *Oxyrrhis* (Dodge and Crawford 1971). More detailed information on various species and the configuration of thecal plates can be found in Dodge (1973, 1985).

The occurrence of surface plates has also been found in some cryptomonadida. These organic structures with upturned margins, sometimes hexagonal or quadrangular (eg, Figure 2.1d), occur either as a single layer below the plasma membrane or as a double layer with the plasma membrane sandwiched between them (Santore and Leedale 1985). The presence of trichocysts lining the walls of the flagellar canal and the occurrence of thecal plates in some species suggest a close affinity with the dinoflagellates, although evidence is not conclusive. The nucleus appears to be a typical eukaryotic form with dispersed chromatin during interphase. Pigmented forms usually have peripheral plastids, and the cytoplasm contains mitochondria, reserve substances, and typical eukaryotic cell organelles.

Considerable research has focused on the cytoplasmic origin of surface scales in species of the Prymnesiida and clarification of their fine structure. Organic or calcareous scales occur on the cell surface in some species, while others possess both kinds. For example, some coccolithophorids (Figure 12.10) produce remarkably elegant surface coats of imbricated, decorated calcite plates (coccoliths) overlying an inner layer of organic scales. Other members of this group, and some Chrysomonadida, possess only organic scales and the fine structure of these aesthetically pleasing cellular products has received substantial attention, especially by Manton and colleagues (Manton and Parke 1960a,b; Manton 1965, 1986), and also by Green and Pienaar (1977), and Pennick (1984). The origin of the scales, both organic and those supplemented by calcite deposition, is consistently within Golgi-derived vesicles, or flattened cisternae of the intracytoplasmic membranes. Differences in details of shape, sequence of development, cytoplasmic location of the scale-bearing vesicles, and the ultimate shape of the scale are distinguished among species. Considerable attention has been given to the origin and mode of calcification of coccoliths by Westbroek and colleagues (eg, Westbroek et al 1986; Van der Wal 1984). Two species have been given particular attention (*Emiliania huxleyi* and *Pleurochrysis carterae*). The origin of the scale-forming vesicles differs in the two species. In *E huxleyi*, the scales originate in a membranous flattened vesicle lying next to the nuclear envelope and derived apparently from a modified Golgi system (Van der Wal 1984). In *P carterae*, the scales are produced within cisternae of the Golgi stack that become detached, and migrate through the cytoplasm as the internal scales mature. In both cases, the mature scales are deposited upon the surface of the cell by exocytosis. A more detailed account of scale formation is given for *E huxleyi* and *P carterae*.

During development of the scale in *E huxleyi*, the coccolith forming vesicle (CV) initially contains a thin organic plate that serves as the substratum for calcite deposition. The vesicle membrane, closely apposed to the surface of the developing scale, and a complex tubular mass of membranes (reticular body) attached to the outer surface of the vesicle, apparently contain an acid mucopolysaccharide that enhances calcium carbonate deposition. The crystals of calcium carbonate are deposited first on the base plate forming a thin basal sheet. Then, the elaborate and architecturally elegant rim of calcite is added by the molding action of the expanding vesicular membrane (Figure 12.10). The close association of the concave side of the scale-forming vesicle with the surface of the nuclear envelope apparently is essential to normal scale development. After maturation of the scale, the reticular body is resorbed and the scale-bearing vesicle migrates to the periphery of the cell, where the scale is deposited on the surface of the cell by exocytosis. The organization of the acid mucopolysaccharide molecules within the membrane surrounding the developing coccolith, and the action of surrounding cytoskeletal components in shaping and organizing the enveloping membrane, may account for the molding action of the coccolith vesicle.

The formation of scales in *P carterae* commences within flattened Golgi cisternae near the nucleus. As the cisternae progressively mature within the Golgi complex, the organic base plate is produced first and is clearly visible within the Golgi vesicle when it is released into the cytoplasm. Subsequently, the vesicle containing the developing scale migrates peripherally through the cytoplasm, where final ornamentation and full development of the scale occurs. Seven stages of development have been detected during the course of the migration of the

218   12. Comparative Microanatomy of Flagellates

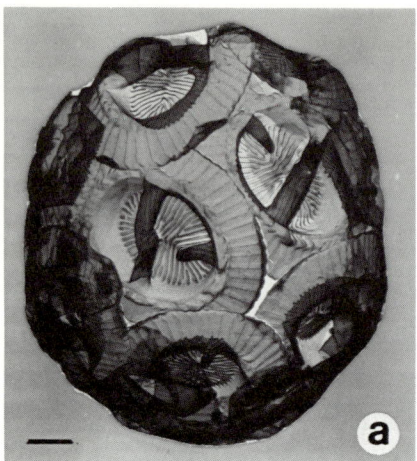

FIGURE 12.10. Coccolith-bearing flagellate (Order: Prymnesiida). (a and b) *Geophryocapsa oceanica*, coccosphere (a) and an individual coccolith (b) showing the ornate organization of these calcite bodies. (c) *Pleurochrysis carterae* diagrammed from an ultrathin section as viewed with the transmission electron microscope. The centrally located nucleus (N) is flanked on either side by massive plastids (Chl). The cell is surrounded by three kinds of surface covering: distally by a single layer of coccoliths (EC), then by several layers of scales (SC), and proximally by columnar material (CM). The coccoliths are synthesized within the vesicles of the Golgi body (G) and occur first as bases (b) and coccolithosomes (CS). As the vesicles migrate toward the periphery, the scales (SC) and coccoliths (CV) mature and are deposited on the surface of the cell by exocytosis. Scale markers = 1 µm. (a and b) Courtesy of Dr. Andrew McIntyre, Lamont-Doherty Geological Observatory of Columbia University. (c) From Westbroek et al (1986). Reproduced by permission of Oxford University Press.

scale-bearing vesicle (Figure 12.10c). The sequence of events commencing at the nuclear region and progressing toward the peripheral membrane as summarized by Van der Wal (1984) is as follows: (1) cisternae bearing incipient scales appear at the distal face of the Golgi array and are composed of two parts, a scale-bearing part with closely apposed membrane and a peripheral lobe without scale; (2) the dilated cisterna contains the maturing scale apposed to the proximal membrane; (3) scale enlargement produces a considerably more dilated vesicle, which distally has two interconnected protrusions that may serve the same function as the reticular body in *E huxleyi*; (4) the lateral and proximal sides of the scale are closely enclosed by the vesicle membrane, and there is an electron-translucent matrix on the distal surface of the scale; (5) a mature coccolith begins to appear, containing a calcite rim; (6) a fully mature form of the coccolith as appears on the periphery of the cell appears within the loosely enclosing membrane; and (7) the vesicle is fully dilated near the plasma membrane in preparation for release of the mature coccolith.

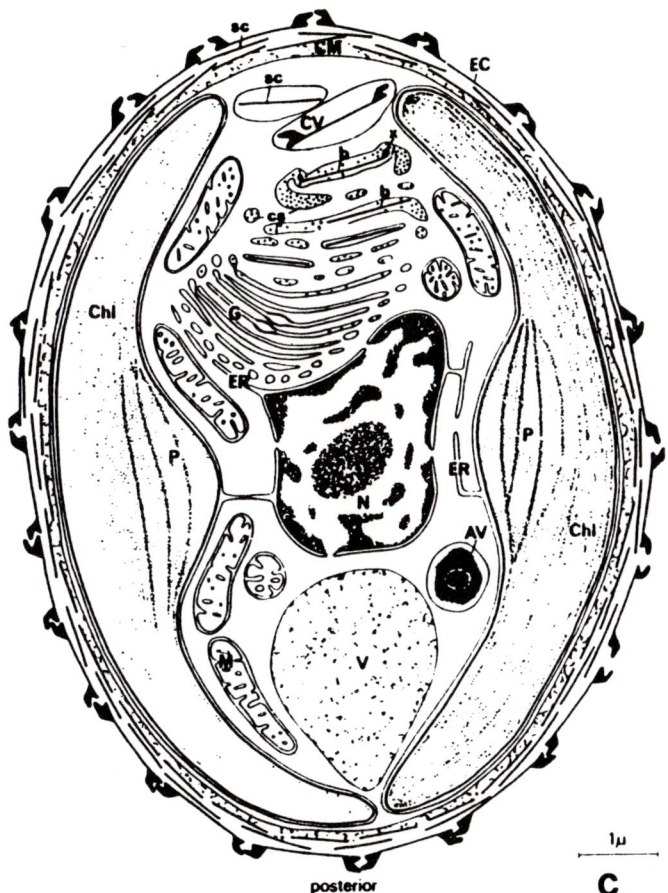

FIGURE 12.10C.

Similar processes of organic or siliceous scale production occur in the Golgi apparatus or membranous vesicles of scale-bearing Chrysomonadida. However, many species are naked or possess only a thin surface layer of organic fibrils. The extensively studied genus *Ochromonas* (eg, *O danica* and *O malhamensis*) is a biflagellated cell with a thin lorica of organic fibrils. One of the flagella is highly active and pantoneme, whereas the other is often trailing and less active. This flagellum may appear in a surface groove of the plasma membrane, and in other chrysomonads is reduced to a flagellar stub enclosed within a membranous pocket in the surface of the cell. A stigma, consisting of osmiophilic granules, occurs within the lobe of a chloroplast immediately beneath the pouch containing the flagellar stub. In *Ochromonas* and other representatives of this group, several plastids are scattered at the periphery of the cell (eg, Figure 16.3). A thin layer of cytoplasm containing the plastids surrounds the nucleus and encloses a

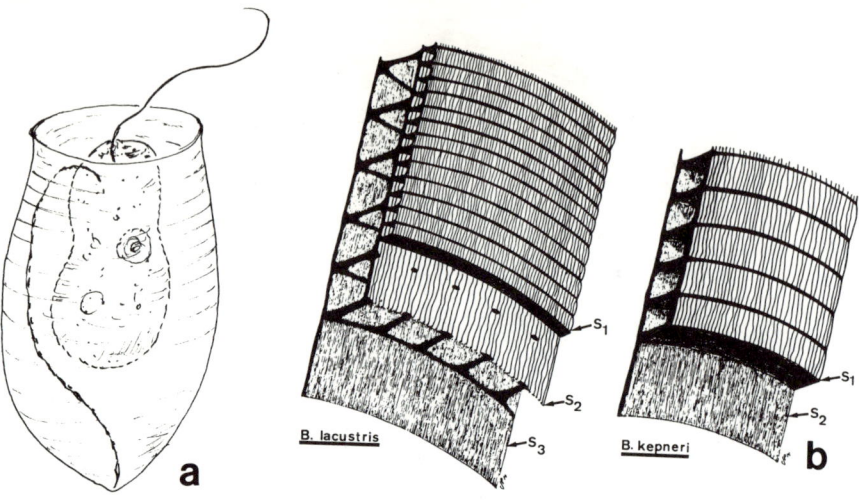

FIGURE 12.11. Diagrammatic representation of the lorica-bearing flagellate *Bicoeca* (a) showing the vaselike lorica and the internal biflagellated organism. The forward-directed flagellum is used for locomotion, and the trailing flagellum attaches to the base of the lorica. Cell size = 10 μm. Fine structure of the lorica (b) exhibits the delicate trabeculate organization of the wall consisting of three layers in *B lacustris* and two layers in *B kepneri*. From Mignot (1974); with permission.

typically large, posterier "storage vacuole" (eg, Anderson and Roels 1967). This vacuole can be converted to a digestive vacuole during phagotrophy of large prey as, for example, during cannibalism when food becomes limited. Golgi bodies and one or more contractile vacuoles occur near the anterior end of the cell.

A very remarkable colorless flagellate, of uncertain taxonomic position, but often assigned to the Chrysomonadida is *Bicoeca* spp (Figure 12.11). This free-swimming organism produces a delicate, conical lorica of organic fibrils embedded within a mucilaginous substance. In some species, the mesh is quite regular and delicate in design (Figure 12.11b). There are two flagella. One is forward-directed, and the other is recurrent and attached to the base of the lorica, thus holding the cell in position. Fine structural details show that there is a band of microtubules emanating from the base of the forward-directed flagellum and curving around the peristome, perhaps as a reinforcement (Moestrup and Thomsen 1976). The major fine structural features are shown in Figure 12.12 (Mignot

▶

FIGURE 12.12. Fine structure features diagrammed for *Bicoeca*, exclusive of the lorica, showing the large excentrically located nucleus surrounded by chondriome and food vacuoles (fv) near the phagocytic canal. The forward-directed flagellum (F_1) bears mastigonemes, while the trailing flagellum (F_2) passes through a depression in the cell surface. A band of microfilaments (Mf) strengthens the trailing flagellum that is used to attach the cell to its organic lorica (see Figure 12.11). From Mignot (1974); with permission.

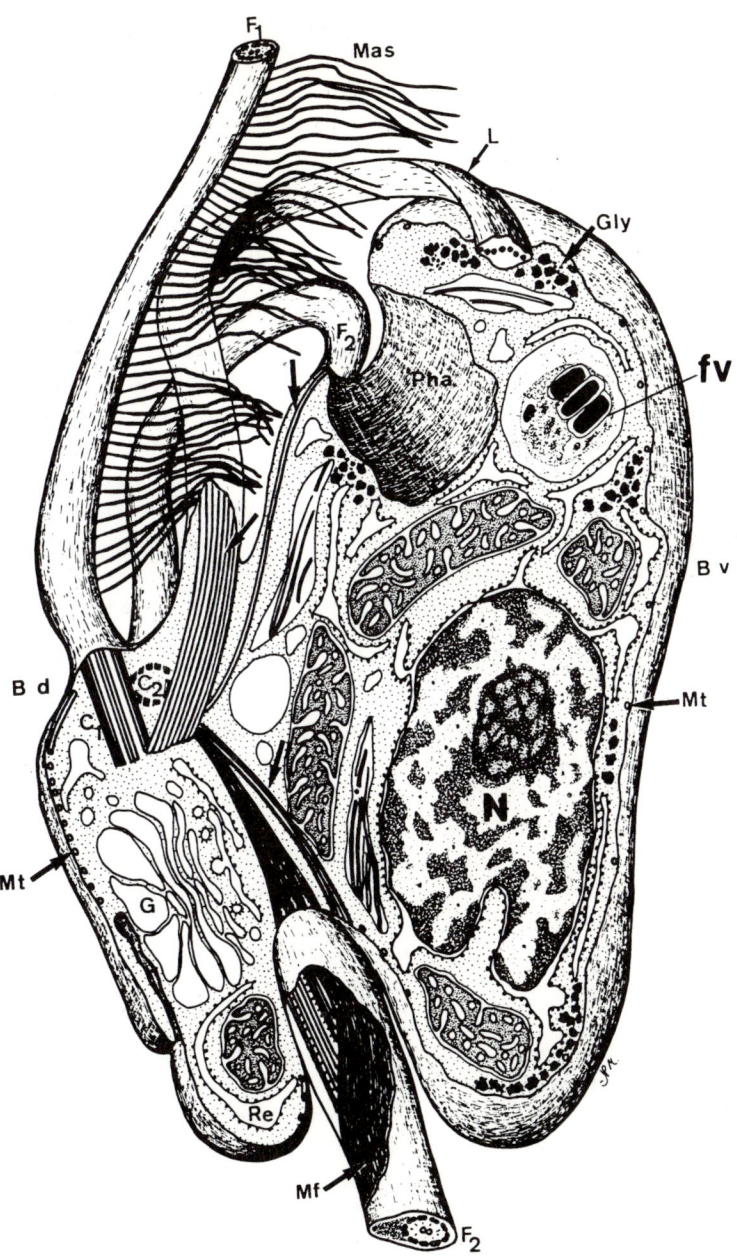

1974; Belcher 1975). The permanent pharynx and peristome are somewhat different from those of other chrysomonads. Additional microtubular reinforcement occurs beneath the plasma membrane, and microfilaments form a thick reinforcing band within the recurrent flagellum, lying within a groove that extends along a narrow portion of the cell body. This groove is adjacent to the site where the flagella bases are inserted. The forward-directed flagellum is haptonematic, with a single file of long mastigonemes projecting from one side. Intracytoplasmic organelles include mitochondria, one or more contractile vacuoles (often with extensive ducts extending into the surrounding cytoplasm), Golgi body, and digestive vacuoles. A loop of cytoplasm forming a lip extends anteriorly from the region of flagellar attachment. The recurrent flagellum attaches to the base of a bell-like lorica (Figure 12.11a). The large, open lorica may aid in concentrating and capturing food particles in addition to having a protective function.

The occurrence of phagotrophic forms among the Chrysomonadida, especially some amoeboid, partially photoautotrophic species such as *O malhamensis*, suggests that this group may be closely linked ancestrally to amoebae. The occurrence of colorless forms among the phytoflagellates, such as *Bicoeca* (with anterior and recurrent flagellum used for attachment), may indicate a close affinity with some zooflagellates that also possess recurrent flagella.

Zooflagellates

The diversity of species and variation in fine structure among the zooflagellates precludes a detailed analysis in a text of this scope. Therefore, only comparative features of some representative species will be considered. The objective is to introduce some of the diverse ways in which the zooflagellates have adapted to their environment to ensure survival amidst the uncertainty of depending on food from external sources. Particular emphasis will be given to representatives of the Trichomonadida and Kinetoplastida.

Trichomonadida

The fine structure and correlated physiology of trichomonads has received extensive attention by Honigberg and colleagues (eg, Honigberg 1978a,b; Honigberg et al 1971) and by Brugerolle (1980). The general fine structural features of this group are illustrated by *Protrichomonas legeri* (Figure 12.13). The pelta-axostylar complex formed by sheets of microtubules and consisting of an apical pelta, juxtanuclear capitulum, and an elongated axostyle dominates much of the cytoplasm. This apparently supportive structure is closely associated anteriorly with the flagellar bases. In *Protrichomonas* there are three free flagella and one intracytoplasmic trailing flagellum. Two large, striated rootlets project posteriorly near the nucleus and in the vicinity of prominent Golgi bodies. The nucleus, partially enclosed by a spathelike portion of the pelta-axostylar complex, is surrounded by hydrogenosomes and intracytoplasmic membranous elements. Digestive vacuoles in various stages of development appear in the peripheral

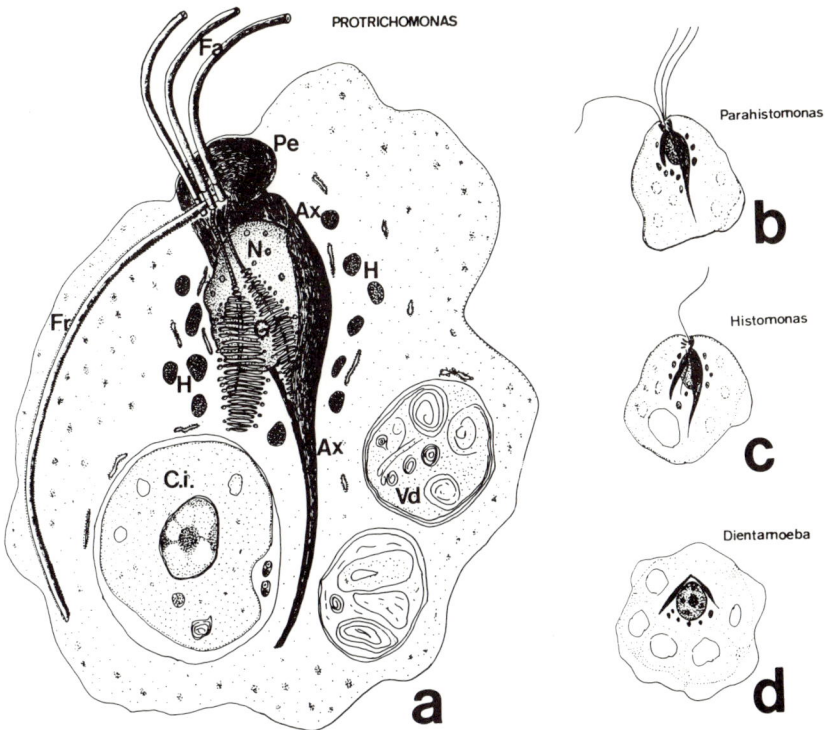

FIGURE 12.13. Diagrams of the major cytoplasmic features of (a) *Protrichomonas* bearing three free flagella and one nonemergent flagellum embedded within the cytoplasm. The pelta (Pe) and axostyle (Ax) lie in close proximity to the nucleus (N), which is near the base of the three flagella bearing rootlets in close proximity to stacks of cisternae of the Golgi bodies (G) and hydrogenosomes (H). Food vacuoles (Ci) and digestive vacuoles (Vd) are distributed throughout the peripheral cytoplasm. Comparative features of (b) *Parahistomonas*, with four emergent flagella, and (c) *Histomonas*, with one emergent flagellum, exhibit the large pelta and axostylar complex characteristic of the Trichomonadida. *Dientamoeba* (d), formerly thought to be an amoeba, possesses characteristic trichomonad cytoplasmic features, and though lacking flagella, is clearly closely related to the other members of this flagellate group. From Brugerolle (1980); with permission.

cytoplasm. *Protrichomonas* ingests *Cryptobia intestinalis*, which also lives in the stomach of the host (*Box boops*, a fish). An early stage of ingestion shows a still intact *Cryptobia* (Ci) within the food vacuole. Masses of glycogen are scattered throughout the cytoplasm. In the same diagram (Figure 12.13), three related genera are shown (*Parahistomonas*, *Histomonas*, and *Dientamoeba*) exhibiting the internal arrangement of the pelta-axostylar complex and the Golgi bodies. These three genera illustrate a progressive change from flagellated forms of trichomonads to the fully amoeboid form *Dientamoeba*. The cytoplasmic details, however, clearly mark these genera as trichomonads. Further comparative details of the fine structure of this group are exhibited by *Tritrichomonas foetus* (Honigberg et al 1971). The prominent pelta surrounds the kinetosomes of the flagella and extends as a channellike sheath near the nucleus and posteriorly as an axostyle (Figure 12.14). Each of the three kinetosomes of the anterior free flagella is supported by filamentous lamellae (F1 through F3). The lamella from flagellum #2 forms a sigmoid sheet extending toward the pelta. Parabasal filaments (PF1 and PF2), and a thickened striated fiber forming the supportive costa (C1) (lying beneath the attachment site of the trailing flagellum) project poster-

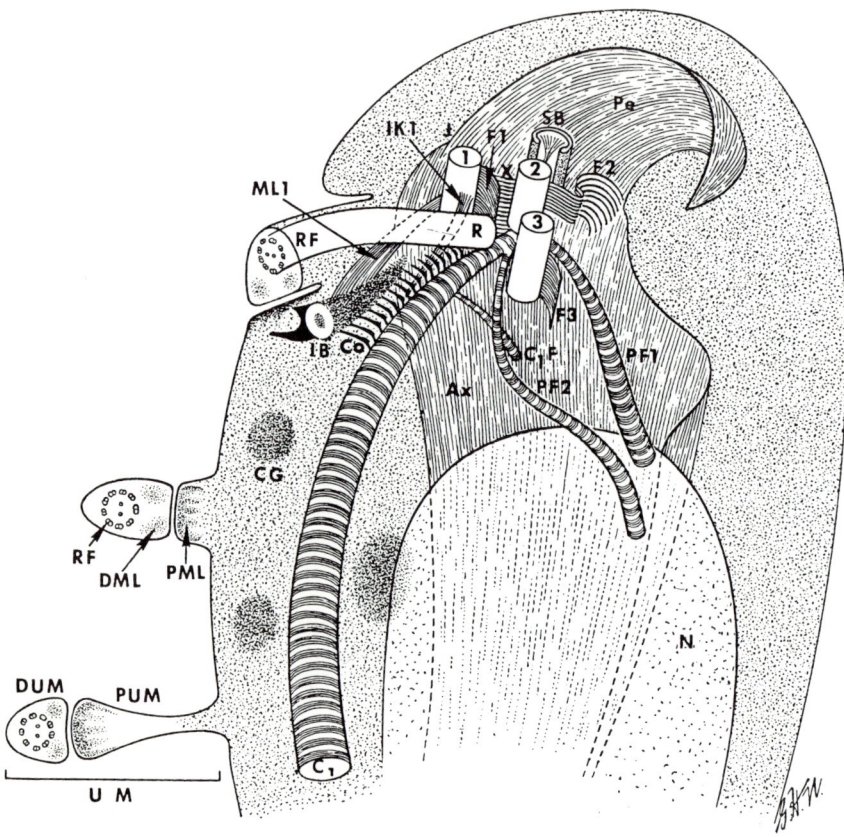

FIGURE 12.14. Fine structural features diagrammed for *Tritrichomonas foetus*. The pelta (Pe) and axostyle (Ax) are composed of sheets of microtubules and partially surround the nucleus (N). The three anterior flagella (1–3) and the recurrent flagellum (RF) arise from a complex association of flagellar bases and striated costa (C_1) and finer parabasal filaments (PF) that extend into the cytoplasm. The distal part of the recurrent flagellum is attached to a fold of the cell surface, forming an undulating membrane (UM). The undulating membrane is strengthened by lamellae (DML and PML) that augment the cytoplasm of the flagellum and fold, respectively. From Honigberg et al (1971); with permission.

iorly within the cytoplasm. The trailing flagellum emerges from the cytoplasm at an oblique angle to the three anterior flagella, and extends posteriorly along the surface of the plasma membrane to which it is attached, at places forming an undulating membrane (UM). When the flagellum is active, it lifts the layer of cytoplasm beneath the attachment sites, and forms the undulating membrane observed as a flickering thin lamina at the edge of the living cell when viewed by light microscopy. The proximal part of the trailing flagellum is a foldlike differentiation of the dorsal body surface which forms the proximal marginal lamella (PML) at the site of flagellar attachment. Further posteriorly, the proximal part of the undulating membrane (PUM) is clearly shown in section as a raised ridge of cytoplasm lifted by the attached flagellum. Among other characteristics, the comblike structure (Co) is a differentiating feature of the family Tritrichomonadinae, distinguishing it from the family Trichomonadinae (including the genus *Trichomonas*) where a costal base occurs near the flagellar basal region. Likewise, the suprakinetosomal body, connected to kinetosome #2 in the vicinity of attachment of the sigmoid filaments, and the infrakinetosomal body, which appears to contribute to the proximal marginal lamella, are structures apparently restricted to Tritrichomonadinae (Honigberg et al 1971). The paracostal bodies, densely staining particles near the costa as observed in light microscopy of *Trichomonas* spp, are hydrogenosomes (as confirmed by electron microscopy).

Kinetoplastida

The Kinetoplastida, encompassing parasitic and free-living forms, have been investigated in some detail. This is probably due to the importance of understanding how to control pathogenic species (eg, *Trypanosoma* and *Leishmania*) as described in Chapter 9. The free-living bactivorous species (*Bodo* and *Rhynchomonas*) provide a generalized view of the organization of the cytoplasm in phagotrophic forms (Figure 12.15) (Vickerman and Preston 1976). In addition to the nucleus, one of the most prominent organelles is the kinetoplast–mitochondrial complex characteristic of this group. The DNA-containing enlarged part of the organelle usually lies close to the flagellar bases, and has a large single mitochondrial segment attached to (Figure 12.15a) and in some species extending around (Figure 12.15b) the kinetoplast. A distinct contractile vacuole occurs near the point of flagellar insertion in both *Bodo* and *Rhynchomonas*, and is situated near the wall of the flagellar reservoir. Bacterial endosymbionts occur sparsely scattered throughout the cytoplasm. Both species also contain some typical eukaryotic organelles including Golgi apparatus, endoplasmic reticulum, and digestive vacuoles in varying stages of development. The cytostomal region differs considerably in the two genera. *Bodo* has a lateral cytostome surrounded by small cytoplasmic projections known as lappets. Subpellicular microtubules, not shown in the diagram, occur beneath the plasma membrane especially at the anterior end of the cell. A long, tapering cytopharynx extends into the cytoplasm toward the opposite side of the cell, where food vacuoles are formed by endocytotic engulfment of the food particles. *Rhynchomonas* has two large

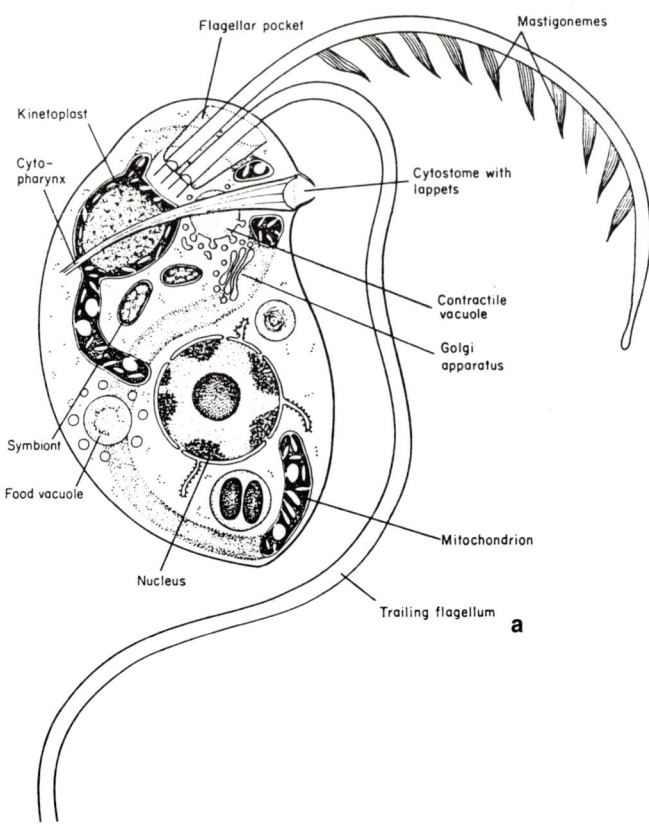

FIGURE 12.15. Fine structural features of two free-living, bacterivorous Kinetoplastida, *Bodo saltans* (a) and *Rhynchomonas nasuta* (b). The prominent kinetoplast, lying near the flagellar bases, possesses a long chondriome with internal cristae resembling those of mitochondria. *B saltans* has a long cytopharynx opening at a cytostome bearing a circle of lappets; whereas *R nasuta* bears two lobes forming a proboscis that is activated by the motion of the shorter flagellum and aids in bacterial ingestion. The cytopharynx opens between the lobes of the proboscis and extends dorsally into the cytoplasm. Both organisms possess a trailing flagellum whereby they glide along surfaces while feeding on bacteria. From Vickerman and Preston (1976); with permission.

anterior lobes forming a proboscis that is attached to the short anterior flagellum. The proboscis lobes enclose the opening of the cytostome, which leads into a long cytopharynx extending posteriorly within the cytoplasm. The vesicles surrounding the cytopharynx are characteristic of many phagotrophic bodonines. The anterior lobes are believed to aid in bacterial prey apprehension and collection into the cytostome. The beating action of the shorter flagellum produces a sweeping action of the lobes. The trailing flagellum (bearing mastigonemes) is typically attached to the substratum during locomotion and feeding, while the free flagellum is used to propel the organism through the water.

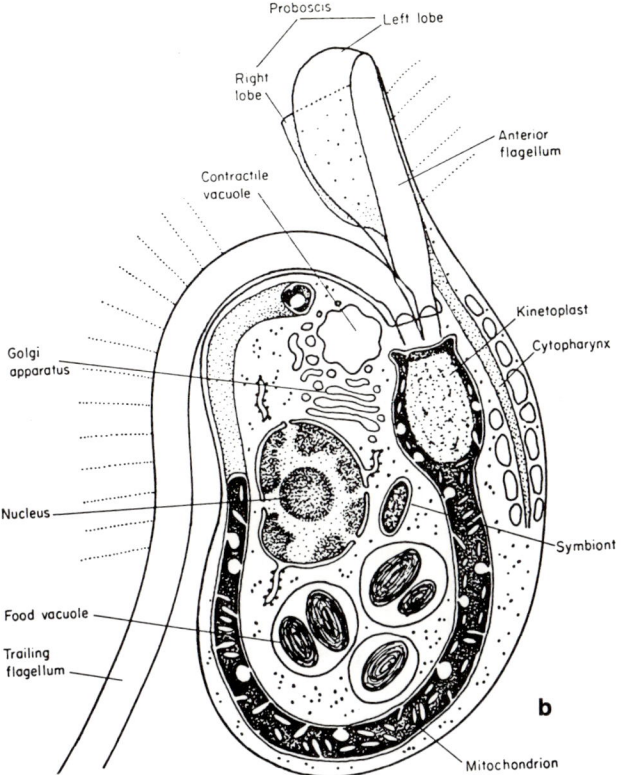

FIGURE 12.15.

The presence of trichocysts beneath the pellicle of the freshwater bodonine *Rhyncobodo* sp (Brugerolle 1985), and other features resembling those of the euglenoids *Entosiphon* and *Isonema*, suggest a phylogenetic affinity between these two groups of flagellates. Other features present in both groups are flagellar paraxial rods and long transition zone in the basal bodies, subpellicular microtubules, three independent fibers at the level of the flagellar pocket, similar fibers in the cytopharyngeal complex, and nuclei with closed mitosis (persistent nuclear membrane) and intranuclear spindle (Brugerolle 1985).

Comparative studies among free-living and ecto- or endoparasitic kinetoplastids (*Bodo* and *Cryptobia*) indicate some close microanatomical features among at least six species: (1) *Bodo* sp (close to *B caudatus*); (2) *Cryptobia branchialis*, attached to fish gills by the flagellum; (3) *Cryptobia dahlii*, living in fish intestines; (4) *Cryptobia intestinalis*, also living in fish intestines; (5) *Cryptobia helicis*, from seminal vesicles of snails; and (6) *Trypanoplasma borreli*, from the blood of *Cyprinus carpo* (the carp fish). All have two flagella inserted within a flagellar pocket, one anterior-directed, and the other recurrent. The kinetoplast and other cytoplasmic organelles resemble those found in other kinetoplastids, but several distinctive structures were also identified (Brugerolle et al 1979). The cytoskeletal system contains some unique microtubular fibers attached to or near the kinetosomes in the region of the flagellar pocket. These structures are not found in other Kinetoplastida. The recurrent flagellum is poorly attached to the ventral surface of the cell in *Bodo* sp, but is closely adherent in *Cryptobia* and *Trypanoplasma*. The trailing flagellum in *Cryptobia branchialis* adheres to the gill surface, but in *Cryptobia helicis* and *Trypanoplasma* it forms an undulating membrane with a fold in the surface of the cell. The use of the recurrent flagellum as an attachment organelle is also observed in *Cryptobia* spp, infecting snail spermatheca. It is also used for attachment of *Trypanosoma vivax* to the labrum in the proboscis of the tsetse fly (Current 1980; Vickerman 1973). In the free-living colonial kinetoplastid *Cephalothamnium*, several cells are united by attachment of their recurrent flagella to a stalk (Hitchen 1974). Thus, it appears that there is an evolutionary tendency within the group toward adaptation of the flagellum as an adhesive or prehensile organelle.

The fine structure of the parasitic trypanosomes resembles that of other Kinetoplastida (eg, Vickerman 1969; and with Preston 1976), especially with respect to the prominent kinetoplast and attached chondriome (Figure 12.16). There is, however, only one flagellum attached to the cell plasma membrane forming an undulating membrane. At the base of the flagellar pocket, a second barren kinetosome marks the site of the undeveloped flagellum. Microtubules lying immediately beneath the plasma membrane form a strengthening pellicular cytoskeleton, and expanded portions of the endoplasmic reticulum occur in close association with the attachment sites of the flagellum to the cell surface. A cytostome probably also occurs in many trypanosomes other than the free-living species (Brooker and Preston 1967). However, detailed accounts are wanting for some species. The form of the cytostome in *T cruzi* and *T conorhini*, during its infection of a vertebrate host was reported first by Milder and Deane (1969). It begins as a funnel-shaped depression lined by the plasma membrane and continues inward as a narrow cylinder penetrating deeply into the cytoplasm and culminating in an expanded cul-de-sac, apparently where the food vacuoles are formed. It appears that the cytostome opens into the flagellar pocket, either at the base or somewhat anteriorly. Delicate fibrils connect microtubules to the membrane lining the border of the cytostome, and this cytoskeletal system may provide the motive force to control opening and closing of the cytostomal aperture.

The highly organized and permanent cytostomal structure in some flagellates, including the remarkably complex structures of representatives such as *Peranema*, *Entosiphon*, *Bodo*, and *Rhynchomonas*, clearly distinguishes them from other protozoa. The flagellum, though of clear functional and taxonomic significance, is less unique, given its occurrence in opalinids and in the flagellated stages of some amoebae (eg, *Naegleria fowleri* and *Naegleria gruberi*).

Given increasing evidence that the flagellates and amoebae may have arisen from a common ancestral group, the acquisition of a permanent cytostomal apparatus during evolution is undoubtedly a significant advance in increasing the cytoplasmic differentiation and functional efficiency of the evolving flagellate

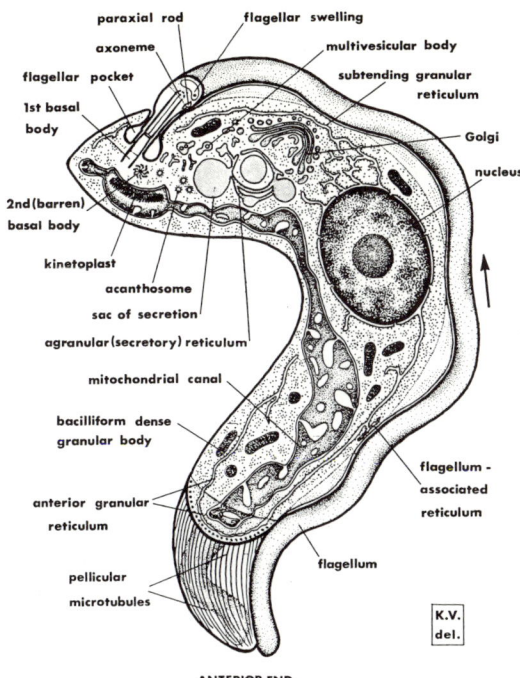

FIGURE 12.16. Fine structure of *Trypanosoma congolense* in the bloodstream phase. The emergence of the flagellum from the posterior end of the cell and its attachment to the cell surface are illustrated with a cutaway view of the internal cytoplasmic structures. The kinetoplast and mitochondrial canal extend along the side of the cell opposite to that bearing the attached flagellum. The cell surface is augmented by longitudinal rows of microtubules. From Vickerman (1969); with permission.

groups. Some of the other major microanatomical features separating flagellates and amoebae are presented in the next chapter.

Summative Perspective

The wide physiological diversity of flagellates, indicated by their range in habitat and functional complexity as shown in Figure 1.2, is reflected in the complexity and diversity of their fine structure. Although most have flagella during all or most of their life cycle, some (eg, *Dientamoeba*) are amoeboid and only the fine details of cytoplasmic organization confirm their position among the flagellates. Their fine structural diversity and range of specialized adaptations are reflected in the complex adaptations observed through electron microscopy. These include variations in flagellar organization, variations in surface structures ranging from scales to mineralized lorica, and diversity of specialized organelles among major groups as reflected in the presence of hydrogenosomes (eg, Trichomonadida), kinetoplasts (Kinetoplastida), and chondriomes varying from individual and scattered small mitochondria to unitary reticulated bodies ramifying throughout the cytoplasm. The lorica of some euglenoids is composed of protein that specifically binds manganese, and to a lesser extent, other metals. The chemical composition of this matrix is not fully determined. It is interesting to note, however, that the shells of some testate amoebae (eg, Ogden 1980) are also rich in manganese as determined by microanalysis. This common fine structural and mineral compositional feature raises intriguing questions about the possible similarity in protein composition or organization that may account for selective binding of manganese. Mineralized scales and walls are produced by flagellates (eg, siliceous tests in chrysomonad spores) and some sarcodinads, as presented in the next chapter. Comparative fine structural and chemical analyses are needed to clarify the mechanisms of these depositional processes, and to determine to what extent the mineralizing matrices are similar in these two groups of organisms (which are believed to be closely phylogenetically linked). Further research on the correlation between fine structure and physiology of the flagellates in relation to their mode of life is needed to elucidate the biological origins of the remarkable complexity of form and function that has evolved in these particular protozoa.

13
Fine Structure of Amoebae and Their Relatives

Comparative Fine Structural Features

Comparative views of the cytoplasm of a naked amoeba, planktonic foraminiferan, radiolarian, and heliozoan are presented in Figures 13.1 and 13.2. Some general descriptions of these four representative cells will be presented before proceeding to more detailed comparative analyses of the various taxonomic groups. The diversity of protozoa within the Sarcodina and related groups, and the broad comparative focus of the chapter necessitate a generalized description of fine structure rather than an intensive analysis of individual species. For purposes of clarity, however, reference to interspecies differences will be made to illustrate general principles of microanatomy and correlated function among the "amoeboid" organisms.

Naked Amoebae

Although variations in organellar form and distribution are to be expected among different species of naked amoebae (Gymnamoebia), and are significant as taxonomic criteria (eg, Page 1976a, 1978, 1980, 1983), the cytoplasmic organization of the marine amoeba presented in Figure 13.1a is fairly representative. The prominent nucleus, with endosome, is immediately surrounded by granular cytoplasm containing mitochondria with tubular cristae, endoplasmic reticulum, some Golgi bodies, food vacuoles and digestive vacuoles in varying stages of maturation, and vacuoles (surrounded by two or more membranes) containing apparently endosymbiotic bacteroids. The abundance of membrane-enclosed organelles, and the density of macromolecular cytoplasmic substituents may account for the optical density of this region (granuloplasm) when viewed with a light microscope. The periphery of the cytoplasm (hyaloplasm), particularly at the anterior presumptively outward-flowing edge, is rich in microfilaments forming an electron-lucent, fine-fibrillar zone immediately adjacent to the plasma membrane. These microfilamentous networks also extend into the occasional pseudopodia protruding from the surface of the cell. No membranous organelles

232   13. Fine Structure of Amoebae and Their Relatives

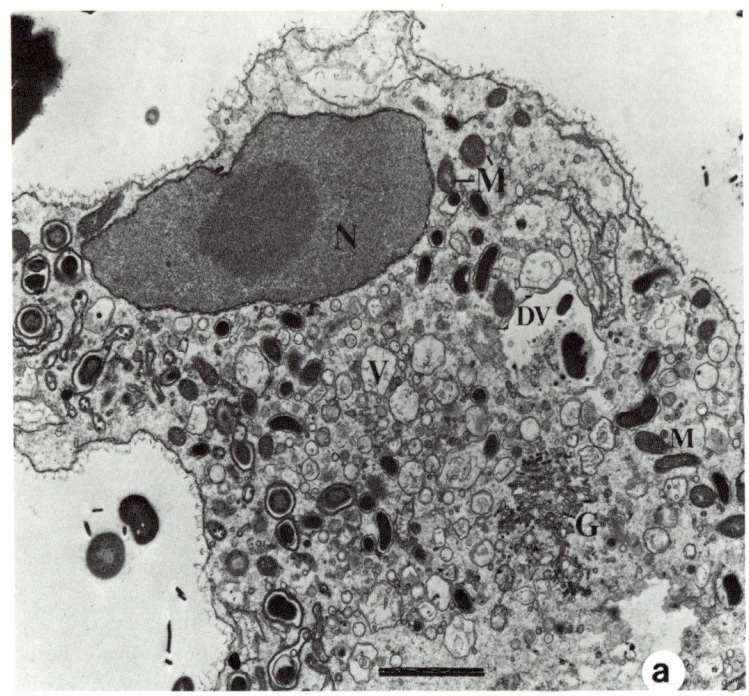

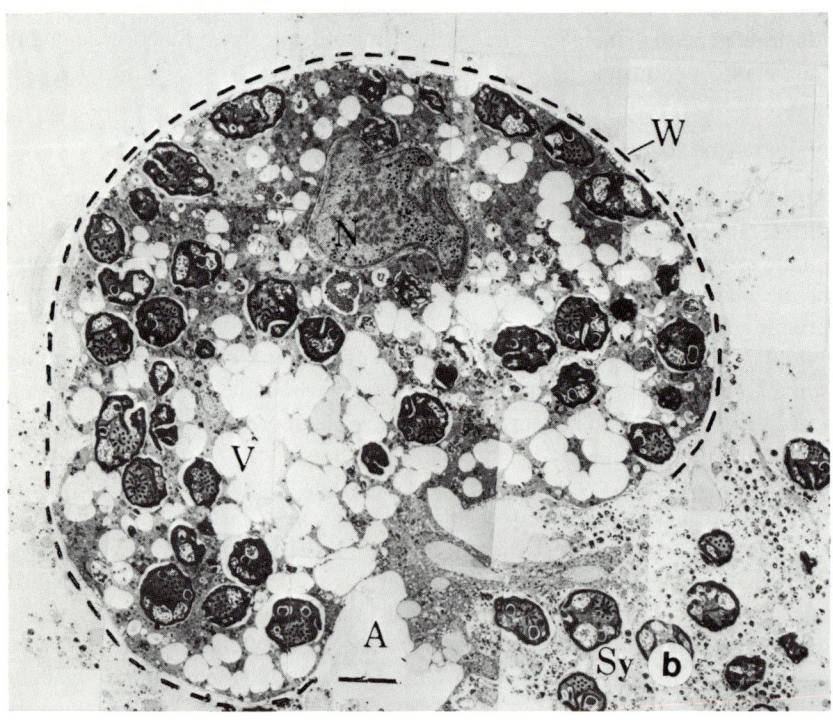

occur in this zone of microfilaments, which may account for its hyaline appearance when living cells are viewed with a light microscope. Although it is not apparent in this low magnification view, the amoeba also has a thin layer of scales coating the plasma membrane. These scales are apparently formed in Golgi-derived vesicles within the cytoplasm. As the vesicles mature and move toward the periphery, the organic scales are formed and eventually released by exocytosis on the surface of the cell. In this respect, the process is apparently very similar to that of scale-bearing flagellates as discussed in the preceding chapter. For purposes of comparison with the naked amoeba, a more detailed view of a testate amoeba (Figure 13.3) exhibits the theca (composed of mineralized scales surrounding the cell). The internal cytoplasm does not exhibit division into granuloplasm and hyaloplasm as found in some naked amoebae, but an array of fine pseudopodia typically emerge in a zone near the aperture that is more hyaline than the inner, more granular cytoplasm.

Foraminifera

The shell-bearing planktonic foraminiferan (Figuare 13.1b) exhibits a cytoplasmic organization similar to that of the granuloplasm of naked and testate amoebae. But, the presence of a calcite shell enclosing and constraining the cytoplasm, and the occurrence of long rhizopodia issuing first as a streamlike mass of cytoplasm protruding from the shell aperture confer a distinctly different profile. The nucleus is located in the cytoplasm within the shell, typically in one of the less peripheral chambers. Mitochondria and lipid storage bodies are abundant in the cytoplasm of the innermost chambers, while additional mitochondria, peroxisomes, digestive vacuoles, and Golgi bodies are increasingly apparent in sections through the more peripheral and hence larger chambers. When the planktonic foraminiferan has symbiotic algae, they are enclosed within cytoplasmic vacuoles occurring throughout the intrashell and extrashell cytoplasm. The presence of a perialgal membrane ensures that the foraminiferan can be isolated cytoplasmically from the foreign symbiotic cells, while also controlling interactions with them. Since the algal symbionts are enclosed within perialgal vacuoles, the foraminiferan can move the symbionts by cytoplasmic streaming into and out of

◄

FIGURE 13.1. Comparative views of the cytoplasm of a marine amoeba (a) and a planktonic foraminifer (b). The amoeba bears organic scales on the plasma membrane and contains abundant amounts of microfilaments in the peripheral cytoplasm beneath the plasma membrane. The shell of the planktonic foraminifer has been removed prior to sectioning and is indicated by a dashed line. Dinoflagellate symbionts (Sy) are enclosed within the intrashell and extrashell cytoplasm, compared to bacteriod symbionts within vacuoles of the amoeba. N = nucleus, M = mitochondria, G = Golgi body, and DV = digestive vacuole. Compare to Figure 13.2. Scales = 2 µm (a) and 20 µm (b). Figure (a) from Anderson (1977b); Figure (b) from Anderson (1984) reprinted with permission from *Marine Plankton Life Cycle Strategies*, copyright CRC Press, Inc., Boca Raton, Florida.

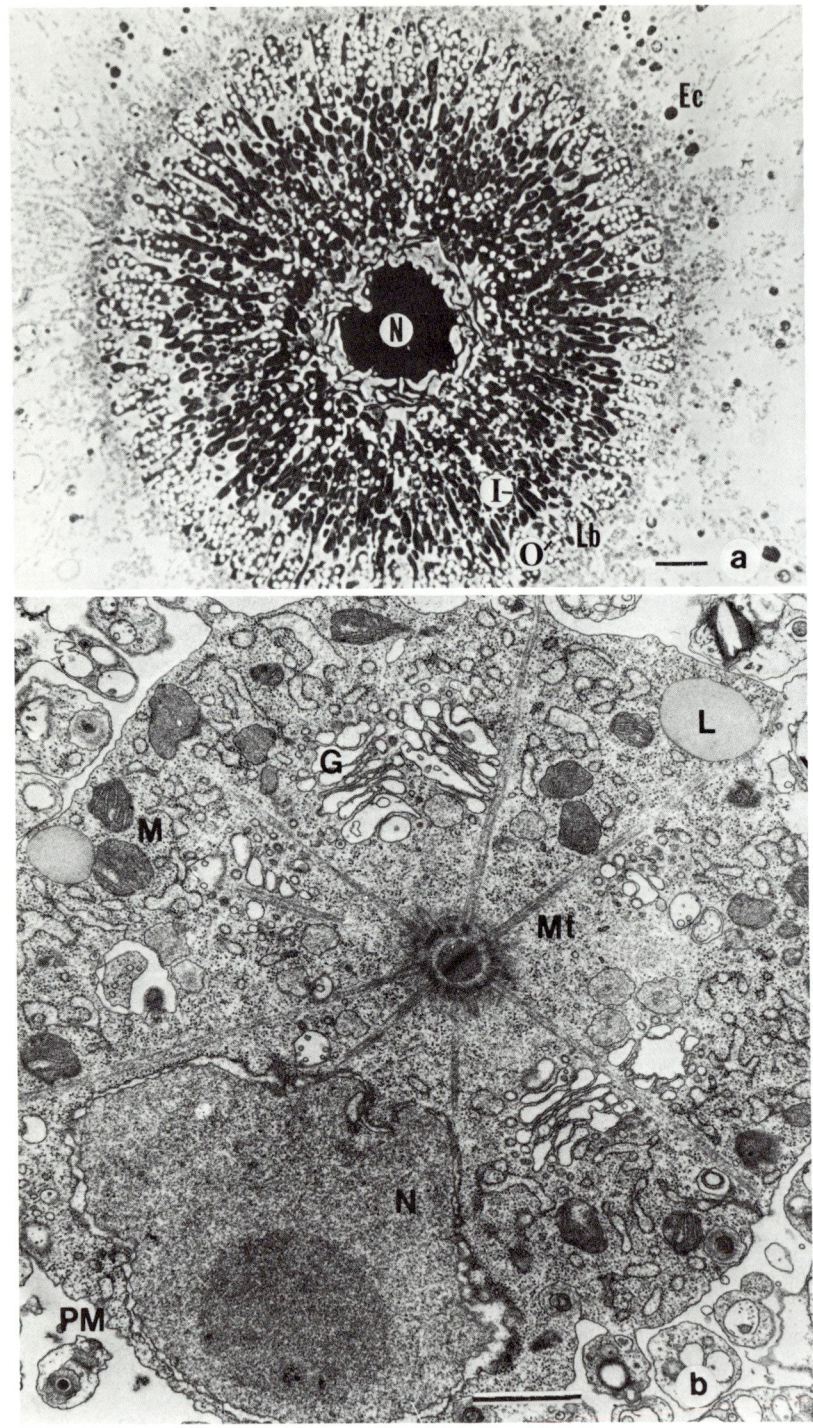

the chambers. Indeed, many symbionts are gathered into the cytoplasm at night and distributed outward along the spines and rhizopodia during the day. This behavior is also observed in some radiolaria, where vacuole-enclosed symbionts also occur.

Radiolaria

The organization of the cytoplasm in a radiolarian, with a spongiose spherical siliceous skeleton, is shown in Figure 13.2a. The cytoplasm is centrally dominated by a large and lobed nucleus surrounded by radially arranged lobes of cytoplasm. The cytoplasmic lobes are bipartite, with an inner slender segment containing dense cytoplasm and an outer broader segment bearing rows of vacuoles. The inner segments of the lobe are rich in mitochondria, endoplasmic reticulum, and Golgi bodies, while the outer segments contain only a fine granular deposit among the vacuoles. Outside the spherical central capsulum, a thin layer of frothy, weblike cytoplasm is penetrated by the stiffened cytoplasmic pseudopodia known as axopodia. Although a radial arrangement of the cytoplasm is common in many radiolaria of the Spumellarida (spherical cytoplasmic plan with a perforated organic capsular wall), some species have a much more compactly arranged mass of radial lobes. In these species, moreover, the subcellular organelles are distributed more homogeneously throughout the cytoplasm surrounding the central nucleus. Lipid droplets or other food storage particles may be arranged in radial arrays marking the geometry of the closely spaced radial lobes where they occur. The Nassellarida, with a prolate cell body and perforations only at one pole of the central capsule, do not typically exhibit lobes within the intracapsular cytoplasm, but possess a rather homogeneous groundplasm containing a nucleus, endoplasmic reticulum, mitochondria, peroxisomes, and a prominent cone of microtubules forming an internal cytoskeleton (Figure 13.4). The broad end of the cone of tubules is oriented toward the base of the capsule,

---

◄

FIGURE 13.2. Comparative views of the cytoplasm of a radiolarian (a) with spongiose siliceous skeleton and a marine heliozoan (b). The radial lobes of cytoplasm in the radiolarian central capsule bear an inner densely granular segment and an outer vacuolated segment. The central capsule is surrounded by an organic capsular wall perforated by thin strands of cytoplasm connecting intracapsulum with extracapsulum. The extracapsulum consists of frothy cytoplasmic masses and radiating axopodia holding algal symbionts. The heliozoan (*Heterophrys marina*) possesses a prominent centroplast containing a dense lamella surrounded by a spherical granular coat. Microtubules (Mt) forming the axonemes radiate from the centroplast and pass peripherally into the axopodia radiating around the cell. Compare with the freshwater heliozoan *Actinophrys sol* (Figure 4.5j). N = nucleus, M = mitochondria, L = lipid droplet, Lb = intracapsular lobes with inner (I) and outer (O) segments, and Ec = extracapsulum. Scales = 20 μm (a) and 1 μm (b); (a) from Anderson (1983); and (b) courtesy of Dr. C. F. Bardele, adapted from Bardele (1975).

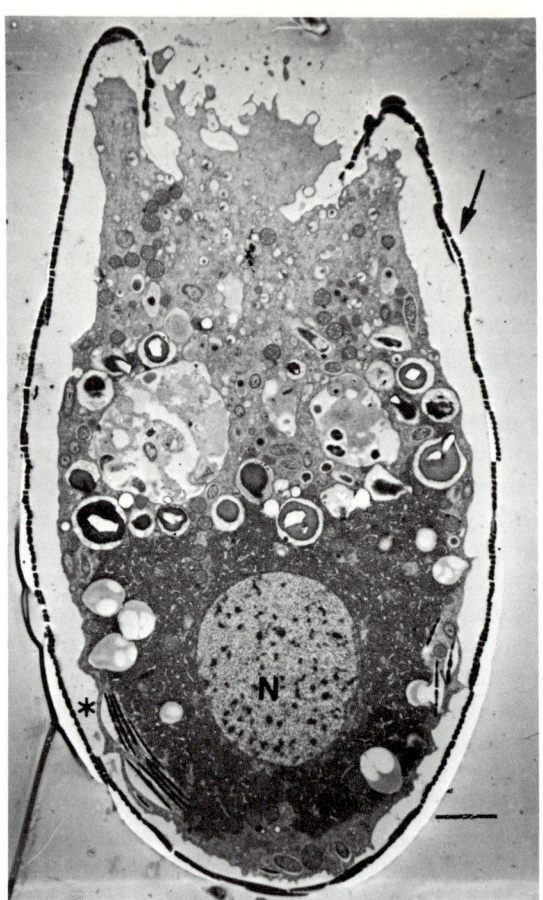

FIGURE 13.3. Fine structure of the cytoplasm of the thecate amoeba *Trinema lineare*. The thin test composed of imbricated siliceous scales (arrow) contains a wide pseudostome with protruding pseudopods composed of finely granular cytoplasm. A prominent nucleus (N), surrounded by dense cytoplasm, containing mitochondria and scale forming vesicles (asterisk), lies at the opposite pole of the cell. At the interface between the dense and more hyaline cytoplasm, there are numerous food vacuoles and digestive vacuoles. Scale = 2 μm. Courtesy of Dr. C. G. Ogden, British Museum of Natural History, London, England.

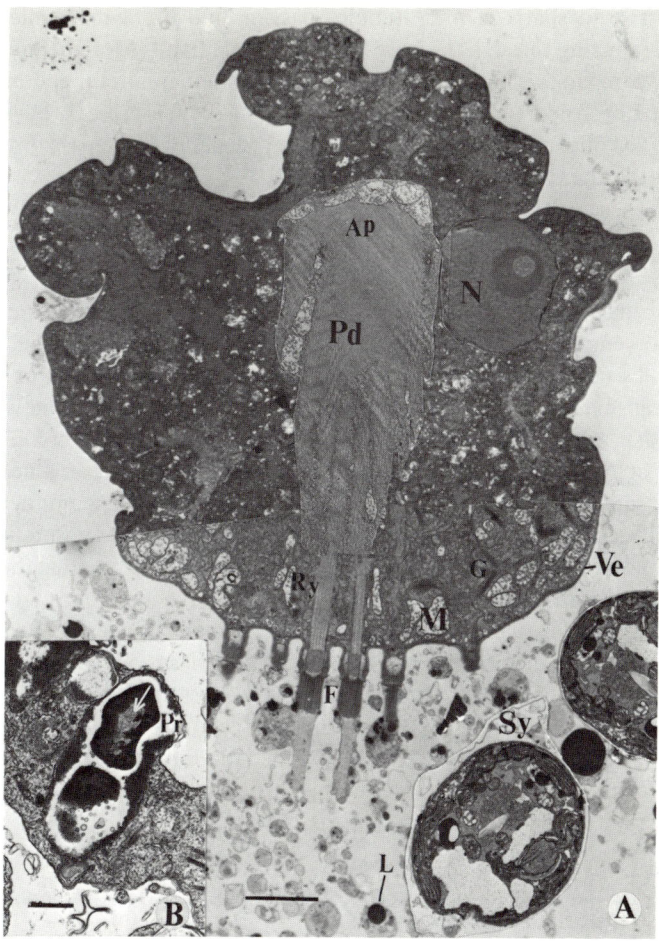

FIGURE 13.4. Fine structure of the cytoplasm of a nassellarian (A) (radiolarian) showing the nucleus (N) and massive microtubular organizing center (AP), with conical rays of microtubular axonemes projecting toward the base of the cell where they emerge through specialized collarlike pores known as fusules and form the array of axopodia extending into the environment. The total conical array is known as the podoconus and the pore field where the axopodia emerge is called the porochorus. Mitochondria (M), and Golgi (G) are distributed throughout the peripheral cytoplasm. Symbionts (dinoflagellates) are held in the peripheral cytoplasm outside the central capsule, and food vacuoles (B) containing bacteria or other small plankton prey are dispersed among the rhizopodia between the axopods. Scales = 0.4 μm (A) and 4 μm (B). From Anderson (1983); with permission.

and gives rise to shafts of microtubules passing into the cytoplasmic strands (fusules) projecting through the pores in the capsular wall. More detailed information on comparative fine structure of amoebae, Foraminifera, and radiolaria is presented by Anderson (1983). Algal symbionts, when present, have always been observed outside the central capsule. They are too large to pass through the small pores in the capsular wall.

Heliozoa

Some heliozoa also typically exhibit a radial arrangement of cytoplasmic components. The centrally located nucleus is surrounded by a sunburstlike array of microtubules (axonemes) that begin at the surface of the nucleus or near an extranuclear microtubular organizing center (MTOC), and project outward into the axopodia. Additional details of the comparative fine structure of MTOCs are presented below in the section on cytoskeletal structures. There is no organic central capsule with fusules as observed in radiolaria. This is a major taxonomic discriminating feature. In some species belonging to the order Desmothoracida, there is, however, an organic or mineralized porous shell surrounding the central cell body. Other organelles include mitochondria, Golgi bodies, and endoplasmic reticulum as found in many eukaryotic protista. In many species, large digestive vacuoles occur near the periphery of the central cytoplasm, where food particles are drawn into phagosomes preliminary to their conversion to digestive vacuoles. The axopodia possess specialized extrusomes for capture of prey and production of mucous on the surface of the cell, as described more fully below. One or more prominent contractile vacuoles occur in the periphery of the central cytoplasmic mass of many freshwater dwelling heliozoans.

Major Organelles

Membranous Organelles

The major characteristics of the fine structure of some cytoplasmic organelles including mitochondria, peroxisomes, Golgi body, and endoplasmic reticulum are presented for a foraminiferan in Figure 11.1a, and in general perspective for an amoeba in Figure 13.1a. The mitochondria of amoebae often contain tubular, anastomosing cristae, although some have discoid cristae. In scale-secreting species of amoebae, the Golgi apparatus may be substantially enlarged and give rise to numerous peripheral vesicles containing scales in varying stages of development. Typically, the scale-containing vesicles become larger and bear more mature scales near the periphery of the cytoplasm. Fusion of the Golgi vesicles with the plasma membrane and exocytosis releases the scales on the surface. This usually produces a uniform, thin layer of scales covering the surface of the cell (Fig. 11.2b). In general, the Golgi bodies are found in the vicinity of the nucleus, but may be distributed peripherally, especially during secretory activity.

Nuclear Fine Structure

The vegetative cells of many amoebae and related organisms contain but one nucleus, which divides to produce two or more daughter nuclei at the time of reproduction. In some colonial radiolaria, however, the nucleus divides in advance of reproduction, filling the cytoplasm with closely spaced nuclei containing cordlike masses of chromosomes (Anderson 1983). The benthic foraminiferan *Allogromia* sp (NF) (Lee and Pierce 1963) contains several small nuclei during most of its life cycle. The heliozoa are mononucleate during most of the vegetative phase of their life cycle with the exception of five genera: *Actinosphaerium, Echinosphaerium, Camptonema, Gymnosphaera,* and *Actinocoryne* (Febvre-Chevalier 1985).

The nuclear fine structure of amoebae, Foraminifera, radiolaria, and heliozoa exhibits, on the whole, a typical eukaryotic organization with a double membrane envelope surrounding the nucleoplasm containing one or more nucleoli, and usually, dispersed strands of chromatin (euchromatin). In some species, clumps of heterochromatin occur within the nucleoplasm or at its periphery. Helices of ribonucleoprotein have been reported in the nucleoplasm of *A proteus*, and at least subportions of the helices pass into the cytoplasm during interphase (Stevens and Prescott 1971; Daniels 1973). In the parasitic amoeba *Entamoeba histolytica*, clumps of chromatin (peripheral chromatin) rich in RNA surround the periphery of the nucleoplasm. There is a prominent centrally located centrosome. Nuclear vesicles may occur within or near this peripheral chromatin layer (Albach and Booden 1978). A similar ring of peripheral chromatin is observed in *E invadens*, but there is a "network-like" structure surrounding the nuclear envelope (Siddiqui and Rudzinska 1965) that is not present in *E histolytica*.

A honeycomblike coat or fibrous lamina occurs on the inner membrane of the nuclear envelope in some "naked" amoebae (Flickinger 1974; Daniels 1973) (eg, *Amoeba proteus, A dubia, A discoides,* and *A amazonas*), benthic Foraminifera (reviewed by Anderson and Bé 1978), testate amoebae (eg, *Lesquereusia*, Harrison et al 1976), and heliozoa (*Hedraiophyrys* and *Sticholonche*) as reviewed by Febvre-Chevalier (1985). The organization of the chromosomes and nucleoli is variable. A large centrally-located spheroidal nucleolus (centrosome) is observed in some amoebae (eg, Bowers and Korn 1969; Daniels 1973; Page 1976a), planktonic Foraminifera, and radiolaria (Figure 13.4) (Anderson 1983). In at least one planktonic foraminiferan (*Globigerinoides sacculifera*), the nucleoplasm is divided into a dense mass of nucleolar material and a lighter region containing dispersed chromatin (Anderson and Bé 1976, 1978). The nucleus of the large amoeba (*Pelomyxa palustris*) contains a large nucleolus and small clumps of nucleolar material (micronucleoli) dispersed near the periphery of the nucleoplasm. Peripheral nucleoli of larger size also occur in the nucleus of *P carolinensis*, while in *A proteus*, nucleoli are scattered throughout the nucleoplasm (Daniels 1973). In many eukaryotic cells, including the amoebae and their relatives, the outer edge of the nuclear envelope may be associated with clumps of nucleoprotein appearing opposite each of the membrane pores. Although the chemical composition of these clumps is not determined, they may be strands of mRNA passing into the cytoplasm. Their occurrence can be sporadic, depending on the physiological state of the cell. In the benthic foraminiferan (*Ovamina opaca*), a complex membranous system of cisternae, vesicles, and tubules surrounds the nuclear envelope and extends peripherally to connect with the endoplasmic reticulum (Dahlgren 1967; Anderson and Bé 1978). However, in many of the representative organisms included here there are no specialized cytoplasmic organelles immediately surrounding the nuclear membrane other than occasional ribosomes located on the outer membrane of the nuclear envelope.

Cytoskeletal Structures

There are four types of cytoskeletal and tensile structures typically found in eukaryotic cells: (1) actin-containing microfilaments (ca 5 nm diameter); (2) a heterogeneous group of intermediate filaments (8 to 14 nm), so named since they are intermediate in size between microfilaments and microtubules; (3) microtubules, typically more rigid scaffolding, rodlike structures (ca 30 nm diameter); and (4) spasminlike filaments in some protists. The microfilaments form tensile elements within the cell, but when cross-linked laterally or bound into crystalline array, they may become quite stiff. The general properties of cytoskeletal and supportive structures are discussed by Bereiter-Hahn (1987).

In general, fine structural studies of amoebae have documented the presence of microfilaments and intermediate type filaments (range 4 to 10 nm) in the groundplasm of naked amoebae (eg, in *A proteus*, and *P carolinenesis* as reviewed by Daniels 1973) and testate amoebae. The presence of substantial amounts of microfilaments beneath the plasma membrane in the hyaloplasm of some species suggests that they serve as cortical support and may be involved in locomotion as described in the chapter on motility. Actin fibrils may also traverse thick cytoplasmic strands, giving additional stability as observed in *Physarum* and *Amoeba* (eg, Bereiter-Hahn 1987).

Microtubules appear to be transitory, and assembled as required to provide support structures in the cytoplasm of Foraminifera. They occur as isolated shafts of microtubules within the intrashell cytoplasm, and particularly as parallel arrays of stiffening components in the central cytoplasm of filopodia (eg, McGee-Russell 1974 and Anderson and Bé 1978, p 137). These microtubules are apparently assembled and disassembled as requirements for support and cytoplasmic dynamic activity change. Among the radiolaria and heliozoa, microtubules form a rather stable assemblage of intracellular scaffolding. In the Spumellarida (radiolaria with a spheroidal body plan), rays of parallel microtubules originate near or within an invagination of the nuclear membrane and radiate outward through the cytoplasm, continuing into peripheral axopodia as the axoneme. The arrangement of the microtubules within these axonematic bundles (Figure 13.5) and their point of origin near the nucleus are of taxonomic significance (eg, Cachon and Cachon 1971; Anderson 1983). Among the Nassellarida (Figure

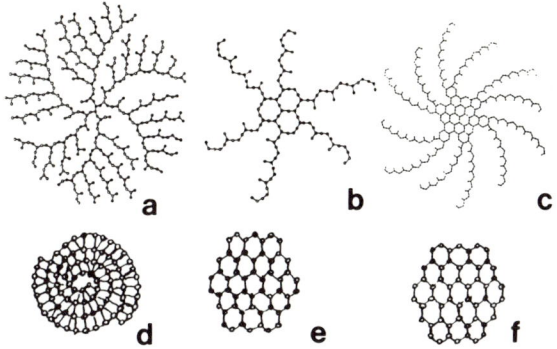

FIGURE 13.5. Axonemal pattern in the axopods of radiolaria (a–c) and heliozoa (d–f). (a) Spumellarida, (b–c) Nassellarida, (d) *Actinophrys* and *actinosphaerium*, (e) *Gymnosphaera*, *Actinocoryne*, and *Hedraiophrys* (axoneme without inner differentiation), (f) *Sticholonche* (with axonemes articulated in pits in the nuclear envelope). (a–c) From Cachon and Cachon (1971), (d–f) from Febvre-Chevalier (1985); with permission.

13.4), an apparently stable conical array of microtubules forms the podoconus. A microtubule organizing center (MTOC) is at the apex of the cone, and shafts of microtubule bundles project down toward the base of the central capsule, where they emerge and pass outward as the axonemes of the stout axopodia. The function of the podoconus is uncertain, but it is interesting that it is oriented with its axis parallel to the long axis of the cell. This suggests that it may serve a cytoplasmic organizing role in determining the shape of the cell. It is clear that such a conical structure has the potential to add considerable strength to the cell, particularly in anchoring the central cytoplasm to the base of the prolate central capsule (Anderson 1983). The axonemes of the axopodia are not permanent structures. They are disassembled when the axopodia are retracted, and only the microfilaments are observed in ultrathin sections. During extension of the axopodia, microtubules are assembled from the tubulin precursor monomers and form the central stabilizing shaft of the extending axopod.

The heliozoa also possess labile microtubule shafts in their axopodia, but these arise from radiating axonemes originating near or upon the nuclear membrane at MTOCs (Febvre-Chevalier 1985). Among the actinophryids (*Actinophrys* and *Sticholonche*), a simple or stratified dense plaque within the nuclear envelope or close to it gives rise to the microtubular bundles. Centroplasts, by contrast, are spheroidal microtubular organizing centers. There are two types. One contains a tripartite, centrally located discoid, 0.2 to 0.4 μm in diameter and 0.1 μm thick, exemplifying a classic form of the centroplast (Bardele 1975). The other, called an axoplast (Febvre-Chevalier 1972), is devoid of central differentiation. The appearance of the internal structure of the centroplast varies with kind of fixation and according to taxonomic group. In *Heterophrys elati*, *H marina*, and *Acanthocystis turfacea*, a tripartite central disc sandwiched between two hemispheres of flocculent or dense material is surrounded by a cortex where the bases of the axonemes are inserted. The axoplast of three genera of heliozoa (*Hedraiophrys*, *Gymnosphaera*, and *Actinocoryne*) lacks a central disc. Only a central sphere of microfilamentous material is present, surrounded by a shell of similar composition bearing the bases of the axonemes (Febvre-Chevalier 1985).

The microtubules in axonemes of radiolaria and heliozoa are organized in regular arrays (Figure 13.5). Radiolarian axonemes in cross section display radially arranged, usually curved sheets of microtubules extending from a centrally located axis, or a set of hexagonally shaped hollow tubes with three microtubules on each side (eg, Cachon and Cachon 1971; Anderson 1983, p 80). Four patterns of microtubular organization occur in heliozoa: (1) double interlocking coils formed by two sheets of microtubules spirally wound around each other as in some actinophryid heliozoa; (2) a simple triangle of microtubules in the helioflagellate *Ciliophrys*; (3) irregular hexagons and triangles packed together (eg, the centrohelids, *Heterophrys*, and *Raphidiophrys*); and (4) irregular hexagons forming an imperfect floor tile pattern as in the centrohelids lacking a tripartite body in their axoplast (eg, *Hedraiophrys* and *Gymnosphaera*). In addition to the differences in axopodial microtubular organization between radiolaria and heliozoa, the organization of the cytoplasm at the point of attachment to the central cell body is different. In radiolaria there is a specialized collar of cytoplasm called a fusule passing through a pore in the wall of the central capsule. Typically, a electron-dense deposit of filamentous or apparently homogeneous matter occurs within the fusule cytoplasm immediately distal to the point where the axopod joins the central cell cytoplasm. In some Nassellaria, a narrow sheath of endoplasmic reticulum surrounds the periphery of the fusule, and provides continuity between the intracapsular and extracapsular endocytotic membrane systems (Anderson 1977a).

Cytoplasmic Specialized Structures

Endosymbionts

Endocytotic symbionts enclosed in a vacuole are commonly observed in a wide range of amoeboid organisms. Bacteroids enclosed within vacuoles are often observed in freshwater and marine amoebae. Algae, including dinoflagellates, prymensiomonads, prasinomonads, and some red algae, have been observed in a few amoebae, but more frequently in Foraminifera and radiolaria. An algal symbiont within a perialgal vacuole is shown in Figure 11.1c. The vacuolar membrane is apparently altered, perhaps by inclusion of a specialized protein that prevents fusion with lysosomes, thus protecting the enclosed symbiont from digestion. In the large amoeba *Pelomyxa palustris*, there are neither mitochondria nor Golgi bodies, thus making it quite different from most other eukaryotic cells. The presence of bacteroids of two kinds (narrow bacteria usually gram positive, and broad ones usually gram negative) may serve the function of mitochondria, but this is not certain. Glycogen bodies (5 to 20 μm diameter spheroidal masses) increase in abundance after feeding and are often surrounded by bacteroid symbionts. *Entamoeba histolytica* also lacks mitochondria, but this is understandable given its anaerobic habitat (human intestine). Golgi bodies have been observed, but apparently rough endoplasmic reticulum has not been seen. The ribosomes occur in the cytoplasm as helical arrays (Albach and Booden 1978). Glycogen occurs as both alpha (rosette), and beta (individual) granules. By contrast, the large amoeba *Pelomyxa carolinensis* contains both Golgi bodies and numerous, up to 300,000, spheroidal mitochondria (1 to 2 μm diameter). The multinucleate species (*Pelomyxa illinoisensis*) is smaller than the other species of *Pelomyxa*, and is replete with mitochondria and Golgi bodies. Crystals are commonly observed in vacuoles within its cytoplasm, and these may be nitrogenous waste products (triuret) as also observed in lesser concentrations in *P carolinensis* and *A proteus*. *Pelomyxa palustris* does not contain membrane enclosed crystals, but there are numerous mineral and sand grains suspended within the cytoplasm of most specimens. Daniels (1973) reports the occurrence of smooth-surfaced spheroidal bodies about 350 nm in diameter in some specimens, and numerous smaller 100 nm electron-opaque spheroids (each encased within an irregular vesicle) in those cells with inward serrations along the nuclear envelope. Species in the genus *Paramoeba* and the parasitic genus *Janickina* possess a juxtanuclear body known as a parasome or *Nebenkörper* (Grell and Benwitz 1970; Page 1983). This is a DNA-containing body, containing a middlepiece which gives a positive Feulgen reaction, and two peripheral (end) pieces. Hollande (1980) concluded that the parasome in *Janickina* at least may be a kinetoplastid symbiont. He named it *Perkinsiella amoebae*. A single parasome per nucleus is typical in many species, but supernumerary parasomes are very common. This is consistent with a symbiont role for the parasomes.

Algal symbionts enclosed in vacuoles are commonly observed in many planktonic and some benthic Foraminiferan species. They are typically, although not exclusively, dinoflagellate symbionts in many spinose planktonic Foraminifera. Dinoflagellate, red algae, and diatom symbionts have been observed in some benthic Foraminifera (Lee 1980a).

Other Membrane-bound Organelles

The cytoplasm of planktonic Foraminifera is distinguished from benthic forms by the presence of vacuoles (Ca 8 to 20 μm) containing a puffy of filamentous organic matter called the "fibrillar system" (Anderson and Bé 1976). The function of these organelles is not known, but they may aid flotation as a buoyancy mechanism. The heavy calcareous shell is clearly negatively buoyant, requiring some compensatory flotation device. The origin of the fibrillar system in the cell is uncertain. But, various stages of formation have been observed, commencing with small vesicles containing electron opaque tubules and progressing toward larger vacuoles containing a fibrillar material apparently derived from the tubular inclusions. In reproductive cells, membranous whorls and annulate lamellae forming stacks of membrane-bound cisternae within the cytoplasm are apparently reserve membranes used to enclose the numerous daughter nuclei that are formed during gamete production (Spindler and Hemleben 1982). These bodies, especially observed in some planktonic Foraminifera, are particularly abundant before daughter nuclei are formed, but decline in abundance after proliferation of the nuclei. A variety of membrane-bound bodies specific to a given species have been reported in some benthic Foraminifera. For example, Hedley and Wakefield (1969) observed spindle-shaped "secretion granules" (2 to 3 µm in length) in the peripheral cytoplasm of *Gromia oviformis*. These bodies situated near the oral region of the shell consist of three regions: (1) a thin rod portion about 0.2 µm in width that extends the length of the granule; (2) concentrically arranged lamellae composed of closely packed granules; and (3) an electron-dense reticular body containing small granules 10 to 15 nm in diameter. The function of the body is not known, but based on the similarity of the internal matrix to surrounding mucous material, Hedley and Wakefield suggested that these granules may be the source of the fibrillar mucous coat surrounding the cell. Membrane-bound vesicles containing either fibrillar material or dense, beadlike aggregates are also observed in the rhizopodia of radiolaria containing a gelatinous coat (Anderson 1983). In some regions, these vesicles appear to have opened releasing fibrillar, apparently gelatinous material into the surrounding space.

Specialized granules are also observed along the periphery of axopodia in heliozoa. These probably account for the granular appearance of the axopodia when viewed by light microscopy. They have been identified as varieties of extrusomes, either secreting mucous (mucocysts) or ejecting barbed threads (kinetocysts) that facilitate prey capture. Their fine structure has been carefully elucidated (eg, Bardele 1970b, 1975). Kinetocysts are oval organelles 0.4×0.3 µm found in centrohelids with a centroplast (*Heterophrys*, *Raphidophrys*, and *Acanthocystis*, and in the desmothoracan heliozoan *Clathrulina*). The membrane-bound organelles probably originate from the Golgi body, and before release are fixed to the membrane of the cell by a proteinaceous body. They contain a two-part dense body consisting of an outer barblike pointed portion and an inner spherical segment. This dense body is surrounded by a less dense matrix with radial and concentric striations. A kind of electron-dense granule has also been observed in Actinophyrida and in the centrohelid heliozoans with an axoplast (Febvre-Chevalier 1985). Freeze fracture examination shows several masses of granules with rectangular arrays of six to 100 particles.

Extracellular Surface Structures

A filamentous coat of proteins, probably derived from Golgi-produced vesicles, is commonly observed on the surface of many naked amoebae including *A proteus*, *A discoides*, and *A amazonas*. However, it is not present in *A dubia* (Flickinger 1974). It may serve to anchor small particles to be ingested, and structurally reinforce the plasma membrane. The organic coat of the rhizopod-bearing protozoan, *Gromia oviformis*, forms a spheroidal chamber with a thickened collar known as an oral capsule (Figure 13.6). This wall is composed of

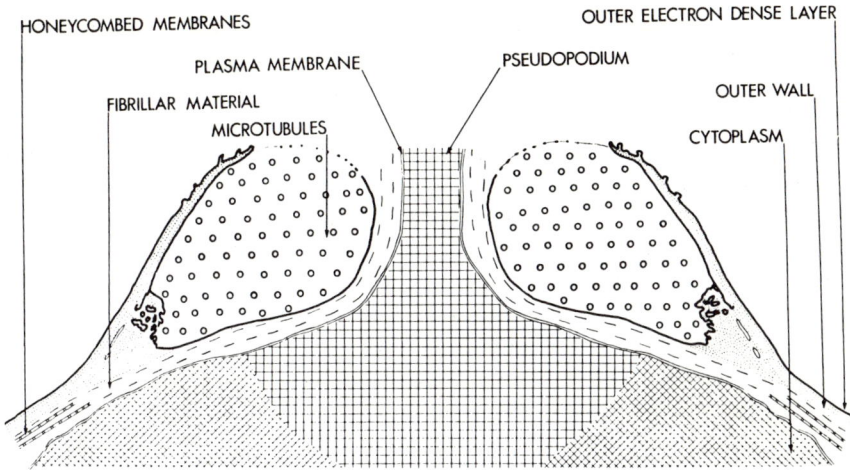

FIGURE 13.6. Diagram of the fine structure of the rhizopod-bearing *Gromia oviformis* showing the oral region of the organic capsule surrounding the cell. Pseudopodia protrude through the oral aperture, surrounded by a massive collar bearing microtubules. The capsular wall is composed of an outer electron dense layer and underlaid by a layer of honeycombed membranes. A thin layer of fibrillar material separates the cytoplasm from the organic wall. From Hedley and Wakefield (1969); with permission.

several layers (Hedley and Wakefield 1969). The outermost layer is a thin, electon-dense deposit overlying a finely stratified outer wall that encloses a complex, honeycombed organic layer. The innermost lamina consists of a fibrillar material deposited immediately adjacent to the plasma membrane. Internal microtubules strengthen the lip of the collarlike oral capsule where the pseudopodia emerge. The honeycombed organic membrane is formed by a remarkably regular array of hexagonally-arranged cylindrical structures (10 to 20 nm diameter), each linked by a narrow septum. Although it is not known how the honeycombed organic layer is produced, it is clear that it can substantially strengthen the wall while remaining lightweight.

Organic or mineralized scales are secreted on the surface of some amoebae (eg, Grell and Benwitz 1966; Anderson 1976; Page 1983; Yamaoka et al 1984) and testate amoebae (eg, Hedley and Ogden 1973, 1974; Ogden 1979, 1980; Ogden and Hedley 1980). These appear to derive from vesicles produced by the Golgi body (Figure 13.3) or from cisternae of the endoplasmic reticulum. The scales are released on the outer surface of the cell by exocytosis. Silicified scales or spicules, observed on the surface of heliozoa, are produced by endoplasmic membranous vesicles or from invaginations of the cell membrane (Bardele 1975; Patterson and Dürrschmidt 1986). The deposition process can be carefully monitored in the siliceous scale-bearing species *Acanthocystis erinaceoides*, by growing the individuals in silicon-free medium to obtain naked cells. Upon replacing the cells in silicon-rich medium, scale formation immediately commences. The scales are deposited within vesicles at the periphery of the cell, and first appear as a spongy, possibly organic state before silicification and eventual release by fusion of the vesicle with the surface membrane of the cell (Patterson and Dürrschmidt 1986).

In some testate amoebae, as exemplified by the genus *Netzelia*, the shell is composed of silicified particles (idiosomes) and intervening organic plaquelike bodies called sieve-plate granules (Netzel 1983). These granules bear slitlike perforations in the center (Figure 13.7), and are surrounded by an apparently adhesive rim that cements the idiosomes and intervening granules together into a spherical, single-layered wall (Figure 13.8). The wall is deposited during formation of a daughter cell at asexual reproduction. The mother cell produces a bulge of cytoplasm through the shell aperture which grows into a shape approximating that of the daughter cell. As the bulge enlarges, siliceous idiosomes contained within vacuolar spaces in the cytoplasm are distributed upon the surface of the bulge. Subsequently, organic sieve-plate granules appear in the cytoplasm, and are probably derived from the Golgi apparatus (Netzel 1983). They are transported to the plasma membrane and inserted between the idiosomes, which become oriented with the broader surfaces parallel to the surface of the bulge. The cement of the granules apparently fuses with the idiosomes and hardens to form the test wall. After deposition of the granules, the cytoplasmic bulge is withdrawn. The mother nucleus divides and one nucleus is carried into the daughter cell cytoplasm. The two cells separate upon completion of reproduc-

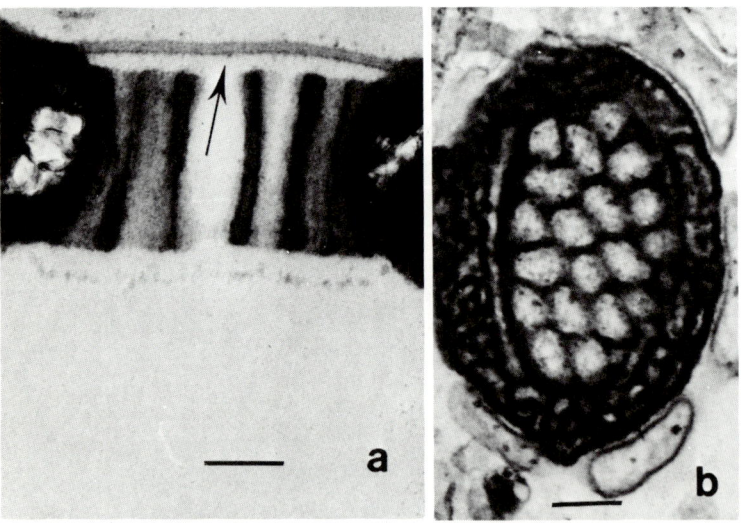

FIGURE 13.7. Adhesive fenestrated plaques cementing siliceous granules within the test of the testate amoeba *Netzelia*. A longitudinal section (a) exhibits the slitlike organization of the grillwork shown more clearly in cross-section (b). A thin organic lamella (arrow) closes the pore of the plaque on the inner surface. Scales = 0.1 µm.

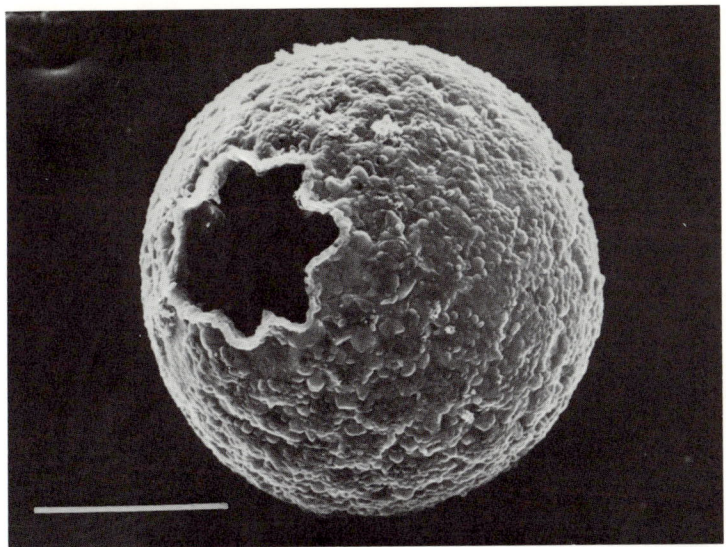

FIGURE 13.8. Scanning electron microscopic view of the test of *Netzelia tuberculata* with a lobate orifice (pseudostome) and granular wall composed of siliceous particles cemented together by numerous organic plaques (Figure 13.7). Scale = 40 μm.

tion, each possessing a completed shell. Although the shapes of the mother and daughter shell are very similar, the individual idiosomes may vary considerably. This is particularly likely in the genus *Netzelia*, as the idiosomes are of at least two kinds: (1) those synthesized within vesicles in the cytoplasm, and (2) those composed of a central grain of foreign matter that has been coated in the cell with a veneer of silica (Anderson 1987). In some cases, organic particles such as starch grains collected from algal prey can be used as a kernel for deposition of the silica veneer. Hollow idiosomes observed in ultrathin sections of the wall may represent silica coats remaining after the inner organic matter has been degraded.

The shells of Foraminifera may be organic, agglutinated detritus fused within an organic or mineralized cement, or composed of calcium carbonate in a variety of chemical or crystallographic forms. The organic wall of the monothalamus (single-chambered) *Allogromia* spp is typically an amorphous, somewhat flexible coat of fibrillar material with a collarlike aperture where the rhizopodia emerge. Agglutinated shells are composed of a variety of mineralized products gathered by the Foraminiferan rhizopodia and cemented together within an organic matrix. Diatom frustules and mineral grains of suitable size can be clearly distinguished within the organic cement that is deposited between them and partially coats their surfaces. The porcellaneous shelled species, such as *Spiroloculina* (representing the *Miliolina* suborder), produce a test composed of high magnesian calcite needles. The needles, enclosed within a thin organic coat, are synthesized within vacuoles in the cytoplasm and deposited at the periphery during wall formation. During new chamber addition, a thin outer organic layer is deposited first (Figure 13.9), forming the outer organic lining (OOL). A layer of needles oriented with their long axes parallel to the surface of the OOL is deposited by action of the streaming peripheral cytoplasm. Immediately subsequent to the deposition of the outer layer of needles, an irregular heap of scattered needles, some grouped into clumps with their long axes parallel to one another, is deposited in the inner part of the wall. This layer is sealed by an inner layer of regularly arranged needles, forming the thin and smooth inner surface of the chamber wall. An inner organic lining (IOL) is deposited immediately next to the cytoplasmic surface (Hemleben et al 1986). The smooth outer surface layer and the underlying mass of irregularly disposed needles produces the porcellaneous appearance of the shell when observed by surface illumination.

Some benthic and many planktonic species of Foraminifera produce a bilamellar calcite wall (Figure 13.9b). This is composed of two calcite layers separated by an inner organic deposit known as the primary organic membrane (POM). During new wall deposition, the POM is formed first, usually in conjunction with a thin osmiophilic layer of cytoplasm that constitutes the mold or anlage for the chamber wall. This structure is produced by an array of rhizopodia that expand outward from the surface of the shell near the aperture during the onset of chamber formation. Thin patches of calcite are deposited on the outer and inner surfaces of the POM, forming the first stages of the bilamellar wall construction. As the patches of calcite grow laterally, they fuse to form a thin layer of calcite enclosed by an organic lining. Additional calcite is added on both surfaces of the wall, but typically more heavily on the distal side. This results in considerable thickening of the wall, varying in amount among species. The surface texture is species-specific and can be ridged, rugose, or smooth. At certain places in the developing wall, strands of cytoplasm traversing the POM mark sites where pores will develop due to lack of calcium carbonate deposition in these regions. Thus, a depression is formed in the wall. A multilayer lining, representing the successive layers of organic linings deposited during each successive layer of calcification, coats the surface of the pore and forms a perforated pore plate across the base of the pore. Calcitic spines in spinose species are anchored within the wall of the chamber and grow inside cytoplasmic sheaths supported by the outward-radiating rhizopodia. Upon completion of a new chamber, the cytoplasmic mold is withdrawn and rhizopodia fill the new chamber. After feeding, cytoplasm is synthesized and the chamber becomes filled with increasingly denser deposits of cytoplasm. Each new chamber possesses one or more apertures penetrated by the major strands of cytoplasm forming the rhizopodial mass surrounding the organism. Not all Foraminifera, however, have shells with a large oral aperture. Some of the benthic species (eg, *Heterostegina*, *Elphidium*, *Sorites*, *Peneropolis*, and others) may possess many smaller apertures either at the rim of the shell, along the sutures, or in the wall closing the apertural region. In *Elphidium*, for example, the pores are distributed along sutures and the final chamber is closed. The pores, moreover, are remarkably sievelike. These lead into internal canals within the wall that open into the main chambers of the

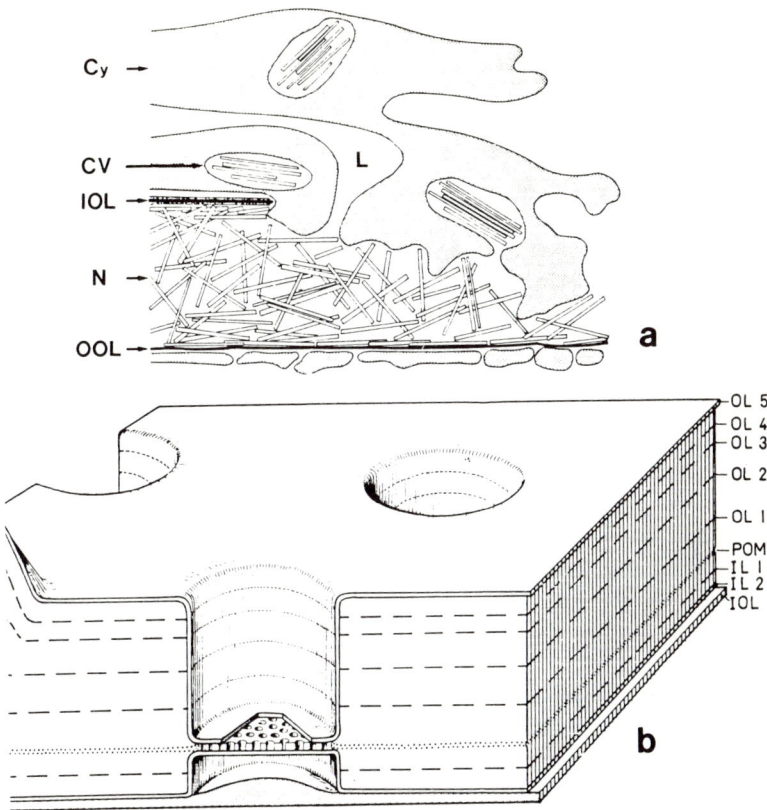

FIGURE 13.9. Calcite wall construction in Foraminifera (a) *Calcituba polymorpha*, a benthic foraminifer, and (b) *Globorotalia menardii*, a nonspinose planktonic foraminifer. *C polymorpha* deposits a wall rich in magnesian calcite. Needles of the mineral occur first within cytoplasmic vacuoles (CV) separated by a lacunar system (L), and are deposited by exocytosis. The first deposited needles are arranged in a parallel fashion upon the inner surface of an outer organic layer (OOL), then randomly organized forming the increasing wall. This inner layer of loosely aggregated needles is covered on the cytoplasmic side by a final layer of parallel oriented needles lined by an organic lining (IOL). The bilamellar wall of many planktonic Foraminifera (b) consists of a primary organic membrane (POM) that serves as an initial calcifying layer and separates the inner calcite layers (IL) from the outer wall layers (OL). The POM is also the site of pore plate deposition containing fine pores within a recessed cylindrical canal within the shell wall. An inner organic lining (IOL) separates the calcite wall from the plasma membrane of the cell. (a) By Dr. W. Berthold, from Hemleben et al (1986), and (b) from Hemleben et al (1977); with permission.

test. Digestion in these organisms occurs outside the test. Diatoms are a common prey. The frustules are opened and the cytoplasmic contents removed. Diatom plastids are particularly selected by the organism and drawn into the test through the small pores. The size of the pores, however, excludes frustules and other mineral debris. thus, the Foraminiferan takes only usable food particles into the intrashell cytoplasm (J.J. Lee, personal communication). This remarkable organization of the shell, including the elaborate internal canal system (Hottinger 1978), permits efficient distribution of the rhizopodial system and rapid ingestion of prey particles (Hottinger and Dreher 1974; Hottinger and Leutenegger 1980; Lee and Hallock, 1987). In general, the shells of Foraminifera are clearly more than protective coverings. In addition to specialized pore structures to aid feeding and exchange with the environment, some of the larger, algal symbiont-bearing benthic species have remarkably sculptured shells with domelike regions to house the symbionts, and possibly increase their light-capturing capacity within the calcite shell. The spinose planktonic Foraminifera clearly use the radiating calcite spines to support an extensive web of cytoplasm used in prey capture, and to distribute symbionts at the periphery during the day. These extensive scaffolding-like structures, as with the complex siliceous skeletons of radiolaria, provide anchorage for the rhizopodia and may enhance mechanical advantage of the pseudopodia during capture of large and motile prey such as copepods and other small crustacea.

Among the radiolaria, the development of the often ornate siliceous skeletons has been carefully documented for a few species (eg, Anderson 1983, 1986). The structure of the skeleton is dictated by a cytoplasmic sheath (cytokalymma) produced by the extracapsular cytoplasm. The cytokalymma (Figure 13.10) is a hollow cytoplasmic structure that forms the "mold" for the skeleton deposited within it. It is remarkably dynamic, exhibiting a flowing and molding motion during the deposition of the amorphous silica. The inner membrane of the cytokalymma immediately surrounding the space where the silica is laid down is undoubtedly a kind of silicalemma. A silicalemma is a membrane that actively secretes silica at sites of silica deposition. The first stage of skeletal formation is the appearance of electron opaque granules (ca 0.3 µm diameter) within the cisterna of the cytokalymma. These granules may serve as silicification centers, since they gradually become augmented by silica-producing siliceous grains that grow and fuse, forming the skeletal structure. Since the growing mass of silica is constrained within the space of the cytokalymma, it assumes a final shape dictated by the moldlike surface of the cytokalymmal cisterna. Thus, the species-specific shape of the skeleton can be explained by the dynamic processes of genetically controlled cytoplasmic streaming and cytokalymmal organization, which dictate the spatiotemporal events during silica deposition.

Current knowledge of mineral structures in a wide variety of protozoa is summarized in a symposium volume on biomineralization in lower plants and animals (Leadbeater and Riding 1986). The remarkable variety of surface structures, shell architecture, and activities of the cytoplasm in producing these extracellular

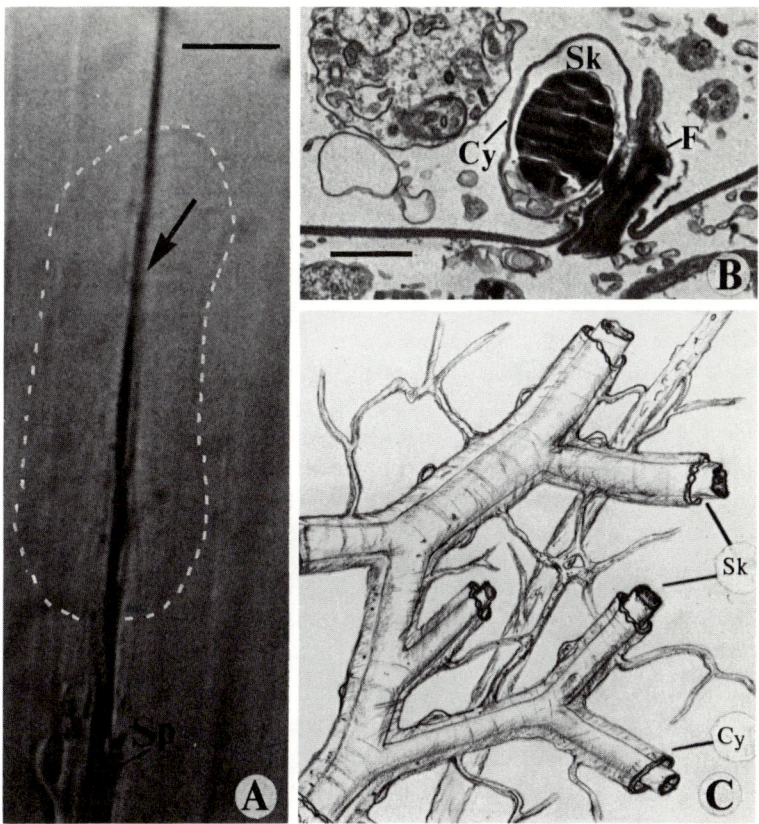

FIGURE 13.10. Silica deposition during skeletal formation in radiolaria. (A) Spine elongation occurs within a thin cytoplasmic sheath (arrow) known as a cytokalymma that serves as a living mold to deposit and shape the silica forming the elaborate shells of this organism. (B) A transmission electron micrograph of the cytokalymma sheath (Cy) enclosing a portion of the skeleton (Sk). The cytokalymma in this portion of the cell is attached to the membrane of a fusule projecting from the capsular wall. (C) Three-dimensional diagram of the organization of the cytokalymma (Cy) and skeleton (Sk) in a radiolarian species with a spongiose siliceous skeleton. Scales = 20 µm (A) and 1 µm (B). From Anderson (1983); with permission.

products is one of the most intriguing and surprising aspects of the study of protozoan cellular biology. The elucidation of the genetic and cytoplasmic developmental processes which account for the elegant, species- specific design of protozoan surface structures will undoubtedly be a major scientific breakthrough in molecular biology.

Summative Perspective

The fine structure of the Sarcodina is relatively conservative compared to the diversity of structures observed among the flagellates. Based on comparative fine structural and physiological data, three major patterns of cytoplasmic organization have been proposed for the Sarcodina (Anderson 1984): (1) diffuse, (2) transitional, and (3) zonal. Diffuse patterns of cytoplasmic specialization are characteristic of many "naked" amoebae. The cytoplasm (eg, Figure 13.1a) is not segregated into distinct regions by organic or mineralized barriers. The cytoplasm may exhibit momentary zonalized distributions of the organelles, but the organization is fluid and changing with time. Transitional forms of cytoplasmic specialization are characteristic of many testate amoebae and Foraminifera (Figure 13.1b). The cytoplasm is partially segregated by a mineralized or organic barrier that forms a test or shell separating intrashell cytoplasm from extrashell cytoplasm. This barrier, however, is perforated or bears an oral aperture through which cytoplasm can flow into or out of the shell. Hence, there is a fluid and variable distinction between inner and outer masses of cytoplasm. A gradation in cytoplasmic organization is typically observed from the intrashell to the extrashell cytoplasm. Therefore, these organisms are considered transitional in organization between the diffusely organized species and those that have a distinct zonal organization. The radiolaria and some heliozoa exhibit zonal organization. In these organisms, there is a definite and stable organic barrier (eg, capsular wall of radiolaria) that segregates major regions of the cytoplasm into structural and functional zones. The intracapsular cytoplasm of radiolaria (eg, Figure 13.2a), where food reserves are stored and major respiratory activity is localized, is organized quite differently from the extracapsular cytoplasm where prey capture and digestion occur. The marked and stable differentiation of these two zones is characteristic of zonal specialized species. Variations in nuclear organization, including patterns of division during mitosis, presence or absence of honeycomb laminated lamellae, and variations in peripheral membrane elaboration (eg, Figure 13.2a) are characteristic of different species. Variations in cytoplasmic organelles such as the fibrillar system found in planktonic Foraminifera, specialized mineralized products produced in the Golgi body, and variations in kinds of endosymbionts are characteristic of some of the major groups of Sarcodina. Further fine structural and physiological research on locomotion, prey capture, and specialized mechanisms for production and secretion of mineralized products may yield significant insights into the diverse ways amoeboid eukaryotic cells have adapted to diverse environments.

14
Fine Structure of Ciliates

General Cytoplasmic Features

A substantial literature exists on the fine structure of ciliates, especially with respect to the taxonomic significance of the surface cortex and subcortical structures, and structural–functional correlates of subcellular organelles. Some of the finer details, especially with respect to the taxonomic significance of comparative fine structure, have been reviewed by Small and Lynn (1985). A comparative survey of some major features of representative organisms is presented here with the aim of summarizing current knowledge and establishing a broad context for the physiological and functional topics discussed in succeeding chapters.

The general organization of the cytoplasm and macronucleus of *Tetrahymena pyriformis* is shown in Figure 14.1. A higher magnification of the periphery of the cell (Figure 14.2) exhibits the cortical structure, consisting of a peripheral pair of membranes forming an alveolate pellicle. A similar organization found in *Paramecium aurelia* is diagrammed in Figure 14.3. A layer of numerous mitochondria and fewer microbodies occurs in the cytoplasm immediately beneath the pellicle. The cilia arise from kinetosomes anchored beneath the pellicle. Between the centrally located nucleus (Figure 14.4) and the peripheral cytoplasm, the most prominent inclusions are numerous vacuoles of varying size, endoplasmic reticulum, masses of glycogen, scattered lipid droplets, and occasional mitochondria and microbodies. Mucocysts, containing mucin deposits and surrounded by a single membrane, are located immediately beneath the pellicle membranes at pore sites usually regularly spaced among the cilia. Variations in the organization of the surface membranes, attached organelles, and the kind and distribution of subpellicular membranous organelles are important taxonomic discriminators.

Cortical Fine Structure

The organization of the microtubules lying within the cortical cytoplasm and associated with the kinetosomes has been extensively studied for many ciliates.

14. Fine Structure of Ciliates

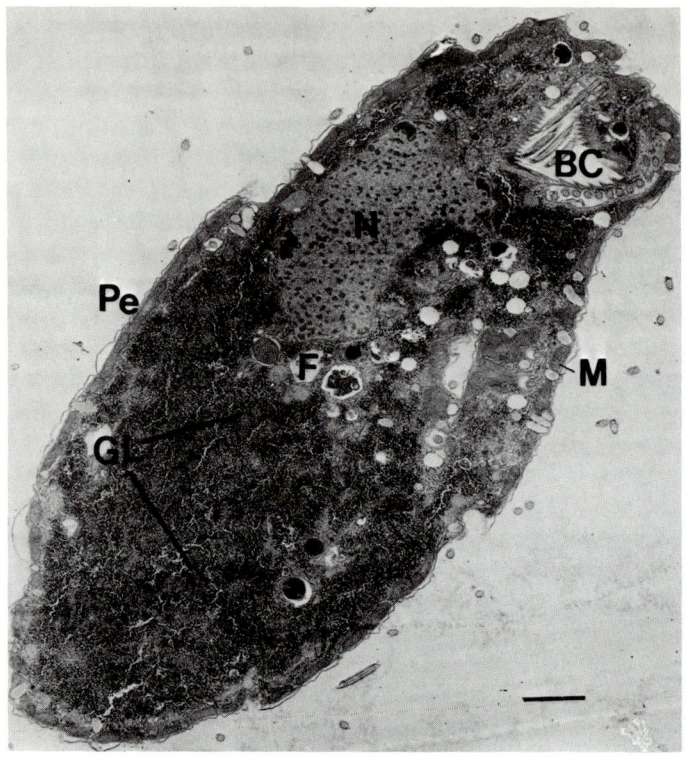

FIGURE 14.1. Fine structure of the ciliate *Tetrahymena pyriformis* containing a large macronucleus (N), food vacuoles (F), buccal cavity (BC) with cilia of the membranelles, abundant stores of glycogen (GL), and peripherally located mitochondria (M) beneath the pellicle (Pe). Scale = 2 µm.

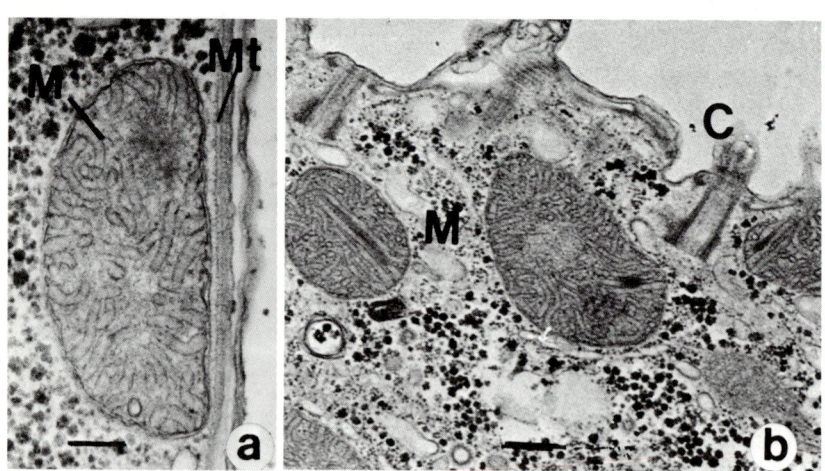

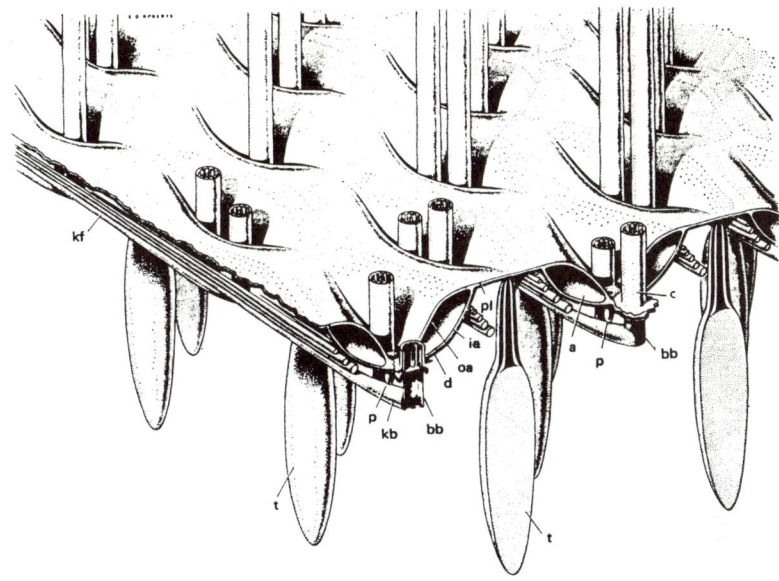

FIGURE 14.3. A three-dimensional reconstruction of the pellicle of *Paramecium aurelia*. The anterior end of the animal is to the observer's upper left, and the ciliary rows are directed diagonally to the observer's left. The kinetodesmal fibers (kf) arise anterior to the basal bodies (bb) and join other fibers running along the (cell's) right margin of the ciliary row. Within the ciliary row, the carrot-shaped trichocysts (t) alternate with the associated parasomal sac (p). The alveolar space (a) between the ciliary rows is surrounded by an outer alveolar membrane (oa), and an inner alveolar membrane (ia) which is associated with a dense granular layer (d). The microtubular bands are omitted from this diagram. From Jurand and Selman (1969); with permission.

A representative diagram for *Tetrahymena* (Figure 14.5) shows the prominent latticework of three bands of microtubules lying immediately beneath the pellicle. The longitudinal bands (lt) are parallel to their kineties at the right. A transverse band (tt) originates at the anterior side and proximal end of each kinetosome, and extends laterally across the meridian where the muciferous bodies occur. It ends abruptly after passing under the adjacent row of longitudinal microtubules. A shorter arcuate set of postciliary microtubules (pt) extends from the posterior and proximal edge of the kinetosome upward toward the point where the transverse band passes beneath the longitudinal band. The longitudinal bands are not continuous, but are formed by overlapping sets of parallel microtubules as shown

◀

FIGURE 14.2. High magnification views of the peripheral cytoplasm of *Tetrahymena pyriformis*: (a) mitochondrion (M) with tubular cristae lying near the microtubular ribbon (Mt) beneath the pellicle; (b) mitochondria and ciliary base (C) at the periphery of the cell. Scales = 0.2 μm.

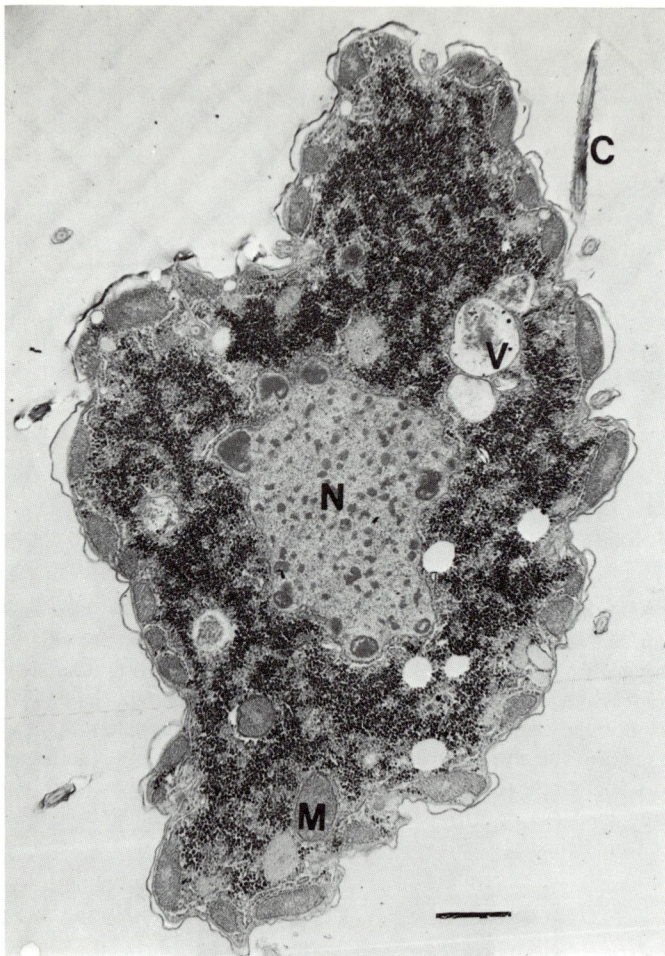

FIGURE 14.4. Cross-section of *Tetrahymena pyriformis*. A prominent macronucleus (N) is surrounded by masses of glycogen, vacuoles of varying size (V), and peripherally located mitochondria (M). Cilia (C) arise from the pellicle consisting of membrane-enclosed alveoli. Scale = 1 μm.

in the diagram. A thin layer of amorphous material (e) forms a continuous lamina surrounding the cell and interspersed between the lateral and longitudinal bands of microtubules. A basal microtubule lies along the anatomical left side (viewer's right) of the proximal edge of the kinetosome. Anteriorly-directed tapering bands of kinetodesmal fibers (kf) extend from the triplet set of microtubules in the kinetosome of each cilium. These banded fibrils are similar to those forming rootlets at the base of kinetosomes in flagellates. A layer of mitochondria and endoplasmic reticulum occurs immediately beneath the microtubular ribbons.

FIGURE 14.5. Three-dimensional view of the structures of the somatic cortex of *Tetrahymena pyriformis*. At the left and right are segments of two kineties showing the basal bodies, and associated bands of microtubules forming laterally directed transverse fibers (tf) and arcuate post ciliary bands (pf). Anteriorly directed kinetodesmal fibers (kf) extend from the triplet microtubules at the base of the cilium. A longitudinal set of microtubules (lt) is underlaid by a thin layer of epiplasm (e) and covered on the outer side by the pellicle. Basal microtubules (bt) lie longitudinally near the basal bodies of the cilia. From Allen (1967); with permission.

Similar patterns with minor variations occur among other members of the Hymenostomatida, including the commonly observed genera *Colpidium*, *Glaucoma*, and *Tetrahymena*. For purposes of contrast, the cortical organization of the multiflagellated *Opalina* previously thought to be a ciliate is shown in Figure 14.6. The pellicle is organized into folds or ribs approximately 50 nm in thickness and 600 nm high, separated from adjacent ribs by grooves 100 nm wide. Each fold is internally reinforced by a single row of 20 to 25 parallel, longitudinal tubules (ca 17 nm diameter) embedded within an amorphous matrix (Noirot-Timothée 1959; Pitelka 1963; Patterson 1985).

In the dikinetid-bearing ciliate *Colopoda* (Lynn 1976), the pair of kinetosomes is linked by fibrillar ribbons (desmoses). A transverse ribbon of microtubules extending upward and laterally beneath the pellicle originates at the base of each of the kinetosomes in the pair. The anterior kinetosome in the pair has a *single* postciliary microtubule; not the arcuate band extending upward toward the longitudinal microtubules as found in *Tetrahymena*, for example. The posterior kinetosome, however, has a postciliary ribbon, and also a fanlike laterally-directed kinetodesmal fibril. The dikinetids of *Loxodes* have a tangential transverse

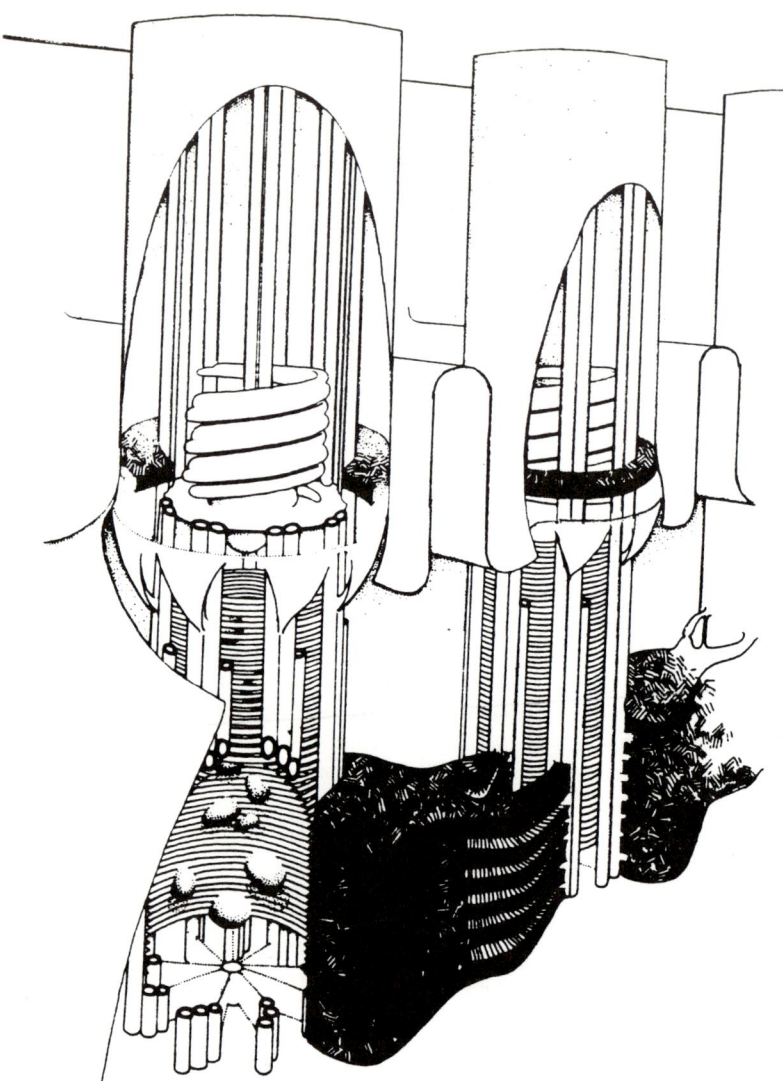

FIGURE 14.6. A three-dimensional diagram of the organization of two kinetosomes and interkinetosomal connective (dense basal structures) in the multiflagellated *Opalina ranarum* previously thought to be related to the ciliates. The organization of the membrane is clearly different than most ciliates, and exhibits folds between the flagella. There are no arcuate post ciliary bands as in some ciliates, and the spiral-like body encircling the central pair of microtubules is characteristic of this group. From Patterson (1985); with permission.

ribbon of microtubules attached to *each* kinetosome. From a comparative perspective, the dikinetids of the following genera possess overlapping postciliary microtubules and have longitudinal contractile myonemes: *Geleia, Kentrophoros, Loxodes, Tracheloraphis, Trachelonema, Stentor, Condylostoma, Blepharisma*, and *Spirostomum* (Lynn 1976).

The fine structure of the contractile spirotrich *Spirostomum* shows that there are two fibrillar arrays in the peripheral region of the cytoplasm (Figure 14.7). A longitudinal fibrillar bundle (LFB) runs lengthwise through the cell, and is associated with the ciliary kinetosomes and the membrane lining the crevice where the kinetids are inserted into the plasma membrane. A contractile fibrillar system (CFS), that accounts for the contractile activity of the cell, is situated more proximally at the boundary between the endoplasm and ectoplasm. The longitudinal fibrillar bundle and the subpellicular fibrils probably have a supportive or strengthening role (Yagiu and Shigenaka 1963). The preoral membranelles of *Spirostomum* form a polykinetid with a robust, fan-shaped fibrillar bundle extending into the endoplasm. The dimensions of the minute fibrils in the

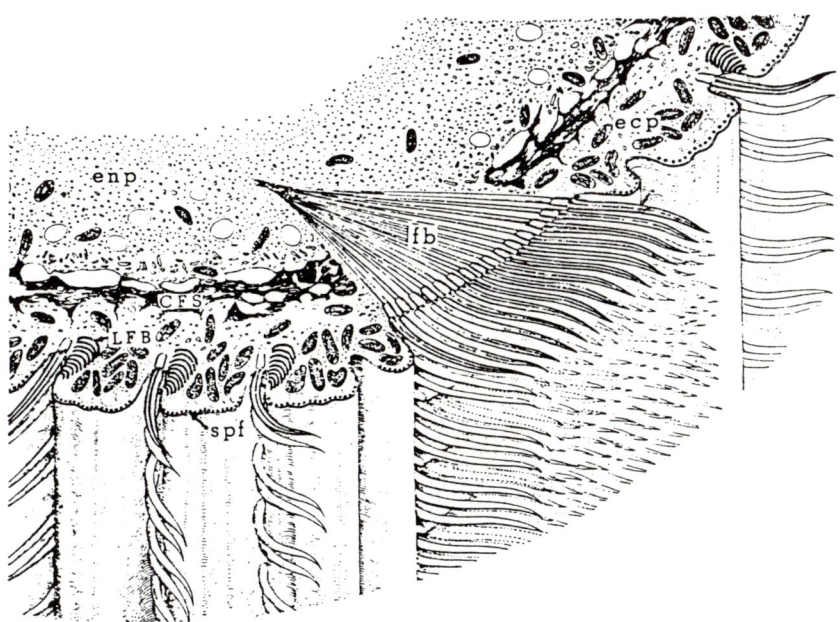

FIGURE 14.7. Three-dimensional view of the peripheral cytoplasm of *Spirostomum ambiguum*. Preoral membranelles consist of the fan-shaped mass of cilia (fb) running longitudinally along the grooved pellicle and rows of kineties containing dikinetids. Subpellicular fibrils (spf) lie immediately beneath the pellicular membranes and longitudinal fibrillar bundles (LFB) run parallel to the basal bodies of the kineties. A contractile fibrillar system (CFS) lies between the ectoplasm (ecp) and endoplasm (enp). From Yagiu and Shigenaka (1963); with permission.

contractile fibrillar system and the associated vacuoles closely resemble the contractile elements ("M bands") (Randall and Jackson 1958) in *Condylostoma*, *Stentor*, and *Vorticella*, which also exhibit pronounced contractile and extensional movements. These "M bands" resemble the contractile fibrils in smooth and striated muscle of higher animals, further strengthening the conclusion that they are the contractile components of the cell. They are lacking in the noncontractile spirotrich *Blepharisma*, which, however, also contains kinetodesmal fibers extending anteriorly in an overlapping manner reminiscent of those in the longitudinal fibrillar bundle of *Spirostomum* (Kennedy 1965). These composite data strongly suggest that the "M bands" in spirotrichs are the contractile system of the cell.

The cortical fine structure of the psammobiotic (sand-dwelling) gymnostomatid *Tracheloraphis dogieli* gives further evidence that the kinetosomal fibrils are probably supportive and not contractile elements (Raikov et al 1975). This vermiform ciliate has dikinetids. The posterior member of the pair has a kinetosome with a backward directed postciliary fiber (PC) contributing to longitudinal overlapping bands of postciliary ribbons (Figure 14.8, SPC) underlying cortical ridges. A kinetodesmal fiber extends anteriorly from the base of the posterior kinetosome. The anterior kinetosome gives rise to a lateral transverse ribbon of microtubules that arch upward and over the bundles of myofilaments (MY) running longitudinally beneath the cortical ridges. These bundles of myofilaments are ensheathed by vesicles resembling the arrangement in spirotrichs, and thus enhance the conclusion that the stacks of postciliary microtubules are supportive, whereas the myofilaments are contractile. Pigment granules and rhabdocysts (extrusomes) lie immediately beneath the plasma membrane. There is no regular pellicular structure as in *Tetrahymena*, but a rather irregular layer of vacuoles appears to serve a similar role as the pellicular alveoli.

The cortical structure of *Lacrymaria* (Fig. 7.5) is of theoretical interest in cell biology, since this organism exhibits pronounced elongation and contraction of the body. The locus of contractile activity is not fully determined, but the cortical organization of the microtubules exhibits some clear differences relative to other ciliates described above (eg, Bohatier 1970; Tatchell 1980). Each kinetosome is associated with a postciliary ribbon consisting of seven microtubules arranged in two ranks. The outer rank contains four parallel microtubules and the inner rank has three. Cross-connections (10 nm long) anchor the outer rank of microtubules at occasional locations to the nearby inner membrane of the pellicular alveolus. The inner rank is linked to the outer rank by two connectors attached to each other in a V to "dogleg" configuration. The transverse ribbon consists of two parts: a band of three microtubules (trio) and a set of six (sextet). Both arise from near the base of the kinetosome, on the left side (as viewed from within the cell), but diverge near the top of the kinetosome. The trio extends only about 1 μm into the interkinetal crest, while the sextet curves anteriorly until it forms a band almost parallel to the line of kineties. The transverse ribbons are also linked to the pellicular membrane. The postciliary and transverse ribbons form a system of overlapping microtubules on both sides of the kinety, the former extending

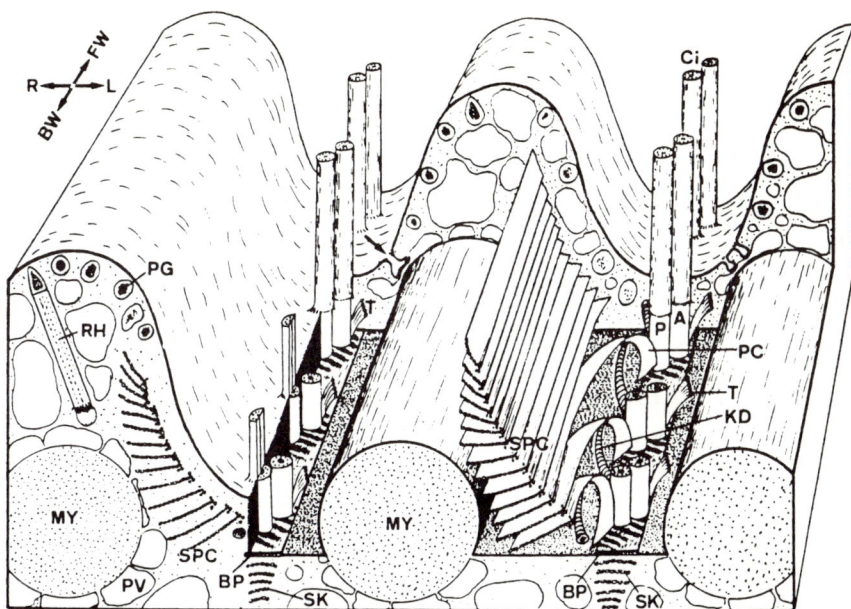

FIGURE 14.8. Pellicular organization of the marine sand-dwelling ciliate, *Tracheloraphis dogieli*, showing the ventral kineties; hence right and left are reversed as the organism is oriented ventral side up. Large bundles of myonemes (MY) run longitudinally and are associated on the left side with bands of overlapping microtubules forming the post-ciliary ribbons (SPC). These are derived from the postciliary fibers (PC) arising from the base of the cilia (ci). Transverse fibers (T) and kinetodesmal fibers oriented toward the anterior (KD) are associated with the ciliary bases. Pigment granules (PG) and ejectosomes (RH) lie beneath the pellicle surface. From Raikov et al (1975); with permission.

posteriad, and the latter anteriad. At the anatomical left-hand of each kinety, a tract of striped fibers (the myoneme) extends the whole length of the somatic kineties. The striped appearance in longitudinal section is due to the distribution of the constituent filaments (ca 4 nm diameter). These filaments do not appear to be ordered in the contracted organism, nor does their diameter change upon extension of the body. A band of filaments of similar diameter (4 nm) extend around the body beneath the pellicle, but these "transverse filaments" are not arranged in a banded pattern. During body extension and ingestion of prey, the neck is rigidly held while the cytostome is formed. Tatchell (1980) suggests that the rigidity of the neck may be maintained by the overlapping anteriorly-directed transverse ribbons that connect to the pellicular membrane by cross-connections. The longitudinal "myonemes" may account for contraction as in other ciliates, but this is not fully confirmed.

The parasitic ciliate *Balantidium caviae* exhibits some interesting subcortical features comparable to free-living ciliates (Paulin and Krascheninnikow 1973). Postciliary bands of microtubules (4 to 5 tubules per band) are attached at the posterior right margin of each kinetosome and course upward and posteriorly. The left transverse ribbon, composed of a single row of four or five microtubules, originates near the anterior margin of the kinetosome and extends medially. Kinetodesmal fibers originate at the anterior right margin of the kinetosome, extend toward the surface, and turn anteriorly below the pellicular membrane. They do not form overlapping arrays of stacked ribbons as, for example, occurs in *Paramecium*. Somatic kinetosomes possess rootlike bands of microtubules extending into the endoplasmic region. A detailed examination of the peripheral cytoplasm of the peritrich (tintinnid) *Cyttarocylis brandti* demonstrated a complex pellicular membranous system called the "Perilemma" (Laval-Peuto 1975). The periphery of the cell is enclosed by a triple layered system consisting of an outer membrane, and an inner vacuolar space connected to a complex intracytoplasmic set of vacuoles and cisternae and the inner membrane of the cisternal spaces. Thus, the pellicle is not the typical alveolate or quiltlike mosaic of cisternae as observed in *Paramecium*, *Tetrahymena*, and other ciliates. The double membrane system apparently extends around somatic cilia and those of the adoral membranelles. The extensive network of intracytoplasmic cisternae to which it is connected also is continuous with the cisterna of the nuclear envelope, thus forming a large and complex vacuome within the endoplasm and extending broadly beneath the surface membrane of the cell. Adoral membranelles are anchored in the subpellicular cytoplasm by bands of microtubules extending downward into the endoplasm, and laterally beneath the vacuome of the perilemma.

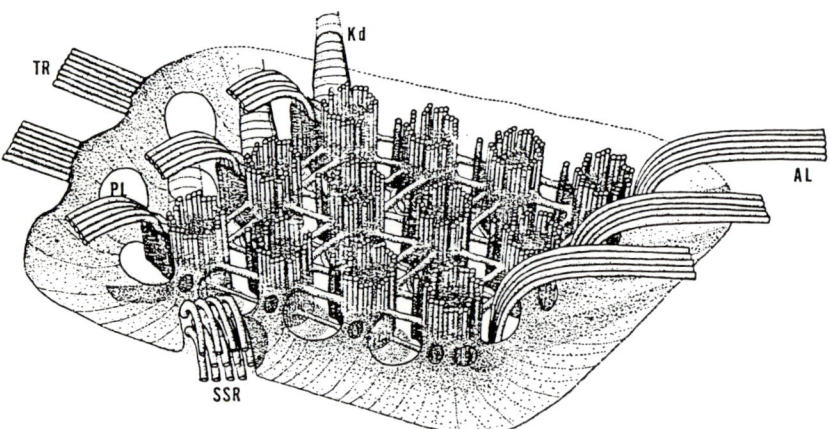

FIGURE 14.9. Fine structure of the cilia in the cirrus of an hypotrich *Gastrostyla steinii* illustrated in abbreviated form showing the kinetodesmal fiber (Kd), posterior bands of microtubules (PL) arising only from the posterior set of cilia, and anterior bands of microtubules (AL) arising from only the anterior set of cilia. Subectoplasmic rootlets (SSR) extend from the basal bodies and penetrate into the cytoplasm. From Grim (1972); with permission.

The subpellicular organization of the closely coordinated cilia forming the cirri of hypotrichs exhibits some specialized features not observed in somatic cilia (Grim 1972). The cilia are interconnected by a complex network of microtubules (Figure 14.9) anchoring the kinetosomes to each other. Postciliary ribbons of microtubules originate only from the posterior set of kinetosomes in the cirral array. Likewise, anteriorly-directed ribbons arises from the anterior set of kinetosomes. Kinetodesmal striated bands are anchored at the base of the posterior four kinetosomes. The total arrangement provides an extensively reinforced and symmetrical array of kinetosomes, with strong anchoring bands of microtubules connected to the surrounding pellicle and subectoplasmic rootlets (SSR) penetrating into the deeper cytoplasm.

An anastomosing network of fine filaments has been observed in many ciliates at the interface between the pellicle and the cytoplasm. This layer, called the epiplasm, is often intimately associated with the inner membrane of the pellicle and the microtubular ribbons extending from the base of the kinetosome. The epiplasm may strengthen the cortex (eg, Hausmann and Mulisch 1981). In at least one hypotrich, *Euplotes vannus*, a unique cortical network replaces the usually fine filamentous layer (Walker 1975). The organization of the meshwork, immediately beneath and also fused with the pellicular membranes, further suggests that it is a cortical strengthening structure. The areolate and netlike organization of the layer may add additional strength against tension and rupture of the lamina.

Specialized Cytoplasmic Structures

Oral Apparatus

The oral ciliature and/or subpellicular cytoskeletal apparatus is highly variable among species of ciliates. Although the cortical ciliature has provided substantial taxonomic discriminatory value, the organization of the oral ciliature or feeding apparatus is often less distinctly species-specific and appears to have followed several lines of adaptive convergence. Thus, the organization of the feeding organelles may be very similar in otherwise highly diverse species. For example, a pharyngeal basket (Figure 14.10) composed of parallel arrays of microtubules, although varying in detail of organization, occurs in such widely diverse genera as *Nassula* (class: Nassophorea) and *Chilodonella* (class: Phyllopharyngea). Broad patterns in organization of the feeding apparatus are useful, however, in making discriminations of major taxonomic groups (eg, Small and Lynn 1985). Eisler (1986) has made an elegant study of the pharyngeal organization of some nassulid ciliates (*Nassula*, *Furgasonia*, and *Pseudomicrothorax* spp) toward elucidating their phylogenetic relationships. The organization of the cilia in the oral region has been carefully determined for several species and some examples are presented to illustrate the structural–functional correlation of these organelles. In the hymenostome *Paramecium*, there are three bands of cilia traversing the wall of the vestibulum: the quadrulus, a band of four cilia extend-

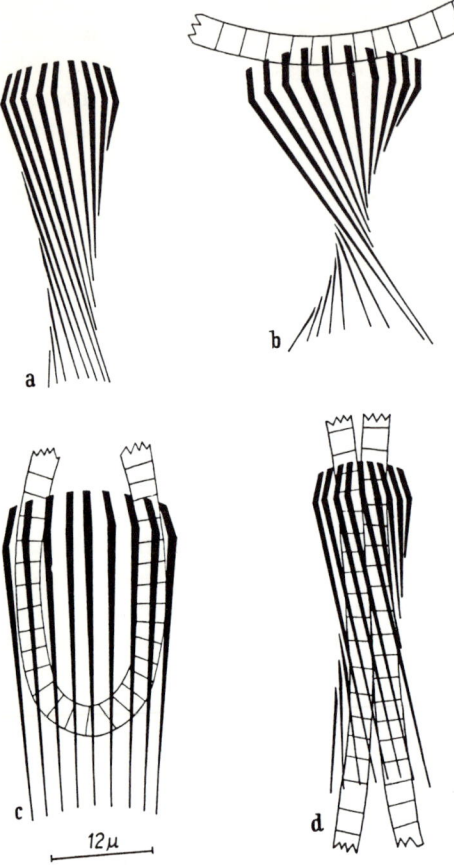

FIGURE 14.10. Pharyngeal basket of *Nassula* showing various stages of ingestion of a filamentous alga (a) in the normal state within a nonfeeding cell, (b) during grasping of the algal filament, and (c) during expansion and engulfment of the filament and ingestion of the strands (d) following partial dissolution by digestive enzymes released from the cytopharynx by lysosomes that stream into the cytopharyngeal opening. From Grell (1973); with permission.

ing largely along the dorsal (left) surface of the vestibulum, but spiralling into the region of the cytopharynx; and two peniculi, the dorsal and ventral extending along the lateral wall of the vestibulum (Figure 14.11a). In *Tetrahymena*, there are four ciliary membranelles, hence the name of the genus meaning four membranes (Figure 14.11b). The undulating membrane (UM) forms a long, curving haplokinety at the margin of the buccal cavity. Three intrabuccal membranelles (M1, M2, and M3) lie to the viewer's right (anatomical left) of the undulating membrane (eg, Williams and Bakowska 1982). The buccal cavity is reinforced and supported within a shroud of intermediate filaments (16.4 nm diameter) that apparently are not contractile. These filaments, somewhat different in size than metazoan intermediate filaments, are connected into a rigid framework capable of maintaining the overall architecture of the buccal cavity in the absence of microtubules (Williams 1986). The kinetosomes of the buccal cilia are interconnected by a complex network of microtubules that extend posteriorly, as also in *Paramecium*, as a tapering array near the base of the cytopharynx. This is where the food vacuole is formed, and apparently where membranous vesicles are

FIGURE 14.11. Oral membranelles of (a) *Paramecium* and (b) *Tetrahymena thermophila*. The vestibulum of *Paramecium* (v) leads into the buccal cavity (BC) with an endoral kinety (EK) at the rim of the buccal opening (BO). Three major bands of membranelles line the wall of the buccal cavity, the dorsal peniculus (DP), ventral peniculus (VP), and quadrulus (Q). These membranelles spiral slightly toward the dorsal surface of the buccal cavity wall and terminate near the cytostome (or mouth) where the extremely thin, short cytopharynx (CP) commences. Food vacuoles (FFV) are formed by the membrane at the base of the cytostome. A preoral suture (PRS) and postoral suture (POS) extend from the margins of the buccal overture. The four membranelles characteristic of *Tetrahymena* (b) consist of a haplokinety forming the undulating membrane (UM), and three membranelles labelled M1, M2, and M3 that extend into the buccal cavity. (a) From Wichterman (1986) and (b) from Williams and Bakowska (1982); with permission.

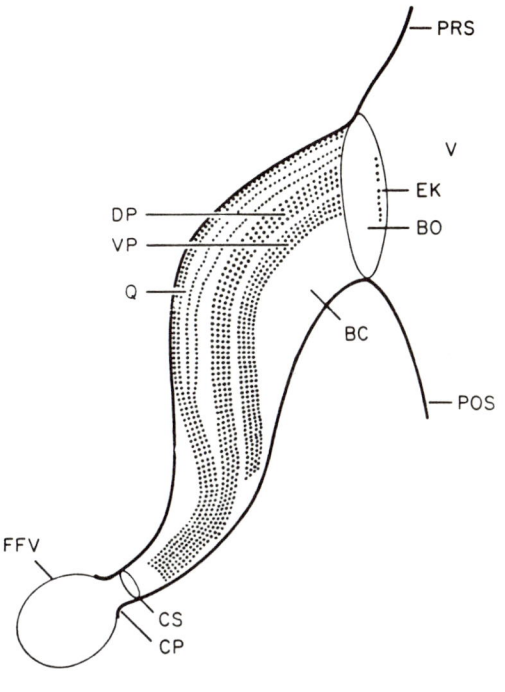

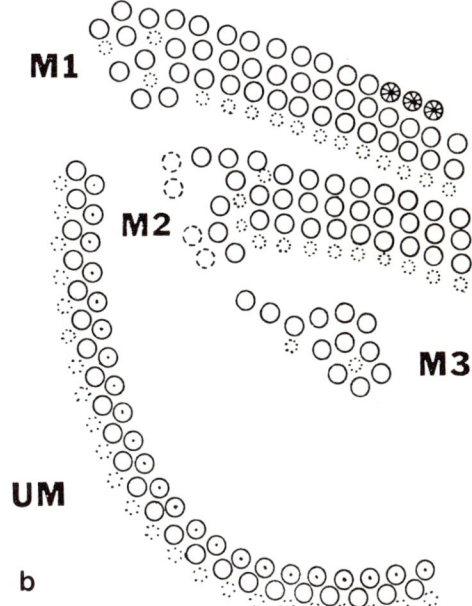

recovered from the cytoplasm to supply the large membrane surface area required to surround the forming food vacuoles. Microtubules in this region may guide these vesicles to the phagosome forming region, and in general provide the scaffolding for directed movement of food vacuoles and its precursors within the peribuccal region of the cytoplasm. The ciliary membranes in hymenostomes are clearly organized to sweep food particles efficiently into the buccal cavity, and to concentrate them at the cytostomal membrane where they are sequestered in food vacuoles.

In the Rhabdophora, possessing a temporary cytostome as characterized by *Lacrymaria*, the conical apical region (Figure 14.12a) contains a fibrous annulus (FA) connected to the transverse microtubules (T) supporting the apex and surrounding the ring of toxicysts (TO) used to capture prey. Rhabdophora with a permanent cytostome (Figure 14.12b) have an inwardly directed basket of membrane-covered microtubules (T) leading to the phagoplasm, where prey is engulfed within cytoplasmic vacuoles.

The marine gymnostomate ciliate *Helicoprorodon gigas* has a highly complex cytopharyngial basket (Figure 14.13) externally surrounded by concentric rings of cilia (C), terminating centrally as the cytostome (Raikov and Kovaleva 1980). The cytopharynx is lined peripherally by nemadesmata (rods of microtubules) lined by an inner (IL) and outer (OL) lamella of longitudinally oriented microtubules. The nemadesmata are arranged in a regular cylindrical array in the distal portion (levels 1 through 5), but at about level 6, they begin to diverge and no longer lie exactly in a circle. By level 8, only a few strands of the proximal microtubules are visible. The outer lamella is more truncated than the inner lamella and terminates at about level 4. Diagonally situated rods of microtubules (T1 and T2) project from the peripheral folds of the circumcytostomal ridges and extend into the endoplasm. Food is drawn into the cytostome and transported proximally into the phagoplasm, where food vacuoles are formed (level 8).

Cytoplasmic Organelles

Macro- and micronuclei of ciliates vary in size, spatial proximity to one another, and chromatin organization. In some species, the macronucleus contains one or more nucleoli, sometimes with peripheral clumps of heterochromatin. The micronucleus, however, may be without nucleoli and contains only scattered masses of chromatin. In some fine structure views, the nuclear envelope of the macro- and micronucleus are in continuity with each other, as well as with the endoplasmic reticulum. This is particularly the case in the hymenostomes and scuticociliatids, where the micronucleus lies in close proximity to or within a concave surface of the macronucleus. This membrane continuity may permit exchange of metabolites (Kaneshiro and Holz 1976). The nuclear envelope is usually porous in both nuclei, and a fringe of filamentous material resembling nucleic acids (perhaps mRNA) is observed in the cytoplasm as clumps opposite the pores.

Specialized Cytoplasmic Structures 267

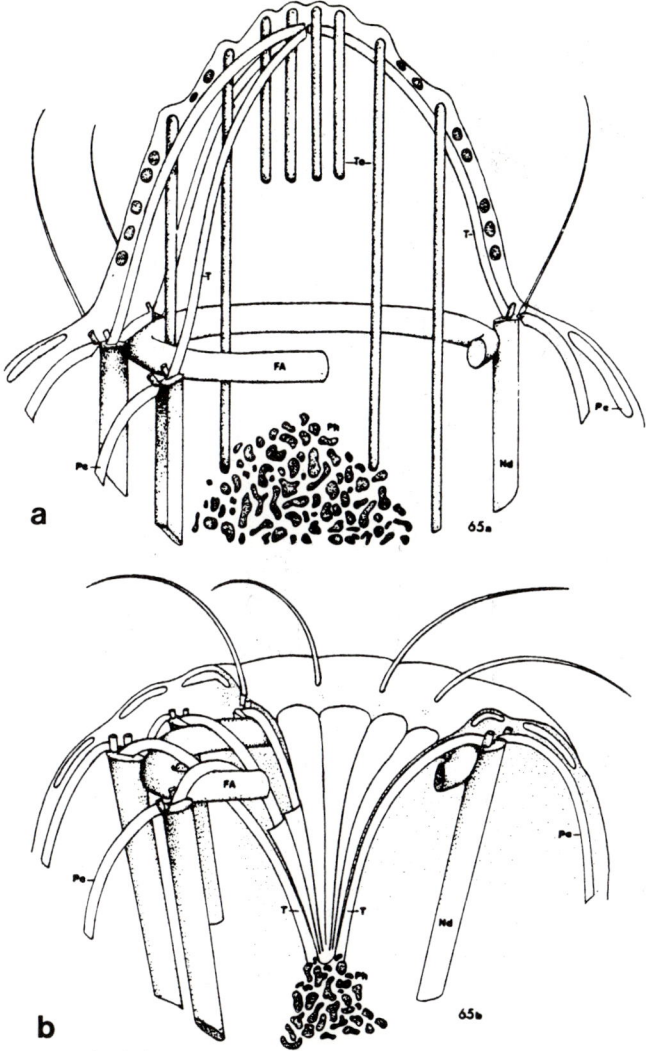

FIGURE 14.12. Schematic diagram of the rhabdos, cytopharyngeal apparatus characteristic of the Rhabdophora. (a) A temporary cytostome formed by a domelike array of microtubules that support the apical region of the ciliate. A stout ring also known as a fibrous annulus anchors the transverse ribbons of microtubules that support the apical cone. (b) A permanent cytostome contains inflexed ribbons of microtubules covered by the cell membrane. Food particles are collected at the base of the invaginated cone (b) or during ingestion when the temporary cytostome forms an endocytotic cavity upon contact with prey (a). From Small and Lynn (1985); with permission.

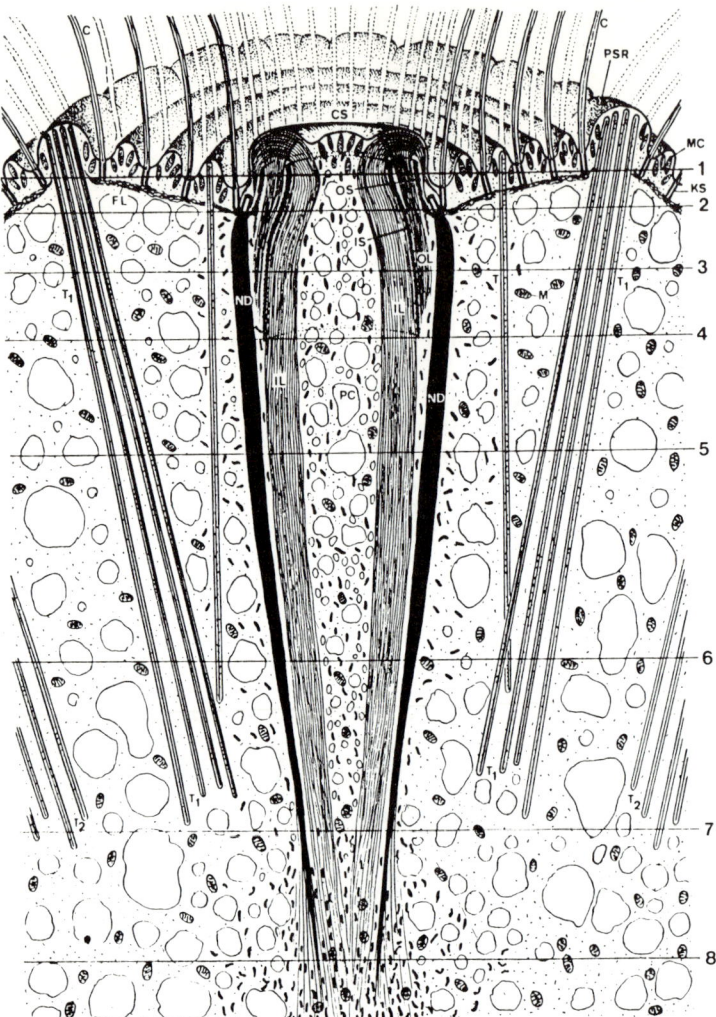

FIGURE 14.13. Feeding apparatus of *Helicoprorodon gigas* diagrammed from an ultrathin section viewed with a transmission microscope. This longitudinal view of the section shows the concentric array of cilia around the oral aperture or cytostome. Horizontal lines demarcate zones of major organizational features. Nematodesmata (ND) are arranged in a regular array in the distal portion of the cytopharyngeal apparatus (levels 1–5), but at level 6 the microtubules begin to diverge, and they no longer lie in an exact circle. An outer lamella (OL) extends from the first to the fourth level, and the inner lamella (IL) forming a circumoral ring extends from the exterior down to level 8. Diagonally situated rods of microtubules (level 1–7) extend from the circumcytostomal ridges into the endoplasm. During feeding, the cytostomal opening expands and food is engulfed in food vacuoles. From Raikov and Kovaleva (1980); with permission.

Specialized organelles, known as Müller vesicles in loxodid ciliates (eg, *Loxodes* and *Remanella*), may be geotactic organelles permitting the ciliates to orient their movements in the water column in response to gravity (Fenchel and Finlay 1984, 1986). These vacuoles (ca 10 µm diameter) contain a membrane covered mineral grain (statolith) and various microtubular bands extending into the vacuole from the overlying kinety (Figure 14.14). The arrangement of the statolith and attached postciliary microtubules (PMT) strongly suggests that this is a geotactic sensory device. The loxodid ciliates are microaerophiles which do not tolerate high oxygen tensions. Therefore, they migrate in the water column to regions of optimum oxygen concentration and prey abundance. The gravity-sensing capacity of the vacuoalar-bound mineral grain may account for the organism's ability to maintain a constant direction of movement when perturbed by water turbulence or human experimental intervention (Fenchel and Finlay 1984).

The mitochondria of ciliates are major centers of energy transformation, ATP production, lipid metabolism, and other metabolic functions as occur in many eukaryotes. They are typically individual, spheroidal, tubular-cristate organelles distributed mainly throughout the peripheral cytoplasm, but varying in organization and distribution in some species according to the physiological state of the organism. For example, the mitochondria of *Tetrahymena* in the log phase of growth are elongate and oriented cortically with their long axis parallel to the cell surface; however, in stationary phase growth, they are smaller, rounded, and distributed deeper in the cytoplasm (Elliott 1963). In the marine scuticociliate *Uronema*, the mitochondrion is a single chondriome with long lobes oriented parallel to the long axis of the cell and lying immediately beneath the pellicle. At the posterior, the lobes join by cross-connections to form a basketlike arrangement (Kaneshiro and Holz 1976). The unique nonmotile posterior cilium of *Uronema* possesses a typical kinetosome and is flanked on each side by a pair of parasomal sacs. A fibrous ring surrounds the complex. As in many ciliates, a complex set of collecting tubules forms the water expulsion apparatus. These membrane-enclosed tubules are concentrated near the posterior of the cell and occur interposed between cisternae of ER and the lobes of the chondriome. This position permits close contact with the ER on one side and the outer membrane of the chondriome on the other side, thus possibly improving water and waste metabolite collecting efficiency. The system of tubules are coated on the cytoplasmic side by electron-dense peglike projections arranged in a helical pattern around the tubule. Some tubules contain an inner tubule, but these internal tubules were never coated. The tubules project posteriorly, anastomose, and are in close association with the contractile vacuole membrane, thus adding additional evidence that these are the nephridial or collecting tubules for the contractile vacuole (Kaneshiro and Holz 1976). Similar coated tubules form the nephridial system of other ciliates as described previously in Chapter 11.

The elimination of water and possibly soluble waste is largely through the expulsion of fluids at the pore of the contractile vacuole. But, in many ciliates

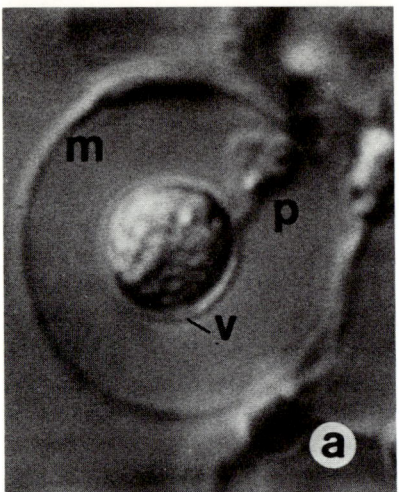 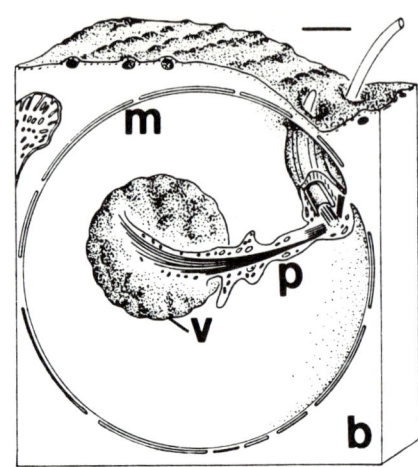

FIGURE 14.14. Light micrograph (a) and schematic drawing (b) of the Müller vesicle in *Loxodes*. The vacuole (ca 7 μm diameter) contains a mineralized granule enclosed by a vesicle membrane (v) and attached to a ciliary basal body by the microtubules of a postciliary fiber (p) that is enclosed within a membrane-bound strand of cytoplasm. The vacuolar membrane (m) surrounds the vesicular apparatus, and is continuous with a membrane-lined canal surrounding the rudimentary cilium shown in (b) as barely protruding from a depression in the pellicle. The movement of the mineralized particle in response to gravity as the organism changes position may provide a stimulus that helps the organism orient during locomotion. Scale = 1 μm. From Fenchel and Finlay (1986); with permission.

there is a specialized region (cytoproct) where residual digestive vacuole waste is excreted. This region, as exemplified by *Paramecium*, is a slitlike fold in the pellicle underlaid by sheets of microtubules. As the residual vacuole approaches the cytoproct, contact is made with the microtubular framework, which apparently strengthens and supports the cytoproct slit region and guides the vacuole to the point on the plasma membrane where membrane fusion and waste expulsion occur.

Microbodies, surrounded by a single membrane and containing granular inclusions, are frequently observed in close association with mitochondria. In some ciliates the microbodies are preferentially situated in the deeper cytoplasm amidst the scattered mitochondria, whereas the majority of the mitochondria are located near the periphery. The microbodies, in conjunction with the mitochondria, play an important role in carbohydrate and lipid metabolism of some ciliates. Their location deeper in the cytoplasm is functionally understandable, as they are not aerobic centers of metabolism to the extent of the mitochondria. Their role in degrading nitrogenous compounds such as nucleic acids, and the production of intermediates in the conversion of lipid to glycogen, among other functions, makes their central cytoplasmic location more understandable. The complex interrelationship between the biochemistry of the microbody and mitochondria, and the role of these organelles in the physiology of the cell during various stages of growth and in response to varying environmental conditions is discussed more fully in the next chapter.

Summative Perspective

The elucidation of the fine structure of ciliates, especially the arrangement of the subpellicular fibrils, has been especially helpful in modern taxonomic studies. Current taxonomic categories have largely been constructed on cytoplasmic features rather than on "gross" cellular features, which often obscure conservative and possibly phylogenetically more significant fine structural features. The examination of the macronucleus and micronucleus during various stages of cellular development and across different species has substantially contributed to understanding the physiology and possible phylogenetic history of the group. Some of the behavioral characteristics of ciliates, such as varying modes of prey apprehension, motility, contractility, and mechanisms of cellular division have been explained by fine structural analyses. A good example is the recent clarification of the organization of the Müller vesicles in *Loxodes* and *Remanella*, and their possible role in geotaxis. The remarkable adaptive radiation of ciliates during their evolution is clearly documented by the diversity of their fine structural features. Variations in ciliary organization, such as formation of cirri and oral membranelles, have been further clarified by analysis of the microtubular and filamentous bands associated with subpellicular structures in these kineties. The correlation of fine structural and physiological variables during growth, variations in metabolism, and reproduction is an expanding area of modern protozoological research which promises to yield significant insights into eukaryotic cellular structure and function. Some of these aspects of research are reviewed in subsequent chapters concerned with the physiology and life cycles of protozoa.

Section III Physiology and Life Processes

15
Basic Biochemistry and Physiology

Cellular Dynamics

The living cell is a dynamic system which constantly undergoes change and reorganization of its molecular components. Moreover, it is not at physical and chemical equilibrium. Chemical reactions, distribution of many ions and molecules across the plasma membrane, and the organization of structural elements are not in the lowest possible energy state; hence, energy is required to maintain the dynamic and peculiarly ordered state of the cell. Heterotrophic organisms obtain energy by taking in organic (carbon-containing) compounds from their environment. These compounds are metabolically degraded to release energy in a form that is widely utilizable throughout the cell (eg, as ATP), or some are stored as food reserves upon transformation (eg, carbohydrate granules or lipid droplets). One of the fundamental processes essential to life is the conversion of carbohydrate or lipid to smaller compounds, with the release and conservation of energy as ATP. Since there is always some inefficiency in natural energy conversion processes, not all of the energy released during metabolism is conserved as chemical energy in the bonds of ATP, and some of the energy loss is in the form of heat. Thus, living systems lose some heat energy to their environment, especially during periods of intense metabolic activity. The major biochemical pathway for metabolic conversion of glucose and fatty acids to smaller molecules and ATP is shown in Figure 15.1. The major events are summarized here, but more detailed information is available in standard physiology and biochemistry texts. Only some concepts pertinent to our comparative analyses of protozoan biology are discussed.

Glucose Metabolism

Glucose, either assimilated directly from the environment or by degradation of storage carbohydrate, is converted during glycolysis to pyruvic acid by way of the Embden–Meyerhof pathway. This process involves the conversion of glucose to glucose-6-phosphate, and the transformation of glucose-6-phosphate to fructose-6-phosphate in order to permit addition of another phosphate on the first carbon

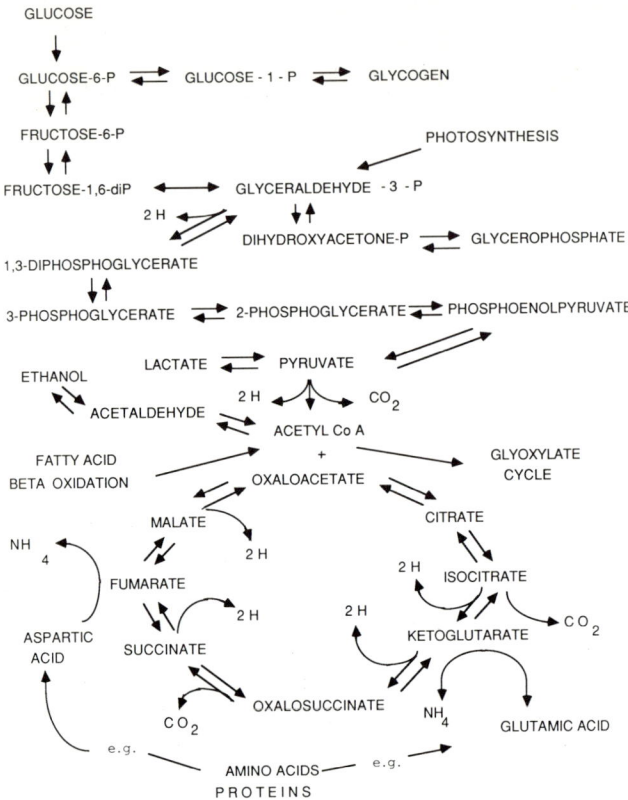

FIGURE 15.1. Some major intermediate compounds in the glycolysis pathway (glucose to pyruvate) and the tricarboxylic acid cycle (TCA cycle) commencing with oxaloacetate that combines with active acetate (acetyl-CoA) to form citrate. Amino acids are synthesized from TCA cycle intermediates by transamination (addition of ammonia from other compounds) as in the production of glutamic acid by amination of ketoglutarate. Likewise, fumarate is converted to aspartic acid. By reversal of the reactions, amino acids can contribute to the intermediates in the TCA cycle. Fatty acid degradation by beta oxidation (Figure 15.2) yields active acetate that can be used in generating TCA cycle intermediates. Or, acetyl-CoA generated by glycolysis can be used in the glyoxylate cycle (Figure 15.3). The breakdown of glycogen is shown as a precursor for glucose. End products of glycolysis in the absence of oxygen (anaerobic metabolism) may include ethanol or lactate, depending on the species. These compounds are also sources of metabolites by reversal of the pathway. Glycerophosphate can be used as a nutrient source by interconversion with dihydroxyacetone phosphate. Reduced compounds (eg, NADH) produced by glycolysis and the TCA cycle are sources of energy for ATP production. Reduced NAD and other cofactors are symbolized as 2H. Some ATP is also produced by substrate level phosphorylation of ADP during the glycolysis reactions.

atom yielding fructose-1,6-diphosphate. This initial step requires two molecules of ATP. These, however, are regained at later steps, along with additional ATP molecules to make the process energetically favorable to the cell. Fructose-1,6-diphosphate is cleaved to yield two 3-carbon compounds, glyceraldehyde-3-phosphate and dihydroxyacetone phosphate. These two intermediate products are interconvertible, thus allowing the total carbon from glucose to be utilized in the metabolic pathway leading to pyruvate through glyceraldehyde-3-phosphate. Alternatively, one of the branches in the metabolic pathway utilizes dihydroxyacetone phosphate to generate glycerophosphate, and by dephosphorylation, glycerol. The specialized process of interconverting glycerophosphate and dihydroxyacetone phosphate (Figure 15.1) in energy metabolism of anaerobic protozoa is discussed at a subsequent point. The glyceraldehyde phosphate is converted by an oxidation step plus addition of inorganic phosphate to produce 1,3-diphosphoglyceric acid. The oxidation of the substrate also converts NAD to NADH. NAD is a coreactant (also known as cofactor) in the process, and thus conserves some of the reducing potential of the aldehyde as reduced NAD (NADH). Transfer of the phosphates from 1,3-diphosphoglyceric acid in two steps to ADP yields 2 molecules of ATP for each molecule of glyceric acid. The mainline product of this two step reaction is pyruvic acid. Note that at each of these steps, 2 molecules of ADP are converted to 2 molecules of ATP. This occurs since fructose-1,6-diphosphate is split into two 3-carbon compounds, each leading to 1,3-diphosphoglyceric acid. Hence, the latter part of the pathway generates 4 ATP molecules and 2 pyruvic acid molecules. Since 2 molecules of ATP were utilized initially, the net gain in the conversion of 1 molecule of glucose to pyruvate is 2 ATP molecules. Under anaerobic conditions, the NADH can be reoxidized to NAD by conversion of pyruvic acid to lactic acid. Hence, one of the end products of anaerobic metabolism (anaerobic glycolysis) is lactate. Under some circumstances, part or all of the lactate can be degraded to ethanol with the loss of CO_2.

In aerobic protozoa with typical eukaryotic mitochondria, the pyruvate is converted to acetate by loss of a carbon in the form of CO_2. A molecule of NADH is simultaneously generated. The acetate is not free, but linked to a carrier, coenzyme A (CoA), as acetyl-CoA. This compound enters the internal space (matrix) of the mitochondrion, where the remainder of the oxidative production of ATP occurs. The two carbons carried as acetate are joined to the 4-carbon compound (oxaloacetate) in the Krebs cycle to yield the 6-carbon, tricarboxylic citric acid. Hence, this cycle is also known as the tricarboxylic acid (TCA) cycle (lower half of Figure 15.1). As the TCA cycle proceeds, two carbon atoms are successively cleaved off as CO_2, and in the process, some of the reducing potential of the compounds in the metabolic process is conserved as NADH. NADH molecules and other reducing compounds (eg, $FADH_2$) enter the final part of the oxidative metabolic pathway in the inner membrane of the mitochondrion. Here, the reducing potential of each NADH and its associated chemical energy, are used to produce 3 molecules of ATP for each NADH and 2 molecules of ATP for each $FADH_2$. In the overall reaction, including events in the TCA cycle, however,

3 ATP molecules are acquired for each $FADH_2$. The additional ATP is produced by a substrate-level production of ATP-equivalent molecules (GTP) at the site in the TCA cycle where $FADH_2$ is produced. There are five NADH or equivalent ATP-producing groups generated within the steps leading to and within the TCA cycle. Since 2 molecules of acetate are contributed to the reactions in the TCA cycle, this gives 10 NADH or equivalent ATP producing groups for each glucose consumed. In the inner membrane of the mitochondrion, 3 ATP are produced for each NADH or equivalent group, thus yielding 30 ATP for this part of the reaction. But, we must remember that 2 NADH were also produced during glycolysis, when phosphoglyceraldehyde was converted to phosphoglyceric acid. This yields an additional 6 ATP. Thus, from the total TCA cycle activity and glycolysis, 36 ATP molecules are generated in the mitochondrion. To this 36, we must add the 2 ATP net yield produced when pyruvate was generated from phosphoglyceric acid. Hence, the total net yield is 38 molecules of ATP for each glucose molecule consumed. Compared to the net yield of only 2 molecules of ATP during anaerobic glycolysis, the 38 ATP yield from aerobic metabolism is significantly more efficient.

Fatty Acid Metabolism

Fatty acids are also broken down in the mitochondrion of many eukaryotic cells to yield acetyl-CoA. This process, known as beta-oxidation, sequentially cleaves two carbon fragments from the carboxyl end of the fatty acid to make acetyl-CoA at each step. In the process of cleaving off the two carbon acetyl fragments, the second carbon from the carboxyl end (beta carbon) is oxidized to a carbonyl group and transferred to another coenzyme A, thus preparing the molecule for the next cleavage step as shown in Figure 15.2. By successive cleavage of acetate from the fatty acid, the total molecule is fragmented. Since most fatty acids in living systems contain an even number of carbon atoms, the process eventually leads to complete degradation of the fatty acid and production of acetyl CoA molecules. Thus, in a fatty acid with N carbon atoms, $N/2$ acetyl-CoA molecules are generated. These acetyl-CoA molecules enter the TCA cycle and generate ATP during aerobic phosphorylation, further providing an energy source for the cell.

Amino Acid Metabolism

As illustrated in the diagram, amino acids can also contribute to ATP production through incorporation into the TCA cycle. In this example, deamination of glutamic acid yields alpha-ketoglutaric acid, a component of the TCA cycle. The deamination can occur by action of an enzyme known as glutamate transaminase. This enzyme transfers the amino group to another keto-containing compound, producing an amino acid while leaving the deaminated compound as a keto-acid. For example, oxaloacetate may receive the amino group from glutamic acid and form aspartic acid. The deaminated glutamic acid in this coupled reaction becomes alpha-ketoglutaric acid. Likewise, reversal of this coupled reaction by

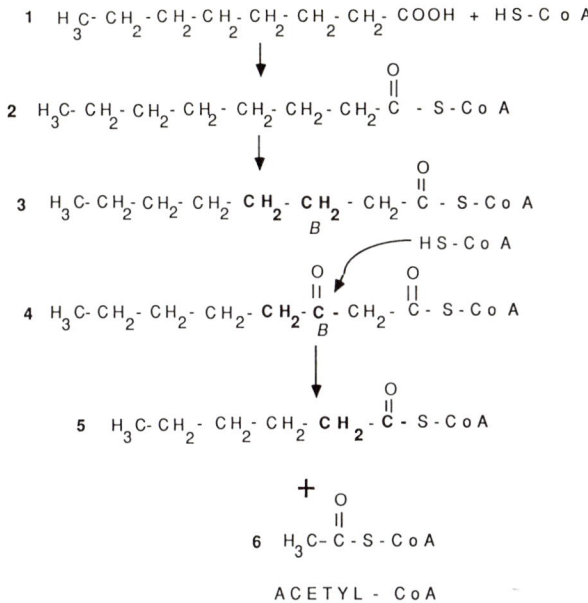

FIGURE 15.2. Major steps in the oxidative degradation of fatty acids yielding one acetyl-CoA molecule for each two carbon fragments cleaved from the fatty acid. In the initial steps, the carboxylic acid group is esterfied to coenzyme A forming an acyl-CoA intermediate. The second carbon from the carbonyl group, known as the beta carbon, is oxidized to a keto group. The terminal two carbon segment is cleaved from the molecule as acetyl-CoA. A coenzyme A molecule is added to the beta carbon, yielding a shorter chain acyl-CoA molecule in preparation for the next phase of the cycle. Each turn of the cycle yields an acetyl-CoA. Since there are usually an even number of carbon atoms in biologically produced fatty acids, all of the carbon atoms are converted to acetyl-CoA during complete beta oxidation.

deamination of aspartic acid produces oxaloacetate, one of the initial components of the TCA cycle.

Thus, amino acids assimilated from the environment by transfer across the plasma membrane or by degradation of cellular protein can be metabolically transformed and contribute to the TCA cycle. The amino acids or protein can also be acquired through ingestion by phagocytic organisms. The metabolic map shown in Figure 15.1 is only a representative one exhibiting some major examples of metabolic interconversions. More detailed examples of variations in the metabolic sequences can be found in standard biochemistry textbooks.

Glyoxylate Cycle and Gluconeogenesis

Conversion of lactate and other alpha-hydroxyacids to metabolically useful compounds such as ketoacids, along with the conversion of alcohol to acetaldehyde, occurs in peroxisomes. These single, membrane-bound organelles contain catalase and also L-alpha-hydroxyacid oxidases mediating the oxidation of the reduced compounds to the metabolically useful forms. Peroxisomes have been documented in a wide variety of protozoa, including flagellates, amoebae, Foraminifera, and ciliates (eg, Hogg 1969; Müller 1969; Anderson and Tuntivate-Choy 1984). These organelles and the morphologically similar microbodies have been found both in plant and animal cells. A significant link in the complex metabolic activity of some protozoa, as well as other eukaryotes, is performed by the microbody (glyoxysome) and the glyoxylic acid cycle (Figure 15.3). The microbody contains enzymes that mediate the conversion of glyoxylate to malate by addition of acetate. The glyoxylate can be derived from the TCA cycle intermediate, isocitrate, through its cleavage by the enzyme isocitrate lyase, yielding 1 molecule of glyoxylate and 1 molecule of succinate. The glyoxylate can combine with acetate in the microbody to yield malate, which also is a component of the TCA cycle. This is an important shunt mechanism in some protozoa lacking enzymes for a complete TCA cycle in the mitochondrion. In these cells, the enzymes mediating the conversion of isocitrate to malate are lacking or defective. Isocitrate is converted to malate by the cooperation of the microbody enzymes, thus completing the TCA cycle. This intracellular cooperation among organelles is an excellent example of how subcellular adaptation through complementary metabolic pathways overcomes enzyme deficiencies in one organelle. Moreover, glyoxylate produced by deamination of glycine (Figure 15.3) provides another pathway for incorporation of amino acids into TCA cycle intermediates; ie, through the production of malate. The formation of malate in the microbody is also a significant step in the process of gluconeogenesis (ie, the production of glucose and glycogen from other intracellular compounds). The malate is converted to oxaloacetate, which is decarboxylated to yield pyruvate. From this point, the glycolysis pathway is reversible to yield glucose. Thus, fatty acids can be converted to glucose by first being cleaved into acetyl-CoA, and through action of the microbody enzymes combined with glyoxylate to yield malate and its degradation product pyruvate. Likewise, the conversion of glycine to glyoxylate and production of oxaloacetate in the microbody provides a route for the conversion of amino acids to glucose (Figure 15.1). The production of glycogen from glucose, however, does not occur by reversal of the glycogen cleavage process. Glucose is released from glycogen polymers by the action of the enzyme phosphorylase. This hydrolyzing enzyme cleaves off the terminal glucose unit and adds a phosphate to yield glucose-1-phosphate. By repeatedly cleaving off the terminal glucose from the polymer, the glycogen is progressively degraded to monomeric glucose. During gluconeogenesis, however, the glucose that has been synthesized by reversal of the glycolysis pathway is incorporated into glycogen by the action of a synthetase, and not phosphorylase. The synthetase, as found in many eukaryotic cells, requires the presence of a precursor strand of glycogen to initiate the polymerization. Also, the glucose must be united with a carrier, uridine diphosphate (UDP). This carrier is a nucleic acid derivative (ie, a diphosphonucleoside) which provides the proper chemical organization for the glucose to be utilized by the synthetase. The presence of two different enzymes

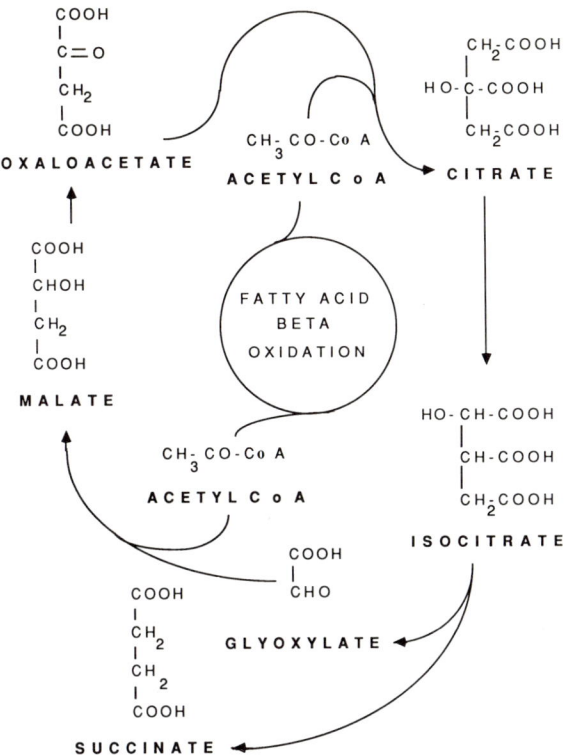

FIGURE 15.3. The glyoxylate cycle. Isocitrate derived from the TCA cycle can be utilized to yield glyoxylate and succinate. Acetyl-CoA, from beta oxidation or other sources, is combined with glyoxylate to yield malate. Since isocitrate and malate are components of the TCA cycle and pass freely across the membranes of mitochondria and glyoxysomes (peroxisomes), it is possible for the glyoxylate cycle to be linked to the TCA cycle. The incorporation of acetyl-CoA into malate by way of the glyoxylate cycle also permits reversal of the glycolysis pathway by utilizing oxaloacetate as an intermediate in the production of pyruvate, and eventually glucose as diagrammed in the reverse steps of Figure 15.1. The amino acid glycine can also be converted to glyoxylate by deamination, and thus contributes to synthesis of intermediates in the cycle.

for the hydrolysis of glycogen (ie, phosphorylase) and for its synthesis (glycogen synthetase) from glucose makes cellular control of these separate pathways more exact. Moreover, further control mechanisms have been observed in aerobic cells to regulate the use of glucose most efficiently. One of these processes, known as the Pasteur effect, suppresses glycolytic fermentation of glucose in the presence of oxygen and favors the production of TCA cycle compounds. This is energetically advantageous, as the aerobic production of ATP as cited above is much more substantial than with fermentation. Hence, by suppressing the fermentation pathway in the presence of oxygen, the cell utilizes the glucose more efficiently in synthesizing as much ATP as possible from each glucose molecule. Clearly, however, in the absence of oxygen fermentation is a useful though less efficient alternative pathway for ATP production. Although some protozoa have been extensively studied to document the presence of these pathways and their control, much additional research is needed to fully document the extent to which protozoan metabolism is similar to or different from other eukaryotic cells.

Hydrogenosomes

Among some anaerobic protozoa, as discussed more fully later, mitochondria are lacking or are not fully active as respiratory organelles. In some of these organisms, a unique organelle known as a hydrogenosome, recently documented in thorough biochemical and fine structural studies by Müller and colleagues (eg, Müller 1980, 1985), serves as a key metabolic center in ATP generation. The name hydrogenosome comes from the observation that protozoa with this organelle generate free molecular hydrogen as a gas during metabolism. The hydrogen gas is generated by reduction of hydronium ions (H^+) to molecular hydrogen at a terminal point in the metabolic pathway. Pyruvate derived from glycolysis is decarboxylated and oxidized to produce acetate in the form of acetyl-CoA (Figure 15.4). During the oxidation step, an intermediate compound (probably a ferredoxin or some other electron transfer protein) accepts the electron. This reduced carrier (shown as Fd^- in the diagram) donates the electron to hydronium ions to produce molecular hydrogen, and is thus reoxidized (Fd). In some of these aerotolerant organisms (capable of existing in an oxgyen atmosphere), the oxygen can be used to reoxidize the electron transfer protein. The acetyl-CoA generated in this part of the process transfers the CoA to succinic acid to make succinyl-CoA and acetate. The energy stored in the succinyl-CoA bond is used to make ATP from ADP. The CoA is transferred to another acetyl group as shown by the recurrent arrow (steps 1 to 5) in Figure 15.4. Thus, the hydrogenosome is a remarkably adaptive metabolic center. Though less efficient than the mitochondrion in generating ATP, it is peculiarly adaptive to function in either anaerobic or aerobic environments. It is an intriguing example of an evolutionary adaptation which permits organisms to exploit environments of varying oxygen availability, while also maintaining an efficient ATP synthesizing cellular machinery.

Intracellular Regulation

The energy stored as ATP is used by the cell in manifold ways to sustain its biosynthetic processes, maintain cellular structure, mediate motility, and establish the unique ionic and molecular composition of its cytoplasm in the presence of a variable and often decidedly different external milieu. For example, the biosynthesis of proteins, lipids, polysaccharides, and other large polymers requires energy activating steps that utilize ATP or its equivalent energy sources. The dynamic properties of the cytoskeletal system and its responsiveness to inter-

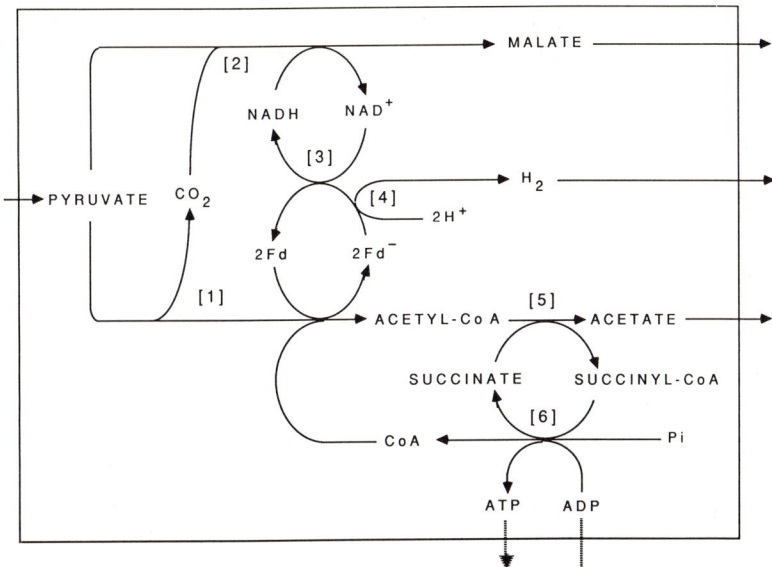

FIGURE 15.4. Some major reactions in the hydrogenosome as found in *Trichomonas foetus* and *T vaginalis*. Arrows indicate the assumed direction of reactions in vivo. Pyruvate is converted by (1) pyruvate/ferredoxin oxidoreductase to acetyl-CoA, or alternatively (2) to malate by malate dehydrogenase (decarboxylating) (NAD). These two reactions are coupled by intermediates (3) that are reversibly oxidized and reduced (NAD and Fd, ferredoxin). Molecular hydrogen (4) is produced during these redox reactions and is detected in the medium of the organism. Hence, the organelle is named an hydrogenosome. The conversion of acetyl-CoA to acetate is coupled to the production of ATP in step (6). Therefore, the reactions are energy conserving and yield ATP to maintain energy requiring processes in the cell. From Steinbüchel and Müller (1986); courtesy of Dr Miklos Müller, Rockefeller University, New York.

nal and external stimuli require energy. Hence, ATP is often used in the continuous redesign of the cellular molecular organization. Ions, and molecules essential to the internal physiological balance of the cytoplasm, are sometimes actively transported across the plasma membrane by carrier molecules that utilize energy stored in ATP. For example, Na^+ is potentially toxic to most cells and is actively transported out of the cell. This process occurs at the expense of ATP that is hydrolyzed to ADP and inorganic phosphate during sodium translocation. Calcium is also translocated outside of the cell by active transport processes. As explained below, some calcium is actively transported across the mitochondrial and ER membranes and stored within membrane-bound spaces inside these organelles. The same carriers may also carry K^+ in the opposite direction during each translocation cycle. These carriers that require ATP as an energy source are designated permeases, or ATPases. The location of the translocating function can be localized or diffusely spread throughout the surface of the cell. There is substantial evidence that in some protozoa, the contractile vacuole is a major site of sodium extrusion. Amino acids, vitamins, and some sugars can also be actively accumulated by the cell. Internal concentrations of some ions are carefully regulated by storage of excess amounts within cytoplasmic organelles. For example, cytoplasmic Ca^{++} is typically maintained at a level of ca 10^{-7} to 10^{-6} M. This is accomplished by actively transporting cytosolic calcium into the mitochondrion and/or endoplasmic reticulum where it is stored. The amount of calcium can become so great in the mitochondrial matrix that it precipitates in the form of calcium phosphate. The level of calcium in the cytoplasm is a critical factor in the chemical control processes of many eukaryotic cells. Evidence is increasingly mounting to show that the concentration of calcium in the cytoplasm is a signal used to regulate cytoplasmic functions. Release of calcium ions from the storage sites in the mitochondrion and endoplasmic reticulum, or by influx from the environment across the plasma membrane, activates a special group of calcium-binding proteins known as calmodulins. These proteins are altered in structure by the binding of calcium to special sites on the calmodulin molecules. Once structurally altered, the calmodulin serves as a signal molecule to regulate other molecular processes in the cell. The binding of the calcium is reversible, so that a reduction in intracytoplasmic calcium levels results in a dissociation of the calcium from the calmodulins. In general, when the calcium ion concentration increases above 10^{-6} M, calmodulins become activated. Although evidence for functions in protozoa are just beginning to accumulate, it appears that the more calcium available in the cytoplasm, the more sites on the calmodulin that are occupied by the calcium. Structural alterations of the calmodulin appear to be correlated with the number of sites occupied by calcium ions; hence, varied signals can be generated by these varied states of calmodulin. This offers a very fine control gradation indeed for coordination of intracellular functions. Some of these control processes involve regulation of motility and contractile mechanisms in protozoan cells as discussed more fully in Chapter 18.

The many reactions characteristic of life are mediated by biological catalysts known as enzymes. These are typically protein molecules, although there is some evidence that RNA may act as a catalyst for modification of other RNAs. The coordinated and closely regulated sequence of chemical reactions in the cell is maintained in part by a system of feedback controls on enzyme activity. Feedback control occurs when chemical products in one part of the metabolic pathway are capable of binding to surface sites on the enzyme and alter its activity. Those feedback regulators that increase enzyme activity are known as activators, and those that decrease activity are designated as inhibitors. In general, the action of activators is called positive feedback control, whereas the action of inhibitors is called negative feedback control. Positive feedback control, resulting in increase of the product catalyzed by the enzyme, is indicated by (+). Negative feedback control, resulting in diminished product formation at the site of the enzyme controlled reaction, is indicated by (−). The effect of feedback control by products within metabolic pathways ensures that a steady and coordinated production of cellular products is maintained during metabolism (eg, Figure 15.7).

Specialized Metabolic Pathways in Flagellates

Many flagellates produce metabolic products and contain enzymes indicating the presence of the Embden-Meyerhof pathway for glucose metabolism (eg, Lindblom 1961; Belsky and Schultz 1962; Honigberg 1978a; Mansfield 1978; Klein and Miller 1981), and many have typical eukaryotic aerobic metabolic pathways. However, substantial variations occur in the further metabolism of carbon intermediates derived from the glycolytic pathway.

The aerobic flagellate *Trypanosoma brucei*, when in the long slender form found in the blood stream, lacks a functional TCA cycle and lactate dehydrogenase. Consequently, no lactate is produced by fermentation. The major end product of glycolysis is pyruvate (Fairlamb 1981; Opperdoes 1981). All of the glycolytic enzymes are sufficiently abundant to account for the observed glycolytic activity. They are localized in three subcellular compartments: the glycosome (microbody), mitochondrion, and the cytoplasm. Two moles of pyruvate are produced from each mole of glucose, and 1 mole from each mole of glycerol. The glycerol can enter the glycolysis pathway by conversion to glycerophosphate and incorporation into dihydroxyacetone phosphate (as shown in Figure 15.1). Since there is no functional TCA cycle in these species, pyruvate is excreted as an end product of metabolism. NAD is regenerated from NADH by a glycerophosphate shuttle linking the glycosome and mitochondrion. Uptake of oxygen oxidizes the intermediates in the electron transport chain of the mitochondrion, thereby providing a sink for the electrons removed from the glycerophosphate. The enzyme mediating this step is glycerophosphate oxidase. No ATP is apparently generated in the mitochondrial uptake of oxygen to form water. In this process, dihydroxyacetone phosphate is regenerated and oxidized NAD is produced. All of the ATP is produced by substrate level phosphorylation of ADP. By contrast, the exoerythrocytic stage of the malarial parasite *P berghei* (a sporozoan, not a flagellate) contains cytochrome oxidase activity, indicating ATP production by respiration. Moreover, several key enzymes in the Embden-Meyerhoff glycolytic pathway and the TCA cycle have been detected using histochemical techniques (Meis et al 1984). The production of nucleic acid intermediates (eg, the pyrimidine thymidine) involves the ubiquinones and oxygen metabolism (Sinden 1986).

The blood trypomastigote and intracellular amastigote forms of the flagellate *Trypanosoma cruzi* all utilize oxygen. Glucose is metabolized to form CO_2 and intermediates for the TCA cycle. The presence of a complete TCA cycle and functional mitochondrion suggests a typical aerobic pathway for carbon metabolism. Amino acids and carbohydrates are the main exogenous energy sources, but triglycerides appear to be the main endogenous energy reserve (Gutteridge 1981). The promastigotes of *Leishmania* are reported to utilize carbohydrates via glycolysis and the TCA cycle. In addition, the amino acid proline can act as a major energy source at least in *L donovani*. Under these circumstances, glycolysis appears to be less significant (Coombs 1981). Transport systems in the plasma membrane for uptake of glucose and several amino acids have been identified and partially characterized.

In addition to organic sources of carbon, dark CO_2 fixation has been demonstrated in several free-living and parasitic flagellates (Figure 15.7). For example, in *Euglena gracilis* Z variations of carbon dioxide uptake during the growth cycle indicate that this fixation of inorganic carbon into organic compounds is an anaplerotic replenishment of components in the TCA cycle (Peak and Peak 1981). Current evidence suggests that CO_2 is joined with pyruvate to form oxaloacetate by action of the enzyme phosphoenolpyruvate carboxylase (PEPC). This enzyme catalyzes the addition of CO_2 to the pyruvate as a carboxyl, thus forming the dicarboxylic acid oxaloacetate, one of the major constituents in the TCA cycle (Figure 15.1). By comparison, CO_2 is also fixed by *T brucei*, but by action of the enzyme phosphenolpyruvate carboxykinase (Klein 1981). No PEPC was found.

The incorporation of CO_2 into TCA intermediates in *E gracilis* is apparently related to nitrogen deficiency of the cell. Cells subjected to ammonia starvation showed marked increases in PEPC activity and incorporation of CO_2 into TCA intermediates when the medium was replenished with NH_3 (Peak and Peak 1981). The increase is more pronounced in heterotrophically grown cells ($10\times$) as compared to autotrophically grown cells ($1.55\times$). The resulting uptake of ammonia and increase in TCA intermediates capable of uniting with the ammonia to form amino acids (Figure 15.1) may provide the source for amino acid biosynthesis in the nitrogen depleted cells. Some proposed feedback regulation loops by metabolic products in the TCA cycle are also shown in Figure 15.7 (Peak and Peak 1981). Similar nitrogen regulation of the amino acid degrading enzyme L-asparaginase in *Chlamydomonas* spp was found by Paul and Cooksey (1981). Cells grown in nitrogen-free medium had a 2- to 3.5-fold increase in activity after 24 to 48 hr. L-asparaginase degrades asparagine to asparatic acid and free NH_4^+, thus providing additional sources of nitrogen at the expense of the cytoplasmic asparagine pool. The increased enzyme activity did not disappear in the presence of high concentrations of nitrogen when added to the medium, indicating absence of an asparaginase-degrading system.

Carbon sources other than glucose can also sustain metabolism in a unique group of protozoa known as the acetate flagellates. Among this group of protozoa are many euglenoids (eg, *Astasia* and *Euglena*), some volvocidans (eg, *Polytomella* and *Polytoma*), chrysomonads (eg, *Ochromonas*), and some parasitic kinetoplastids (ie, *Trypanosoma cruzi*; Docampo et al 1979). The free-living acetate flagellates are capable of utilizing organic acids, especially acetate, and some of the intermediates in the TCA cycle as substrate. These compounds are often toxic to other protozoa, particularly at lower pH, due to the ease with which protonated forms of the acid (RCOOH) penetrate the plasma membrane and produce toxic concentrations in the cytoplasm. The acetate flagellates are more resistant to low pH effects, but are also sensitive below a critical pH threshold. Some of the acetate flagellates apparently utilize membrane transporting mechanisms to maintain the proper intracellular concentration of the substrate. The metabolic pathway for acetate incorporation, as exemplified by *Astasia longa*, is probably by conversion of acetate to acetyl-CoA (Hunter and Lee 1962). The essential enzyme, acetic-thiokinase, is present and active in cell-free extracts. Condensing enzyme mediates condensation of acetyl-CoA and oxaloacetate to yield citrate, thus initiating TCA cycle metabolism of the acetate carbon. Exogenous sources of TCA intermediates and some amino acids including alpha-ketoglutarate, glutamine, malate, aspartate, serine, citrate, and succinate were assimilated, and induced respiratory stimulation ranging from 6 to 21% as measured by oxygen consumption. However, DL-alanine, *cis*-aconitate, isocitrate, and oxaloacetate were inert (Hunter and Lee 1962). In general, stimulatory effects of assimilated exogenous organic acids were greater at a lower pH (5.0) than at a higher pH (7.0), suggesting again that permeability of the membrane is greater for the protonated form of the substrate. The metabolic fate of acetate in resting populations of *Euglena gracilis* was reviewed by Collyard and Danforth (1970). They report that when acetate is the sole source of carbon and energy, 42% of the acetate carbon is oxidized to CO_2 and the remaining 58% is assimilated, mainly by conversion to carbohydrate reserves. During starvation, *Euglena* oxidizes carbohydrate reserves at a rate corresponding to an O_2 uptake of about 8 to 10 µl/10^6 cells per hour. Oxidation of the reserves continues when acetate is supplied to the starved cells, but evidence points to a separate compartment for metabolism of carbohydrate reserves and utilization of exogenous acetate. This is also true for the metabolism of ethanol and pyruvate, which appear to be oxidized separately from carbohydrate reserves. The rather constant proportion of exogenous carbon oxidized to carbon dioxide, compared to carbohydrate synthesized, suggests that these two processes are dynamically and tightly linked by cellular control processes (Collyard and Danforth 1970). One plausible hypothesis is that the oxidation of acetate is enhanced by the presence of ADP (a product of assimilation of acetate into carbohydrate), and that carbohydrate production is stimulated by the presence of ATP (a product of acetate oxidation). Thus, the prevailing balance of ADP to ATP in the cell would govern the relative rates of acetate oxidation versus acetate assimilation into carbohydrate. Since the two processes are linked by the interconversion of ATP to ADP, a steady ratio of the two processes would be expected to develop in the cell leading to the .42/.58 ratio. Large quantities of carbohydrate (paramylon) are produced at the beginning of exponential growth, when *E gracilis* is cultured in a medium with DL-lactate as the sole carbon source (Briand and Calvayrac 1980). C-14 labeled CO_2 is also incorporated during dark growth, and the labeled carbon appears in paramylon in varying amounts depending on the physiological state of the cells. It is most intense in the exponential phase of heterotrophic growth. The carbohydrate reserves, accumulated within transformed and vesiculated mitochondria during carbon depletion, can be utilized as a major source of energy production. A fully active Embden–Meyerhof pathway has been shown to be present in many euglenoids, including *Astasia longa* (Barry 1962).

Some current knowledge of "acetate metabolism" in *Polytomella* has been succinctly reviewed by de la Cruz and Gittleson (1981). Acetate-grown cells can be adapted to utilize propionate and butyrate as substrates, and there is good evidence that at least propionate is actively transported into the cell, since its uptake is ATP-dependent (Cantor and Withroe 1970). A similar utilization of propionate has been demonstrated in *E gracilis* strain Z (Hosotani et al 1980). Illuminated cells grew with propionate as the sole carbon source, but could not utilize propionate when grown in the dark. Labeled carbon from propionate was first recovered in the amino acid fraction and gradually transferred to protein and paramylon during subsequent incubation of the cells. In the amino acid fraction, aspartic acid was most densely labeled initially, followed by glutamic acid, indicating that propionate is metabolized by way of succinate. The illuminated cells apparently assimilate propionate through propionyl-CoA and methylmalonyl-CoA into tricarboxylic acid cycle intermediates for growth. Ethanol-grown cells of *Euglena gracilis*, when placed in acetate, frequently adapt over an 80 min period, gradually assuming metabolic characteristics of cells that have been maintained solely on acetate as the carbon source (Danforth and Wilson 1957). These characteristics include a higher acetate oxidation rate at pH 7.0 than for ethanol grown cells. The rate of acetate oxidation in acetate-grown cells is relatively independent of pH, whereas that of ethanol-grown cells is markedly increased at low pH. Moreover, at pH 5.5 the respiratory rate of acetate-grown cells is independent of acetate concentration above 4 mM, but for ethanol-grown cells, the respiratory rate can be increased by increasing acetate concentrations up to 18.4 mM. Among the exogenous substrates utilized by *Polytomella* are the organic acids butyrate, valerate, succinate, alpha-ketoglutarate, and lactate; the alcohols ethanol, butyl alcohol, and hexyl alcohol; and intermediates in the glycyolysis pathway including phosphoenolpyruvate and pyruvate under controlled growth conditions (Wise 1955; de la Cruz and Gittleson 1981). Substantial evidence indicates that the acetate, as acetyl-CoA, can be utilized in the microbody to yield succinate that is incorporated in cell syntheses. In acetate medium, two molecules of active acetate combine to yield one molecule of succinate. Although key enzymes in the glyoxylate cycle are present in colorless and pigmented volvocidan acetate flagellates, current evidence suggests that these enzymes are present in the mitochondrion rather than in the microbody (Lembi 1980). Mitochondrial localization of glycolate dehydrogenase has been confirmed cytologically in the pigmented species *Chlamydomonas reinhardtii* (Beezley et al 1976). The presence of microbodies in *Chlamydomonas*, *Chlorogonium*, *Polytoma*, and *Polytomella* has been confirmed ultrastructurally (eg, Lembi 1980), but although catalase is present, it does not appear to be specialized in function and is not associated with either glycolate or glyoxylate metabolism in *Chlorogonium* (Stabenau and Beevers 1974). Catalase is an enzyme that converts hydrogen peroxide to water. Clear evidence indicates the presence of glyoxylate cycle enzymes in the glyoxysomes of *Euglena* (Graves et al 1972) as occurs in the microbodies of some higher eukaryotes. The glyoxysomes of *Euglena*, however, do not contain catalase as found in microbodies of other eukaryotes. The beta-oxidation enzymes catalyzing the degradation of fatty acids are abundant in the glyoxysome of *Euglena* grown heterotrophically on the 6-carbon fatty acid hexanoate. Under these conditions, the beta-oxidation enzymes (see Figure 15.2) are 20 times more active in the glyoxysome than in the mitochondria (Graves and Becker 1974). This suggests that the beta-oxidation of fatty acids occurs almost entirely in the glyoxysome of *E gracilis* when grown heterotrophically on hexanoate. Therefore, the entire portion of the glyconeogenic pathway from fatty acid to succinate is localized in the glyoxysome of *Euglena*. This is in marked contrast to the mitochondrial localization of fatty acid degradation in higher eukaryotic cells, and further exemplifies that acetate (acetyl-CoA) assimilation during gluconeogenesis undoubtedly occurs in the glyoxysome.

Further studies on the effects of glucose versus acetate metabolism in *Euglena gracilis* indicate that the glyoxylate cycle operates at a high level in the presence of acetate, but is suppressed in the presence of glucose (Heinrich and Cook 1967). Large transients in CO_2 fixation occur during glucose metabolism, but not when acetate or enthanol replace glucose as the carbon source. The enhanced activity of the glyoxylate cycle in the absence of glucose, and possible replenishment of the TCA cycle intermediates by this pathway, probably spares the necessity to incorporate CO_2 into oxaloacetate as a source of TCA intermediates. TCA cycle intermediates and 3-phosphoglycerate inhibit PEP carboxylase and block this route of CO_2 fixation (Lloyd and Cantor 1979). Although glucose caused a moderate depression of many of the enzymes of the glyoxylate cycle, it neither repressed nor inhibited malate synthase, the enzyme that produces malic acid from glyoxylate and acetate. A stimulatory effect of acetate uptake by glucose has been observed, however, in some volvocidans. Cassin (1974) reported that the acidophilic flagellate *Chlamydomonas acidophila*, which grows well at pH 2.0 but not at pH 7.0 or 8.0, can assimilate acetate only in the presence of glucose. Her data also suggest that the acid tolerant growth of this species can be explained by H^+ exclusion from intracellular compartments, possibly by a barrier at the membranous envelope of the peripherally located plastids. Similar mechanisms of H^+ exclusion in pigmented algae inhabiting acid waters of strip mine pits have been suggested by Brock (1973).

Chrysomonads and prymnesiomonads possess a wide range of capacities to assimilate inorganic and organic sources of carbon for energy and biosynthesis (Droop 1974; Aaronson 1980). Although at least three chrysomonads have been shown to be obligate phototrophs, since they cannot grow in the dark either on glucose or on acetate (*Mallomonas epithelattia*, *Microglena arenicola*, and *Paascherella tetras*), many are capable of utilizing organic acids, carbohydrates, and alcohols as a carbon source (Aaronson 1980). For example, the marginally phototrophic chrysomonad *Ochromonas malhamensis* assimilates acetate, pyruvate, glucose, fructose, sucrose, ethanol, and glycerol when grown in the dark. The volvocidan *Pandorina morum* incorporates acetate carbon into insoluble lipoidal and carbohydrate reserves when grown in the light, but not when dark-grown (Palmer and Togasaki 1971). Although a strain capable of growing in the dark incorporated acetate carbon into polysaccharide reserves, both strains exhibited only limited conversion of acetate to carbohydrate even when grown in light. The prymnesiomonads *Pavlova gyrans* and *Syracosphaera carterae* utilize glycolate when grown in light (Aaronson 1980). By comparison, the euglenoid *Euglena gracilis* can utilize glycolate and the amino acids glycine and serine as sole carbon source in CO_2-free atmosphere when grown in light, but not when grown in the dark (Murray et al 1970). In the presence of glucose, however, dark-grown cells were capable of assimilating glycolate. Of several related metabolic compounds, it is of interest to note that only glycolate was utilized as the sole carbon source in light under "anaerobic" conditions. Glycerol is utilized by some prymnesiomonads, including *Prymnesium parvum* grown in the dark. By comparison, only one cryptomonad, *Chroomonas salina*, has been shown to grow on glycerol in either light or darkness (eg, Antia et al 1979). Photoheterotrophic growth on glycerol was greater than phototrophic growth in the absence of glycerol, or in chemoheterotrophic growth in darkness with glycerol. During extended chemoheterotrophic growth, the cells become cream-colored, suggesting a reduction of the photosynthetic pigment concentration (reviewed by Gantt 1980). Glycolate enzymology of nine species of prymnesiomonads indicates the presence of glycolate dehydrogenase as occurs in some volvocidans. However, carbohydrate photosynthates are different among most prymnesiomonads and volvocidans. The latter produce largely sucrose, fructose, glycerol, and glucose; whereas the former contain principally mannitol, although glucose is also present in some species (Norris 1980). Carbohydrate storage products also discriminate among some of the commonly cited acetate flagellates. For example, euglenoids as represented by *Euglena* spp contain paramylon, a beta-1,3 glucan as opposed to amylose, an alpha-1,4 glucan commonly found in green monads. Amylopectin is a branched polymer containing an alpha-1,6 glycosidic linkage at each branch point, in addition to the linear alpha-1,4 linkage. It is also a common reserve product in some green monads. The major carbohydrate reserve of chyrsomonads and prymnesiomonads is leucosin, also a beta-1,3 glucan, but of different polymer organization than paramylon.

Acetate metabolism, indicated by a substantial increase in respiration in the presence of acetate, has been documented in a blood stream form of *Trypanosoma cruzi* (Docampo et al 1979). Acetate oxidation occurred via the TCA cycle and respiratory activity was inhibited by the mitochondrial blocking agent antimycin A, suggesting that terminal metabolism of the acetate occurs in the mitochondrion. There was a 6 to 7% residual oxygen uptake that was not inhibited by antimycin concentrations. Acetate is the end product of metabolism in a group of hydrogen-producing anaerobic flagellates containing the unique cytoplasmic organelle known as a hydrogenosome (Müller 1985). Although these organelles were discovered in protozoa, more recently they were shown to also be present in anaerobic fungi (Yarlett et al 1986). Some anaerobic and aerotolerant flagellates, such as those found in the termite gut, rumen alimentary canal, or as pathogens in mammalian systems, use hydrogenosomes (Müller 1981, 1985) as the final electron acceptor. Trichomonads excrete a certain amount of glucose carbon as acetate formed by activity of the hydrogenosome (Figure 15.4). Moreover, the evolution of molecular hydrogen in low oxygen environments is clear evidence for the presence of this organelle. Approximately 50% of glucose carbon is converted to succinate and 50% to acetate, suggesting that at least half of the glucose carbon enters the hydrogenosome. Lactic acid is the predominant end product of anaerobic glycolysis (Doran 1959). *Trichomonas vaginalis*, one of the most widely studied hydrogenosome-containing trichomonads, is aerotolerant and can subsist in oxygen concentrations up to at least 20% by volume. Under these circumstances, it is possible for the oxygen to serve as the electron acceptor (as shown in Figure 15.4). The aerotolerant intestinal flagellate *Giardia lamblia* produces ethanol, acetate, and CO_2, either anaerobically or aerobically, using exogenous or endogenous glucose sources. But, molecular hydrogen is not produced. It can respire in the presence of O_2 and endogenous respiration is stimulated by glucose, but not by other carbohydrates and TCA cycle intermediates (Lindmark 1980). The current biochemical evidence suggests the occurrence of glycolysis, energy production by substrate level phosphorylation, and a flavin-based, iron–sulfur protein mediated electron transport system. Results of studies with enzyme blocking inhibitors indicate the absence of cytochrome-mediated oxidative phosphorylation and a functional TCA cycle (Lindmark 1980). In the absence of hydrogen evolution by *Giardia*, it is questionable if these organisms possess a fully functional hydrogenosome as found in the trichomonads. Hydrogenosomes have been conclusively demonstrated in three trichomonad species (*Tritrichomonas foetus*, *T vaginalis*, and *Monocercomonas* sp) (Müller 1985). The production of hydrogen gas, acetate, and carbon dioxide by obligate intestinal flagellates (eg, Hypermastigida of termites and wood roaches) may also indicate a hydrogenosome contribution to their metabolism (Müller 1980). They also possess intracellular organelles resembling hydrogenosomes in fine structure. It is interesting to note that the production of molecular hydrogen as a metabolic end product is also characteristic of some bacteria of a clostridial type. Hydrogen-producing bacteria are also known to inhabit anaerobic environments such as the intestinal tract. The similarity in biochemistry of these monerans and the hydrogenosome, coupled with the fine structural observation that the hydrogenosome is probably enclosed by a complex double membrane, suggests that the hydrogenosome may have arisen by a bacterial endosymbiosis during evolution of the flagellates (Müller 1980).

Some Specialized Metabolic Pathways in Amoebae

The elucidation of metabolic pathways requires sufficient cell mass to permit chemical analyses, and an axenic and preferably adequately defined medium to ensure unambiguous conclusions. Clearly, cells grown in the presence of other organisms, even those consumed as food (nonaxenic culture), may not be sufficiently free of the cooccurring organisms to permit unambiguous detection of metabolic pathways that are specific to the protozoan. The soil-dwelling amoeba *Acanthamoeba*, and the parasitic amoeba *Entamoeba* have been especially productive sources of information on metabolism, as they have been cultured axenically in sufficient quantities to permit refined biochemical analyses of whole cells and substituent organelles (eg, Diamond 1961; Fulton 1969; Reeves et al 1977; Albach and Booden 1978). Clearly, there is much uncharted territory to be discovered in elucidating the metabolic pathways of many free-living as well as parasitic species of amoeboid organisms.

Acanthamoeba and other hartmanellid amoebae possess an Embden–Meyerhof pathway and can utilize glucose as a sole carbon source, or in some strains, sucrose (eg, Adam and Blewett 1967; Dolphin 1976). *Naegleria gruberi*, however, while exhibiting aerobic metabolism, does not assimilate exogenous glucose or polysaccharides. Apparently, amino acids form the main carbon source during all growth phases (Weik and John 1977). *Acanthamoeba castellanii* can be grown on a defined medium containing only five amino acids when glucose is the sole carbon source (Adam 1964; Dolphin 1976). These are: (1) arginine (amino side group); (2) methionine (sulfur-containing side group); (3) leucine; (4) isoleucine; and (5) valine (hydrocarbon side groups). When, however, acetate, is the sole carbon source, glycine must also be provided (Dolphin 1976). The most common free intracellular amino acids, when the amoeba is grown on glucose, are alanine, proline, glycine, glutamic acid, and serine. These can be derived from the intermediates of the Embden–Meyerhof (glycolytic) pathway when glucose is catabolized. The requirement for glycine when acetate is the sole carbon source suggests the following metabolic concept (Dolphin 1976). *Acanthamoeba* has a functional glyoxylate pathway that supports glucose production by gluconeogenesis, hence generating intermediates in the Embden–Meyerhof pathway that can be used to synthesize required amino acids. However, this pathway, which is based on acetate metabolism, may be too slow to furnish an adequate pool of free amino acids. The addition of glycine could supplement the glyoxylate cycle (Figure 15.3) and enhance gluconeogenesis contributing to the synthesis of adequate pools of the major free amino acids. In this free-living amoeba, therefore, some of the major metabolic pathways for energy production and sources of carbon for amino acid synthesis can be explained by appealing to classic examples of eukaryotic glycolytic pathways and anaplerotic (replenishing) functions of the glyoxylate cycle.

The metabolism of pathogenic amoebae has been investigated with particular interest in the hope of finding unique pathways that could be blocked by pathogen-specific inhibitors. Thus, these inhibitors may be administered as drugs to eradicate the pathogen without harming the host. *Entamoeba histolytica* is of particular interest. In contrast to *Acanthamoeba*, where mitochondria with a functional TCA cycle are demonstrable (eg, Waidyasekera and Kitching 1975), no mitochondria have been reported in *E histolytica* (Albach and Booden 1978). And, neither TCA cycle nor respiratory chain activity have been observed. The cells can grow, however, in oxygen and produce acetate with consumption of O_2. Among a wide variety of substrates examined, including carbohydrates, carboxylic acids, fatty acids, and amino acids, only glucose and L-serine were found to stimulate oxygen uptake (Weinbach and Diamond 1974). The anaerobic breakdown of glucose to CO_2 and pyruvate indicates an active glycolytic pathway. Also, clear evidence of all intermediates in the Embden–Meyerhof pathway have been found in the Laredo strain of *E histolytica*, which grows at a lower tempera-

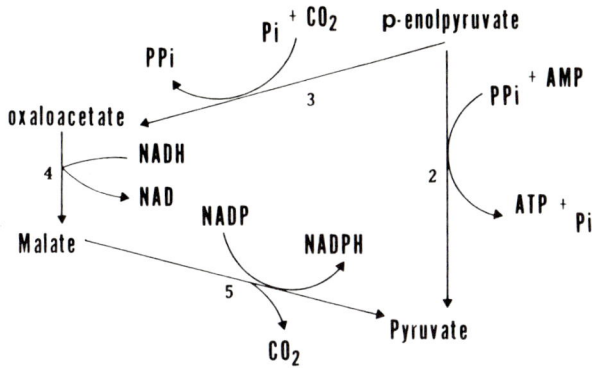

FIGURE 15.5. Four-enzyme pathway of *Entamoeba histolytica*. Pyruvate is a common end product of all reactions commencing either with oxaloacetate or phosphoenolpyruvate. NADPH and ATP are generated during steps (5) and (2), respectively. From Reeves et al (1977) and McLaughlin and Aley (1985); with permission.

ture than do the normal strains of the parasite (Bragg and Reeves 1962). This organism ferments glucose to CO_2, acetic acid, and ethanol as main products (McLaughlin and Aley 1985).

The pathway for production of acetate in the parasitic strains of *E histolytica* is of particular interest. Current knowledge (Reeves 1970) suggests a four-enzyme step pathway for pyruvate production (Figure 15.5), and further oxidation of pyruvate to acetate by the combined action of two enzymes. In the four-step pathway leading to pyruvate production, a direct pathway (2) catalyzed by the enzyme pyruvate phosphate dikinase (PPDK) converts phosphoenolpyruvate to pyruvate, and generates one molecule of ATP. Alternative pathways lead to oxaloacetate (3) with the incorporation of CO_2. This step is mediated by the enzyme phosphopyruvate carboxylase. In step (4), NAD is regenerated from NADH by production of malate from oxaloacetate. The enzyme is malate dehydrogenase. In step (5), pyruvate is produced from malate by loss of CO_2 and generation of a reducing equivalent NADPH through the action of malic enzyme.

Pyruvate is converted to acetyl-CoA by oxidative decarboxylation of the pyruvate. Two pathways are possible. Pyruvate oxidase (Takeuchi et al 1975) converts pyruvate in the presence of O_2 and CoA to acetyl-CoA + CO_2 and H_2O_2. Pyruvate synthase catalyzes the conversion of pyruvate to acetyl-CoA with elimination of CO_2, and the release of two electrons that can be accepted by flavin adenine dinucleotide (FAD), flavin mononucleotide (FMN), riboflavin, and ferredoxin, but not NAD or NADP (Reeves et al 1977). The two reactions are as follows:

$$\text{pyruvate} + \text{CoA} + O_2 = \text{acetyl-CoA} + CO_2 + H_2O_2 . \quad (1)$$

$$\text{pyruvate} + \text{CoA} = \text{acetyl-CoA} + CO_2 + 2e^- . \quad (2)$$

The acetyl-CoA is converted to acetate and ATP by action of the enzyme acetyl-CoA-synthetase (Reeves et al 1977) as follows:

$$\text{acetyl-CoA} + \text{ADP} + P_i = \text{acetate} + \text{ATP} + \text{CoA} , \qquad (3)$$

where P_i is inorganic phosphate.

Earlier reports that *E histolytica* produces hydrogen in the presence of bacteria as prey and contains a hydrogenase have been shown to be incorrect (McLaughlin and Aley 1985). The results were undoubtedly due to contamination with food bacteria, illustrating the importance of having axenic cultures for biochemical analysis. No evidence of hydrogenosomes has been found in carefully prepared samples of *E histolytica* (Müller 1980). Further details of metabolic pathways in *Entamoeba* can be found in the comprehensive review by McLaughlin and Aley (1985).

Some Metabolic Pathways in the Ciliate *Tetrahymena*

The development of a chemically defined medium for culture of *Tetrahymena* (eg, Kidder and Dewey 1951; Hutner et al 1972; Holz 1973) has permitted extensive and carefully documented biochemical analyses of metabolism in this ciliate. It is probably one of the most extensively studied ciliates from a biochemical perspective. Some current knowledge of carbon metabolism comparable to the main themes developed above for flagellates and amoebae will be summarized here.

Glucose is readily assimilated by *Tetrahymena* and substantial evidence indicates a fully functional Embden–Meyerhof pathway (eg, Shrago and Elson 1980). Under anaerobic conditions, pyruvate can be converted to lactate, providing the main source of ATP. The enzyme converting pyruvate to lactate (lactate dehydrogenase) is not in the cytosol where it is found in many eukaryotes, but in the mitochondrion. There is considerable evidence that a significant part of the glycolytic pathway, commencing with fructose-6-phosphate, is duplicated in the mitochondrion of *Tetrahymena* (eg, Shrago and Elson 1980). Hence, all of the necessary enzymes for anaerobic metabolism of fructose-6-phosphate to lactate are present in the mitochondrion. This is clearly different from most eukaryotic cells, where glycolysis occurs substantially if not totally in the cytoplasm.

In *Tetrahymena*, approximately 80% of the glucose carbon is converted to glycogen, and very little carbohydrate is used for fatty acid and protein synthesis via intermediates of glycolysis and the TCA cycle (Borowitz et al 1977; Shrago and Elson 1980). *Tetrahymena* utilizes acetate avidly, both in the natural environment and in laboratory cultures. A functional glyoxylate cycle, including all of the essential enzymes for initial steps in gluconeogenesis, has been demonstrated in peroxisomal (glyoxysomal) organelles (eg, Hogg 1969). The compartmentation of enzymes in membrane-bound organelles of glycogenically active *Tetrahymena* was determined by Müller et al (1968). They found that the mitochondria possess about 20% of the cellular protein, and contain the bulk of the respiratory enzyme activity and most of the enzymes mediating the TCA cycle. The microbodies contained the bulk of the catalase, alpha-hydroxy acid oxidase (converting alpha-hydroxy acids such as lactate to keto-acids), glyoxylate oxidase, glycolate oxidase, isocitrate lyase, and malate synthase activities. The isocitrate lyase, catalyzing the cleavage of isocitrate to succinate and glyoxylate, provides the source of carbons combined with acetyl-CoA to produce malate (by action of malate synthase). Thus, as with some flagellates and amoebae, TCA cycle intermediates can be channeled into gluconeogenesis through the glyoxylate cycle. Moreover, during anaerobiosis, substantial amounts of lipid (fatty acids) is converted to glycogen by way of the glyoxylate cycle. Some of this activity is incorporated in autophagosomal vacuoles ("glycosynthesomes") containing precursor glycogen, along with mitochondrial and glyoxysomal enzyme activities essential to glycogen synthesis (May et al 1982). Large masses of glycogen are observed in the vacuoles and also in the surrounding cytoplasm during later phases of gluconeogenesis. The extensive production of glycogen from fatty acids during anaerobiosis provides a carbon source that can be utilized by glycolysis to ensure continued ATP production.

Carbon dioxide incorporation into oxaloacetate occurs by way of phosphoenolpyruvate carboxylase (Shrago and Elson 1980), and not by the enzymatic action of pyruvate carboxylase as observed in many microorganisms and higher animals. In this respect, *Tetrahymena* resembles some acetate flagellates (eg, *Euglena*) and *Entamoeba histolytica*, which also utilize PEP carboxylase in metabolic fixation of CO_2 (eg, Figures 15.6 and 15.7). The relative proportion of glucose carbon appearing in metabolic intermediates during anaerobiosis and CO_2 fixation was determined by Ryley (1952) as follows:

$$2 \text{ glucose} + CO_2 = 2 \text{ succinate} + 1 \text{ lactate} + 1 \text{ acetate} .$$

The incorporation of CO_2 into succinate is apparently a necessary step in anaerobic supplementation of the TCA cycle intermediates, since in the absence of CO_2 glycogen is converted only to lactate. A possible metabolic pathway leading from glycogen to succinate is as follows (Shrago and Elson 1980):

Glycogen → phosphoenolpyruvate

Phosphoenolpyruvate + CO_2 → oxaloacetate

Oxaloacetate + NADH + H^+ ↔ malate + NAD^+

Malate ↔ fumarate

Fumarate + NADH + H^+ + ADP → succinate + NAD^+ + ATP .

During gluconeogenesis, the amino acids tyrosine, leucine, isoleucine, and alanine are effectively incorporated into glycogen. However, glutamic acid and serine are relatively poor precursors, and aspartic acid is not utilized (Mavrides 1973; Shrago and Elson 1980). In the absence of acetate, these amino acids are significant sources of TCA intermediates funneled into the glyoxylate cycle. Tyrosine, for example, is first deaminated by an enzyme, tyrosine transaminase, that transfers the amino group to a receptor keto-acid, converting it into an amino acid, and transforming the tyrosine into a keto-containing compound. By further reorganization of the molecule and opening of the benzene ring, an intermediate

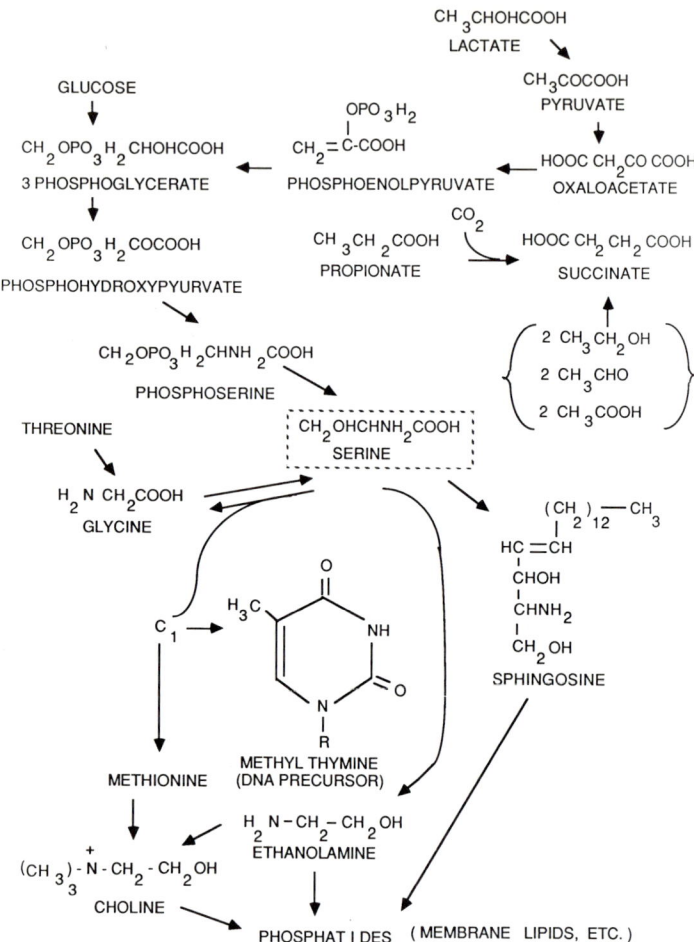

FIGURE 15.6. Some reactions leading to the synthesis of the amino acid serine and its contribution to other compounds of biological significance. Glucose, lactate, pyruvate, or oxaloacetate are utilized by way of a common intermediate, 3-phosphoglycerate, to produce serine (within dashed-line box). The amino acids threonine or glycine can also serve as sources. Products derived from the intermediate production of serine include the amino acid methionine, the DNA base precursor methyl thymine, and large molecular weight compounds such as sphingosine and phosphatides used in cell membrane synthesis. Based on data from Dewey and Kidder (1960) and Kidder (1967); with permission.

(fumarylacetoacetic acid) is produced. This is hydrolytically cleaved to yield fumaric acid and acetoacetate. The fumaric acid is an intermediate in the TCA cycle. Leucine is converted to acetyl-CoA and acetoacetate by a pathway commencing with a deamination and incorporation of CO_2. This fixation of CO_2 during leucine catabolism may be as significant quantitatively as the fixation into oxaloacetate by PEP carboxylase (Borowitz et al 1977).

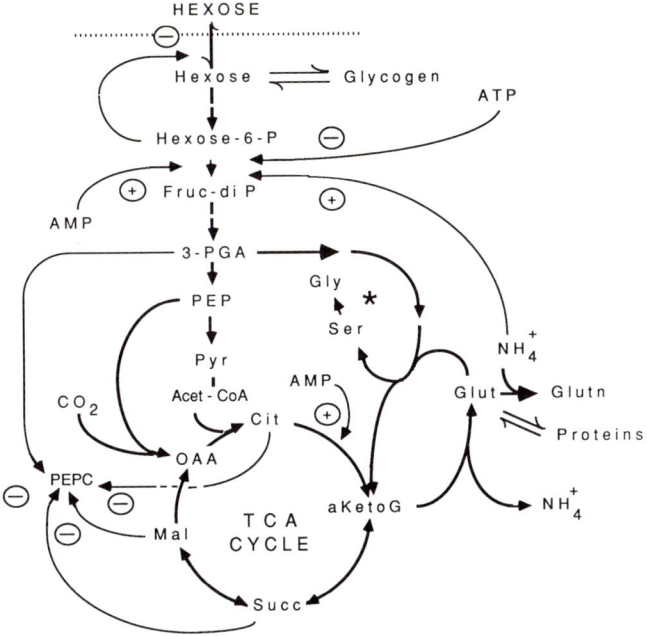

FIGURE 15.7. Regulation of metabolic pathways. Coordination of cellular function by feedback regulation is shown for some aspects of the glycolytic pathway, TCA cycle, anaplerotic incorporation of CO_2 into the TCA cycle, and nitrogen metabolism. Negative feedback (−) inhibiting enzymes at key points in the metabolic pathways and positive feedback (+) activating enzymes are shown as light lines. Metabolic pathways are shown as bold lines and arrows. Uptake of hexoses (6-carbon sugars) is regulated in some cells by negative feedback effects of hexose-6-phosphate (eg, glucose-6-phosphate). This control mechanism ensures that only sufficient hexose is transported into the cell to meet requirements of the pathway. If too much phosphorylated intermediate builds up, the uptake of hexose is inhibited. Likewise, intermediates in the TCA cycle and three phosphoglyceric acid (3-PGA) have negative feedback influences on the enzyme phosphoenolpyruvate carboxylase (PEPC). PEPC catalyzes incorporation of CO_2 into phosphoenol pyruvate (PEP), producing the TCA intermediate oxaloacetate (OAA). High levels of AMP, indicating possible depletion of high energy adenylates, as well as build up of ammonium indicating a likely depletion of amino acids or proteins, stimulate the glycolysis pathway. This occurs at the step producing fructose diphosphate (Fruc-di-P). ATP, however, has a suppressive effect. AMP also stimulates the formation of the TCA intermediate alpha-ketoglutarate (a-Keto-G), and thereby enhances ATP production by oxidative phosphorylation. When the cell has been grown under nitrogen-limiting conditions, the cell can become depleted in amino acids. Addition of ammonium to the medium can stimulate conversion of TCA intermediates to amino acids, thus reducing the pool of TCA compounds. This in turn stimulates CO_2 incorporation into PEP. The role of serine in buildup of complex molecules (*) is illustrated further in Figure 15.6. Based on data from Dewey and Kidder (1960), Sols (1967), and Peak and Peak (1981).

Fixation of CO_2 may also be an important step in the complex pathway leading to the conversion of cytoplasmic NADH to NAD^+ in *Tetrahymena*. During glycolysis in the cytoplasm, NADH is generated and must be reoxidized to NAD to sustain the glycolytic pathway. In many protozoa, this is accomplished anaerobically by lactate dehydrogenase. In *Tetrahymena*, however, this enzyme is bound in the mitochondrion, and assuming the mitochondrial membranes are impervious to NAD as in other cells, then an alternative oxidizing source for NADH must exist in the cytoplasm. The following cytosolic reactions may satisfy the requirement (Shrago and Elson 1980). The NADH-generating step during glycolysis is shown at the outset.

Glyceraldehyde-3-phosphate + NAD^+ + P_i →
1,3- diphosphoglycerate + NADH + H^+

Phosphoenolpyruvate + CO_2 → oxaloacetate

Oxaloacetate + NADH + H^+ → malate + NAD^+

Other Ciliates

In comparison to the anaerobic metabolism of *Tetrahymena*, the rumen ciliates, though studied less extensively, exhibit substantial differences (eg, Dehority 1986). These organisms inhabiting the alimentary canal of herbivorous mammals are of potential economic importance, as they may contribute substantially to the nutrition and vitality of domesticated cattle, especially during their early development. There are two main groups of rumen ciliates: (1) those in the order Entodiniomorphida characterized by a firm pellicle and the absence of cilia except on the peristome, and (2) the "holotrichous" group belonging to the order Trichostomatida, with a uniformly ciliated body and a mouth situated at characteristic locations according to species. The first group has been studied extensively by Coleman (eg, Coleman 1979) who has maintained them in culture using reducing gases (ie, carbon dioxide or nitrogen, or a mixture depending on the species). Bacteria appear to be indispensible to the robust growth of this group, both as ectocommensals and as food organisms. The carbohydrate metabolism of *Entodinium* spp studied in vitro is supported by exogenous maltose and glucose. Intracellular stores of starch are hydrolyzed to maltose and then glucose, and each product inhibits the enzyme that produced it, thus effecting negative feedback control of the degradative process. Consequently, when these compounds are exogenously supplied, the rate of intracellular starch hydrolysis is proportionately depressed, and the cytoplasmic pool of maltose and glucose remains fairly constant. Both sugars are incorporated into storage polysaccharide at equivalent rates. However, as with *Tetrahymena*, apparently little carbohydrate carbon from starch, glucose, or maltose is incorporated into protein (Coleman 1979). Indirect evidence suggests the presence of a classic Embden–Meyerhof glycolytic pathway. In *Entodinium caudatum*, the final products of starch fermentation are acetic and butyric acids, with small amounts of lactic, propionic, and formic acids, and CO_2 (Coleman 1979).

Little is known about the details of carbohydrate utilization in the second group (holotrichous Trichostomatida), but members of the Isotrichidae (eg, *Isotricha* and *Dasytricha*) ferment glucose, fructose, sucrose, raffinose, and pectin. *Dasytricha* can also ferment the disaccharides cellobiose (cellulose constituent) and maltose (amylose constituent). *Isotricha*, however, can ingest and utilize small vegetable starch grains. The major end products of carbohydrate metabolism by these holotrichs are hydrogen, CO_2, acetate, butyrate, and lactate. The production of hydrogen gas and acetate suggests that hydrogenase is present, and possibly that hydrogenosomes mediate anaerobic oxidative metabolism (Müller 1980) in addition to an Embden–Meyerhof pathway leading to CO_2 and lactate production. By comparison, in the termite gut where acetate is a major metabolic product excreted into the lumen and utilized by the host as its major source of carbon, the ciliates very likely also contain hydrogenosomes. This is indicated by electron microscopic evidence of microbody-like organelles and the production of hydrogen gas (Müller 1980). If this is confirmed, then the remarkable symbiosis between termite and ciliates (as well as flagellates), and the production of acetate as an essential carbon source for the host, is sustained largely by the presence of this unique subcellular organelle. The principal storage product of intestinal ciliates is amylopectin, or paraglycogen, as it is sometimes called.

Although many ciliates store either glycogen or amylopectin, electron microscopic and cytochemical evidence suggests that there are species-specific differences in the amount and kind of these carbohydrates accumulated in the cytoplasm (Verni and Rosati 1980). Among 10 species examined, three major storage forms of the carbohydrate were detected: (1) alpha glycogen (rosettes of granules), (2) beta glycogen (individual granules), and (3) masses of amylopectin. The major storage products for the 10 species coded as aGL (alpha glycogen), bGL (beta glycogen), and Ap (amylopectin) are as follows: *Paramecium bursaria* (bGL); *Blepharisma japonicum, Oxytricha bifaria, Oxytricha falax, Gastrostyla steinii, Stylonichia* sp, *Diophrys scutum*, and *Diophrys* sp (all Ap); and *Euplotes crassus* and *E eurystomus* (aGL). An examination of storage products reported in the literature suggests that the Kinetophragminophora as a group are more heterogeneous, and the Oligohymenophora and Polyhymenophora are more conservative in the range of carbohydrate storage products.

Some Selected Topics on Comparative Protozoan Metabolism

In the foregoing sections, some fundamental comparative data on carbon metabolism in major groups of protozoa were discussed. More general topics are addressed here, including some comparative data on nitrogen metabolism, lipid composition and biosynthesis, and other pertinent data as a context for the succeeding chapter on protozoan nutrition. The range and magnitude of research on the biochemistry of protozoa necessitates some selection of topics in a general survey text of this kind. Comprehensive reviews of protozoan metabolism and general biochemistry consider topics in greater detail (eg, Morris 1973; Levandowsky and Hutner 1979, 1980; Aaronson 1980; Antia 1980; Shrago and Elson 1980).

Nitrogen Metabolism

Among the flagellates, especially the phytoflagellates, the source of inorganic nitrogen is nitrate or ammonium, with ammonium preferentially assimilated. Indeed, the acetate flagellate *Euglena gracilis* does not utilize nitrate as a sole nitrogen source (eg, Lloyd and Cantor 1979). Among the more "typical" flagellates there are some exceptions to the preferential uptake of ammonium over nitrate. Among 38 freshwater 'chlamydomonads' studied by Cain (1965), the following did not preferentially consume ammonium: *Chlamydomonas gloeopara, C microsphaerella, C peterfi, Gloeocystis gigas*, and *G maxima*. In general, growth is roughly equivalent with either nitrogen source, but in the photosynthetic flagellate, *Dunaliella tertiolecta*, Paasche (1971) noted that growth was 30% faster when ammonium was the nitrogen source, as compared to nitrate. He suggested that a higher content of photosynthetic enzymes may be responsible for the enhanced growth on ammonium–nitrogen. Two enzymes appear to account for the total reductive assimilation of nitrate into metabolically assimilable ammonium: (1) nitrate reductase, that catalyzes the reduction of nitrate to nitrite; and (2) nitrite reductase, catalyzing the reduction of nitrite to ammonium. The presence of nitrate reductase in the aerotolerant, but typically anaerobic-dwelling ciliate *Loxodes* (Finlay and Fenchel 1986), adds additional evidence of the diversity of protozoa that possess this fundamental enzyme for the initial step in assimilation of nitrogen for synthesis of organic compounds. There is good evidence that nitrate reductase in some phytoflagellates (eg, *Chlamydomonas*) and algae is regulated by negative feedback inhibition when ammonium is assimilated (Morris 1973). It appears that ammonium itself is not the regulator molecule, but some product of ammonium metabolism. Amino acids and amides come to mind as the most likely regulators, but no inhibition has been demonstrated in cell-free preparations of the enzyme. The conversion of nitrate to nitrite appears to induce the genetic transcription of mRNA for protein synthesis of nitrite reductase. Thus, the quantity of this enzyme appears to be regulated at the genetic level according to the availability of substrate.

Organic sources of nitrogen can also be utilized by some protozoa. For example, Birdsey and Lynch (1962) found that eight species of chloromonads were able to utilize urea, and five assimilated uric acid and xanthine as nitrogen sources. *Euglena gracilis*, however, was unable to utilize these three compounds. Some chrysomonads and cryptomonads show excellent growth on urea, and at least one cryptomonad, *Hemiselmis virescens*, grows well on some urea derivatives (alloxan, parabanic acid), but does not grow on others (including biuret, guanidine, hydantoin, and allantoin) (Antia 1980). A wide range of amino acids, urea, hypoxanthine, and uric acid support growth of some species of chrysomonads and cryptomonads as summarized by Antia (1980). The assimilation of urea appears to be mediated in most cases by the enzyme urease, which converts urea to ammonia and CO_2. Lui and Roels (1970) report that *Ochromonas malhamensis* excretes urea when grown on ammonium as the nitrogen source, and reutilizes the urea for growth on exhaustion of the inorganic nitrogen. Urease activity is increased when urea is the sole nitrogen source and when the cells are nitrogen starved. It is interesting to note by contrast that no urease is present in strains *E*, *S*, and *W* of *Tetrahymena pyriformis*, nor is there any evidence of a urea cycle. None of the strains was able to grow on either citrulline or ornithine in place of arginine (Dewey et al 1957). The marine volvocidan *Dunaliella tertiolecta* incorporates glutamate into soluble protein, accounting for about 14% of the total glutamate uptake, while < 1% of the amino acid carbon was incorporated in chlorophyll (Precali and Falkowski 1983). The production of amino acids by *Tetrahymena*, as exemplified by serine biosynthesis (Figure 15.6), has been studied by Dewey and Kidder (1960) and Kidder (1967). In addition to transhydroxymethylation of glycine, serine is produced from products of carbohydrate metabolism. There is evidence from nutritional experiments that serine can be synthesized from such compounds as acetaldehyde, fatty acids (short chain), and ethanol. Anaplerotic incorporation of CO_2 into pyruvate to yield oxalacetate also leads to serine synthesis by way of 3-phosphoglycerate (Figures 15.6 and 15.7). Serine, in turn, is a significant source of other major organic compounds including thymine, choline, sphingosine, and other large lipid molecules, some of them used in membrane biosynthesis.

Teterahymena is also capable of assimilating amino acids from growth media, and there is good evidence that although rumen ciliates can take up amino acids from the rumen fluids, they obtain most of their amino acids from particulate protein ingested as food (Coleman 1979). Interestingly, these anaerobic ciliates are proteolytic and release amino acids into the medium (Warner 1956). Ammonia is also an end product of metabolism in some species, as is generally thought to be the case for most protozoa. Likewise, some flagellates effectively utilize exogenous amino acids as a nitrogen source; for example, *Prymnesium parvum* assimilated arginine, aspartate, isoleucine, lysine, phenylalanine, and valine. *Pavlova gyrans*, however, could not utilize any of the tested amino acids. Although *E gracilis* does not take up arginine from its growth medium, this amino acid is accumulated intracellularly in both free and peptide forms as a major nitrogen reserve in the cells when they are grown in medium with excess nitrogen (Park et al 1983). Other flagellates exhibiting varying degrees of amino acid-stimulated growth include *Hymenomonas* sp, *Ochrosphaera neapolitana*, and the coccolithophorid *Emiliana huxleyi*. The latter was most restricted and utilized only arginine, histidine, and lysine (Antia 1980).

Uptake of the complex, nitrogenous compounds comprising the nucleic acids has been documented in several protozoa. The parasitic amoeba *Entamoeba histolytica*, for example, can utilize various preformed purines and pyrimidines and the corresponding nucleoside for RNA and DNA synthesis (eg, Albach and Booden 1978). Purine and pyrimidine nucleosides are apparently incorporated preferentially over the free bases when supplied in chemically defined medium. Using radioactively labeled compounds, Gibbs et al (1974) demonstrated that thymidine was incorporated into DNA of the flagellate *Ochromonas danica*. Further studies showed that adenine, thymidine, and uracil were incorporated into DNA in decreasing order; and in *Ochromonas malhamensis*, radioactively labeled uridine, cytidine, and orotate were all incorporated into the pyrimidines of DNA and RNA (reviewed by Aaronson 1980). In general, the nucleic acid biosynthesis of *Euglena gracilis* appears to be typical of most eukaryotic cells (Wolken 1967).

Comments on Lipids in Protozoa

The presence of phospholipids with unsaturated fatty acid moieties and sterols in the membranes of most eukaryotic cells has been well documented, and many protozoa have been shown to contain a variety of widely occurring sterols in cell extracts or membrane preparations. For example, sterols in the chrysomonad *Ochromonas danica* include brassicasterol, clionasterol, ergosterol and in the main, poriferasterol. *O danica* is also of interest, as it contains a long chain sulfolipid characterized by Haines (1973). In *O malhamensis* and *O sociabilis*, poriferasterol was the major sterol. The distribution and composition of lipids in other chyrosomads is discussed by Aaronson (1980). Exogenous cholesterol fulfills all lipid requirements for growth of the anaerobe *Entamoeba histolytica*, but only suboptimal growth was obtained with betasitosterol (Fulton 1969). By contrast, the anaerobic rumen ciliate *Entodinium caudatum* requires betasitosterol for growth (Broad and Dawson 1976), and probably utilizes it and other sterols for the synthesis of stigmastanol and campestanol, which comprise 73 to 80% of the sterols present (Coleman 1979). *Tetrahymena* cells, however, do not produce any typical sterols such as cholesterol, ergosterol, and stigmasterol, but rather synthesize a pentacyclic triterpenoid tetrahymanol (II). Tetrahymanol (Figure 15.8) is commonly found in membrane lipids extracted from the plasma membrane and a wide variety of intracellular organelles (Nozawa and Thompson

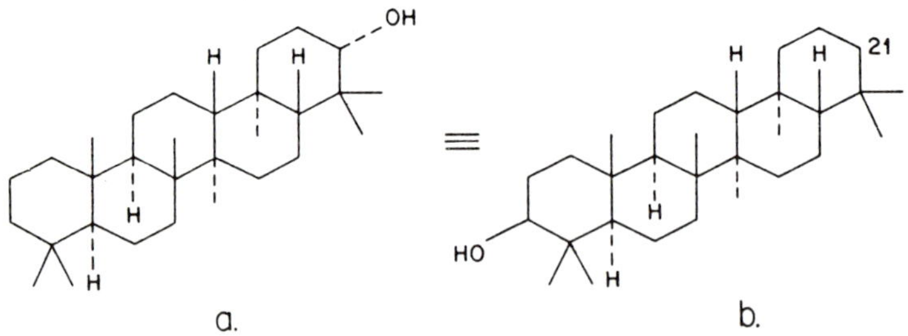

FIGURE 15.8. The structural formula for tetrahymanol, a product of the metabolism of the ciliate *Tetrahymena*. This compound is included in the membranes of the ciliate. From Holz and Conner (1973); with permission.

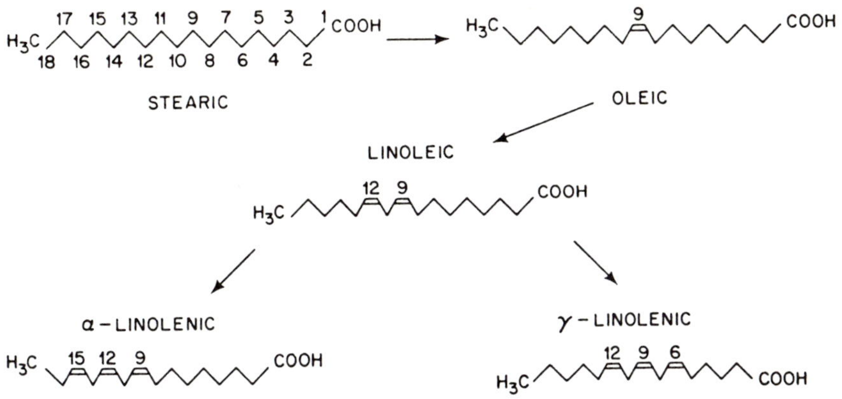

FIGURE 15.9. Polyunsaturated fatty acid synthesis in *Tetrahymena pyriformis*. The formation of a double bond (unsaturation) at position 9 converts stearic acid to oleic acid. Further double-bond additions yield linoleic acid, which is an intermediate for alpha-linolenic or gamma-linolenic acid formation. From Holz and Conner (1973); with permission.

1979), and comprises 0.14% of the dry weight of cells (eg, Holz and Conner 1973). Although sterols are not synthesized by *Tetrahymena pyriformis*, they are accumulated from the growth medium and appear in cellular membranes. When cholesterol is added to the culture medium, approximately 60% is accumulated by the cells regardless of the concentration in the medium. There appear to be three separate phases in the uptake: (1) a rapid, temperature-independent initial phase, probably involving adsorption; (2) a lag phase when no appreciable detection is possible; and (3) a temperature-dependent, hence metabolic phase. The cumulative data suggest that the sterol is ingested by phagocytosis, and subsequently, as with other sterols, is accumulated and transformed to a common derivative (Holz and Connor 1973). Cholesterol or one of its metabolic derivatives also blocks the conversion of precursors (especially mevalonic acid) to tetrahymanol, and thus limits its biosynthesis (Wilton 1983). Some species of *Tetrahymena* require trace amounts of exogenous sterol when grown in chemically defined medium, but this requirement can be spared or bypassed by appropriate supplementation with other lipids (Holz and Connor 1973).

The unsaturated fatty acids found in many protozoa are assimilated from the growth medium, or produced from saturated fatty acids either synthesized in the cell or accumulated from the medium. An enzyme known as a desaturase adds double bonds at specified points in the long saturated hydrocarbon chain. For example, in *Tetrahymena*, stearic acid is converted to oleic acid by addition of a double bond at carbon 9 in the chain (Figure 15.9). Further desaturation successively produces linoleic (diunsaturated) and linolenic (triunsaturated) derivatives. Several species of *Tetrahymena* are capable of growing in a chemically defined medium without addition of fatty acids: all strains of *T pyriformis*, *T corlissi* Th-X, *T setifera* HZ-1, and *T paravorax* RP. Only glucose and amino acids were supplied as a carbon source (Holz and Connor 1973). Acetate and other short chain organic acids are readily assimilated by these species and converted into long chain fatty acids. But, the quantity and composition of the products varied with the kind of carbon source. *T pyriformis* grown at elevated temperatures (ca 40°C) show abnormalities in cell morphology and ultrastructure; the membranes are deranged. Biochemical analyses of membrane extracts confirmed an abnormally high amount of saturated fatty acids. The presence of saturated fatty acids, though anomalous relative to lower temperatures, has adaptive value at higher temperatures. Saturated fatty acids are more cohesive and reduce membrane fluidity, making the cell membranes more resistant to thermal denaturation. Considerable attention has been given to the biochemical explanation for this temperature-induced response. Apparently, at elevated temperatures, the desaturase step is blocked, perhaps by alteration of the enzyme. This was substantiated by showing that cells grown at elevated temperatures are normal if their medium is supplemented with normal unsaturated fatty acids obtained from cells grown at lower temperatures. In general, cells grown at 35°C or lower are able to assimilate long chain exogenous fatty acids, and either shorten or desaturate them to produce unsaturated intermediates and linolenic acid as a major end product. Gamma-linolenic acid is also characteristically found in chrysomonads, but is lacking in prymnesiomonads, thus further strengthening the taxonomic separation of these otherwise closely related groups of organisms (Aaronson 1980).

Synthesis of fatty acids in chrysomonads, as in other phytoflagellates, is via acetyl-CoA, and monounsaturated acids are synthesized by way of a desaturase enzyme requiring oxygen and NADPH associated with a particulate enzyme complex:

$$\text{Stearoyl-CoA} \xrightarrow{O_2 \ NADPH} \text{oleoyl-CoA}$$

Particulate enzyme complex

Some comparative data on fatty acid composition among chrysomonads and prymnesiomonads is presented by Aaronson (1980). A comparative analysis of metabolic pathways for lipid synthesis in the chrysomonad flagellate *Ochromonas malhamensis* and the free-living amoeba, *Acanthamoeba castellanii* indicated the presence of desaturases typical of animal and plant pathways (Pollero et al 1976). The supernatant, but not the particulate fraction of the cell, contained both types of desaturating enzymes. In both organisms, alpha-linolenic and arachidonic acids could be synthesized. When *A castellanii* was grown in the presence of C-14 labeled linoleic and alpha-linoleic acids, carbon from these compounds was incorporated mainly in triglycerides and phospholipids. The major fatty acids of *Acanthamoeba* spp include linoleic acid, and as with higher animals, polyunsaturated 20-carbon fatty acids (Korn 1964). *Acanthamoeba* is in an unusual position, in that in addition to this animallike quality, it possesses the plantlike characteristic of biosynthesizing long chain fatty acids de novo from nutrients in its medium. Korn (1964) proposed the following pathway acetate → stearate → oleate → linoleate → polyunsaturated long chain fatty acids. There is also evidence that the unsaturated fatty acid, oleate, is converted to the monounsaturated long chain fatty acid, 11-eicosenoate.

Discussions of variations in fatty acid composition of *Tetrahymena* and *Paramecium* under various growth conditions, including changes in proportion of saturated to unsaturated fatty acids with culture age and among species, is presented by Holz and Conner (1973), Kaneshiro et al (1979), and by Nozawa and Thompson (1979). The lipid composition of amoebae has been investigated by Halvey and Finkelstein (1965) and Costas and Griffiths (1984), who concluded that among several genera examined, including *Acanthamoeba*, *Hartmanella*, and *Mayroella*, the presence of long chain (14- to 20-carbon chain length) fatty acids varied so greatly with different growth conditions and with culture age that no clear separation of taxa could be made on this basis. Differences in lipid composition between *Dictyostelium discoides* and *Acanthamoeba castellani* included the absence of a 15-carbon saturated fatty acid and several 19- and 20-carbon atom fatty acids (some unsaturated) in *D discoides* that were present in *A castellani* (Costas and Griffiths 1984). The characterization of *Acanthamoeba* strains, using a variety of nutrients and enzyme inhibitors, shows promise of a refined physiological basis for discrimination among organisms that otherwise exhibit very similar morphological characteristics (Costas and Griffiths 1986).

Digestive Enzymes

Subsequent to the discovery of the lysosome by deDuve and colleagues, substantial attention has been given to characterizing the digestive enzyme composition of various protozoa. These data have been useful in understanding the nutritional and feeding habits of many protozoa, and in clarifying at a biochemical level why certain ingested substrates are utilizable and others are not. In general, it is noteworthy that most of the major macromolecules (proteins, carbohydrates, nucleic acids, mucopolysaccharides, and some lipids) are substrates for these enzymes. Thus, many phagocytic protozoa possess the capacity to ingest and degrade a wide variety of biologically synthesized macromolecules as sources of nutrition. Some interesting or unusual enzymes have been reported for organisms living in specialized environments. For example, some rumen ciliates and termite gut protozoa possess cellulolytic enzymes (cellulases) and can digest wood or wood products. Among these, *Polyplastron multivesiculatum* hydrolyzes cellulose, pectin, and a number of oligosaccharides; and *Eudiplodinium maggii* degrades cellulose to cellobiose and then to glucose, which is fermented principally to acetic and butyric acids (Coleman 1979). A cellulase is also found in the soil and the freshwater-dwelling amoeba (*Acanthamoeba*), but it is not certain if this serves in digestion or the metabolism of cyst wall cellulose. Collagenolytic (collagen hydrolyzing) activity has been found in pathogenic strains of *Entamoeba histolytica*, but not in nonpathogenic strains. It is possible that the hydrolysis of collagen has something to do with the invasion and pathogenicity of the virulent strains, especially their ability to penetrate host tissue (Munoz et al 1984). Variations in host susceptibility may involve differences in tissue collagen composition and its susceptibility to amoeboid enzymatic hydrolysis.

Summative Perspective

The metabolic activity of protozoa exhibits wide variations, from plantlike activity found in some phytoflagellates to parasitic heterotrophs that are completely dependent on their host for major preformed organic compounds. Many of the phytoflagellates, and some alga-containing symbiotic protozoa, are capable of synthesizing most if not all of their organic requirements from simple inorganic sources including nitrate or ammonia, carbon dioxide, and sulfate or small sulfur-containing compounds. In intermediate groups, varying degrees of dependence on preformed sources of nitrogenous compounds are observed. In some cases, certain amino acids are required, or, in a few instances discovered thus far, some lipids must be assimilated from the environment. In general, it appears that more advanced protozoa have become increasingly reliant on exogenous sources of preformed larger molecules. Although the precise evolutionary pathways leading to this dependence are not known, it appears logical to assume that during evolution these organisms lost certain key enzymes necessary for biosynthesis of major compounds. This resulted in the energy-saving step of assimilation of these compounds from the environment. The major pathways of biosynthesis of intracellular organics is understood only for a few organisms; largely those that can be cultivated in quantity in the laboratory. Some of these, while interesting laboratory specimens, are not necessarily the most widely distributed nor ecologically significant among the wide diversity of protozoan species. Clearly, much remains to be done to elucidate the biochemical and physiological mechanisms that account for the nutritional and ecological roles of major occurring protozoa in the natural environment.

16
Nutrition

Basic Perspectives

Protozoa have developed a wide range of feeding mechanisms, probably in response to the diverse environments and wide range of food available in varying habitats. Some species living in organic-rich media, either as parasites within organisms or in eutrophic natural environments, assimilate dissolved nutrients by absorption through the plasma membrane. In bacterially–enriched environments, further specialized modes of small particle ingestion may also occur. When larger prey (including protista and metazoa) are prevalent, as in some oligotrophic environments, there are additional strategies for snaring and engulfing these larger prey. Some of these adaptations have been reviewed earlier by Hall (1967) and more recently by Nisbet (1984) and others (eg, Sieburth 1979; Anderson 1983; Wichterman 1986; Elliott 1973). An overview of general principles of food capture and digestion are presented here. Following a discussion of trophic behavior and general principles of intracellular digestion, some detailed descriptions of nutritional requirements and predatory activity are presented for major groups of protozoa. Particular attention is given to nutrients required in laboratory cultures of protozoa.

The elucidation of nutritional requirements of an organism requires careful and extensive laboratory investigation to determine major sources of carbon, nitrogen, phosphorus, sulfur, and mineral nutrients. In addition, other growth factors such as vitamins or auxotrophic substances which must be acquired from the environment should also be determined. Further insights into patterns of nutrition come in part from knowledge of the environment of the organism, existing biochemical information on metabolism, and knowledge of nutritional needs of closely related organisms. In turn, new knowledge generated by nutritional studies often helps to clarify the metabolism and biochemistry of the organism, thus contributing to our broader understanding of the molecular basis for the trophic adaptations of the species.

Feeding Strategies

Feeding strategies in protozoa range from assimilation of dissolved substances from the natural environment by absorption across the cell membrane (osmotrophy) to engulfment of particles of macromolecular size (or even whole organisms) by a process of phagotrophy. Osmotrophy in a strict sense applies only to absorption of soluble substrates. This can occur by diffusion across the plasma membrane or by the activity of porters or permeases (membrane carrier molecules). In some cases, however, colloidal-sized particles can be taken up by pinocytosis. This is an active process of accumulating suspended particles into fine introcytoplasmic channels formed by invagination of the plasma membrane. The pinocytotic invagination ends in a cul-de-sac, where small pinocytotic vacuoles are formed to receive the food particles. These pinocytotic vesicles are carried by cytoplasmic streaming into the central cytoplasm, where fusion with lysosomes converts them into microdigestive vacuoles. Small molecules and membrane-permeable substances may immediately pass into the cytoplasm, before substantial digestion commences. The digestion products of larger molecules (including sugars, dipeptides or free amino acids, and other small macromolecules) are subsequently passed into the cytoplasm, where they are utilized for metabolism. Phagotrophy involves the engulfment of food particles into substantially larger vacuoles, often at a specialized site on the cell surface. The food particles are engulfed into phagosomes (food vacuoles). This process is either by simple membrane invagination, forming a concave depression where the particle is absorbed (eg, amoeboid feeding), or by accumulation of prey in a food canal or vestibulum as occurs in some flagellates and ciliates.

Flagellate Feeding Behavior

Many phagotrophic flagellates utilize simple engulfment of the food particles at one or more sites on the plasma membrane. There is no specialized food gathering canal, although there may be a particular site on the surface of the cell where food is likely to be ingested. A specialized region for food engulfment called a feeding pouch or slit occurs in some cryptomonads, euglenoids, and other particle-gathering flagellates. Phagotrophic dinoflagellates have varied mechanisms for prey capture including a pseudopodial net, deployed from a pore in the thecal wall, that surrounds and digests the prey (Jacobson and Anderson 1986), or a highly extensible fingerlike peduncle that snares the prey and serves as a feeding organelle (Spero 1982). The peduncle is strengthened by a set of overlapping sheets of microtubules, which, according to Spero, resemble those in the tentacle of suctoria. He suggests these may be homologous structures indicating similar feeding mechanisms. In the pseudopodial feeders, the thin saclike pseudopodium, called a pallium, is produced by extrusion of plasma membranes from the intrathecal cytoplasm. Lysosomal vacuoles produced in the cytoplasm stream into the pallium and commence digestion of the prey. Digestive vacuoles containing prey matter are carried by cytoplasmic streaming into the theca, where further digestion and metabolism of the food occurs (Jacobson 1987). The pallial reservoir is supported internally by a complex, conical array of microtubules forming a contorted funnel-shaped canal, lined on one side by membrane, and oriented with its narrow end at the pallial pore. The broad end opens in the vicinity of the nucleus.

Peranema, a colorless euglenoid, has a remarkable feeding apparatus as explained in Chapter 12. A stiffened rod of microtubules is contained in the cytostomal pocket, which opens on the wall of the flagellar reservoir near the aperture. The rod-organ, as it is known (Chen 1950; Nisbet 1974), is extensible and aids in prey capture by projecting outward upon contact with the prey. The cytostome is enlarged and held open by the mechanical action of the protruding rod. There is no convincing evidence that the rod punctures the prey (Nisbet 1974). During phagocytosis of the prey (which can be as large as another flagellate or as small as a micron-sized organic particle), euglenoid movement of the *Peranema* aids proper orientation and engulfment of the food (Figure 12.7). In *Entosiphon* (Figure 12.6b), the feeding apparatus consists of a long conical tube (siphon) formed by ribbons of microtubules. During feeding, the siphon is thrust outward from a membrane-lined cavity and expands in diameter (Triemer and Fritz 1987). An organic flap, or cap, closes the siphon when withdrawn, but is displaced to one side as the siphon tube is extended, thus permitting engulfment of food particles. Flangelike blades, formed by sheets of microtubules, extend inward from the periphery of the siphon, forming a cartwheel arrangement in cross section. When feeding is completed, the siphon is withdrawn, the cap closes the open end, and the whole structure is enclosed beneath the plasma membrane.

Sarcodina Feeding Strategies

Sarcondinads, including the amoebae and their relatives, exhibit various means of entangling and engulfing prey. The naked amoebae, though often considered simple in their cytoplasmic organization and feeding behavior, exhibit some remarkably diverse adaptations for food engulfment. Food particles may be engulfed at the anterior end of the cell, where phagocytic vacuoles are produced either by invaginations at the leading edge of the cytoplasm or by outward growth and encircling of prey by lobopodia. Food particles attached to the plasma membrane in the uroid region at the posterior of the cell can also be engulfed in small vacuoles by phagocytosis. The pseudopodial capture of prey varies with the size and physiology of the amoeba. Some smaller monopodial or discoid species engulf food particles largely at the leading edge of the advancing cell. Other species form pincherlike pseudopodia that can partially enclose and constrict large prey (eg, ciliates), reducing their mass to suitable proportions for engulfment. Alternatively, a rather complex structure known as a feeding cup is formed. The feeding cup, as the name implies, is a cuplike depression formed in the surface of the cell upon prey contact. It flows outward, surrounds the prey, and eventually fully encloses it in a phagosome. Upon passage of the phagosome into the endoplasm, it is converted to a digestive vacuole by fusion with primary lysosomes. The testate amoebae, some Foraminifera, and axopod-bearing sarcodinads such as radiolaria exhibit highly diverse pseudopodial activity in apprehending and ingesting prey. For example, in testate amoebae and benthic Foraminifera, peripheral rhizopodia exhibit highly selective prey apprehension, and by rhizopodial streaming, carry suitable food particles to the shell aperture where digestive vacuoles are formed in the intratest cytoplasm. Remarkably dextrous behavior is exhibited by some rhizopods in gathering diatom prey. The frustule of the diatom is separated into the epitheca and hypotheca by rhizopodial streaming and tension. The soft, digestible cell mass is engulfed in digestive vacuoles, but the nondigestible, siliceous thecae are discarded at the periphery of the cytoplasm. In the benthic Foraminiferan *Elphidium* there is no major aperture, but only small pores on the surface of the shell. Captured diatom prey is carried by cytoplasmic streaming toward the surface of the shell, but due to the sieving action of the pores, only the soft cytoplasmic contents of the prey are drawn into the intrashell cytoplasm. Prey selection and its translocation toward the cell body of some rhizopod-bearing organisms can involve elaborate manipulation. The food particles may be rotated and oriented during rhizopodial streaming, or transferred from one strand to another, or even elevated on the tip of a rhizopodial strand. It is not known why some of these very elaborate motions are executed during prey apprehension, but it may be part of the chemosensory processes in selecting particles to be ingested. Radiolaria and planktonic Foraminifera often exhibit very elaborate feeding behavior (Anderson and Bé 1976; Anderson 1983). Radiolaria snare algae, protozoa, and small metazoa (copepods and other plankton) as prey (eg, Swanberg and Anderson 1985), using their pseudopodia as a net and invasive apparatus. An algal prey is apprehended at the surface of sticky rhizopodia or axopodia, and engulfed into small food vacuoles where it is transported proximally toward the cell body. The engulfed algal prey can be distinguished clearly from algal symbionts by the thick sheath of cytoplasm formed around it. Algal symbionts are enclosed in only a very thin cytoplasmic envelope. Cytoplasmic streaming carries the engulfed prey to the central capsular region, where digestive vacuoles are formed in the frothy cytoplasmic layer immediately next to the capsular wall. There is some evidence that a complex selection process occurs in accepting or rejecting the algal prey. Two types of prey selection have been observed (Anderson 1978). In a type I response, the algal prey is rapidly assessed and either accepted immediately upon contact with the rhizopodia or rejected. In a type II response, the algal prey is first engulfed in vacuoles, but after a short duration of one or more minutes unacceptable prey are rejected. The reason for the delay in type II response is not known, but it may involve a protracted process of chemosensation to determine the acceptability of the prey, or it may simply represent a period when the total mass of food is assessed and any excess is rejected. Larger prey such as copepods and other metazoa can be as large as or larger than the radiolarian. The prey is first apprehended by the sticky axopodia and rhizopodia. As the prey struggles, it is often further entangled in the web of feeding rhizopodia. Furthermore, the rhizopodia immediately become mobilized and begin streaming toward and around the prey. Strands of rhizopodia attach to the surface of the prey and exert tension, resulting in rupture of the exoskeleton at weaker points. These are often thin areas at joints in the appendages or where the exoskeleton is less robust. The feeding rhizopodia invade the tissue spaces and selectively attach to suitable prey tissue. Some of the rhizopodia penetrate beneath the tissue and separate it from the inner surface of the carapace, or selectively attach to and translocate masses of muscle or lipid taken from the prey. These are engulfed in digestive vacuoles in the sarcomatrix layer near the capsular wall. Nondigestible matter is carried to the tips of the rhizopodia by cytoplasmic streaming and discarded.

Ciliate Feeding Strategies

The ciliates have developed some of the most complicated and efficient mechanisms for filter feeding. Prey particles are swept by ciliary beating into a food canal (eg, vestibulum) leading to a region known as the cytostome, where food vacuoles are formed. The organization of the canal and the location of the cytostome varies among species. Comparative structural features of the feeding apparatus of various ciliates are discussed in Chapter 14. The pattern of water currents generated by ciliary beating in the oral region of a vorticellid (Figure 16.1) illustrates the variations in hydrodynamic behavior of these diverse groups. A generalized diagram (Figure 16.2) of the sequence of events following food vacuole formation in ciliates illustrates the four stages of digestive vacuole development (Fok et al 1982; Allen 1984; Allen and Fok 1984). During the transformations of the intracellular vacuoles, the membrane topography and molecular composition (especially of the glycocalyx) changes. The pH of the food vacuole (stage I) is neutral to alkaline, but becomes acidic in subsequent stages prior to initiation of digestion. In stage II, the food vacuole becomes more rounded and contracts by formation of surface vesicles that are shed into the cytoplasm. The vacuole becomes acidic by fusion with acidisomes (vesicles with acid pH). The vesicles produced during contraction of the food vacuole are accumulated at the oral region to be recycled in food vacuole formation. The condensing food vacuole also loses water, and the surrounding membrane is closely appressed against the enclosed food particles. Primary lysosomes fuse with the condensed food vacuole and supply hydrolytic enzymes to form stage III. As digestion proceeds, the digestive vacuole enlarges and the periphery assumes a wavy outline. Further expansion and formation of surface evaginations leads to stage IV, where digestion is in its final stages. Some of the residual enzymes are enclosed in the surface vesicles and released into the cytoplasm, forming dense bodies that can act as primary lysosomes supplying enzymes to other food vacuoles during subsequent phases of ingestion. The "spent" digestive vacuole, known as a residual body, is carried to the cytoproct or anal pore, where

312   16. Nutrition

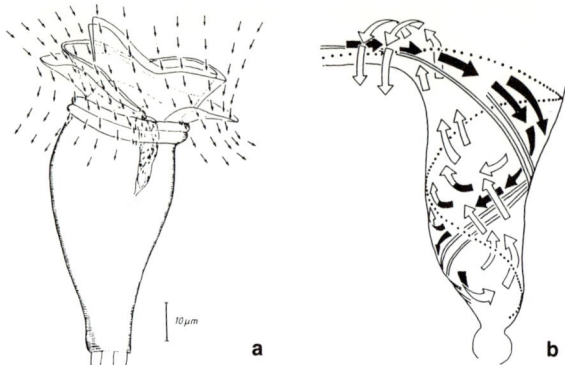

FIGURE 16.1. Water currents in the region of the oral membranelles (a) and within the infundibulum (b) of suspension feeding ciliates. (a) Feeding currents of *Zoothamnium* sp. (b) Incoming currents (black arrows) within the conical infundibulum are drawn along the right side of the polykineties and are eventually driven through the paroral membrane (white arrows exiting). From Fenchel (1980); with permission.

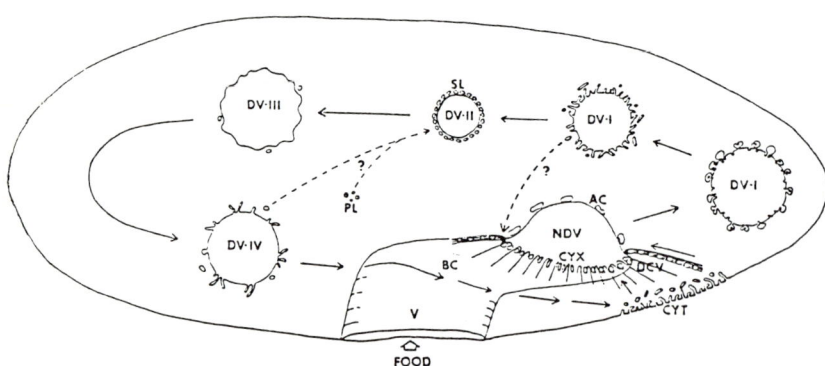

FIGURE 16.2. Schematic drawing of vacuolar changes during feeding and digestion in *Paramecium caudatum*. The four stages of the vacuole are indicated by DV-I to DV-IV. AC = acidosomes; BC = buccal cavity; CYT = cytoproct; CYX = cytopharynx; DCV = discoidal vesicles; NDV = nascent digestive vacuole; PL = primary lysosomes; SL = secondary lysosomes; V = vestibulum. Acidosomes with low pH fuse with the newly formed food vacuole, lowering its pH. Subsequently, fusion of the food vacuole with primary lysosomes (DV-II) initiates enzymatic digestion of the prey. Following major digestive stages, the vacuole becomes alkaline once again, and blebs off vesicles possibly containing residual digestive enzymes. Before defecation of the last stage (DV-IV), the vacuole fuses with the cytoproct membrane, and residual fragments of the spent digestive vacuole are recovered as vesicles that are carried by microtubules to the region of the cytopharynx where they are recycled in formation of new food vacuoles. Dashed lines show proposed paths of recovery of membrane constituents from the phagocytic vacuoles. From Allen (1984); with permission.

defecation occurs. The membrane of the residual vacuole fuses with the plasma membrane and contributes to the cell surface. The increase in cell surface area produced by this addition of membrane can be compensated for by intracytoplasmic invagination of vesicles at other sites on the membrane, or in some cases, the membrane fragments may be recovered by resorption at the site of defecation.

Although many ciliates have an oral ciliature to collect and concentrate food particles, others use specialized cytoplasmic structures to grasp and engulf food. The cyrtos or pharyngeal basket formed by a conical or tubular array of nematodesmata (rods of microtubules) as occurs in *Pseudomicrothorax dubius* is of particular interest. This cylinder of microtubules is lined on the inner surface by a membrane, and opens distally at the cytostome on the ventral surface of the cell. Algal strands or filaments of cyanobacteria are engulfed in the cyrtos and drawn into food vacuoles formed at its base (Hausmann and Peck 1978, 1979; Peck and Hausmann 1980). The feeding behavior of *Pseudomicrothorax* in the presence of the filamentous cyanobacterium (blue-green algal) *Oscillatoria formosa* is a complex, two phase process (Peck 1985). During the first phase, the ciliate makes a complex swimming motion over the surface of the filament with its ventral cilia attached. This is called contact swimming. In the second phase, phagocytosis begins with attachment of the cytostome to the filament. Lysosomal streaming is observed in the cytopharyngeal-cytostomal region, whereby digestive enzymes are deposited at the mouth of the cyrtos and upon the surface of the algal filament. The filament is ruptured or weakened, and drawn into the membrane-line cytostome where food vacuoles are formed and digestion begun. If the feeding is interrupted experimentally before the algal strand is engulfed, lysosomal vacuoles appear and accumulate at the orifice of the cytostome. These are apparently the excess lysosomes that would have been used to aid algal strand ingestion. Rather large segments of algal filaments can be ingested, forming a substantial mass within the intracytoplasmic vacuolar system.

Suctoria with radiating cytoplasmic tentacles take up colloidal-sized food particles by pinocytosis at the surface of the cell body, or snare larger prey such as other ciliates at the tips of the tentacles. The tentacles can be of several types, either pointed and needlelike, or with swollen and knoblike tips. When a suitable prey encounters the distal end of the tentacle, it becomes attached by the immobilizing action of the haptocysts that discharge toxic substances into it. The pellicle membranes of prey and the tip of the tentacle fuse, forming a cytoplasmic bridge between predator and prey. The cylinder of peripheral microtubules within the cytoplasm of the tentacle moves upward toward the prey and splays outward forming a trumpetlike configuration, while the plasma membrane invaginates to form a channel carrying prey cytoplasm into the body of the predator. Inward directed streaming of the plasma membrane lining the tentacle channel is thought to provide the translational force for movement of prey cytoplasm into the cell of the suctorian, where food vacuoles are formed. The detailed events during prey attachment, immobilization, and cytoplasmic engulfment have been documented by Rudzinska (1965) and Bardele (1972, 1974).

Some of the nutritional requirements of representative genera of major groups of protozoa will be presented with special reference to some of the biochemical and physiological concepts introduced in the preceding chapter.

Flagellate Nutrition

Inorganic Nutrients and Vitamins

Representatives of the general group known as acetate flagellates will be discussed first, followed by some illustrations of dinoflagellate, trichomonad, and kinetoplastid nutritional requirements. Our understanding of flagellate physiology and nutrition has been substantially enhanced by the innovative and sustained efforts of specialists in protozoan culture techniques and physiological research (eg, Pringsheim 1912; Hutner 1936; Hutner and Provasoli 1951; Provasoli and Pintner 1953; Hutner et al 1972). Some inorganic and organic sources of nutrition will be discussed in a comparative format (Table 16.1). Minimal mineral requirements (expressed as mg/l) for heterotrophic culture of dark-grown *Euglena gracilis* Z using L-glutamic acid as sole carbon and nitrogen source are: KH_2PO_4 (44.0), $MgSO_4 \cdot 7H_2O$ (11.7), $CaCl_2 \cdot H_2O$ (0.0734), $MnCl_2 \cdot 4H_2O$ (0.024), $ZnCl_2$ (0.0042), Thiamine HCl (0.005), and vitamin B_{12} (0.0002) at pH 3.17 (Kempner and Miller 1972). Under these growth conditions at 27.5°C, the doubling time was 13.2 h and the final yield of cells was 1 to 2 × 10^8 cells/ml. Inorganic phosphorus at concentrations > 1 mg/ml inhibits growth in terms of cell density per ml and generation time. The source of this inhibition is unknown, but similar inhibition of growth of strain SM-L1, *Euglena*, occurs during "overfeeding" with sulfur and acetate (eg, Buetow and Schuit 1976). Maximum growth of streptomycin-bleached *Euglena gracilis* maintained in a defined medium occurs with an inorganic phosphate concentration of 4 to 5 µg/ml (Buetow and Schuit 1968). At lower concentrations, the cells apparently deplete the phosphorus medium, but are able to survive for 6 to 7 days. This may be attributed to the phosphorus-accumulating capacity of the cells yielding an intracellular reserve as also occurs in the volvocidans, *Pandorina* and *Volvox*. Moreover, in *Euglena gracilis* SM-L1, phosphorus limitation induces formation of a nonspecific acid phosphatase. This enzyme hydrolyzes phosphate from organic compounds, and may provide an additional source of phosphate from intracellular or extracellular organophosphate compounds. In the SM-L1 strain, carbon deprivation or aging of the culture also produces an increased in a wide range of acid hydrolyases (Baker and Buetow 1976) degrading proteins, carbohydrates, and mucopolysaccharides. A marked increase in the production of these enzymes occurs as the cultures proceed from a midexponential to late stationary phase of growth, with a particularly marked increase for beta-glucosidase. During carbon deprivation, enzyme activities expressed on a per cell basis decreased for DNase, while those of beta-galactosidase, cathepsin D, and RNase increased. The increase in these digestive enzyme activities appears to be an adaptation of the *Euglena* to varying metabolic demands occasioned by changes in exogenous

TABLE 16.1. Some major nutrients required for laboratory culture of protozoa.

				Flagellates				Amoeba
Nutrients	Astasia longa	Peranema trichoph.	Polytom. caeca	Ochrom. malham.	Ochrom. danica	Crypth. cohnii	Euglena gracilis	Acanth. spp
Organic								
Na acetate	82	30	200	—	—	—	—	—
DL Malic acid	—	20	—	—	—	—	—	—
Succinic acid	—	20	—	—	—	—	—	—
$(NH_4)_2$ Citrate	—	—	—	120	—	—	—	1 g
Citric acid	—	—	—	—	—	100	—	—
K_3 Citrate	—	60	—	—	—	100	—	—
DL-Alanine	—	—	—	—	40	—	—	140
L-Arginine HCl	—	—	—	50	1 g	100	300	—
Glutamic acid	—	—	—	300	10	—	—	200
Glycine	—	—	—	—	40	+ +	—	—
L-Histidine	—	50	—	50	—	—	—	15
DL-Methionine	—	—	—	60	—	—	—	75
L-Isoleucine	—	—	—	—	—	—	—	75
L-Leucine	—	—	—	—	—	—	—	27
DL-Valine	—	—	—	—	—	100	—	—
Betaine HCl	—	—	—	—	—	500	—	—
Glucose	—	—	—	1 g	1 g	100	—	1.8 g
Sucrose	—	—	—	—	—	—	—	—
Cholesterol	—	00.05	—	—	—	—	—	—
Methyl linoleate	—	1 g	—	—	—	—	—	—
Lecithin	—	1.0	—	—	—	—	—	—
Inorganic								
KH_2PO_4	10	—	—	30	30	10	4.4	80
Na_2HPO_4	—	—	—	—	—	—	—	80
$MgCO_3$	—	—	—	50	40	—	—	—

Flagellate Nutrition   315

TABLE 16.1. (Continued.)

	Flagellates							Amoeba
Nutrients	Astasia longa	Peranema trichoph.	Polytom. caeca	Ochrom. malham.	Ochrom. danica	Crypth. cohnii	Euglena gracilis	Acanth. spp
$CaCO_3$	—	—	—	15	5	50	—	—
$CaCl_2$	—	5	—	—	—	—	0.01	53
NH_4Cl	—	—	20	40	50	8	—	—
$(NH_4)_3PO_4$	10	—	—	—	—	—	—	—
$(NH_4)_2SO_4$	—	20	—	—	—	—	—	—
KH_2PO_4	10	20	50	—	—	—	—	—
KCl	—	—	10	—	—	80	—	—
$MgSO_4$	30	80	10	—	100	360	1.2	80
Metal Salts								
Fe, Co, Mo, Mn, etc.	+	+	+	+	+	+	+	+
Vitamins								
Biotin	—	+	—	+	+	+	—	+
Thiamine	+	+	+	+	+	+	+	+
B_{12}	+	+	—	+	—	—	+	+
Riboflavin	—	+	—	—	—	—	—	—
Pantothenate	—	+	—	—	—	—	—	—
Pyridoxamine	—	+	—	—	—	—	—	—
Nicotinic acid	—	+	—	—	—	—	—	—
Medium pH	6.6	2/3	6–7	5	5	6–6.2	—	6.5

Genera: *Astasia longa*, *Peranema trichophorum*, *Polytomella caeca*, *Ochromonas malhamensis*, *O danica*, *Crypthecodinium cohnii*, *Euglena gracilis*, and *Acanthamoeba*. All quantities are expressed as mg/100 ml medium unless noted otherwise. A + or + + indicates required compounds in concentrations at less than the mg range. Refer to the original sources for complete media including Hutner and Provasoli (1951). Compiled from Wise (1959), Aaronson and Baker (1959), Band (1962), Barry (1962), Allen et al (1966), Keller et al (1968), Kempner and Miller (1972), and Byers et al (1980).

sources of nutrients, or by accumulation of metabolites and other factors associated with changes in age of the cultures (Baker and Buetow 1976). Ammonium can serve as an inorganic source of nitrogen in the absence of organic sources, and appears to regulate anaplerotic incorporation of CO_2 into organic molecules.

Carbon Sources

Among a variety of carbon sources supplied to *Euglena* and related colorless genera (eg, *Astasia*), acetate is preferentially utilized. When supplied with glucose, acetate is consumed first, and glucose is utilized only after a period of physiological adaptation requiring about 200 h at pH 6.8 to 7.1. If preadapted cells are used as an inoculum in the glucose medium, or if 0.1% glycine is added, the lag is shortened to 70 to 100 h. Glycine stimulation of growth is maximal at pH 7.0. Only trace concentrations are required (ca 0.03%), suggesting that it acts as a stimulatory or "sparking" substance (Hurlbert and Bates 1971). A number of TCA cycle acids and amino acids were also found to stimulate glucose carbon uptake at neutral pH. Adaptation of cultures to glucose as a carbon source was apparently due to the appearance of genetically altered organisms (mutants), as also has been observed with *Astasia longa* (Barry 1962). Barry found that the parent strains in her cultures were unable to utilize exogenous sources of C-14 labeled glucose, whereas the mutant strains incorporated the label into metabolic CO_2. This was demonstrated in intact cells as well as in cell extracts. Since the extracts from the parent strains possessed enzymes for metabolism of the glucose, Barry concluded that the lack of glucose utilization by the intact cell is due to membrane impermeability rather than a deficiency in glucose-assimilating enzymes. The utilization of glucose is pH-dependent, whereas ethanol is not.

Ethanol Metabolism

Ethanol-grown *Astasia longa* reaches cell densities of 5 to 6.6×10^6 cells/ml, while those grown on acetate reach densities approximately an order of magnitude lower (5 to 6×10^5) (Buetow and Padilla 1963). The dry weight of ethanol-grown *Astasia* was 23% higher than acetate-grown cells. This was due to a larger carbohydrate content of cells metabolizing ethanol. Lipid is a major storage product in acetate-grown cells, but decreases in ethanol-grown *Astasia*. Since carbohydrates have a greater density than does lipid, the mass of the cell increases during ethanol metabolism. Ethanol-grown cells contain ca 72% more carbohydrate than comparable acetate-grown cells. Moreover, protein accumulation was 17% higher in ethanol-grown *Astasia*. Both *Euglena* and *Polytomella caeca* grow to high densities with ethanol as a carbon source in the presence of high thiamine concentrations (ie, 0.012%), suggesting similar metabolic pathways in the two organisms (Buetow and Padilla 1963). During growth with ethanol as a carbon source, the pH of the medium declines from an initial value of 6.7 to 3.5 or less. By contrast, a moderate increase in pH is observed with cultivation on acetate. In general, a lower pH can be detrimental when short chain organic acids are present in the medium due to the greater permeability of the membrane to the nonionized forms. *Euglena gracilis*, for example, is killed in the presence of acetic, formic, propionic, butyric, and fluoracetic acids at low pH, but not in the presence of fumaric, malic, succinic, or pyruvic acids (Bates and Hurlbert 1970). The differences in toxic effects may be explained by differences in membrane permeability to the acids, and consequently the extent that the pH of the cytoplasm is lowered by release of protons from the accumulated organic acids.

Amino Acid Metabolism

Amino acids are excellent sources of organic nitrogen for bleached *Euglena gracilis* (bacillaris variety, strain SM-L1), and *Astasia longa* (Buetow 1966). At concentrations of 10^{-5} M, *E gracilis* used the following amino acids as sole nitrogen source in a defined medium: glycine, alanine, valine, leucine, isoleucine, serine, threonine, and glutamic acid. Aspartic acid was used at 10^{-2} M. Glutamine and asparagine were also used, but at 10^{-3} M. They were better nitrogen sources than their parent dicarboxylic amino acids, glutamic and aspartic acids. Growth was not supported by phenylalanine, tyrosine, tryptophan, cysteine, methionine, proline, hydroxyproline, histidine, arginine, lysine, or taurine as sole nitrogen source. *Astasia longa* (Jahn strain) was more restricted than *E gracilis* and utilized only asparagine and glutamine as nitrogen sources for growth (Buetow 1966). The limitation on utilization of amino acids as metabolites in this strain of *E gracilis* raises interesting questions about the chemistry of amino acid metabolism. There is good evidence that the plasma membrane of *E gracilis* is permeable to amino acids, but it is not known if some additional internal membrane barriers or lack of necessary enzymes may prevent utilization. Methionine and cysteine serve as sulfur sources, although they are not good nitrogen sources. Variations in strains may account for some differences in utilization. *E gracilis* strain Z incorporates leucine, alanine, and perhaps valine intact into protein, although they are not metabolized; whereas they are metabolized by a bleached strain. Phenylalanine is not utilized by the bleached strain. However, it is incorporated into proteins and is considered a good protein tracer in strain Z (Kempner and Miller 1965). The amino acids that are not utilized by the bleached strain can be categorized according to their chemical composition: (1) aromatic amino acids, (2) those containing secondary amino groups or sulfur in the molecule, and (3) those containing a single carboxyl group and more than one nitrogen atom (the basic amino acids). By contrast, as previously reported, *Astasia longa*, a nonpigmented euglenoid, assimilated only two amino acids with complex amino-containing side groups.

Organic Acids

Growth of *E gracilis* on 10 mM succinate at pH 3.62 yielded densities of 6×10^5 cells/ml at inception of the stationary phase of growth, while cells cultured in succinate at a pH of 7.62 produced 3×10^6 cells/ml at the inception of stationary phase. Factors regulating diffusion of organic acids into cells include the concentration gradient across the membrane, and the proton concentration gradient between the inside and outside of the cell. An inward diffusion of the acids is favored by higher outside concentrations and lower external pH (greater hydronium ion concentration outside the cell). Assuming that uptake of organic acids is by diffusion of the nonionized (protonated) form across the membrane in response to the concentration gradient and the difference in pH across the membrane, the following equation of Jacobs (1940) can be used. This equation predicts the concentration ratio of a monocarboxylic acid inside (C_{in}) to outside (C_{out}) the cell:

$$\frac{C_{in}}{C_{out}} = \frac{1 + 10^{pH_{in} - pK_a}}{1 + 10^{pH_{ex} - pK_a}},$$

where pH_{in} and pH_{out} represent, respectively, the internal and external pH. Examination of the formula shows that if the internal and external pH are equivalent to the pK_a of the acid ($pK_a = -\log K_a$) where K_a is the dissociation constant of the acid, then the concentration of the acid inside the cell equals the concentration outside. On the whole, as the internal pH becomes larger than the pK_a, acid tends to become more concentrated inside the cell. However, an increase in external pH relative to the pK_a tends to favor retention of the acids outside the cell. Evidence from culture studies tends to support the appropriateness of Jacob's model, but with some deviations (Votta et al 1971).

By comparison, nutritional studies of *Polytomella caeca* indicate that fatty acid uptake is probably mediated by an active transport mechanism (Wise 1959). The fatty acids may be bound to coenzyme A during their translocation across the membrane. This would imply active transport by a CoA pump. A CoA pump is an active transport process that requires the substrate to be bonded to coenzyme A. Only those compounds that can form appropriate ester linkages to the CoA are translocated. The occurrence of a CoA pump in *P caeca* is further supported by the fact that nutritionally useful fatty acids are all capable of complexing with CoA, while molecules of similar size and lipid solubility are not utilized (eg, propionate and butyrate are utilized, but fumarate is not). *Polytomella* utilizes fatty acids (up to 5-carbon chain length) and alcohols (up to 6-carbon chain length) as carbon sources. The pH for optimum uptake indicates that fatty acids are utilized in dissociated form (ionized), whereas succinate and pyruvate are utilized in undissociated form. Alcohol availability is proportional to chain length, decreasing with increasing number of carbon atoms. Since longer chain alcohol molecules are restricted in their diffusion across the membrane, this suggests that the alcohols are accumulated by diffusion across the plasma membrane and not by active transport. If there were active transport, one would not necessarily expect to find an inverse correlation between uptake and chain length. Among the fatty acids, acetate concentrations less than 0.2% do not support maximum populations, but concentrations of 0.2 to 1.0% are adequate (Wise 1959). The closely related acetate flagellate *Polytoma obtusum* also grows on acetate as sole carbon source, but not on butyrate nor glucose or other hexoses. Carbohydrate exclusion is apparently due to membrane impermeability. Butyrate is accumulated in the cell against a concentration gradient. This indicates that the membrane is sufficiently permeable, but the absence of key enzymes may account for its lack of metabolism (Chapman et al 1965).

Butyrate metabolism is inducible in *Polytoma uvella* (Cirillo 1957) and depends upon two mechanisms: (1) induction of the enzyme fatty acid oxidase, which precedes growth; and (2) a long-term physiological adaptation spanning several cell generations, resulting in an increase in initial rate of growth. The generation time with butyrate as a substrate for unadapted cells is initially about 24 h, but is reduced to about 10 h after the second subculture in butyrate medium. Adaptation to butyrate (4-carbon acid) also results in a simultaneous adaptation to caproate (6-carbon acid). The inability of *Polytoma obtusum* to assimilate glucose cannot be due to a simultaneous requirement for a dicarboxylic acid and nitrogen source, as occurs in *Euglena* (Hutner 1956), since succinic acid and ammonium chloride were included in the medium.

Carbohydrates

Although glucose is not utilized as a substrate by many "acetate flagellates," it is a major carbon source in culture media for some parasitic protozoa and free-living heterotrophic flagellates. For example, glucose, galactose, and the disaccharide maltose are all utilized individually or in combination by the trichomonad *Tritrichomonas gallinae* in laboratory culture. This organism can be grown in Cysteine-Peptone-Liver extract medium (CPL) with one or more of the above carbohydrates (Matthews and Daly 1975). Variations in glycogen content and metabolic patterns occur as a function of the kind of carbohydrate in the medium. When cells are grown on glucose, then washed and tested for the effects of various carbohydrates on the rate of gluconeogenesis, replenishment with glucose caused greater glycogen synthesis than did addition of galactose. However, the two sugars showed equal gluconeogenic stimulation when added to washed galactose-grown cells (Matthews and Daly 1975). Neither glucose nor galactose had a definite glycogen sparing effect when added to washed cells grown in maltose or in a combination of glucose and maltose.

Glucose-supplemented complex media containing amino acids and other derivatives of natural products support growth in laboratory cultures of Kinetoplastida, including *Leishmania donovani* and *L braziliensis* (Steiger and Steiger 1977), *Trypanosoma cruzi* (Yoshida 1975), and a *Leptomonas* from insects (Guttman 1966; Roitman et al 1972). Colloidal protein is assimilated by pinocytotic uptake in bloodstream forms of *Trypanosoma brucei* (Langreth and Balber 1975). Iron-containing colloidal protein (ferritin) is engulfed in pinocytic, spiny-coated vesicles produced by invagination of the membrane lining the flagellar pocket. The vesicles become united with straight tubular extensions of a complex, mostly tubular collecting membrane systems where the ferritin is concentrated. Other spiny-coated vesicles, apparently primary lysosomes, congregate around the flagellar pouch and fuse with the collecting membrane system. The concentrated ferritin appears next in digestive vacuoles, possibly pinched off from the tubules of the collecting membrane system. A slightly different mode of pinocytosis occurs in short, stumpy forms where the tracer protein eventually appears in multivesicular bodies. These data suggest that carbohydrate and protein polymers can be assimilated intact in *Trypanosoma brucei* without prior extracellular degradation.

A carbohydrate-free soluble medium has been developed for the insect trypanosomatid *Crithidia fasciculata*, which grew well at pH 3.8 to 6.3 in a defined medium containing arginine (an essential amino acid), proline, and glutamic acid (Tamburro and Hutner 1971). Further addition of succinate stimulated growth and matched that with carbohydrate. Tamburro and Hutner suggest that the capacity to grow on noncarbohydrate substrates may be an adaptive feature of this trypanosomatid and others to permit multiplication in the insect gut or hemocoel and salivary gland. At pH 6.9 to 7.5, growth in the defined medium without carbohydrate was slight, but addition of carbohydrates (sucrose or sorbitol) permitted good growth (Tamburro and Hutner 1971). Further insights into the nutrition of Kinetoplastida can be found in Hutner et al (1979).

Among some of the free-living flagellates that have been extensively studied, glucose or sucrose are good carbon sources for the colorless dinoflagellate *Cryptothecodinium cohnii* (Keller et al 1968) and the chrysomonad *Ochromonas* (eg, Aaronson and Baker 1959). The "Ochromonads" are of particular interest, as they exhibit a broad range of trophic variations from *O malhamensis*, which is only partially phototrophic to *O danica* with substantially more chlorophyll and fully capable of phototrophic growth in mineral medium. Both species, however, are able to grow heterotrophically in the dark with adequate organic nutrients. *O malhamensis* requires vitamin B_{12}, whereas *O danica* does not. The B_{12} and thiamine requirements rise steeply with increasing growth temperatures, particularly above 35°C (Hutner et al 1957). The dinoflagellate *Gyrodinium cohnii* also requires vitamin B_{12} when grown at 35°C, but not at lower temperatures (Gold and Baren 1966). Vitamin B_{12} requirement is not uncommon in phototrophic unicells. Among the growth requirements of the green algae reviewed by Provasoli and Carlucci (1974), 37% required one or more vitamins, and of these 64% required only vitamin B_{12}.

Particulate Nutrition

When exogenous nutrients are decreased, both species of *Ochromonas* exhibit phagotrophy, but especially *O malhamensis*. Particulate organic matter, small algae, and even other ochromonads are ingested (eg, Stoltze et al 1969; Daley et al 1973). The particles appear in the large vacuole accompanied by acid phosphatase activity, indicating digestive vacuole formation. This has also been confirmed for *O danica* (Figure 16.3) feeding on the cyanobacterium *Microcystis aeruginosa* (Cole and Wynne 1974). Feeding is initiated by chance contact of the predator and prey. If the prey bacterium contacts the posterior end, it is carried

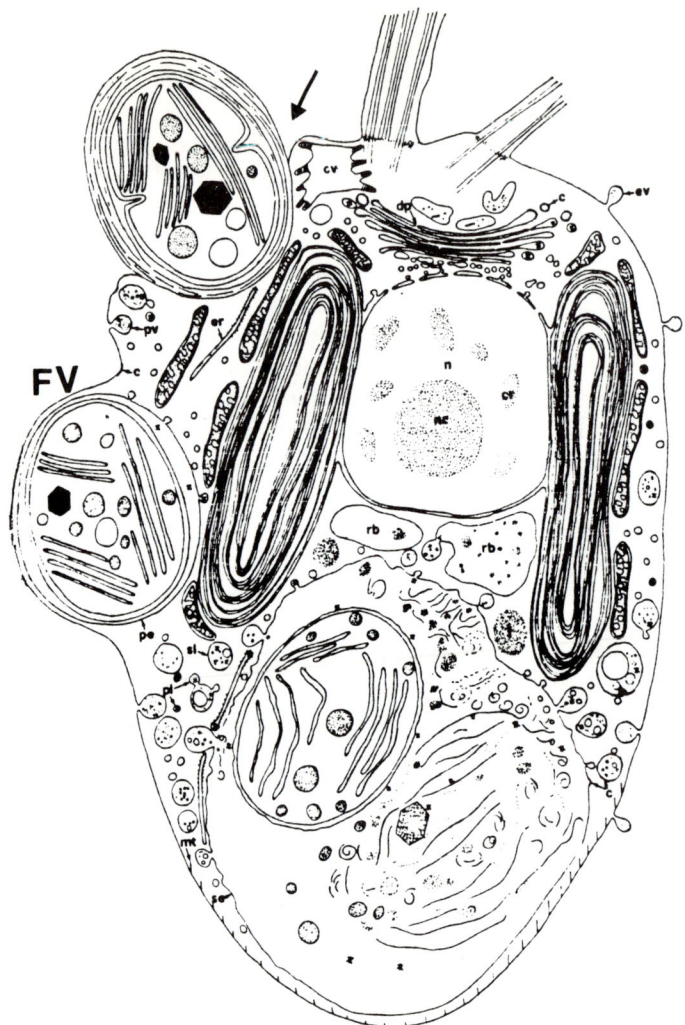

FIGURE 16.3. Diagrammatic interpretation of cellular events during endocytosis of algal prey by *Ochromonas danica*. The prey is first engulfed near the flagellar bases (arrow) and enclosed in a food vacuole (FV) which migrates to the large posterior vacuole, where digestive enzymes are deposited from lysosome and major lysis and degradation of the prey occur. Small vesicles containing digested products are released into the cytoplasm from the large vacuole and carry the nutrients throughout the cytoplasm. From Cole and Wynne (1974); with permission.

to the anterior end by water currents set in motion by the beating flagella. Ingestion occurs on the surface membrane immediately posterior to the flagella. The food vacuole is formed within the peripheral cytoplasm, but migrates posteriorly and eventually fuses with and empties into the large vacuole where digestion is completed. Secretion of membranous bodies containing cytoplasmic granules into the large vacuole also suggests that this is a site of storage and/or catabolic conversion of cytoplasmic-derived organic substances (Anderson and Roels 1967). The ingestion of bacteria may be a source of vitamins in the natural environment. Bird and Kalff (1986) have calculated that a bacterivorous chrysomonad (*Dinobryon*), growing in high biomass in freshwater lakes, consumes more bacteria from the water column than do crustacea, rotifers, and ciliates combined. The marine chrysomonad *Paraphysomonas imperforata* grazes on a wide variety of phytoplankton as well as on bacteria, and, as occurs with *Ochromonas*, resorts to cannibalism when food sources become scarce (Goldman and Caron 1985).

Nutrition of Amoebae

Soluble Nutrients

Some general information on the feeding habits and range of prey for Sarcodinads has been presented in the chapter on ecology (Chapter 5). Therefore, only some illustrative information on nutritional requirements and physiology of feeding will be presented here. Among the free-living amoebae, *Acanthamoeba* has been extensively studied as it is readily cultivated in the laboratory on chemically defined media. Originally isolated from soil samples (eg, Neff 1957) by subculturing on bacteria, this amoeba can be inoculated in axenic cultures by inducing cyst formation through dessication. The bacterial prey dies during drying, but the cysts of the amoeba persist and can be washed and transferred into bacteria-free media. Byers et al (1980) report that *Acanthamoeba castellanii* can be maintained in an axenic medium consisting of six amino acids, but grows more luxuriantly with nine amino acids. In the absence of acetate or glucose synchronous encystment occurs, resulting in 70 to 80% cyst formation. As reported in the chapter on metabolism, when acetate is the sole carbon source, glycine must also be included in the medium (Dolphin 1976). Glucose, ethanol, citric acid, and acetate can serve as major carbon sources (eg, Table 16.1). By contrast, *Acanthamoeba polyphaga* will grow on the same minimal amino acid composition as does *A castellanii* (ie, arginine, methionine, leucine, isoleucine, and valine), but does not utilize acetate as a sole carbon source even in the presence of glycine (Ingalls and Brent 1983). Glucose serves as a major carbon source, and citric acid is also included in the complete medium. Seven amino acids are required by the soil amoeba *Hartmanella rhysodes* in an axenic medium also containing mineral salts, vitamin B_{12}, thiamine-HCl, biotin, and a carbon source such as glucose (Band 1962). Acetate can also be substituted for glucose. Glycine is not required in the medium. The amino acids are arginine, methionine, leucine, isoleucine, valine, lysine, and threonine. The first five amino acids are identical to those required by *Acanthamoeba*, but lysine and threonine are additionally needed for *H rhysodes*.

The pathogenic strains of *Entamoeba histolytica* can be cultivated in axenic media (Diamond et al 1978), and glucose, maltose, or starch can serve as major carbon sources (Segura and Mata 1980). A beta-amylase activity has been found in cell extracts of *E histolytica* with a pH optimum of 6.5. This enzyme hydrolyzes amylose, amylopectin, and glycogen, yielding maltose as a reaction product (Werries and Nebinger 1984). This may help to explain the utilization of starch by *E histolytica* as a carbon source.

The amoebo-flagellate *Naegleria* can be grown in axenic culture on a nondefined medium containing yeast extract, peptone, yeast nucleic acid, folic acid, and hematin (Laverde and Brent 1980). A fully defined axenic medium has been developed for the cultivation of pathogenic and high temperature tolerant species of *Naegleria* (Nerad et al 1983). A chemically defined medium adequate for *N fowleri* and *N lovaniensis* contains a complex mixture of 11 amino acids (arginine, glycine, histidine, isoleucine, leucine, methionine, phenylalanine, proline, threonine, tryptophan, and valine), six vitamins, the nucleic acid precursor guanosine, glucose, salts, and metals. Addition of glutamic acid enhanced population growth and reduced the mean generation time. Glucose could be eliminated from the medium only if glutamic acid were present, indicating that this amino acid can serve as a major carbon source. As these pathogenic and free-living amoebae exhibit remarkable cytoplasmic transformations during the transition from the amoeboid to flagellate stage, a full understanding of their nutrition and the physiological variables accompanying their transformation is a major point of biological interest.

Particulate Nutrition

Many amoebae are phagotrophic, ingesting particulate matter by phago- or pinocytosis. In addition to bacterivorous feeding, many amoebae in culture feed phagotrophically on algae. Ho and Alexander (1974) found that amoebae differ substantially in their ability to prey on algae. Four species of amoebae were examined: *Amoeba discoides*, *Amoeba radiosa*, *Hartmanella castellanii*, and *Tetramitus rostratus*. Algae consumed by all of the amoebae (or at least by three of the four species) included *Chlamydomonas*, *Pandorina*, *Anabaena*, *Ourococcus*, *Ankistrodesmus*, and *Gloeocystis*. However, strains of *Staurastrum* and *Chlorella* did not support growth of the amoebae. The rate of phagotrophic feeding was influenced by temperature, pH, and age of the algal prey. A wide range of amoebae have been reported to consume filamentous fungal species as food (eg, Old and Chakraborty 1986). Among the mycophagous soil amoebae are *Acanthamoeba*, *Cashia*, *Mayorella*, *Saccamoeba*, *Thecamoeba*, *Trichamoeba*, *Vampyrella*, and *Arcella*. Feeding occurs in three stages; ie, attachment to the fungal propagules, complete or partial engulfment into food vacuoles, and digestion of cell walls and protoplasts (Old and Chakaborty 1986). In some cases, the spore wall is perforated after attachment, producing discrete holes, and releasing the protoplast to be consumed by the amoeba.

The molecular events triggering endocytosis (phagotrophy and pinocytosis) have been investigated rather intensely in cultures of free-living and pathogenic amoebae. Pinocytosis is inducible by exogenous factors including a wide range of positively charged solutes (eg, K^+, Na^+, and macromolecular cations). These cations apparently bind to surface receptors on the membrane and alter its permeability to calcium. The increase in intracellular calcium induces cytoplasmic events leading to inward cupping of the membrane and channel formation. The binding of the inducer to the membrane also apparently displaces calcium from the glycocalyx and increases the inward flux of the cation. The calcium is probably actively transported across the plasma membrane. The localization of the calcium beneath the plasma membrane stimulates contraction of microfibrils that mediate membrane invagination, formation of channels, and vesicle separation (Prusch and Hannafin 1979; Prusch 1980a,b).

Inhibition of protein synthesis by starvation or use of drugs prevented induction of pinocytosis by Na^+, but not by K^+ in *Amoeba proteus* (Johansson and Josefsson 1984). The addition, however, of Ca^{++} to the growth medium restored sodium-induced pinocytosis. These data suggest that interventions suppressing protein synthesis deplete the membrane of structures that are necessary for normal Ca^{++} functions during induction of pinocytosis by Na^+-like inducers. The tendency for amoebae to pinocytotically take up positively charged macromolecules, even those of no nutritional value or of toxic properties, was used by Prusch (1981) to investigate solute uptake-specificity in *Amoeba proteus*. Cells which were administered the toxic dye alcian blue took up the dye in pinocytotic channels, but expelled a fraction of it in bulk form through extrusion channels. This response was not elicited with other solutes taken up by pinocytosis, implying that if any selectivity exists in particle uptake, it is by specific solute extrusion mechanisms in the cytoplasm.

Phagocytosis, apparently, is also a carefully regulated and quantitatively determined process. *Acanthamoeba castellanii* distinguishes between nutritive yeast particles and nonnutritive polystyrene beads during phagocytosis (Bowers and Olszewski 1983). When cells were allowed to ingest yeast to capacity, endocytosis stopped, and further presentation of particles (beads or yeast) did not stimulate further uptake. If, however, the cells were allowed to ingest polystyrene beads to capacity, and a second dose of particles (either yeast or beads) were presented, the amoebae exocytosed the internal particles and took up new ones. Yeast cells, chemically altered to make them indigestible but nontoxic, were taken up and treated by the amoebae in the same way as nondigestible polystyrene beads. This indicates that an internal distinction is made between vacuoles containing digestible yeast and those containing nondigestible particles. These data are consistent with the hypothesis that the presence of digestible material within the vacuoles prevents exocytosis. The volume of material consumed by either pinocytosis or phagocytosis is apparently regulated by the volume of intracytoplasmic vacuoles produced (Bowers 1977). Cells of *Acanthamoeba castellanii* which were fed either protein to induce pinocytosis or latex beads to induce phagocytosis, required energy to endocytose the particles. Inhibitors of energy production or treatment with cold stopped endocytosis by both means. However, the total amount of particles consumed by either means was the same. Based on the volume of exogenous material consumed, Bower calculated that the maximum volume of the cytoplasmic compartment regulating rate of uptake was about 500 μm^3/cell, or approximately 15% of the cell volume. The intracytoplasmic events triggering phagotrophic ingestion of particles may involve the formation of adenosine monophosphate (AMP). Uptake of killed yeast was accompanied by relatively small changes in the levels of ATP and ADP, but by large (5- to 7-fold) changes in the AMP levels (Edwards and Doulah 1982). Calcium ion-induced pinocytosis is also stimulated by extracellularly administered cyclic AMP (Josefsson 1975). Induction of phagocytosis in the giant amoeba *Chaos carolinensis* during feeding on ciliates is apparently triggered by membrane surface receptors in contact with chemical substances on the cilia (Lindberg and Bovee 1976). Glutaraldehyde-fixed preparations of *C carolinensis* with prey cilia attached on their surface are engulfed by other *C carolinesis* cells, but are not ingested. Fixed *Chaos* without attached cilia are not ingested, indicating that the cilia are the triggering substance for phagocytosis. Moreover, when increasing amounts of cilia isolated from *Tetrahymena pyriformis* were added to cultures of *Chaos*, the rate at which prey ciliates were ingested was decreased, further suggesting that the cilia block contact sites normally available to initiate phagocytosis of the ciliate prey. Scanning electron microscopic observation of *Amoeba proteus* during phagocytosis shows that unlike freely motile cells, phagocytic cells lose their surface ridges running parallel to the long axis of the cell, and become more firmly attached to the substratum over a wider area of contact (Jeon and Jeon 1976). Broad pseudopods are formed to surround the prey, and this results in the formation of food cups. By contrast, pinocytotic cells exhibit a sharp decrease of contact area with the substratum, suggesting that the induction of phagocytosis may have a different effect on the cell periphery than that of pinocytosis (Opas 1981).

Dramatic changes occur in the phagocytic and pinocytic vacuoles of amoebae shortly after endocytosis. The pH drops abruptly (decreasing by 2 or more pH units) within 5 min after phagosome formation and before fusion with primary lysosomes (Heiple and Taylor 1982), and pinosomes reach a minimum of approximately 5.6 in about 10 min (McNeil et al 1983). The acidic transformation becomes complete at approximately 20 min. Phagosomal pH then slowly recovers to more neutral values over the next 2 h. According to Heiple and Taylor, these early pH changes may be a necessary step in preparation for fusion with primary lysosomes and may be involved in regulation of lysosomal enzyme activity. A substantial body of evidence indicates that protozoa digest food particles within digestive vacuoles formed via pinosomes or phagosomes, as is the case in most eukaryotic cells. Earlier reports that some benthic Foraminifera (ie, *Allogromia laticollaris*) may have a lacunary digestive system in the rhizopodial and intraapertural cytoplasmic region have been shown to be incorrect (Bowser et al 1985). With proper fixation, typical phagocytic and digestive vacuoles are found with prey of varying size, including bacteria and yeast. Most of the digestive vacuole activity is in the intrashell cytoplasm. No lysosomes were found in the reticulopodia. The remarkable capacity of the reticulopodia to invade and shear food-containing gelatin and other organic surfaces must be due to mechanical activity, and not to lysosomal, chemical effects (Bowser 1985). Some differentiation of vacuolar spaces in amoebae is, however, becoming apparent, as exemplified by *Entamoeba histolytica*. Two vacuolar compartments have been detected (Aley et al 1984). The first compartment consists of large vacuoles (> 2 µm diameter) and are formed in rapid equilibrium with the external medium. They are not acidic and apparently never acquire digestive enzymes. Their rapid exchange with the fluid in the surrounding environment suggests that they serve as a shuttle system cycling surface membrane into the cell and back again to the plasma membrane. The second compartment consists of small vesicles (< 2 µm diameter) with acidified contents of endocytic origin. Their pH is 5.2, and they probably represent digestive vesicles.

The detailed events during feeding in heliozoa illustrate some of the variations in feeding modes exhibited by sarcodinads (Patterson and Hausmann 1981; Linnenbach et al 1983). The feeding behavior of *Actinophrys sol* during predation on ciliates is a complex sequence of capture and engulfment. The ciliate is caught by adhesion to the arms of the heliozoan. Within 20 min the prey is enclosed by a funnel-shaped pseudopodial cup that progresses over the prey by cytoplasmic streaming in its differentiated leading edge. This membranous system is produced by fusion of the extrusomes with each other and with the plasma membrane. During this process, a presumably lytic material is secreted. The prey is lysed and coagulates within the digestive vacuole formed by the complete fusion of the feeding funnel. After this time, the perinuclear Golgi region is active and numerous lysosomes fuse with the food vacuole. As a result, the food is hydrolyzed. Within 4 h of feeding, much of the fluid is removed from the food vacuole and the membrane becomes condensed around the coagulated food mass. Simultaneously, vesicles filled with food material are pinched off from the periphery of the vacuole. During this process, the peripheral region of the heliozoan cell becomes vacuolated. The appearance of the cell and of the digestive vacuole remain the same for a period of about 12 h. The number of vacuoles in the cytoplasm gradually decreases. At this time, any undigested remains of the prey are egested. Although the formation of the feeding funnel is a unique adaptation of the heliozoa, many of the subsequent stages of food vacuole condensation, fusion of the primary lysosomes with the food vacuole, and appearance of smaller vacuoles in the peripheral cytoplasm are similar to the events occurring in other sarcodina.

The diversity of prey consumed by representative species of major groups of sarcodinads has been cited in the chapter on ecology and has been reviewed in considerable detail in other publications by Lee (1974, 1980) on benthic Foraminifera, Anderson (1983) on radiolaria, and by Nisbet (1984) for a broad group of protozoa.

Feeding and Nutrition in Ciliates

Soluble Nutrients

Only a few ciliates have been successfully cultivated with high yields in chemically defined media. Among these, *Tetrahymena pyriformis*, *Paramecium* spp, and *Glaucoma chattoni* will illustrate some of the current accomplishments and problems in developing adequate media. Species of *Tetrahymena* will grow on a medium containing amino acids, glucose, nucleic acid precursors, citric acid, vitamins, and inorganic nutrients. The L-amino acids required are arginine, histidine, isoleucine, leucine, lysine, methionine, phenylalanine, threonine, tryptophan, and valine. Most species also require serine, and a few need glycine, proline, alanine, aspartic acid, glutamic acid, and tyrosine (eg, Holz 1973). Amino acids are used by *Tetrahymena* as protein precursors, as sources of energy, and as intermediates in metabolism for synthesis of other essential organic molecules. The amino acid requirement can be met by free amino acids or their polymers: peptides, polypeptides, and proteins. Consequently, maintenance media can be prepared from undefined protein-rich media such as peptone extracts. Some of the amino acid requirements can be substituted by other nutrients. For example, proline, ornithine, and citrulline spare the requirement for arginine (Dewey et al 1957). The ornithine-sparing effect may be attributed to ornithine inhibition of arginine deiminase, thus protecting the pool of arginine against hydrolysis to citrulline and ammonia. A purine and pyrimidine (nucleic acid precursor) are required, and can be satisfied by guanine and uracil. Several carbohydrate sources are adequate for *Tetrahymena pyriformis* including glucose, mannose, maltose, fructose, dextrin, and starch (Holz 1973). Acetate is effectively consumed by *T pyriformis* as a carbon source, and is metabolized in part via the glyoxylate cycle as explained in Chapter 15. Organic acids and alcohol alone, however, have little nutritional value for *Tetrahymena*. Some species require sterols for maximum growth and phospholipids have been stimulatory in the growth of other species, but the biochemical basis is not known. *Tetrahymena patula* can be grown in a chemically defined medium if supplemented with a lipid such as high purity phospholipids; mono-, di-, and triglycerides; and free fatty acids (Luksas and Erwin 1983). Cholesterol or other sterols can be substituted for fatty acid containing lipids, and either ethanolamine or choline permits suboptimal growth in a lipid free medium. An amino acid containing defined medium with citrate and lecithin permitted growth of *T pyriformis* at 0 to 5°C (Cox 1970), and provides a useful medium for study of the effects of low temperature on ciliates as representatives of eukaryotic cells. A practical and inexpensive medium for maintenance of *Tetrahymena rostrata* and *T pyriformis* has been developed to permit high yields in batch culture. It consists of a solution of 1% powdered skim milk and 0.5% yeast extract in distilled water (Saliba et al 1983).

Glaucoma chattoni has been cultivated in a modified medium previously used for *Tetrahymena pyriformis* and *Colpidium campylum* by changing the amino acid, nucleotide, carbon source, and other components to obtain maximum growth (Meskill 1970). The medium contains 17 amino acids, nucleic acid components, fatty acids (linoleic and oleic), stigmasterol, sodium acetate, a mixture of B-vitamins, and several inorganic salts. It lacks citrate and carbohydrate, which were included in an earlier medium devised by Holz et al (1961). The modified growth medium contains 10 essential amino acids required by 15 out of 16 strains of *Tetrahymena pyriformis*; namely arginine, histidine, isoleucine, leucine, lysine, methionine, phenylalanine, threonine, tryptophan, and valine.

A small marine hymenostome ciliate, *Parauronema acutum*, can also be grown on a similar medium containing amino acids, purine derivatives, vitamins, lipids, and artificial seawater (Soldo and Merlin 1977). The following amino acids were required: arginine, histidine, isoleucine, leucine, lysine, methionine, phenylalanine, proline, serine (or glycine), threonine, tryptophan, and valine. The lipid is cephalin, and neither carbohydrate nor organic acids were needed in the minimal medium. Addition of glutamic acid or aspartic acid enhanced growth. Carbohydrates such as glycogen, starch, and glucose-1-phosphate, tested individually, were as effective as aspartic acid or glutamic acid in enhancing growth.

Axenic cultivation of *Paramecium* spp requires a sterol (stigmasterol), fatty acids, phospholipids, amino acids, nucleic acid precursors, vitamins, acetate, and inorganic salts (Soldo and Van Wagtendonk 1969). The basic medium is satisfactory for growth of *P aurelia* complex. *Paramecium multimicronucleatum* also requires carbohydrates as a major carbon source (eg, Johnson et al 1980). Glycogen and mannan gave good growth; dextrin and amylopectin gave only fair growth; whereas glucose, maltose, and amylose did not sustain growth. Ovalbumin or concanavilin A satisfy a protein requirement and apparently stimulate endocytosis. Some details of the historical development of the complete media, variations required for different species, and the rationale used in devising the media can be found in review articles (eg, Van Wagtendonk 1974; Wichterman 1985). An inexpensive medium using skim milk as the major source of carbon and nitrogen has been developed by Schönefeld et al (1986) for the mass culture of *Paramecium tetraurelia* stock 51s and the double mutant *pawn A/pawn B*, which exhibits aberrations in motility. The yield in a 250 l bioreactor was 750 g (wet weight) and 5×10^9 cells. The uptake of soluble phosphorus by the freshwater ciliate *Paramecium multimicronucleatum* is almost completely dependent on the presence of particulate food (as bacteria), and the phosphorus uptake of *P bursaria* with algal symbionts was substantially enhanced in the presence of bacteria as food (Buechler and Dillon 1974). The authors suggest that the relatively high metabolic rate of these organisms and other protozoa may account for the rapid turnover rate of phosphorus by these organisms, thus contributing substantially to nutrient regeneration in the natural environment. The attractive, pink-pigmented spirostome ciliate *Blepharisma* can be cultivated axenically only in the presence of killed bacteria or a bacterial extract (Giese 1973). A satisfactory medium consists of yeast extract, lyophilized bacteria, vitamins, and mineral nutrients. There is some evidence that *Blepharisma* lacks lipases and may therefore require exogenous lipids for maximum growth (Giese 1973). It also appears that some ciliates may require particulates in their environment to form food vacuoles and acquire sufficient nutrients to maintain growth.

Particulate Nutrition

The process of particle gathering by filter-feeding in ciliates has been studied in several species, with the aim of documenting the stimuli that initiate feeding, the role of cilia in moving fluid masses carrying food particles into the cytostome, and the hydrodynamics of fluid flow in and around the feeding apparatus. Fenchel (1980, 1986b) summarized the basic principles of food gathering in suspension-feeding ciliates. In the higher ciliates, food particles are moved by ciliary action in a direction parallel and posterior to the membranelles. Consequently, the food particles are driven into the food canal and toward the cytostome. In most oligohymenophora (hymenostomes and peritrichs), the stream of food particles is drawn through the paroral membrane at the mouth of the infundibulum. Among the polyhymenophores (spirotrichs), the membranellar bands propel water out of the buccal cavity, creating a spiral stream that enters at the lip of the cavity, travels down the right side of the membranelle, and is forced upward and out of the infundibulum. The ribbon of membranelles also acts as a filter. Food particles that are larger than the space between the cilia are retained and eventually collected at the base of the spiral in the region of the cytostome. The filtering activity of the ciliary apparatus is capable of concentrating food by a factor of several thousand. Most of the hymenostome ciliates can collect particles down to a size approximately 0.2 μm, while the larger spirostomes rarely retain particles < 1.0 μm. For some size ranges, collection can be 100%. That is, all of the particles swept into the buccal cavity are collected.

The process of particle gathering in the filter-feeding ciliate *Climacostomum virens* has been studied using cinematography (Fischer-Defoy and Hausmann 1977). The buccal apparatus consists of a funnellike depression in the anterior end of the cell. It contains adoral membranelles, buccal opening, buccal cavity, buccal tube, cytostome, and cytopharynx. Ingestion of several different types of prey has been observed. The green flagellate *Chlorogonium elongatum* is ingested into food vacuoles of varying size, depending on the concentration of food offered. *C virens* can take in 140 *Chlorogonium* cells within 2.5 min. By comparison, *Poteriochromonas stipitata* cells are often assembled within the enlarged buccal tube before a large food vacuole containing the prey is pinched off. The buccal tube is elastic and exhibits wavelike contraction during feeding, which may help to move food particles into the food vacuole-forming region. Large quantities of vacuolar membrane exceeding 1000 μm^2/sec are generated in forming the food vacuoles. Yeast can be ingested, and is vigorously consumed in such quantity that 5 to 10 min after ingestion *C virens* becomes a nearly motionless, globe-shaped body, slowly rotating around its axis. A sudden defecation of nearly all yeast cells occurs after 30 min. The ciliate assumes its normal ovoid shape, moves more rapidly, but momentarily refuses further nutrition.

The stimulus for feeding in carnivorous species has been examined using natural prey and small vibrating models to determine the role of mechanical stimuli in the induction of feeding (eg, Karpenko et al 1977; Orlovskaja et al 1984). Karpenko et al (1977) used as a model of protozoa prey very small magnetic particles driven to oscillate by a magnetic field. They found that the vibrating particle was attacked, but not ingested by the carnivorous ciliates such as *Dileptus anser* and *Didinium nasutum*. This suggests that the attack is initiated by at least a mechanical stimulus, and subsequent ingestion requires additional stimulation, perhaps of a chemosensory kind. Orlovskaja et al (1984) found that feeding response in *Dileptus anser* and *Didinium nasutum* varies with different kinds of chemosensory agents, including adrenaline, some amino acids, phospholipids, and neutral detergents. Particles of neutral absorbents such as carbon or starch, soaked in solutions of the chemical inducing agents, are actively eaten by predators. There may be a membrane chemosensory receptor mediating the response, since treatment of the ciliate with pronase and other proteolytic enzymes disturbs the feeding behavior (Esteve 1981). Molecular receptors may be situated in the membrane glycocalyx which is hydrolyzed by the enzymes, thus accounting for the blockage of sensitivity by the enzymes. In mechanical feeding studies (Karpenko et al 1977), bacteria-feeding ciliates such as *Paramecium caudatum*, *Stylonchia mytilus*, and *Spirostomum ambiguum* react to the stimulus of the vibrating particle, but apparently the mechanical energy or other factors inhibit ingestion. The hydrodynamic perturbations produced by the oscillating particle serve as a source of stimuli inducing a positive response. *Amoeba proteus* approaches the prey, but does not form a food cup, whereas *Amoeba dubia* exhibits a full feeding response including ingestion of the oscillating particles. The combined data suggest that mechanoreception accounts for prey identification at a distance, and that finally chemoreception determines ingestion. It may be that chemical gradients are too ambiguous to permit long-range detection of prey, and that mechanical stimulation is a more predictable source of information for prey location (Karpenko et al 1977).

The attack behavior of *Dileptus anser* with natural prey (*Colpidium campylum*) occurs during a period beginning shortly before dawn and continues until bright daylight, terminating abruptly between 8:30 and 9:00 AM (Miller 1968). Little or no feeding occurs at other times in a 24 h cycle. The proboscis of *D anser* contains trichocysts used to penetrate and hold the prey. Eventually, the prey is drawn into the cytostome by cytoplasmic streaming and carried toward the rear with formation of a digestion cavity. The cytostome simultaneously opens broadly to accommodate passage of prey, often in advance of its arrival. During unsuccessful attempts to strike a prey item with the proboscis, the proboscis may become attached to the surface of the surrounding environment, and portions are sacrificed as it is pulled free. With repeated attacks, the proboscis may be reduced to a stub. But, it is regenerated.

Fine structure of prey capture and ingestion of *Paramecium* by *Didinium nasutum* have been studied in fair detail by transmission electron microscopy (Wessenberg and Antipa 1970). Capture and ingestion of *Paramecium* by *Didinium nasutum* occur by chance collision and proceed with great speed. Upon contact of the conical proboscis with the prey, two types of extrusomes participate in prey capture. Centrally located pexicysts discharge into the prey and become dislodged, adhering to the surface of the prey cell. Longer extrusomes known as toxicysts, arranged around the periphery of the proboscis, penetrate into and anchor the prey to the proboscis of the *Didinium*. The proboscis becomes inverted by inward-directed cytoplasmic streaming, forming an enlarged, funnellike cytopharynx. The prey is drawn into the cytopharynx at the point of attachment to the toxicysts, and is eventually engulfed in a large food vacuole formed by the enclosing membrane of the cytopharynx, which has fused to surround the prey. Other surface extrusomes called cyrtocysts (ca 9 µm long) are also discharged from the pellicle of the *Didinium* during attack. These organelles are not found in the proboscis area of *Didinium*.

Prey capture by *Lacrymaria olar* (eg, Figure 7.5), a ciliate with a long necklike proboscis tipped by an apical dome, involves reorganization of the orgal region to form a cytostome from which food vacuoles are formed. Upon contact with the prey, toxicysts discharge from the apical dome, penetrating and holding the prey. There are two stages to the ingestion process: (1) the ciliated crown expands and (2) the apical dome collapses into the cavity created by this expansion. During the inversion, the microtubular ribbons associated with the apical dome remain attached to it and become the lining of the cytostome surrounding the ingested prey (Tatchell 1981).

There is a clear contrast between the feeding strategy of these ciliates, consuming "macro-prey" usually as one massive particle at delayed intervals, versus the filter-feeding ciliates that consume smaller particles in a more steady stream of trophic activity. The rate of food particle filtering has been estimated by fluid hydrodynamics (Fenchel 1980) and by observations of the amount of prey particles consumed per unit time. Fenchel (1980) estimates that the water transport rate in *Stylonichia* is 4.9×10^{-3} ml/h. Experiments with latex beads yield a maximum clearance of 5.5×10^{-3} ml/h. The maximum clearance of *Colpidium campylum* was 0.433 ml/h. Estimates of grazing rates by tintinnids and their selectivity in feeding have been determined in laboratory cultures of field-collected specimens by Heinbokel (1978), Stoecker (1981), Rassoulzadegan (1982), and Capriulo (1982). When offered a mixture of prey containing dinoflagellates and nondinoflagellates, the tintinnid *Flavella* selectively consumes dinoflagellates, while accepting very few if any cryptomonads, haptomonads, chrysomonads, diatoms, prasinomonads, or chloromonads (Stoecker et al 1981). Maximum ingestion by several tintinnids isolated from coastal waters of California was equivalent to 10 to 20% of their body weight (Heinbokel 1978). In general, ingestion rate increased with increasing food supply until the maximum rate was reached. The ingestion and filtration rates of seven species of field-collected tintinnids were determined by Capriulo (1982). Ingestion rates of 0.05×10^6 µm³ to 1.3×10^6 µm³ per tintinnid per day were found with maximum filtration rates of 2 to 65 µl per tintinnid per h. Filtration rates were independent of temperature, and the highest rates were obtained with the larger food particles. In the natural environment, tintinnids are significant herbivores removing as many as 0.06 to 87 algal cells per predator per h, depending on the season and the type of alga ingested (Capriulo and Carpenter 1980). Their filtration rates varied from 1.03 to 84.7 µl per tintinnid per h, and the magnitude of this predatory activity was of the same order of magnitude as that for copepods. In general, tintinnids have very high feeding rates in proportion to their weight which may be several orders of magnitude higher than those of other zooplankters including copepods. For example, the Mediterranean neritic tintinnid *Stenosemella ventricosa* feeds on nannoplankton with particle diameters from 1.3 to 27 µm and more intensively on particles in the 3–12 µm range. Maximum feeding is as high as 4×10^5 µm^3 (wet food vol.) per tintinnid per day; but under steady food supply, feeding rate reaches a stable mean value of 4.6×10^4 µm^3. This rate is equal to 66% of the fresh weight of the tintinnid. The rate of feeding on particles of a given size is directly dependent on their concentration and *S ventricossa* appears to be able to switch its feeding pressure to particles in other size ranges after depleting a given group of prey (Rassoulzadegan and Etienne 1981).

Some predatory ciliates secrete chemical substances (morphogens) that induce changes in the morphology of the protozoan prey. The altered morphology reduces the probability that the prey will be ingested by the predator. When the hypotrich *Euplotes octocarinatus* is placed in the presence of the carnivorous hymenostome *Lembadion lucens*, some morphogen released by the predator induces a change in the morphology of *Euplotes* within a few hours (Kuhlmann and Heckmann 1985). The cells develop ridges and prominent lateral wings that give them an almost circular profile. Similar responses can be induced by substances released from the predators *Urostyla grandis*, *Stylonichia mytilus*, and *Dileptus anser*. The morphogen has not been chemically identified, but it is a relatively heat-stable substance that is not inactivated by deoxyribonuclease, ribonuclease, or proteinase, but is inactivated by pronase. This indicates that it is probably not DNA, RNA, or a simple protein. The elucidation of the chemical composition and site of biological activity of the morphogens could contribute substantially to our understanding of the complex web of predator prey relationships, and broaden our knowledge of the genetic basis of cellular morphological transformations.

Symbiosis

Symbiosis, the close association between two or more organisms of different genetic composition, is widespread among protozoa and assumes highly diverse forms. Bacteria are endosymbionts of divergent groups of protozoa including endonuclear symbionts in dinoflagellates, cytoplasmic endobionts in amoebae and heliozoa, and "killer-trait" particles in the cytoplasm of some *Paramecia*. Algae, especially photosynthetic dinoflagellates, are by far one of the most common endosymbionts dwelling within the cytoplasm of sarcodina. The sarcodina include amoebae, benthic and planktonic Foraminifera, acantharia, and heliozoa. However, even here diversity is common. Some benthic Foraminifera contain eukaryotic algae, including *Chlamydomonas* and diatoms (eg, Lee 1980a,b). Endosymbionts in radiolaria encompass dinoflagellates, prasinomads, prymnesiomonads, and red algae (Anderson 1983). Only one kind of symbiont usually is associated with a given individual radiolarian. The presence of *Chlorella* within *Paramecium bursaria* is perhaps one of the most widely celebrated and intensively studied endosymbiotic associations among ciliates. However, the colorful, red-pigmented ciliate *Mesodinium rubrum* (also called *Myrionecta rubra*), with bifurcated oral tentacles on the anterior end, also harbors algal endosymbionts (Lindholm 1985). It can be a major source of primary productivity in nutrient-enriched marine environments, especially in upwelling areas (Smith and Barber 1979).

In most cases of algal symbiosis, the endosymbionts are enclosed by vacuolar membranes that separate the host cytoplasm from the foreign cells. This maintains a "respectful distance" between them, and can control exchange between the host and symbiont (Figure 16.4). In general, the coassociation between these diverse organisms results in a benefit to both species. Carbohydrates (maltose) and at least two amino acids (glutamate and alanine) are translocated to the host from the symbiont. Ammonium uptake by the ciliate is stimulated in the presence of the algal symbiont. This nitrogen source is transferred to the alga in addition to products of host metabolism. The exchange of carbon compounds is shown in the upper part of the diagram in Figure 16.4, and of nitrogenous compounds in the lower left part.

In many of these associations, the endosymbiont is frequently sequestered within a favorable environment, protected from predators or adverse stimuli, and often assimilates host-produced substances (sometimes metabolic wastes) that promote growth. In turn, the symbiont provides benefits to the host. Autotrophic endosymbionts share some of their photosynthetic products with the host, and thus contribute to a stable coassociation. In some cases, the presence of the symbionts either directly or indirectly contributes to maintenance of host metabolic processes. In Foraminifera, there is good evidence that for some species at least, the photosynthetic activity of the symbionts provides energy or some stimulatory substance which promotes the energy-requiring process of calcite shell construction. When the symbionts are removed, or photosynthesis is stopped by photosynthetic inhibitors, calcification is suppressed (eg, Duguay and Taylor 1978). Moreover, there is a regular diel cycle of cytoplasmic streaming in planktonic Foraminifera and radiolaria. At daybreak, or within an hour before daylight, the symbionts are moved out of the inner cytoplasm onto the spines or into the peripheral cytoplasm in nonspinose species of radiolaria. At evening, they are withdrawn into the inner cytoplasm. When the symbionts are subjected to photosynthetic inhibitors, the diel cycle is interdicted. It is not known how this close coordination between symbiont and host is achieved.

On the whole, however, there is compelling evidence in a variety of symbiotic protozoa to show that the symbionts provide nourishment for the host, and may be one of the major stabilizing factors in maintaining a close association between these otherwise highly diverse organisms. In some cases, this has been demonstrated by placing symbiont-containing protozoa in food-free (starvation) medium. One group is placed in the light and another in darkness. The illuminated group far outlives the dark control, indicating contribution of nourishment

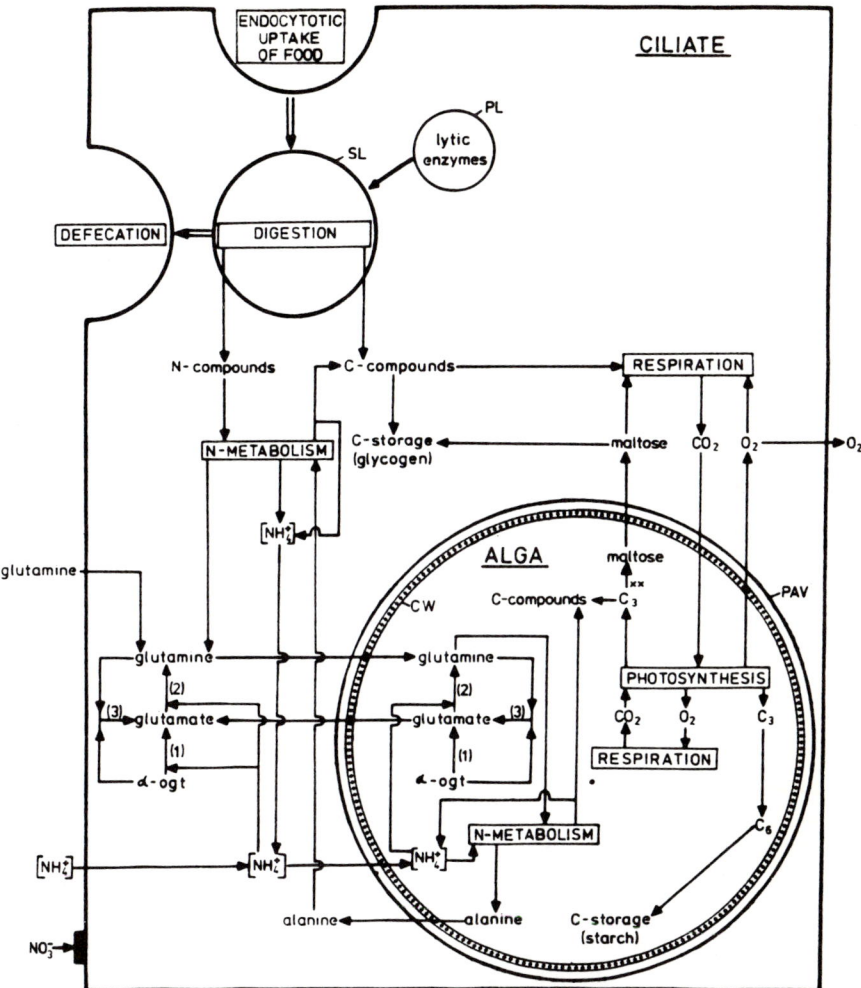

FIGURE 16.4. Host–algal interactions in a symbiont-bearing freshwater ciliate. The algal symbiont (ALGA) is enclosed within a perialgal vacuole (PAV) produced by the host. The vacuolar membrane regulates exchange of materials between the host and symbiont. Arrows indicate the flow of compounds. Digestive processes shown in the upper left produce nitrogenous and carbon compounds used by the host for metabolism and respiration. Some of these products, such as CO_2 and ammonium, are utilized by the alga for nitrogen metabolism and photosynthesis, respectively. Although nitrate is not taken up across the host membrane, the presence of the algal symbionts produces a change in the host membrane, permitting uptake of ammonium ions. These are passed into the algal vacuole for symbiont metabolism. Glutamine transported into the host cell is transferred to the symbiont. In turn, the amino acids, glutamate and alanine, synthesized by the symbiont are released to the host. The carbohydrate, maltose, is consumed by the host in glycogen synthesis and respiration. From Reisser (1986); with permission. See also Lee et al (1985b).

to the host by the photosynthetic activity of the endosymbiont. Among the radiolaria, kinetic studies using radioactively labeled carbon as a tracer have shown that photosynthetically fixed carbon first appears in the symbionts, and then some is gradually transmitted to the host (Anderson et al 1985). The many complex metabolic and adaptive responses required to maintain the host–symbiont association among a variety of freshwater and marine organisms have been extensively analyzed in review articles (eg, Anderson 1983; Corliss 1985; Lee et al 1985b; and Reisser 1986). In overall perspective, it is clear that in many cases the host–symbiont association is highly specific, and only one symbiont species is adopted. This indicates an ability by the host to selectively associate with a given symbiont and to protect it against the usual degradative processes that attack foreign material engulfed by the cell. Evidence is accumulating to show that the membranes of symbiont-enclosed vacuoles are altered during stable symbiotic associations to prevent fusion of lysosomes with the vacuole. Hence, the symbiont is protected against host digestion. The presence of unique high-molecular weight proteins within the membrane of the symbiont-enclosed vacuole suggests that this protein may block lysosomal fusion (eg, Reisser 1986). Whatever the controlling factor may be, it must also permit proper exchange of beneficial substances between the symbiont and host, and likewise allow the host to regulate the density of the symbionts to prevent an overpopulation of the cell. In general among planktonic Foraminifera and radiolaria, the number of symbionts increases as the host matures and increases in size. It is not known how this balance is maintained, but there is good evidence in radiolaria that the host can digest some of the algal symbionts in addition to acquiring soluble photosynthates from them. Thus, it may be that by digesting some symbionts or ejecting excess ones, the proper balance between symbiont and host biomass is maintained. It is also possible that the host regulates the rate of binary fission of the symbiont by some form of hormonelike secretion, but this requires additional experimental verification. In addition to possible host direct control over symbiont abundance, there are environmental affectors known as ecological control factors. These include the amount of light available to sustain symbiont productivity and growth, and sources of food for the host that enhance its growth and possibly stimulate symbiont proliferation (eg, Reisser 1986). Although much remains to be done in clarifying the host–symbiont relationship, it is clear that this unique coassociation between plant or bacteria and a protozoan host provides remarkable benefits to both associates. In the case of symbiotic associations with photosynthetic symbionts, it ensures that the host has a permanent "agricultural" food supply as long as appropriate sunlight and mineral nutrients are available.

Summative Perspective

The diversity of feeding strategies and modes of nutrition presently documented in protozoa clearly rivals the range of behaviors and nutritional requirements established for higher organisms. The remarkable range of habitats of protozoa, and their divergent evolutionary history has produced a wide distribution in nutritional behavior. Some species are largely osmotrophic, requiring a suitable concentration and mixture of soluble nutrients to sustain metabolism. These species are sometimes parasitic and dependent on the unique mixture of compounds found in the host body. Others are free-living and invade aquatic environments rich in organic nutrients. Some of these, known as the acetate flagellates, possess specialized mechanisms in their plasma membranes to regulate the intake of potentially toxic organic acids that can be used as needed in the proper quantity for metabolism. Others combined osmotrophy with various forms of particulate feeding known as phagotrophy. A variety of particulate, food-gathering mechanisms have evolved among all three major groups of protozoa (flagellates, amoeboid organisms, and ciliates). In some cases, the mechanisms suggest convergent evolution as exemplified by the occurrence of food-collecting canals in some flagellates and ciliates, although the architecture is quite different and the method of engulfing particles very divergent. A range of carbon sources has been exploited by protozoa including carbohydrates, amino acids, organic acids, short chain alcohols, and lactic acid (a waste product of many higher organisms). Nitrogen is assimilated as nitrate by species spanning groups as diverse as autotrophic flagellates to ciliates living at the oxic/anoxic boundary of lakes. A variety of organic sources of nitrogen can be utilized by one or more species, including amino acids and urea, an end product of nitrogen metabolism in many organisms. Some protozoa are plantlike (autotrophs), while others are facultative autotrophs, and many intergrade toward or become obligate heterotrophs. Innovative research in protozoan nutrition clarifies some of the common sources of energy broadly utilized by eukaryotic cells, but also highlights the novel aspects that signify the unique evolutionary pathways "explored" by the protozoa. These "byways" in evolution include various forms of symbiosis with clear implications for host nutrition and altered metabolism, occasioned by the inclusion of the symbiont within the host cytoplasm. It is clear that protozoa are not simply typical eukaryotes, especially if that is taken to mean representative of metazoan cells. Many species do possess "typical" glycolysis pathways and utilize aerobic metabolic reactions found in some eukaryotic cells. However, their particular modes of food gathering and some unique nutritional requirements reflect novel energy-transforming metabolic pathways, and further substantiate the peculiar cellular biology that characterizes some species of protozoa. In some cases, knowledge of nutritional requirements can yield practical applications, as for example, the use of vitamin B_{12}-requiring species in bioassay techniques (Hutner et al 1956; Baker et al 1986) and the employment of protozoa in water purification and industrial applications. Given the diversity of protozoan species and the wide range of possible substrates they can utilize as nutritional sources, a broad frontier remains uncharted in this intriguing arena of research, which lies somewhere at the boundaries between biochemistry, bioengineering, and behavioral biology.

17
Respiration and Osmoregulation

Energy Budgets and Growth Efficiency

The energy obtained through nutrition is never used with 100% efficiency by a living system. Similarly, not all of the carbon ingested is retained to synthesize living substance. Energy loss through heat and carbon loss as CO_2 during respiration contribute to the inefficiency of the system. Additional quantities of organic matter are also lost through excretion. The remaining useful energy is used to drive the numerous processes of life including biosynthesis of protoplasmic components, motility, and the maintenance of a balanced physiological state relative to changes in the natural environment. Some fundamental principles of energy physiology and osmoregulation (the maintenance of osmotic balance) are considered in this chapter.

Energy Budgets

A complete account of energy utilization by protozoa can be assessed by the following equation developed by Heal (1967) and applied in several studies (Laybourn 1975; Laybourn and Finlay 1976):

$$\underset{(C)}{\underset{\text{food ingested}}{\text{Amount of}}} = \underbrace{\underset{(P)}{\underset{\text{protozoa produced}}{\text{Amount of}}} + \underset{(R)}{\underset{\text{in respiration}}{\text{Amount lost}}}}_{\underset{(A)}{\text{Amount assimilated}}} + \underset{(E+E)}{\underset{\text{excreted}}{\text{Amount}}} \qquad 17.1$$

The mass of food consumed is accounted for by the amount of protoplasm produced, the amount lost during respiration, and the amount excreted as wastes. The amount assimilated during metabolism is equivalent to the gain in protoplasmic mass plus the amount used in respiration. Protoplasmic mass (P) is calculated based on the increase in numbers of individuals in a protozoan population. Respiration is usually measured as the amount of CO_2 produced, and where possible, excretion is measured by the mass of waste products. With protozoa, it is not possible to directly measure the amount of waste matter since it is not possible to collect the excreted material. Therefore, the excreted products are usually computed indirectly by determining the mass of food ingested and subtracting the amount assimilated (A). The assimilation efficiency of an organism is given by A/C, and can be expressed as a percent. The gross growth efficiency is given by P/C, the ratio of the protoplasmic increase to the amount of food consumed. This is an expression of the efficiency of the organism in converting prey biomass into protozoan biomass. To account for energy utilization, it is necessary to express each of the terms in the equation as energy units. The standard unit presently used is the Joule, or for organisms as small as protozoa, the microjoule (μJ). The energy content is assessed by combusting the material in a microbomb calorimeter and determining the amount of heat released from a known mass of the substance.

The gross growth efficiencies for some protozoan species are as follows: *Tetrahymena pyriformis* (50%) (Curds and Cockburn 1968), *Colpoda steinii* (78%) (Proper and Garver 1966), *Colpidium campylum* (9 to 11%) (Laybourn and Stewart 1975), *Stentor coeruleus* (63.7 to 81.6%) (Laybourn 1976), and *Acanthamoeba* sp (37%) (Heal 1967).

Respiration

The respiratory activity of protozoa, as with other organisms, varies with temperature and also in relation to the kind of substrate metabolized. Some illustrative data are presented, but are not intended to be exhaustive. Within a range of 8.5° to 20°C, maximum respiration occurred at 15°C for the ciliates *Frontonia leucas*, *Spirostomum teres*, and *Vorticella microstoma*. However, *Paramecium aurelia* and *Tetrahymena pyriformis* showed increasing respiration rate with increasing temperature (Laybourn and Finlay 1976). Cowling (1983, 1984) has used a gradient diver microrespirometer to measure oxygen uptake by protozoa. For the testate amoeba *Corythion dubium*, the mean oxygen consumption rate was 27.1 µl/amoeba/h, and for *Tetrahymena pyriformis* 64.2 µl/individual/h.

Oxygen consumption in *Tetrahymena pyriformis* is stimulated by addition of only a few kinds of L-amino acids to the growth medium (Roth et al 1954). Among 39 L-amino acids tested, L-phenylalanine increased oxygen uptake 50 to 60%, whereas L-tyrosine (saturated solution) and L-cysteine were somewhat less stimulatory. In further studies, respiration rate, measured as oxygen uptake, was assessed in *Tetrahymena pyriformis* using a range of phenylalanine- or tyrosine-containing peptides or their derivatives as substrates (Roth and Eichel 1961). Peptides containing L-phehylalanine or L-tyrosine were effective, or more effective, in stimulating oxygen uptake than the free amino acids. Oxygen uptake stimulation varied from 59.6% for the dipeptide L-leucyl-L-phenylalanine to 93.8% for L-phenylalanyl-L-phenylalanine. Increased uptake for individual amino acids were as follows: 61.4% for L-phenylalanine, and 26.8% for L-tyrosine.

The respiratory CO_2 production for the flagellate *Trichomonas gallinae* was assessed by Matthews and Daly (1975) when grown anaerobically or aerobically with different substrates. They found that under anaerobic conditions, the amount of CO_2 produced in $\mu l/10^8$ cells/h was 62 for metabolism of endogenous substrate, 131 with exogenous galactose, 124 with glucose, and 75 with maltose. Under aerobic conditions, CO_2 production expressed in the same units was 69 for endogenous substrate. For exogenous substrates, CO_2 production expressed in the same units was 84 for galactose, 107 for glucose, and 130 for maltose. Respiration measured as oxygen consumption in *Acanthamoeba castellanii* (Waidyasekera and Kitching 1975) varied from 0.283 to 0.486 $\mu l/10^5$ amoebae/h at 24°C.

Osmoregulation

Contractile Vacuolar Activity

Part of the energy produced by respiration is used by the protozoan cell for osmoregulation. Most freshwater protozoan cells are hyperosmotic relative to their environment. Hence, water is accumulated by osmosis and must be removed at a sufficient rate to prevent swelling and lysis of the cell. The contractile vacuole is the main organelle for expulsion of water and perhaps also some soluble wastes. The cyclical expansion and contraction of the vacuole is energy requiring, and there is also increasing evidence that the surrounding vacuolar membrane may also utilize active transport to adjust the ionic composition of the vacuolar fluid. The frequency of vacuolar contraction is inversely related to the ionic strength of the surrounding aqueous medium. With increasing solute concentration (tonicity) of the external medium, there is less water absorbed by the cell, and the contractile vacuole is less active. Indeed, most marine organisms lack a contractile vacuole, since the surrounding seawater is very similar in tonicity to the protoplasm. Several lines of evidence support the conclusion that the contractile vacuole is fundamentally a water regulation device. If freshwater ciliates are subjected to low temperatures or metabolic inhibitors, contractile vacuolar activity stops, and the protozoan gradually swells and eventually bursts (eg, Elliott 1973). Further evidence of the osmoregulatory role of the contractile vacuole is obtained by examination of cellular response to hypotonic media. When *Blepharisma* is acclimated to a solution containing 30 to 40 times the concentration of ions used in its standard growth medium, and then placed in distilled water, the contractile vacuole is incapable of coping with the influx of water. The cells burst within 15 min. (Giese 1973). If, however, the hyperosmotic-acclimated cells are placed in a solution of standard osmolarity (normal culture medium), they swell but do not burst. Eventually, they recover normal size due to contractile vacuolar expulsion of the excess water. Apparently, the small amount of salt in the culture medium is sufficient to prevent lysis until the contractile vacuolar activity compensates for the water influx. When variations in the osmotic pressure of the external environment are not too large, ciliates (eg, *T pyriformis*) are capable of regulating their water expulsion rate to compensate for the influx. When a cell is subjected to media with high osmotic pressure and replaced in a medium of lower tonicity, the shrunken cell is capable of restoring its normal size by adjusting the rate of water loss (Elliott 1973). Thus, during this recovery, the output of fluid by the contracile vacuole must be slower than passive influx of water until the original cell volume is attained. Likewise, when an inflated cell produced by exposure to a hypotonic medium is placed in a medium of normal tonicity, the output of the contractile vacuole is enhanced until the passive influx of water is equilibrated to the output. Elliott (1973) suggests that both the regulation of cell volume (maintenance of constant volume) against a constant osmotic gradient and recovery of cell volume after passive volume changes indicate a feedback system that relates cell volume to contractile vacuolar output. This means that the cell has some mechanism for sensing deviations of the cytoplasmic volume from some predetermined "normal" level, and adjusting contractile vacuolar activity to compensate for these deviations. The variables determining the rate of fluid discharge from the contractile vacuole are two-fold: (1) frequency of contraction, and (2) volume of the vacuole immediately before discharge. In simple terms, the amount of water expelled is directly related to the rate of pumping (contraction) and the volume of fluid that the vacuole can hold at each discharge. Both of these variables are altered when the cell is subjected to osmotic imbalances. Thus, when the cell is exposed to hypertonic media, the interval between successive contractions increases and the volume of the vacuole at each discharge decreases. However, these changes can occur somewhat independently of one another. For example, when a cell is transferred to a higher osmotic medium than is normal for that cell, the rate of contractile vacuolar activity decreases immediately, but the decrease in vacuolar volume requires five to six vacuolar cycles before it reaches a minimum (data of Stoner and Dunham reported by Elliott 1973). A predictable relationship exists between protozoan size and contractile vacuolar activity (Lynn 1982). In a sample of protozoa ranging in length from 25 to 150 µm, a strong negative correlation exists between frequency of contraction and cell size, but there is a positive correlation between volume output of the vacuole and cell size. That is, species with larger cell size have less rapid vacuolar pulsations than do species of smaller size, but, the larger species have more fluid expelled at each systole. The cyclic behavior of the contractile vacuole is more complex than a cursory examination with light optics may imply. Patterson (1977) identified seven phases in the pulsation cycle of the contractile vacuole in *Paramecium caudatum*. The filling of the vacuole during diastole involves two phases. The first is rapid and is due to expulsion of fluid into the vacuole from the surrounding ampullae (Figure 11.4). The second phase is slower and is terminated by the rounding up of the vacuole. At this point, the vacuole remains unchanged for a short time, and then some fluid is passed from the vacuole to the ampullae. After a further short period, the vacuolar fluid is expelled to the environment through the pore. There is a short rest period between the end of the expulsion phase and the beginning of the next diastole. In *P caudatum*, the contractile vacuole appears to have contractile elements. Earlier cinematographic studies of contractile vacuolar activity in *Paramecium multimicronucleatum* (Organ et al 1968) indicated that once the expulsion pore is opened, the vesicle is invaginated by adjacent cytoplasm and is emptied due to pressure from the surrounding cytoplasm. The expulsion may also be aided by pressure of the fibrils that anchor the ampullae to the excretory canal. There is no permanent pore to the vesicle, but it is closed by a sealing of the ruptured membrane where it is in contact with the pellicular excretory canal. At the onset of expulsion during systole, the membrane across the basal opening of the excretory canal is ruptured along one edge. This forms a semicircular slit, with the remainder of the membrane anchored to the opposite rim of the pore. As the fluid rushes out, the free portion of the membrane is displaced up against the opposite wall of the pore as a flap. At the end of the expulsion phase, a constriction of the vesicular and cell membranes at the base of the excretory canal reseals the opening. Obervations by R. D. Allen (reported in Wichterman 1986, p 128) indicate that the membrane closing the pore in *Paramecium* is composed of two cytoplasmic membranes. During vacuolar diastole, the pore is closed by the continuous plasma membrane covering the outer surface of the cell and the vacuolar membrane. These two membranes fuse temporarily during the vacuolar systole and provide a delicate, but durable and precise device that functions as an automatic self-opening and self-closing valve at the pore.

Intracellular Organic Compounds

In addition to the osmoregulatory function of the contractile vacuole as a water expelling device, accumulation of organic intracellular compounds is related to the tonicity of the external medium. With increasing tonicity of the environment, glycerol, bicarbonate, amino acids, or free sugars have been observed to accumulate in a variety of protozoan cells. The euryhaline flagellate *Dunaliella tertiolecta* regulates internal osmolarity by production of glycerol. Cells grown in media of increasing tonicity (0.25, 0.50, and 1.00 M NaCl) contained intracellular glycerol concentrations of 0.12, 0.41, and 1.12 mg/10^8 cells, respectively, as measured by gas chromatography (Jones and Galloway 1979). The quantity of the glycerol produced was also influenced by the wavelength and intensity of light during growth. Further regulation of ionic balance is indicated by adaptive responses of *D tertiolecta* to increased NaCl concentrations. When adapted to medium containing 0.5 M NaCl, transfer to higher salinities produces a lag in growth, suggesting an adaptation period (Latorella and Vadas 1973). However, there is no difference in the intracytoplasmic concentration of Na^+ with the higher salinity of the medium, indicating a mechanism for extrusion or exclusion of the cation. Increased salinity of the medium is accompanied by increased uptake of H^+ suggesting the possibility of a H^+/Na^+ exchange mechanism. Thus, Na^+ is pumped out and H^+ is simultaneously pumped into the cell. Light is required for this process, which may indicate a light-driven active transport mechanism. Cells adapted to a higher salinity also exhibit higher carbonic anhydrase activity. This enzyme mediates uptake of HCO_3^- or CO_2, and may provide a mechanism for enhanced accumulation of the bicarbonate anions to balance the influx of H^+ cations. Loeblich (1970) also reported that carbonic anhydrase increases in *Dunaliella salina* with increased salinity.

Some flagellates subjected to osmotic stress (placed in hypertonic media) react by increasing the intracellular pool of free amino acids. For example, the intracellular concentration of proline has been found to increase in marine algae subjected to hypersaline solutions, and the addition of proline to the growth medium of *Chlamydomonas reinhardii* reduced osmotic stress and increased growth (Reynoso and Gamboa 1982). A linear relation was found between proline concentration added to the medium and resistance to osmostic stress. The precise role of proline in protecting the cell against osmotic stress is uncertain. It may increase the internal tonicity of the cell, thus reducing the osmotic gradient, or alter the cellular water structure, thus protecting against loss of hydration of cytoplasmic constituents due to high external salt concentrations.

Intracellular free amino acid concentrations in *Acanthamoeba castellanii* are also directly related to the tonicity of the growth medium (Drainville and Cagnon 1973). The amount of amino acids and their composition vary with different media of varying osmolarity. The osmolarity of a solution is the molarity of the solute multiplied by the number of osmotically active particles released by the molecule. Thus, a one molar solution of glucose is also 1 osmolar, but a one molar solution of NaCl is 2 osmolar, since each NaCl molecule releases two ions, Na^+ and Cl^-. In general, cell water decreases with increasing osmolarity of the external medium. But, this reduction in water volume cannot account for the increase in internal amino acid concentration. An increase in osmotic pressure of the external medium by 460 mOsmoles decreases the cell water content from 85 to 73%, while the intracellular free amino acid content increases from 7.2 to 178.9 mM. These data, together with variations in composition of the amino acids with differing composition of the external medium, suggest that there is an active regulation of amino acid concentration by the cell in response to changes in osmotic properties of the growth medium. Similar responses have been found in ciliates. The amino acid content of *T pyriformis* increases from 56.5 to 114.9 mM per kg of cells when the osmotic pressure of the medium is increased by adding 100 mM sucrose (Stoner and Dunham 1970). Alanine, glycine, and glutamic acid account for 34.5% of the amino acid pool. Likewise, in the marine ciliate *Miamiensis avidus*, the amino acid pool increased from 75 to 317 mM/kg of cells (Kaneshiro et al 1969). In this ciliate, alanine, glycine, and proline account for 73% of the amino acid pool. In *Acanthamoeba castellanii*, however, alanine, proline, and gamma-aminobutyric acid are the most important amino acid osmotic regulator molecules.

In the flagellate *Ochromonas malhamensis*, the internal osmotic pressure is regulated by two physiological mechanisms: (1) the activity of the contractile vacuole, and (2) regulation of internal carbohydrate concentrations (Aaronson 1980). When external osmotic pressures are below 75 mOsmoles, the flow of water into the cell is regulated by the water expulsive activity of the contractile vacuole. Based on microscopic data, the expulsion of the fluid from the anteriorly located contractile vacuole occurs by fusion of the vacuole with the plasma membrane, followed by a transient collapse of the surface of the cell at the point of expulsion (A. Repak, Z. Dubinsky, and S. Aaronson reported in Aaronson 1980). Above external osmotic pressures of 75 mOsmole, the osmotic regulation is by adjustments in carbohydrate concentration regulated by a series of reactions involving carbohydrate metabolism, and by changes in other small molecules (Kauss 1977). When *O malhamensis* is subjected to osmotic stress due to salts, sugars, or polyols in the external medium, internal concentrations of a small carbohydrate molecule called isofloridiside (IF) is increased. IF is a complex carbohydrate identified by Kauss (1977) as O-α-D-galactopyranosyl-(1 $\rightarrow$ 1)-glycerol (Figure 17.1). When the external osmotic pressure was reduced, the IF was converted to a larger molecular weight storage carbohydrate (leucosin). The complex metabolic pathways mediating the interconversion of IFP and leucosin as conceptualized by Kauss (1977) are shown in Figure 17.1. Isofloridoside is synthesized from galactose and glycero-3-phosphate. UDP acts as a carrier for the galactose. Degradation of IFP to leucosin proceeds by production of glycerol and galactose. The latter is converted to glucose-1-phosphate, which is polymerized to form leucosin. Exogenous glucose consumed as a carbon source and photosynthetic products serve as sources for synthesis of glucose-1-phosphate. This is converted to UDP-glucose and UDP-galactose as precursors for IFP synthesis during osmotic stress. Many of the chrysomonads are remarkably euryhaline, and may adapt to strong variations in osmolarity of their medium by production of intracellular osmolytes similar to those of *O malhamensis* (Aaronson 1980). For example, the heterotrophic chrysomonad *Paraphysomonas vestita* was able to withstand marked salinity changes when transferred from distilled

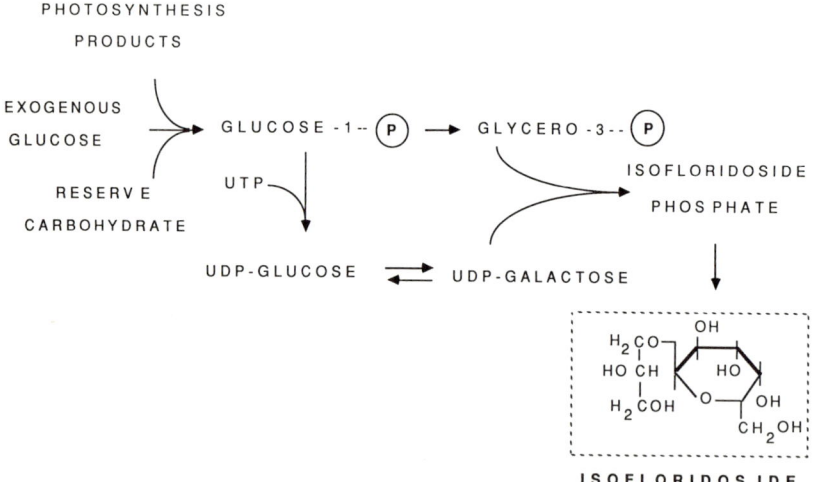

FIGURE 17.1. Metabolic fate of sugars (photosynthetic products and exogenous glucose assimilated by the cell) and leucosin (reserve carbohydrate) in *Ochronomonas malhamensis* in response to changes in external osmotic pressure. The production of isofloridoside in the cytoplasm increases the osmotic pressure of the cell and balances external osmotic pressure. After Kauss (1977).

water to saline solutions with concentrations 2.6 × seawater (Lee 1978). In *O malhamensis*, approximately 70 to 80% of the internal osmotic activity of the cell was attributed to increase in IF, while 10% was due to increase in free amino acids, mainly alanine (Aaronson 1980). The remaining 10 to 20% of osmotic balance was attributed to an increase in K^+ and its undetermined counter-ion.

Intracellular Ionic Balance

Internal pools of potassium and sodium appear to be regulated by active processes in many protozoa. Regulation of sodium ion concentration and its extrusion from the cytoplasm occurs by way of the contractile vacuolar fluid, as observed in amoebae including *Amoeba proteus* and *Chaos chaos*. A similar mechanism has been found in the ciliate *Tetrahymena pyriformis* (eg, Dunham and Child 1961; Elliott 1973). Strong evidence in support of this conclusion is obtained by subjecting cells to solutions containing sufficiently high concentrations of sucrose (100 mM) to suppress contractile vacuolar activity. The intracellular Na content increases from 9 to 27 mM/kg cells (Elliott 1973). These data, combined with the analysis of contractile vacuolar fluid showing a substantially higher Na concentration in the fluid compared to that in the cytoplasm, clearly point to a Na-concentrating and expulsion role for the contractile vacuole. Further evidence for active extrusion of Na ions is obtained from kinetic studies of the rate of Na ion expulsion in relation to the external Na ion concentration. When cells of *T pyriformis* are immersed in media with Na ion concentrations between 2 and 20 mM, the internal Na concentration remains relatively constant at 5 mM/l cells. However, when the external medium contains concentrations of Na greater than 20 mM, the internal Na ion concentration increases linearly with increasing external Na concentration (Figure 17.2). That is, at external concentrations above 20 mM, the extrusion mechanism can no longer compensate for inward flux, and internal Na ion concentrations increase concomitant with increase in external Na ion concentration. This suggests that there is an active transport mechanism which becomes saturated or operates maximally at 20 mM external Na (Dunham and Child 1961). Adaptations of ciliates to high concentrations of NaCl in their growth media apparently involve enhanced ability to actively extrude Na ions. When *T pyriformis* is grown for extended periods of time (many months) in solutions with elevated NaCl concentrations, the internal Na concentraton is relatively low (21 mM/l cells), but when they are briefly equilibrated to high NaCl solutions the internal concentration of Na is much higher (105 mM/l cells). The adaptation is gradual, and indicates a metabolic adaptation to extrude sodium in greater quantities.

The intracellular concentrations of potassium, sodium, and chloride have been studied rather extensively in several species of protozoa. Total content of the cation-forming elements K, Na, Mg, and Ca in *Acanthamoeba castellanii* was assayed at approximately 360 mM/kg dry weight (Sobota et al 1984). Chloride was 173 mM/kg dry weight, and phosphorus was estimated as 492 mM/kg dry weight. Correlations between quantities of P and K in the cells, and the similar

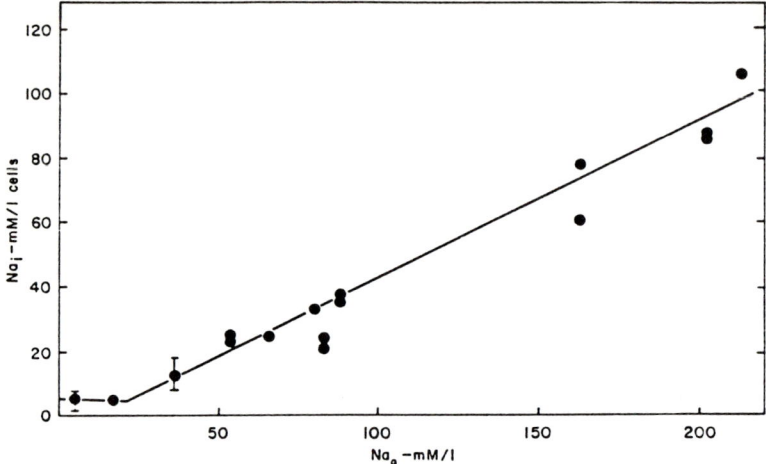

FIGURE 17.2. Relationship between intracellular sodium ion concentration, Na_i (ordinate), and external sodium ion concentration, Na_o (abscissa). When *Tetrahymena* cells are allowed to equilibrate with various external Na concentrations, the Na_i is less than Na_o above concentrations of 5 mM. Between 2 and 20 mM Na_o, the internal concentration is constant at about 5 mM/liter of cells. Internal concentrations increase, however, when Na_o exceeds 20 mM. This indicates that an active process exists for Na extrusion from the cell and is saturated or operates maximally at a level of 20 mM Na_o. From Dunham and Kropp (1973); with permission.

nonuniform distribution of these elements within the cytoplasm of the cell, suggest that the two ions are linked in their distribution within the cell (Sobota et al 1984). Indeed, patterns of distribution of phosphate as a counter-ion for K^+ may also determine the location and pattern of distribution of potassium within the cell. The K concentration in *T pyriformis* appears to be typical of many protozoa (Elliott 1973), and is generally between 20 and 35 mM/kg cells. Sodium concentrations vary more widely (2 to 105 mM/kg cells), depending on the salinity of the external medium, and in general is similar to that of *Paramecium* but is much higher than that of freshwater amoebae. By contrast, intracellular concentrations of chloride (2 to 50 mM/kg) are much lower than in freshwater amoebae (Elliott 1973). In most growth media, the intracellular K concentration is much greater than the Na concentration. This has also been confirmed in the ciliate *Blepharisma* (Giese 1973). Intracellular K concentrations in *Stentor coeruleus* measured by a microelectrode averaged 15.2 to 15.8 mM. The concentration of K in the external medium was 1.3 mM (Fabczak 1983). The distribution of K ions across the cell membrane of *Stentor* is far from being in equilibrium, and suggests that the ion is accumulated by active uptake using metabolic energy.

Potassium accumulation in *T pyriformis* also appears to be an active process. Exposure of the ciliate to low temperature (6°C), hypoxia, or metabolic inhibitors reduced the intracellular K by 30 to 50% (Andrus and Giese 1963). Removal of the metabolic inhibitors resulted in complete recovery of the potassium pool, with accumulation against a sizable concentration gradient, with a 5-fold to 10-fold higher concentration within the cell compared to the outside. The accumulated K within the cytoplasm appears to be segregated into at least three different compartments or states. Approximately 30% of the intracellular K is in a mobile compartment sensitive to cold or metabolic inhibitors. When exposed to these treatments, 30% of the intracellular K leaks from the cell. An additional 20% is sensitive to low temperature only. Thus, in the presence of low temperatures, 50% of the intracellular K is lost. A third compartment (50%) is unaffected by any treatments (Andrus and Giese 1963). The 20% of K in the mobile compartment that is sensitive to cold only is maintained by coupled transport to Na extrusion. This may be an active transport process of pumping Na out of the cell and pumping K in, perhaps by a dual carrier that imports potassium at each cycle of exporting sodium. A similar process appears to occur in *Blepharisma*, where a metabolically coupled transport of Na out of the cell and K into the cell is indicated (Giese 1973). When the ciliate is chilled to 5°C, the internal Na level rises and the internal potassium level falls. When warmed to 23°C, however, Na extrusion resumes and K is accumulated. Calcium ion regulation may also be an active process in many protozoa. The control of intracellular levels of calcium is essential to balanced and coordinated cellular activity, since calcium is an important intracellular signaling molecule. Among other functions, it increasingly appears to control motility and coordination of metabolic activity. Prusch (1980b) reported that cytoplasmic levels of Ca^{++} in *Amoeba proteus* are maintained at very low levels by active extrusion across the plasma membrane. Regulation of calcium, magnesium, and phosphate concentrations in *Tetrahymena pyriformis* involves a complex physiological process whereby the ions are stored in cytoplasmic granules. These mineral grains, ca 0.2 to 2 μm diameter, consist almost entirely of a hydrated, insoluble complex salt of Ca, Mg, and pyrophosphate (Elliott 1973). The granules are surrounded by a cytoplasmic membrane which appears to contain a pyrophosphatase (an enzyme cleaving pyrophosphate into two orthophosphate molecules). Deposition of the granules is apparently a highly controlled process, since Ca and Mg are always in equimolar concentrations in the granule independent of the concentrations of the ions in the growth medium. There is approximately 1 mole each of Ca and Mg for each mole of pyrophosphate in the granules. When *T pyriformis* is subjected to phosphate deprivation, the granules disappear. This is apparently due to some action of the pyrophosphatase in the membrane which degrades the pyrophosphate and releases it into the cytoplasm concomitant with release of Ca and Mg ions (Rosenburg and Munk 1969). Under these conditions, Ca is secreted into the medium to maintain a proper internal calcium concentration. The calcium efflux ceased after all of the granules were gone, and only at this point did Mg efflux commence. This pattern of ion release is probably part of a regulatory mechanism to mobilize phosphate in the face of phosphate deprivation. Calcium is an inhibitor of the membrane-bound pyrophosphatase, and would be excluded rapidly to prevent enzyme inhibition and consequent reduction of phosphate mobilization.

Magnesium is an activator of the enzyme, and would be conserved until all of the granules were consumed as a means of stimulating the pyrophosphatase activity. Thus, the control of intracellular concentrations of Mg and Ca may be essential to regulation of phosphorous metabolism in *T pyriformis*.

Complicated metabolic adjustments to external variations in calcium ion concentrations have also been reported in the flagellate *Astasia longa* (Tolmashova 1981). When subjected to media with excess calcium chloride (0.5%), the flagellates become physiologically adapted to living in media with enriched calcium concentrations. This adaptation becomes enhanced and persists for at least 2 weeks after the cells are transferred into calcium-free media. After the 2 week period, however, the calcium tolerance gradually declines and reaches the initial level. These long-term modifications in calcium tolerance and the physiological polymorphism of the flagellate in tolerating increased calcium concentrations indicates the physiological plasticity of these flagellates. It also highlights the varied mechanisms that protozoa have developed to maintain proper intracytoplasmic balance of calcium ions as required for coordinated cellular activity. More detailed information on ionic balance and osmoregulation can be found in reviews by Kitching (1967), Elliott (1973), and Giese (1973).

Summative Perspective

Protozoa are remarkably efficient organisms. Energy and matter are utilized to sustain life processes, with gross growth efficiencies up to 70% or more. Some of the inefficiency in energy utilization is due to heat loss. Carbon is lost as excretory products and carbon dioxide. Remaining sources of energy and matter are used to synthesize cell structures and drive the numerous metabolic reactions that must be in a controlled and balanced state to sustain life. Some of these metabolic processes were described in greater detail in Chapter 15. In addition to motility, biosynthetic activity, reproduction, and other cellular dynamics, the maintenance of a balanced internal ionic state and osmotic equilibrium with the external environment is a major challenge for aquatic organisms. Ionic balance is maintained by active membrane processes regulating the concentration of internal ions, probably largely by way of contractile vacuole activity in ciliates, and also by plasma membrane transport in other forms. Protozoa exhibit varied means for regulating internal osmotic balance, including use of energy driven processes to alter organic and/or inorganic substances within the cytoplasm. Thus, they are able to adjust cytoplasmic osmolarity to balance environmental osmotic pressure. In freshwater ciliates especially, a constant influx of water must be balanced by expulsion (largely through contractile vacuolar activity). If metabolism is inhibited or energy sources are otherwise depleted, the cells swell and burst. The several physiological processes reviewed in this chapter illustrate a fundamental principle of life, namely that energy must continually be utilized to maintain balanced growth and activity in living systems, and only the most reduced metabolic states, as occur in some resting stages of protozoa, are attended by minimal or no expenditure of energy. Much energy obtained by metabolism is used in motility and control of locomotion, as discussed in the next chapter.

18
Motility

Sensation and Coordination

The many cellular activities of protozoa, clearly visible at the light microscopic level, exhibit a degree of order and organization that indicate a high level of coordination. Motility, feeding, contractile vacuolar activity, and cytoplasmic streaming are among some of the well-documented activities that have been observed by light microscopy. These activities and other physiological responses are becoming increasingly understood at the molecular level. Much of this gain can be attributed to combined evidence from electron microscopy and biochemistry that clarifies structural-functional relationships at the subcellular level. Our knowledge of the molecular basis for cellular mechanics, including motility, responses to environmental stimuli, and coordination of physiological events, is one of the newest and most rapidly developing fields in cellular biology. In spite of many recent strides in understanding the molecular basis for cellular functions, our knowledge is only scanty and much remains to be discovered (eg, Sleigh 1984a,b; Bereiter-Hahn et al 1987). Considerable progress has been made, however, in explaining the many varied activities of protozoa from the perspective of general cellular biological theory. And, protozoa have proven to be very useful tools in examining some of the basic cytomechanical properties of eukaryotic cells. The clearly identifiable motile organelles of some protozoa, including cilia and flagella, and the relative ease of isolating cells and observing cytoplasmic activity as in ameboid movement, have provided additional tools to the cell biologist in verifying observations obtained with other cell types.

Protozoa, as other eukaryotic cells, exhibit irritability. That is, the cells are capable of responding to environmental stimuli. Mechanical, chemical, heat, and light stimuli in varying degrees, with varying capacity among different groups of protozoa, evoke cytoplasmic responses. Chemosensory responses by protozoa were summarized as early as the beginning of the 20th century (Jennings 1906) and have been demonstrated in a wide variety of species. Among the ciliates, *Tetrahymena thermophila* is attracted to complex peptide mixtures and amino acids (Leick and Hellung-Larsen 1985), and *Paramecium tetraurelia* is attracted to acetate (Van Houten 1977). The flagellates *Bodo* (a Kinetoplastid),

Gymnodinium fungiforme, and *Crypthecodinium cohnii* (nonphotosynthetic dinoflagellates) are attracted to organic compounds including amino acids and protein extracts (Engelmann 1882; Hauser et al 1975; Pfeffer 1886; Spero 1985). The microflagellate, *Pseudobodo tremulans*, is attracted to soluble products from prey bacteria (including ammonium, histidine, glycine, and threonine), but is not attracted to washed bacterial cells (Sibbald et al 1987). This substantiates an hypothesis by Goldman (1984) that heterotrophic microflagellates could locate enriched microenvironments by a chemosensory mechanism. For autotrophic flagellates, however, phototaxis (motile responses to light stimuli) is also a significant survival response. Light stimuli are particularly effective for those organisms with light sensing organelles (eg, the eyespot of some flagellates). Other stimuli are sensed at the surface of the cell membrane. The plasma membrane bears an electrical potential (voltage) difference across its surface. This electrical potential difference is produced by the asymmetrical distribution of ions across the membrane. Sodium and calcium ions are continually transported out of the cell. Chloride and bicarbonate ions, and anionic protein molecules are in greater concentration within the cell. Consequently, this asymmetry in ion distribution across the membrane produces a difference in voltage. The large quantity of anions (negatively charged ions) inside the cell produces a net negative charge within the cell. Usually, the inside of the cell is approximately 70 to 90 millivolts (mV) negative to the outer surface of the cell. In some protozoa, values as low as -68 mV have been reported (eg, Febvre-Chevalier et al 1986). This is called the resting potential of the membrane. The voltage difference can be measured by inserting a very fine electrical-conducting reference electrode into the cytoplasm, and placing a sensing electrode on the outer surface of the cell. When the cell is sufficiently stimulated above the threshold level required to evoke a response, the potential difference across the membrane is momentarily abolished. This response is at first localized at the site of stimulation, but spreads rapidly over the entire surface of the cell. The loss of charge difference across the membrane, called depolarization, produces further intracytoplasmic changes that lead to changes in cellular activity. As an example, membrane depolarization accompanying contraction of the stalked heliozoan *Actinocoryne contractilis* (Febvre-Chevalier et al 1986) is elicited by mechanical stimuli or injection of electrical current into the cell. The heliozoan is sessile, possessing a stout stalk topped by a spheroidal head bearing numerous axopodia. When the organism is stimulated by a brief mechanical stimulus, the membrane of the cell undergoes a rapid depolarization. The internal negative charge is rapidly abolished, and influx of positive ions produces a complete reversal of polarization, causing a momentary positive internal charge (as much as 50 mV positive). This depolarization is followed within 50 to 60 ms by a rapid contraction of the heliozoan stalk. When stimuli of increasing intensity are applied to the cell, the membrane response is graded, exhibiting increasing intensity of depolarization. Thus, the membrane must have receptor potentials that respond to stimuli of varying intensity. That is, when mild electrical shocks of increasing intensity are applied to the cell surface, the degree of depolarization of the membrane grows with increasing intensity of the shock. If the shock is below a certain intensity, the membrane exhibits a minor depolarization, but does not produce a complete depolarization. A complete depolarization produces a signal in a sensing electrode known as an action potential. After an action potential has been evoked, the membrane gradually becomes repolarized as the asymmetry in distribution of the ions across the membrane is reestablished. The cell consequently resumes its "normal" relaxed state. Similar membrane responses have been detected in free-swimming ciliates. *Paramecium* and *Stylonichia* exhibit graded membrane potential responses when stimulated at the anterior region, resulting in ciliary reversal and backward swimming. This allows avoidance of noxious stimuli. Stimulation of the posterior region of the cell, however, evokes hyperpolarization. That is, stimulation of the posterior end produces an increase in the potential difference across the membrane. This hyperpolarization accelerates forward ciliary beat, producing greater forward motion of the ciliate and avoidance of the posterior noxious stimulus (Naitoh and Eckert 1969, 1973; dePeyer and Machemer 1978). In *Paramecium caudatum*, the distribution of mechanoreceptors (receptors sensitive to mechanical stimuli such as pressure) has been studied using fine needles as a probe (Ogura and Machemer 1980). A shift in tendency of the membrane to undergo hyperpolarization as opposed to depolarization was again noted as the region of stimulation progressed from the anterior to the posterior region of the body. However, mechanoreceptor currents cancelled out in the midposterior region of the cell, where neither reversal nor enhanced ciliary beat of the cilia occurred. Maximal depolarizing sensitivity was found slightly posterior to the front end of the cell. Stimulation of the dorsal surface of the cell produced greater membrane response than stimulation of the ventral surface at the same latitude along the axis of the cell. Quantitative analysis of the data indicated that the receptor potential in *Paramecium* is due to activated Ca and K mechanoreceptor channels distributed over the somatic cell surface in a pattern of overlapping gradients. These graded responses and change in direction of polarization are undoubtedly fine adaptations to environmental stimuli that allow these motile organisms to respond positively to attractants and avoid noxious stimuli. The sessile ciliate *Stentor*, as with *Actinocoryne*, produces only depolarization (not hyperpolarization) of the membrane when stimulated (Wood 1982).

Role of Calcium Ions

Increasing evidence is accumulating to show that responses of the cells to depolarization of the membrane involves the inward movement of calcium, and in some cases also potassium, into the cytoplasm. If calcium is not present in the external medium, then the normal response of the cell is interdicted. The role of calcium as a signaling or coordinating molecule in cellular activity has been widely established for many eukaryotic cells, spanning a wide range of organisms from protozoa to mammals. It is possible for transient increases in calcium ions to act as intracytoplasmic signals, since the cytoplasm is normally low in calcium (as described in Chapter 17). As a result of active transport of calcium out of the cytoplasm, the normal intracytoplasmic level is usually about 10^{-7} molar. If, however, calcium concentration increases to about 10^{-6} or slightly greater, the cell responds to the influx of calcium ions by making an adjustment in its activity. The form of the adjustment varies according to the kind of cell and its physiological state. For example, as explained above, when the stalked heliozoan is stimulated, external calcium is required to elicit a contraction response. Apparently, the influx of calcium depolymerizes cytoskeletal microtubules in the stalk, making it more flaccid. The role of calcium in controlling microtubular assembly and disassembly has been documented (Schliwa 1976), and it is known that an increase in calcium concentration induces dissociation of the tubulin monomers producing microtubular disassembly. The presence of microfilaments at the base of the stalk suggests that these contractile elements are activated and account for the massive retraction of the flaccid stalk (Febvre-Chevalier et al 1986). After membrane repolarization and calcium outward transport, the microtubular cytoskeleton is reassembled, leading to extension of the stalk. Likewise, in ciliates, a noxious stimulus at the anterior end produces a depolarization of the membrane followed by influx of calcium ions resulting in a reversal of ciliary beat.

Calcium Gated Channels

Although the exact mechanisms of this kind of molecular control are not fully understood, some of the events accompanying the ciliary reversal have been clarified. The resting (unstimulated) membrane surrounding the ciliate is impermeable to calcium. There are, however, protein molecules in the membrane, especially in the ciliary plasma membrane, that act as gated channels for calcium. These proteins are transmembrane channels that can be held closed to calcium movement or opened to allow calcium to diffuse into the cytoplasm. The movement of Ca^{2+} into the cytoplasm is affected by the pH of the cytoplasm (Umbach 1982). The intracellular pH of resting *Paramecium caudatum* is 6.80 ± 0.05. Intracellular alkalinization enhances the inward flow of Ca, while internal acidification depresses the Ca current. The alkalinization of the cytoplasm to pH 7.15 converts the normally graded quasiregenerative Ca responses into all-or-none action potentials. That is, the normally gradually intensifying transmembrane flow of Ca concomitant with increasing stimulation does not occur, and instead, an all-or-none full action potential is elicited. These results suggest that Ca channel permeability is blocked by intracellular protonation (H^+ ion bonding) of a single site on membrane proteins, thus blocking the transmembrane Ca flow. At acid pH, the channels can be fully blocked by the ion, producing a suppressed response. While at alkaline pH, the hydronium ion concentration is so low that the channels are fully open when stimulated (all-or-none state).

In the resting state, the calcium channels of a ciliate membrane are closed, but when the membrane is depolarized, the channels are momentarily opened and calcium diffuses into the cytoplasm. The increase in calcium concentration triggers ciliary reversal. When the membrane repolarizes, and the channels close once again, outward active transport of calcium restores the normal low level of calcium in the cytoplasm and the cilia resume forward beating. It is interesting to note that a mutant *Paramecium* known as a pawn (after the chess piece that can only move in one direction) lacks functional calcium gated channels. This mutant, therefore, cannot reverse ciliary beat when stimulated, and can only swim in a forward direction. The genetics of this anomaly is discussed more fully in Chapter 20. The gene coding the production of the calcium channel protein is defective in the pawn mutant, and therefore the membrane lacks competent calcium transport proteins. However, the protein deficiency can be overcome by injecting cytoplasm from normal ciliates into the mutant cell (eg, Haga et al 1984). Apparently, the ingested molecule is incorporated into the membrane, thus restoring normal calcium channel activity. The molecule effecting the recovery is not a nucleic acid. Therefore, it is not a gene that compensates for the abnormal condition. Rather, the molecule is a heat-labile, sulfhydryl-(SH-) containing protein that appears to affect existing mutant channels on the ciliary membrane. The protein is of low molecular weight ($<$ 30,000 daltons), acidic in chemical composition, soluble in water, and does not have calmodulin activity. These molecular biological discoveries lend additional credence to the hypothesis that gated control of calcium influx is the basis for membrane mediated control of ciliary reversal during noxious stimulation. The exact intracytoplasmic mechanism of this triggering of reversal and recovery of normal forward beating is unknown, but increasing evidence points to the role of cytoplasmic calcium-binding proteins as mediators in the response.

Calmodulins

One group of calcium-binding proteins collectively known as calmodulin occur in many eukaryotic cells, including ciliates (eg, Nozawa and Nagao 1986). These are typically small, low molecular weight proteins localized within specific sites in the cell. For example, the calmodulin isolated from *Tetrahymena pyriformis* is a protein containing 147 amino acids (Nozawa and Nagao 1986). Localization of calmodulin in *Paramecium tetraurelia* using an immunofluorescent stain indicated three distinct regions of specificity (Maihle et al 1981): (1) large, spheroidal cytoplasmic organelles, possibly food vacuoles, (2) along the entire length of oral and somatic cilia, and (3) along a linear punctate sequence corresponding to the kineties (ciliary basal bodies) of the cell. Calmodulin activity associated with proteins extracted from *P tetraurelia* indicated localization in the cell body and cilia (Walter and Schultz 1981). The protein is a low molecular weight polymer (ca 17,000 daltons). The amount of calmodulin in the cell body was 75 µg/g, and in the cilia 50 µg/g. Its presence in cilia suggests a regulatory role in ciliary action. Calmodulin and similar calcium-activated proteins bind calcium at specialized sites on the molecule, resulting in changes in the conformation (shape) of the protein. In some species, as many as four calcium ions can bind to each calmodulin molecule. When the calmodulin changes configuration, it can act as a signal molecule stimulating activity in enzymes or other responsive macromolecules in the cell. For example, in ciliates, calcium-activated calmodulin induces the formation of cyclic guanosine monophosphate (cGMP). This nucleotide contains one phosphate molecule covalently linked to the oxygen on the third and fifth carbon atom in the sugar (ribose) molecule of the nucleotide. Hence, the phosphate is anchored at both ends to the sugar moiety of the molecule and forms a bridging (cyclic) link between the third and fifth carbon atoms. During hyperpolarization of a ciliate membrane as occurs during enhanced forward swimming, cyclic AMP (cAMP) is produced in the cytoplasm (Hennessey et al 1985). This is a derivative of ATP, and as with cGMP, is produced by cleavage of pyrophosphate from the triphosphorylated nucleotide and cyclization of the remaining phosphate across the third and fifth carbon atoms of the ribose molecule. Cyclic GMP and AMP can act as messenger molecules triggering cellular responses. The studies of Hennessey et al (1985) with *Paramecium* show that the increase in cytoplasmic cAMP does not induce enhanced forward swimming unless the membrane is hyperpolarized. If the membrane is kept at a constant resting potential and cAMP is injected into the cell, there is no increase in ciliary forward beating. If, however, the membrane is not stabilized at a constant resting potential, injection of cAMP results in hyperpolarization and increased forward swimming. Although the presence of calmodulin and its calcium-dependent enhancement of cGMP during membrane depolarization have been documented in a variety of protozoa, our knowledge is incomplete as to how the cellular responses evoked by the altered calmodulin are generated at the molecular level. As a means of establishing a context to discuss these issues, some fundamental knowledge of how protozoa move is presented. Mechanisms of ciliary and flagellar activity are presented first, followed by a discussion of ameboid motion.

Ciliary Motion

The basic structures of cilia and flagella have been discussed in the chapter on fine structure. Both structures exhibit a remarkably similar internal organization, but differ in length. Their diameter is about 0.2 μm, signifying a constant internal organization, while the length of a cilium is in the range of tens of microns, and flagella are hundreds of microns long. Cilia exhibit an oarlike motion. During the power or effective stroke, the stiffened cilium propels water lateral to its surface and parallel to the cell surface. For structures as small as cilia, the water is highly viscous (eg, Yates 1986) and the cilium exerts considerable propelling force on the surrounding water as it undergoes an oarlike motion (Figure 18.1). However, during the recovery phase when the cilium is returning to a stiffened configuration, it gradually unfolds toward the tip. The length of the cilium is gradually drawn lengthwise toward the straightened position, thus avoiding the heavy resistance by the surrounding viscous water that would occur if the stiffened cilium was returned oarlike to the initial position. Thus, each

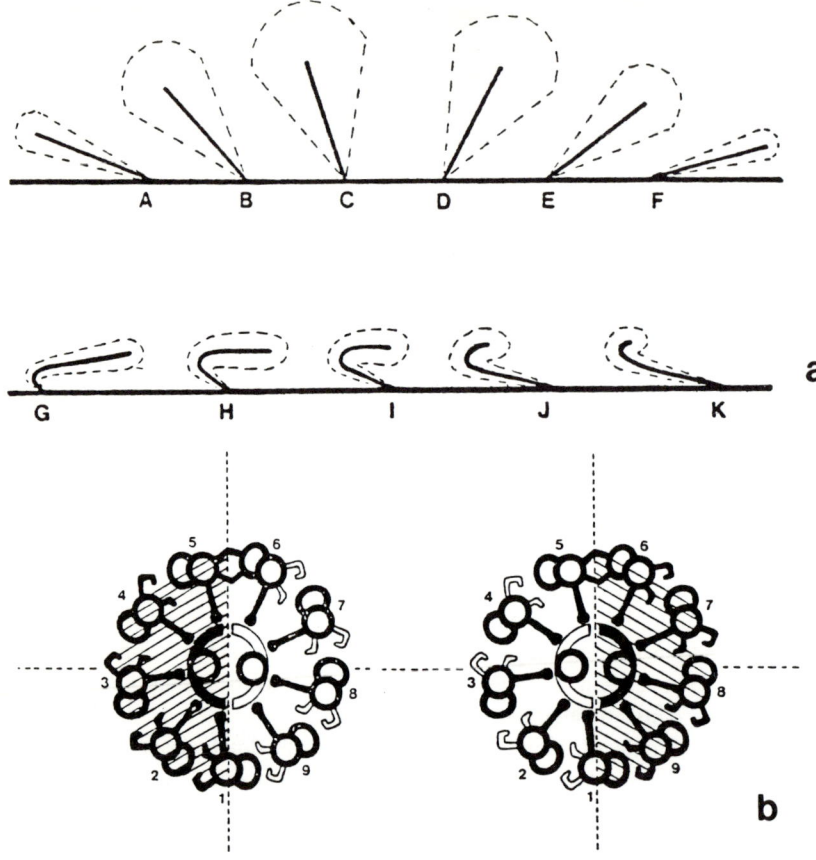

FIGURE 18.1. Ciliary action during the stroke cycle (a) and internal response of the microtubule doublets (b) during an individual ciliary beat. The power stroke (effective stroke) of the cilium (A-F) occurs with the cilium in a stiffened mode that entrains a relatively large amount of water (dashed lines). This effective stroke propels the organism through the water. The recovery stroke (G-K) occurs by gradual folding upward of the ciliary tip while close to the surface of the cell. This mode of recovery entrains less water and subjects the cilium to less drag, thus increasing the efficiency of the ciliary beat cycle. The cyclical bending of the cilium producing the oarlike action is the result of sliding motion of the internal microtubule doublets (b). Alternative sliding motion on the left and right sides of the cilium (shaded regions) produces the regular pattern of bending motion characteristic of the ciliary beat. (a) From Sleigh (1984a) and (b) from Satir (1984); reproduced by permission of the Society of Protozoologists.

cycle of ciliary action involves an oarlike propelling phase and a gradual unfolding recovery phase. Due to the viscous properties of the surrounding water, the action of the individual cilia become coordinated. The water molecules provide a liquid matrix that constrains the motion of the cilia. Thus, eventually the individual activities of the cilia become coupled, increasing efficiency and con-

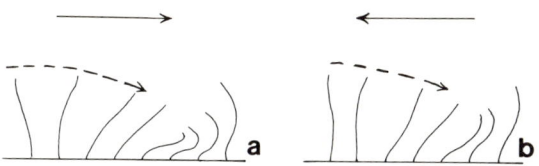

FIGURE 18.2. Relationship between the effective stroke of a cilium and the direction of the metachronal wave rippling across the surface of the ciliary mass. In symplectic motion (a), the effective stroke (dashed arrow) is in the same direction as the direction of the metachronal wave. In anaplectic motion (b), the power stroke of the cilium is in a direction opposite to the direction of apparent movement of the metachronal waves. For more detailed information see the text and Sleigh (1984a).

serving energy. The net result of this hydrodynamic coupling is to generate metachronal synchrony of the cilia. That is, the cilia become coordinated in their action such that each cilium moves slightly in advance of the nearest neighbor cilium, giving the appearance of a successive wave of activity spreading over the surface of the cell. The action of each cilium is slightly out of phase with that of the adjacent cilium, resulting in the metachronal wave effect. This metachronal synchrony was once thought to be caused by cytoplasmic control of the ciliary beat; perhaps by a signal carried out by the subpellicular fibrils interconnecting the ciliary basal bodies (eg, Foissner 1981). Although this cannot be entirely ruled out, some elegant studies by Sleigh and colleagues (eg, Sleigh 1973, 1984a,b) have shown that the viscous coupling by the water is sufficient to explain the effect.

The pattern of ciliary action has been conceptualized into four categories by Knight-Jones (1954) based on the direction of the metachronal wave relative to the direction of the power stroke during beat of the cilia (Figure 18.2). These categories are as follows. (1) In a symplectic mode of coordination, the power stroke of the cilia is in the same direction as the apparent direction of the metachronal wave. That is, the power stroke of the cilium is directed along a line coincident with the direction of the metachronic ripple across the surface of the cilia. (2) Antiplectic action occurs when the power stroke is directed opposite to the direction of the travelling metachronal wave. Further categories relate to direction of the power stroke when it is perpendicular to the direction of the metachronal wave. (3) A laeoplectic pattern occurs when the power stroke is directed toward the left side of the line of direction of the metachronal wave. (4) A dexioplectic pattern occurs when the power stroke of the cilium is directed toward the right side of the directed line of motion of the metachronal wave.

Flagellar Motion

Flagella typically exhibit a sinusoidal motion in propelling water parallel to their axis. In some cases, several sinusoidal waves appear to simultaneously travel up the flagellum. The undulating action of the flagellum either propels water away from the surface of the cell body (pulsellum) or draws water toward and over the cell body (tractellum). A pulsellum produces cellular motion very much like that of a fish tail. The propelled water leaving the flagellum produces an equal and opposite force on the cell, causing it to move in a direction opposite to the flow of water. Thus, the flagellum, located on the posterior part of the cell, pushes the cell body through the water very much as the posteriorly located tail of a fish propels it through the water. The action of a tractellum "pulls" the cell forward through the water, and hence is usually located on the anterior part of the flagellate. Sessile flagellates (stalked flagellates attached to a substratum) are immobile and use flagellar action to move water across the cell surface, thus increasing exchange with the surrounding medium and also bringing food particles in contact with the surface of the cell where ingestion takes place. In some choanoflagellates (collar flagellates), the flagellum is a pulsellum moving fluid away from the cell body and drawing food-laden streams of water through the collar-like corona of microvilli surrounding the site of flagellar attachment (Figure 18.3). In other attached flagellates such as *Bicoeca* (with a meshlike lorica), the flagellum is a tractellum drawing food-laden streams of water toward the cell body and passing into the cavity of the conelike lorica (Figure 18.3).

Mechanisms of Ciliary and Flagellar Motion

Internal Organization

The varied undulating motions exhibited by cilia and flagella can be explained by bending action of the organelle based on internal fine structure. A detailed diagram of the cross section of a cilium or a flagellum is presented in Figure 18.1b. The nine doublets of microtubules at the periphery are each composed of two microtubules. One microtubule of the pair is a complete cylinder (subfiber A), while the other (subfiber B) is incomplete and attached to subfiber A. Radial spokes connect the doublets to the central sheath surrounding the pair of microtubules at the center of the axoneme. Each subfiber A bears two lateral arms that project toward subfiber B of the neighboring doublet. These arms are composed of a specialized protein known as dynein, and exhibit ATPase activity. Calcium-bonded calmodulin activates the ATPase in the arms and is believed to be an essential step in the initiation of movement of the organelle. Additional elastic protein strands (interdoublet links), composed of the protein nexin, connect adjacent doublets. The two central microtubules are connected by a central microtubule bridge. The arrangement of the microtubular doublets and their interconnections with one another and the central shaft provide a stable yet flexible association among the component structures.

Mechanics

Current evidence (Satir 1984) indicates that the bending motion of cilia and flagella is produced by sliding of the adjacent microtubular doublets relative to one another (Figure 18.1b). A bending action in a plane (as occurs during the

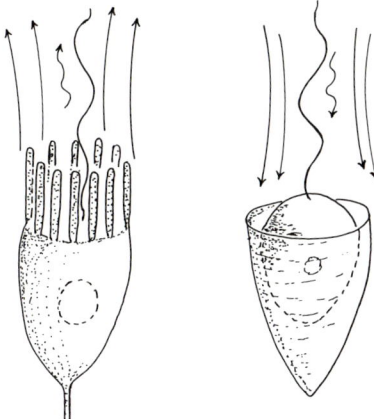

FIGURE 18.3. Types of flagellar motion based on the direction of the movement of water relative to the body of the cell. (a) A pulsellum propels water away from the cell body as occurs in some choanoflagellates. The currents of water in this sessile organism carry food particles that are gathered on the corona of microvilli surrounding the point of flagellar insertion. (b) A tractellum as occurs in *Bicoeca* moves currents of water toward the cell body. For free-swimming flagellates, the pulsellum pushes the cell through the water much as occurs with the swimming action of a fish or tadpole. A tractellum pulls the cell through the water, and therefore is usually oriented ahead of the cell body as the organism locomotes through the fluid environment.

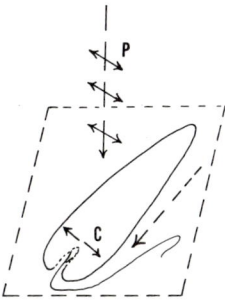

FIGURE 18.4. Diagrammatic interpretation of the organization of the photoreceptor molecules in *Euglena* and their interaction with polarized light based on the research of Creutz and Diehn (1976). When a ray of plane polarized light (P) is passed through a culture of *Euglena*, the cells gradually orient with their long axes perpendicular to the ray of light. Hence, they swim in a direction perpendicular to the light ray (dashed arrow). This phenomenon is explained by assuming that the light-receptive molecules in the eyespot have an orientation such that the transition vectors (direction of electron excitation) are oriented perpendicular to the long axis of the cell (C). As shown in the diagram, the cells gradually orient themselves to maximize absorption of the light. This requires that the transition vectors be aligned parallel to the vectors of the plane polarized light ray (P) as illustrated. For nonpolarized light, with vectors arranged radially around the light ray, the cells will orient toward the light (parallel to the light source), as this will present the transition vectors parallel to the vectors of the light ray. A maximum number of molecules will then be stimulated, thus accounting for the positive phototactic response of the *Euglena*.

ciliary power stroke) can be generated if a set of doublets on one side of the cilium is shifted toward the tip of the cilium, while the ones on the opposite side either remain stationary or are shifted toward the base of the cilium. This differential sliding action can cause extension on one side of the cilium and contraction on the other, thus producing a bend in the cilium away from the elongating side. More complex patterns of sinusoidal activity, as observed in flagella, can be generated by various combinations of sliding patterns among the nine doublets at the periphery of the axoneme. In addition to the sliding motion, there must be some component that provides resistance to shear so that the sliding action can be converted into bending motion rather than lateral displacement of the microtubules. The radial connections to the central tubule sheath and the nexin bridges between doublets appear to satisfy this requirement. The major question of what causes the sliding motion between adjacent doublets appears to be answered by the ATPase activity associated with the dynein side arms. According to current theory, the side arms projecting from an A subfiber are in contact on the opposing surface of a B subfiber of the adjacent doublet. Dynein is a complex molecule composed of globular subunits and a fingerlike projection that contacts the adjacent B subfiber. This finger like projection is apparently capable of flexing when ATP is hydrolyzed to ADP and orthophosphate. For each ATP cleaved, the dynein arm moves the adjacent doublet by one incremental unit. Thus, repeated cleavage of ATP and the succession of dynein-driven incremental displacements of the adjacent doublet produces a net translational motion of the adjacent doublet. The action of the flexing dynein arm has been compared to the flexing action of the myosin cross-links between actin filaments in muscle, which produces sliding action of the filaments relative to one another, and hence contraction of the muscle fibrils. Although the sliding doublet model can adequately explain the bending motion of cilia and flagella, it is unknown how these activities are coordinated to bring about the particular pattern of motion characteristic of different motile states and specific to each species. The discovery that calmodulin potentiates the ATPase activity of dynein provides an additional link to the calcium-induced changes in ciliary beat pattern. But, it is not certain what the role of calmodulin is in the very fine gradations of locomotory behavior observed in ciliates and flagellates.

Contractile and Locomotory Behavior

The varied motile activities of protozoa encompass a wide range of behaviors including cyclosis of cytoplasm, food intake mechanisms, contractile responses of the cell body, and various locomotory responses. Some of the current knowledge about these varied behaviors and the molecular basis for their action has been summarized by Sleigh (1984b), and the biophysics of water movement during filter feeding has been reviewed by Fenchel (1986b). Microfilaments containing actin, and presumably their association with myosin, produce contractile and cytoplasmic streaming responses. Translocation of fluids produced by myosin motile activity on the surface of the actin filaments appears to be involved in cyclosis of cytoplasm. Contractile activity of microfilaments produces movement of food into the cytopharyngeal basket of *Pseudomicrothorax*, and contraction of the tentacles in suctorians. Firm evidence for the presence of actin-containing microfilaments in the collar and other tentacles of choanoflagellates was presented by Leadbeater (1983). His microscopic evidence indicates that the changes in morphology and range in functionally significant movements of these organelles can be explained by myosin-associated, actin-produced events. The arrangement of microfilaments in choanoflagellate tentacles exhibited striking parallels with those in microvilli of some other phyletically remote eukaryotic cells. These findings suggest that there may be some fundamental molecular processes governing movement and contraction in microvilli and tentacles across diverse groups of organisms, and may represent an early phylogenetic adaptation of eukaryotic cells.

Body contractions of ciliates appear to be based on a different set of filaments organized into myonemes. The organization of some of these systems has been presented in Chapter 14. In *Stentor* and *Spirostomum*, there are two sets of contractile elements that can account for the contraction and elongation of the cells. Myonemes, as observed in *Stentor*, consist of 75 to 100 microfilaments, each of about 4 nm diameter arranged parallel to the long axis of the ciliate. Curved bands of microtubules (km fibers) attached to basal bodies of the cilia extend longitudinally. During contraction of the cell, the myonemes become shortened and thickened, forming aggregations of randomly oriented tubular filaments of 10 to 12 nm diameter (Huang and Pitelka 1973). This indicates that the microfilament-mediated contraction is not a simple sliding mechanism. The km bands consist of 21 microtubules arranged in an overlapping pattern. In the extended state, there are five overlapping microtubules in the band, but in the contracted state as many as 40 microtubule ribbons overlap, and the basal bodies to which they are attached are more closely spaced, suggesting that the assemblage has contracted by a sliding action. Apparently, extension is not by microfilament elongation, since the microfilaments expand ahead of the microtubule bands and become folded. It appears that sliding of the microtubules upon one another produces elongation. This is further supported by the observation that cross-links between the microtubules are broken during contraction, but are reestablished during elongation. Contraction of the myonemes is controlled by calcium concentrations. This has been clearly demonstrated in *Spirostomum* by Etienne (1970), who used an intracytoplasmic light-emitting dye, aequorin, stimulated by calcium, to show that calcium is released during contraction. The source of the calcium must be intracytoplasmic, since there is no membrane depolarization during contraction, and hence no influx of external calcium ions. The best evidence at present is that the calcium is released from the cisternae of nearby endoplasmic reticulum. It has been shown that this is a site of calcium storage. By contrast, contraction of the stalk in sessile peritrichs such as *Vorticella* is produced by a single, dense, filamentous fiber, known as a *spasmoneme*. This large fiber runs the length of the stalk and is typically positioned asymmetrically within an elastic sheath. Stiffening fibers called batonnets are arranged helically, surround the sheath, and produce the pronounced helical shape of the stalk when it contracts. The spasmoneme binds calcium during contraction. The large spasmoneme of *Zoothamnium geniculatum* can be removed by dissection and chemically analyzed. Contracted spasmonemes contain 3.81 g of Ca/kg dry mass, while extended ones contain 2.12 g/kg dry mass (Routledge et al 1975). Additional details of protozoan contractile responses can be found in Stebbings and Hyam (1979).

Locomotory responses are largely due to the beating action of cilia and flagella produced by microtubular sliding motion as discussed in the preceding section. Considerable attention has been given to locomotory responses of protozoa to a variety of stimuli including geotaxis (response to gravity), phototaxis (response to light), chemotaxis (response to chemical agents), and thigmotaxis (response to mechanical stimuli). Some representative studies are summarized for ciliates and flagellates.

Geotaxis

Geotaxis is nominally a significant response for organisms which must orient in the water column in relation to chemical gradients or depth-stratified variables essential to their survival. Thus far, several well-documented geotactic behaviors have been described and their mechanism explained with fair precision. The ciliate *Loxodes* possesses geosensing organelles known as Müller's vesicles of variable number but always arranged in one or two rows along the aboral surface of the cell. These bodies, as explained in greater detail in Chapter 14, contain a barium mineral deposit suspended on a thin cytoplasmic thread within a vacuole. An apparently sensory hairlike array of microtubules projects into the vacuole and contacts the barium granule. The orientation and pressure of the barium granule in relation to the sensory "hair" results in ciliary activity that orients the movement of *Loxodes* toward gravity. This behavior brings the ciliate into deeper water, where food sources and appropriate low oxygen concentrations favor survival (Fenchel and Finlay 1986). Geotactic responses in flagellates have been demonstrated in *Chlamydomonas* and *Euglena*. *Chlamydomonas* is negatively geotactic (swimming against gravity). Bean (1977) reported that the negative geotactic response involves a gradual adjustment in the swimming pattern of the flagellate involving long, gradual turns that require an uninterrupted swimming path of at least 200 µm. If the organism is deterred by collision with obstacles, the orienting responses are inhibited. This is attributed to random swimming reactions elicited by collisional contacts. The mechanism of normal geotactic orientation, therefore, involves long, slow reorientation maneuvers (from net downward to net upward) that require hundreds of micrometers of free swimming space. A metabolic inhibitor, sodium azide, produced inhibition of the orientation behavior. The effects of this compound were specific to the reorienting response and did not influence general swimming activity. This differential effect suggests that negative geotaxis in *Chlamydomonas* is an active process, and not attributable to hydrodynamic effects associated with the shape of the cell. Winet and Jahn (1974) developed a "gravity propulsion" model that explains geotactic reorientation as the physical consequence of a torque (rotational producing force) produced on actively swimming cells which have an asymmetric shape and a gyrational swimming path. Although this model may apply to some flagellates, the inhibitory effect of sodium azide indicating active physiological involvement appears to make it less plausible in the case of *Chlamydomonas*. *Euglena gracilis* strain Z also exhibits a geotactic swimming response by orienting its body perpendicularly to the force of gravity (Creutz and Diehn 1976).

Phototaxis

Chlamydomonas and *Euglena*, among other phytoflagellates, are well-noted for their positive phototactic responses. They swim toward a light source and accumulate in illuminated regions of the water column. The phototactic response of *Chlamydomonas* and *Euglena gracilis* is regulated by a biological clock apparently based on a biochemical oscillation, usually of a 24 h period, and entrainable to external light and dark cycles (eg, Bruce 1970). The circadian clock in *Chlamydomonas* not only influences the magnitude of the phototactic response in nondividing cells, but also may control both phototaxis and growth in dividing cultures. During the illuminated phase of the light/dark cycle the cells are much more photoresponsive, and although the magnitude of the response could be altered by changing the physiological conditions of the environment, the length of the rhythm period has been shown to be remarkably constant (Bruce 1970). The sign and magnitude of the phototactic response varies with light intensity (Feinleib and Curry 1971a,b). Accumulation in a light beam becomes increasingly more positive until the intensity is ca 10^3 ergs/cm²/sec. Higher intensities produce a net weaker response of individuals in a population, and the response becomes negative (avoidance) at intensities greater than 10^4 ergs/cm²/sec. Research by Smyth and Ebersold (1970) and Hirschberg and Stavis (1977) with mutant strains exhibiting negative phototaxis indicate that the light response is under genetic control. These cells exhibit avoidance reactions to light intensities that elicit positive phototaxis in normal cells. Three separate nuclear mutations are present in the mutant cells. Two of the mutant genes are linked (present on the same chromosome), while the third is on a separate chromosome (Hudock and Hudock 1973). Optimum motility in *C reinhardtii* occurred in a medium containing either Ca^{2+} or Mg^{2+}, while phototaxis required Ca^{2+} and either K^+ or NH_4^+ (Stavis and Hirschberg 1973). Similar ion effects have been observed in *Euglena*, and substantial effort has been made to clarify the molecular basis for motile responses in this widely studied flagellate, especially by Diehn and colleagues (eg, Diehn 1973). Further information on historical and broad scientific perspectives can be found in Wolken (1967) and Buetow (1982).

Light-Sensing Organelle

The anteriorly located flagellum of *Euglena* produces a power stroke that directs water currents toward the posterior of the cell, thus driving the organism forward. Under constant illumination, the organism swims with a helical motion in a relatively straight direction toward the light source. The light-sensing organelle is an enlarged part of the flagellum (paraflagellar body), located near its base. The cells exhibit a positive phototactic response to light up to an intensity of ca 3 W/m^2, but above this threshold they are negatively phototactic. The organisms respond to white light, and also for experimental purposes, to actinic light in the blue band (472 nm). Chemical evidence indicates that the light-trapping molecules are flavins, and that they are ordered in such a way that the transition moments (direction of optimal electron jump during excitation) are perpendicular to the long axis of the cell (Figure 18.4). In other words, the molecules are organized in the sensory organelle with the transition moments of the molecules oriented perpendicular to the long axis of the *Euglena*. Thus, the molecules will be maximally excited when the electromagnetic vectors of the light beam are oriented perpendicular to the long axis of the cell. Under these conditions, the electromagnetic vectors of the light beam will lie parallel to the transition moments of the photoreceptive molecules, resulting in maximum absorption of the light energy. Evidence for this conclusion comes from observations of phototactic responses of *Euglena* when illuminated with polarized light (Creutz and Diehn 1976). The organisms swim in a direction perpendicular to the plane of polarization. In this orientation, the photoreceptive molecules are aligned with their transition moments parallel to the vector of the polarized light, and thus are maximally stimulated. When the cells are exposed to nonpolarized light (ie, with force vectors organized radially around the direction of the light path), they orient with their long axis directed toward the light source. That is, the long axis of the cell is oriented parallel to the ray of light. In this position, the transition moments of the light-absorbing molecules are oriented in such a way that most of them will be parallel to the radially arranged vectors of the light beam. If the cells deviate from this orientation, then the photoreceptive molecules will be less efficient in trapping light energy. Gradual adjustments are made in the swimming activity until maximum stimulation is once again restored. The orientation of the cells can be explained as a constant adjustment in flagellar motion until the photoreceptors are aligned to maximally absorb light. At this point, flagellar motion becomes uniform and the cells are oriented almost totally in the direction of the light. The major assumptions in this model are summarized by Creutz and Diehn (1976) as follows: (1) the photoreceptor molecules are organized so that their transition moments are aligned perpendicularly to the longitudinal axis of the cell; (2) light stimuli presented from the posterior direction are shaded from the photoreceptor molecules by the cytoplasm of the cell, and any movement away from this orientation will increase light absorption and be favored; and (3) the mechanism controlling cell movement is such as to cause maximal light absorption by the photoreceptive molecules (at least for intensities below 3 W/m²). Although prior models of positive phototaxis assumed that the orienting response to light was due solely to a shading effect of the body on the photoreceptor, this seems less plausible given the current data on polarized light effects.

Modulation of Photosensory Responses

When *Euglena* is stimulated with variations in light intensity, a marked and characteristic change in motility is observed. If the organism has been adapted to a uniform light intensity of sufficiently low intensity to avoid irritation, either a sudden drop in intensity (step down stimulus) or a sudden increase in intensity (step up stimulus) induces a change in motility called a photophobic response. For example, on perception of reduction in blue light intensity, forward swimming motion abruptly stops and the *Euglena* rotates in place (tumbles) for a finite period of time (step down photophobic response). More generally, if a culture is illuminated with a beam of actinic light, the organisms become concentrated within the illuminated zone. This is partially a result of photophobic responses that occur when the *Euglena* approaches the boundary of the light beam where the intensity begins to drop. The tumbling motion prevents further locomotion out of the illuminated area, and upon recovery, the organism resumes locomotion into the direction of the light. The organisms are said to become isolated in an actinic light trap. However, in addition to light variations, ion composition of the growth medium also influences the photophobic reactions. Under constant light intensity, strong motility can be maintained by Mg^{2+} alone, but optimum motility is found in the presence of Mg^{2+}, Ca^{2+}, and K^+ (Doughty and Diehn 1979). Ca^{2+}, Co^{2+}, Mn^{2+}, and Ba^{2+} induce a concentration-dependent increase in the rate at which the cells change swimming direction. When a photophobic tumbling response is induced by a step down stimulus, the duration is enhanced in the presence of divalent cations. The series of ions in decreasing order of effect are $Ca^{2+} > Ba^{2+} > Mn^{2+} > Co^{2+} > Mg^{2+} = Ni^{2+} = 0$. Both magnesium and nickel inhibit response. Interestingly, the composition of the anions in culture also determines motility behavior and sensitivity to changes in light intensity (Doughty and Diehn 1984). Persistent motility was maintained with chloride or sulfate, but not with acetate, nitrate, or propionate as the only added anions. Cells in Cl^--containing media showed constant photophobic response duration following repeated stimulation. In general, the foregoing data of Doughty and Diehn (1979) indicate a role for Ca^{2+} and Mg^{2+} in control of flagellar activity modulated by anion composition of the growth media. The exact mechanism for this effect is not fully understood. Research data indicate that the photophobic response is elicited by a change in Ca^{2+} concentration in the near vicinity of the axonemal 9 + 2 structure within the flagellum. How the change in light intensity induces this alteration in Ca^{2+} concentration is not known. There appears to be no photoreceptor membrane depolarization associated with the response as observed in plasma membranes of *Paramecium* when ciliary reversal is induced by influx of Ca^{2+} ions. Nonetheless, the research data indicate that the source of the Ca^{2+} ions is from the external medium and not from intracellular spaces. Whatever the mechanism of Ca uptake, it appears to be an indirect one. There is no evidence of a gated Ca^{2+} channel in the cell membrane regulating diffusion of Ca^{2+} into the cell, as occurs in some cilites. One hypothesis suggested by Doughty and Diehn (1979) is that the blue light, for example, effects an active change in the photoreceptor structure of *Euglena*. This change could influence Ca^{2+} fluxes into the intraflagellar space and may be regulated by electrochemical gradients of other cationic or anionic compounds. This is made more likely by evidence of a photoreceptor membrane-located transport system for Na^+/K^+ as a key component in control of the intraflagellar free Ca^+ levels (Doughty et al 1980). These changes in intraflagellar Ca^+ could determine the photobehavior of *Euglena* mediated by changes in flagellar activity. It is becoming increasingly clear that variations in calcium concentration regulate a wide variety of ciliary and flagellar motile activities, but the physiological mechanisms mediating changes in calcium ion concentration in response to environmental stimuli are apparently varied among different species. Comparative research studies on flagellar activity and cellular phototactic responses between *Chlamydomonas* and *Euglena* may help to clarify some of the unique features of these two flagellates, as well as common mechanisms that account for their positive phototactic behavior.

Amoeboid Locomotion

The mechanisms of cytoplasmic flow and protrusion of pseudopodia during amoeboid motion have long been of interest to protozoologists and cell biologists. More recently, the phenomenon has acquired increasing significance with the recognition that cell motility is an important factor in understanding the spread of malignancies within the body of higher organisms, and in understanding natural defense mechanisms of vertebrates mediated by phagocytic wandering cells. Two aspects of amoeboid motility are discussed here: (1) the flowing type motion exhibited by amoebae that extend pseudopodia; and (2) the reversible cytoplasmic flow, known as shuttle streaming, observed in plasmodia of slime molds.

Cytomechanical Models

The major cytoplasmic events accompanying motion of amoebae have been well-documented by high resolution light microscopy. The cell is partitioned into two distinct regions, a relatively viscous hyaline ectoplasm and a less viscous, granular endoplasm. The ectoplasm forms a gellike sheath surrounding most of the cell. During protrusion of pseudopodia, the granular cytoplasm streams forward into a bulgelike protrusion of the cell edge. In monopodial species, a clear, fountainlike flow of granuloplasm occurs as the central stream of low viscosity cytoplasm moves forward and spreads laterally toward the edges of the cell. This stream is eventually directed toward the posterior of the cell, where recruitment occurs in the uroid region. Electron microscopic examination of amoebae fixed during locomotion shows that a layer of microfilaments is associated either with the plasma membrane, or separated from it and forms a boundary between the granular and hyaline parts of the cytoplasm (Stockem et al 1982; Grebecki 1986). Two types of filaments have been identified, a thin filament 5 to 8 nm diameter and a thick filament ca 16 nm diameter (Stebbins and Hyams 1979). In general, thick filaments occur in the uroid region and appear to be myosin-containing filaments. Thin filaments containing actin are distributed continuously over the entire surface of the cell. Filaments in the intermediate and uroid regions form a nonoriented meshwork, while those in the frontal region lie parallel to the plasma membrane. There are no organized microfilaments in the granuloplasm, since the actin is in the nonpolymerized form (Grebecki 1986). Some current models explaining amoeboid motion are presented and their relative merits are discussed. Two major theories for locomotion are (1) the ectoplasmic tube contraction model, and (2) the frontal contraction model. The ectoplasmic tube contraction model attributes cytoplasmic streaming to a pressure gradient in the cytoplasm induced by contraction of the ectoplasm. There are several variations of this model. One of the earliest formulations assumed that contraction in the uroid region produced the pressure gradient (Mast 1925). Modern versions assume that the total cortex of the cell is capable of contraction, producing a uniform internal pressure that can be released at regions where pseudopodial projections occur (Grebecki 1986). The frontal contraction model (Allen 1961) assumes that flow is produced by conversion of the fluid endoplasm into a gel-like state at the frontal zone. Thus, the transformation of the sollike endoplasm into a gellike frontal ecotplasm produces a constant pull on the endoplasm moving it forward.

Evidence for the frontal contraction model over that of the uroid contraction model of Mast was presented by Allen et al (1971), who hypothesized that if the uroid were the site of pressure production, then by applying a suction to this region, flow should stop. They applied a small capillary under vacuum to the uroid and demonstrated that the cytoplasm sitll flowed forward. Also, destruction of the uroid cytoplasm by energy from a laser beam did not stop forward flow of the granuloplasm. Moreover, Allen and colleagues point out that the birefringence of endoplasm is lost when it flows into the frontal region and becomes part of the ectoplasmic tube, suggesting a gellike condition.

Recent work by Grebecki (1986), however, lends strong support to a modified form of the ectoplasmic contractile tube model called the cortical contraction theory. He proposes that the microfilaments surrounding the cell produce a pressure in the underlying granuloplasm. During locomotion, it contracts in volume isotonically (changing fiber length but maintaining constant tension) in the posterior of the cell, and isometrically in the anterior parts (maintaining constant length, while increasing tension in the filament). This general contraction around the cell periphery produces an internal hydrostatic pressure. However, frontal tips of pseudopodia are in a relaxed state occasioned by the separation of the microfilaments from the plasma membrane and accumulation of hyaloplasm between them. Thus, separation of the microfilaments from the plasma membrane allows the pressurized hyaloplasm to expand the plasma membrane and flow forward, filtering through the mesh of microfilaments. Hence, a hyaline cap is formed as the granular material of the cytoplasm is retained behind the microfilament mesh, and the frontal tip of the pseudopodium expands outward. During filtration through the microfilament mesh, polymerized actin is retained, but monomeric actin (G-actin) passes through the mesh. Upon polymerization against the plasma membrane, the G-actin reconstitutes the membrane-associated fibrils. This process of microfilament detachment, forward flow of G-actin-containing hyaloplasm into the protruding pseudopodium, and reconstitution of the membrane-bound microfilament layer can occur cyclically. This may explain the repeated formation of pseudopodial protrusions during forward advance of many amoebae. Evidence in support of this hypothesis comes from application of chemical and light stimuli to regions of the cell. Korohoda (1977) demonstrated that application of benzene to the surface of *Amoeba* induced psuedopod formation concomitant with dislocation of the microfilaments from the membrane. Similarly, Grebecki (1980, 1981) showed that application of a narrow beam of light into the granuloplasm enhanced streaming and locomotion rate, while localized shading of the cytoplasm produced a relaxing effect. Moreover, by applying a localized spot of shade onto the periphery of the cell, the relaxing effect induced psuedopod formation. By selective, successive application of the shaded spot to various parts of the cell, the *Amoeba* could be led to locomote in any prescribed direction. If a circular shaded area is projected onto the path of an amoeba, it becomes "trapped" in the shaded region. As it moves toward the edge of the shadow, light suppresses pseudopodial formation. Consequently, new pseudopodia are produced only toward the shaded side of the cell. The net result is that the cell moves only within the region of the circular shadow (Klopocka 1982). In general, the activity of a motile front is strongly promoted by its local relaxation induced by shade, and inhibited or hampered by the contracting effects of light (Grebecki 1986). By employing a shade stimulus in conjunction with orientation of the surface to which the locomoting amoeba is attached, Klopocka (1982) demonstrated a weak negative geotaxis in *Amoeba proteus*. She noted that there was a tendency for the amoeba to follow an upward-directed branch in the pathway over a downward-directed branch. These data, moreover, highlight once again the strong regulatory effect of light on the membrane-associated microfilaments and consequent control of pressure gradient effects in membrane expansion and psueodopodial extension.

By contrast, extension of filopodia and control of bidirectional streaming in benthic Foraminifera is largely controlled by microtubules (eg, Bowser et al 1984; Bowser and Rieder 1985). Bidirectional streaming of cytoplasm occurs exclusively on the surface of microtubules, and transport of both organelles and bulk hyaline cytoplasm occurs bidirectionally along microtubular fibrils. Extension of filopodia occurs by elongation of microtubules pushing against the plasma membrane (Travis et al 1983), forming axial cytoplasmic extensions.

Role of Membranes

The role of the membrane in amoeboid movement has been of interest for some time. Earlier theories presumed that the plasma membrane was the site of contractile forces produced by surface tension effects. This is not widely accepted at present, however, in view of research data showing that contractile microfilaments and not surface tension effects can account for cytoplasmic flow. Other theorists (eg, Goldacre 1952) postulated that amoeboid movement was accomplished by extension and synthesis of the plasma membrane at the anterior end, and resorption at the posterior end through vesicularization and endocytosis of membrane vesicles in the uroid. The research of Grebecki, however, makes this cyclic process of membrane synthesis and expansion followed by endocytotic vesiculation less likely. Stockem (1972) has shown, moreover, that the membrane surface is remarkably constant during locomotion and vesicular endocytosis. This can be attributed to a balanced state of endocytosis at the uroid region and exocytosis at the psuedopodial surfaces. The decrease in surface area produced by invagination and production of intracytoplasmic vacuoles is balanced by fusion of Golgi-derived vesicles with the plasma membrane during exocytosis at some frontal areas of the cell. This is a relatively small and intermittent membrane surface exchange, and cannot account for the rate of locomotion of the cell during psuedopodial expansion and forward movement. Furthermore, careful cinematographic analyses of the movement of marker particles on the surface of moving amoebae (Hülsmann and Haberey 1973) show that the membrane rolls forward during locomotion and is not resorbed at the uroid region. In other words, a marker particle on the surface of the cell beginning at the dorsal side moves forward toward the frontal edge and then is carried under the cell onto the ventral side. Subsequently, it moves posteriorly until once again it is carried upward and forward onto the dorsal surface. This indicates that the membrane is moving in a tractorlike motion as the amoeba locomotes. In the small amoebae *Naegleria gruberi* and *Acanthamoeba castellani*, which lack a substantial glycocalyx, attachment to the substratum at the ventral surface during locomotion is by two types of attachment zones (T. M. Preston in Grebecki 1986). Broad areas of a loose "associated contact" form weak attachment sites. From these platforms protrude adhesive micropseudopodia responsible for much closer focal contacts. The associated contact platforms are very mobile during locomotion and are easily separated from the substratum, while the focal contacts are stable and leave "footprints" on the substratum after separation.

Forms of Amoeboid Locomotion

Based on the organization of pseudopodia and the morphodynamics of locomotion, Grebecki and Grebecka (1978) proposed a new terminology to describe various forms of *Amoeba proteus*. In the older terminology, cells with a single dominant pseudopodium were classified as monopodial and those with several major pseudopodia as polypodial. This simplified classification neglects important differences in the morphodynamics of locomotion. Grebecki and Grebecka propose four categories: (1) monotactic, (2) orthotactic, (3) polytactic, and (4) heterotactic. Some major descriptors of these three types are summarized here and more detailed explanations are available in the source (Grebecki and Grebecka 1978). Monotactic forms have a single pseudopodium, circular in cross section, never contractile, but always leading and forming a vesicular frontal cap. The latter is very important. In the monotactic form, the frontal hyaline cap is formed by vesicles that migrate to the frontal region and fuse to form the hyaline zone. The locomotory path follows bending and looping paths. Orthotactic forms have a single pseudopodium with a bell-like, noncontractile, leading cross section, and lack vesicles in the hyaline cap. Locomotion is unidirectional. Polytactic forms have several pseudopodia with a circular cross section when free, and bell-like when attached. The frontal cap is hyaline. The locomotory path follows a zigzag pattern with frequent changes in general direction. The polytactic form responds to directional stimuli by transforming into an orthotactic type, and into a heterotactic type after general shock. The orthotactic pattern previously described responds to directional or localized stimuli by reorientation. The final form, heterotactic, has several pseudopodia with circular cross section. The frontal cap is hyaline and locomotion is by somersaulting movements. This form responds to directional stimuli by transforming into the orthotactic type. The monotactic and orthotactic forms exhibit an unbranched cytoplasmic streaming pattern. The direction and velocity of streaming are stable in the monotactic form. Fountain streaming sometimes occurs. Velocity is variable and there is no fountain streaming in the orthotactic form. A branched streaming pattern is present in polytactic and heterotactic forms. It is regularly reversing in the pseudopodia, but unidirectional in the trunk and uroid of polytactic forms. A fountain is occasionally observed. In heterotactic forms, the cytoplasmic streaming reverses irregularly everywhere. A fountain pattern of streaming is common in attached pseudopodia.

Ionic and pH Effects

Extracellular pH and cation concentration influence the form and locomotory behavior of amoebae (Bratz-Schade and Stockem 1972; Opas 1975). During pH changes, the rate of locomotion of amoebae is zero at roughly pH 5.0, rises to a maximum at pH 6.7 to 6.8, falls to an intermediate minimum at pH 7.0, reaches a secondary maximum at pH 7.4 to 7.5, and finally falls to zero at roughly pH 8.5. In general, as the pH of the medium increases the pseudopodial activity also increases. An inverse relationship was found between speed of locomotion and pseudopodial activity. Within the pH range of 5.0 to 8.5, there is a constant locomotive activity which can be interpreted as a balanced state between speed of locomotion and rate of pseudopodial formation. Motility is blocked by addition of thorium dioxide at pH 6.0, or by the addition of an iron-containing protein, ferritin, at pH 4.5 (Braatz-Schade and Stockem 1972). Below pH 3.5, the amoebae retract all of their pseudopodia and round up into spheres, regardless of whether thorium or ferritin is added to the culture medium. The role of calcium ions in amoeboid locomotion requires additional research. As often occurs early in a research endeavor, there are some apparently conflicting reports with regard to the role of Ca^{2+} in cytoplasmic streaming in amoebae. Some events such as pinocytosis appear to be regulated by Ca, while intracellular probes with calcium-detecting dyes do not always support a calcium signal role in locomotion (Gawlitta and Stockem 1980). It is known that conditions of high ionic strength (as well as calcium ion concentrations below 10^{-7}) disassemble myosin and actin filaments and indicate a role for calcium regulation of contractile activity.

Cytochemical fluorescent staining techniques show that in locomoting *Chaos carolinense* and *Amoeba proteus*, actin is uniformly distributed throughout most of the cytoplasm except in the tail ectoplasm and in the plasma gel sheets, where *distinct* actin structures were observed (Taylor et al 1980). Moreover, formation of distinct fluorescent actin structures is correlated with contractile activity. The use of cytochemical probes that mark the presence of actin and its state of polymerization shows that in *Acanthamoeba* actin polymerization is regulated, at least in part, by a protein prolifin, which binds to actin monomers, and by a "capping protein" which nucleates polymerization (Pollard et al 1982). By microinjection experiments using myosin inhibitors, it has also been shown that myosin is involved in amoeboid locomotion. Contractile actinomyosin fibrils, possibly regulated by changes in calcium concentration, can also account for shuttle-streaming activity of slime mold plasmodia.

Shuttle Streaming in Slime Molds

Slime mold plasmodia exhibit a characteristic form of motility known as shuttle streaming. This has been documented and especially studied in the large yellow-pigmented species *Physarum polycephalum*. The plasmodium consists of a network of veins (ca 1 mm in diameter) that terminates at the forward moving end as a broad fanlike mass of cytoplasm. Streams of cytoplasm within the veins exhibit a cyclic pattern of reversible flow, proceeding forward for a short duration and then flowing in the opposite direction. The activity is clearly visible under a light microscope, as the numerous nuclei and other cellular inclusions are readily observed streaming with the cytoplasmic flow. When the plasmodium is advancing, the net movement of cytoplasm in the forward direction is greater than in the reverse direction, thus producing a net forward movement of the protoplasmic mass. The remarkable rhythmicity of cytoplasmic flow and the velocity of motion of the stream has long engendered interest by cell biologists and protozoologists.

Cytomechanics

Current fine structural and cell biomechanical research has begun to clarify the processes involved in the cyclical streaming (Grebecki 1979; Grebecki and Cieslawska 1984; Stockem 1979; Wohlforth-Bottermann 1979). The flow of the cytoplasmic stream is currently interpreted as a pressure gradient phenomenon.

Strong evidence points to rhythmic contraction of the ectoplasm in the veins, increasing the pressure of the cytoplasm and forcing it into regions of lesser pressure. The shuttle streaming is a result of oscillations in contraction between the posterior parts of the cytoplasm and the anterior parts. Thus, contraction in the posterior part of the plasmodium and dilation in the anterior part produce a pressure gradient favoring flow toward the anterior region of the cell. Reversal occurs by dilation of the ectoplasm in the posterior of the plasmodium concurrent with constriction in the anterior part. Consequently, the pressure gradient is reversed and the cytoplasm flows in the opposite direction. In general, a more sustained contraction at the posterior part of the plasmodium compared to that in the anterior of the cell results in a net mass transfer of cytoplasm in the forward direction. During simultaneous recording of contraction and streaming at two points along the plasmodial veins and frontal channels of *Physarum polycephalum*, the contraction:relaxation time ratio was 48:52 (Grebecki and Kolodziejczyk 1983). The forward:backward streaming time ratio was 55.5:44.5. These proportions varied insignificantly between different control sites. As predicted from a pressure flow model based on contraction and relaxation processes, contraction of the veins is correlated with relaxation of the frontal zone. Likewise, there is a weak but positive correlation between contraction at a site on the veins and the streaming developed anterior to the site. This suggests that mass forward movement of cytoplasm is by a pressure differential between the veins and frontal zone. Further support comes from studies by Grebecki and Cieslawska (1984), who showed that dissecting the frontal zone of the plasmodium from the posterior mass of veins suppressed forward movement. Frontal regions continued to move only if they were attached to some veins, and the more veins remaining, the greater the motive force. This further substantiates the pressure flow interpretation of cytoplasmic streaming mediated by contractile processes. The mechanism of contraction has been investigated, among other means, by isolating small segments of plasmodial cytoplasm and observing their behavior (Stockem 1979). Small portions of cytoplasm removed from the plasmodium assume a dumbbell shape. These segments of cytoplasm continue to exhibit rhythmic activity. When one lobe of the segment contracts, the other expands, and vice versa. Contractile filaments in the ectoplasm composed of actin and myosin are believed to produce the constriction of the cytoplasm during contraction. A reversible process of polymerization of actin from G-actin (soluble monomeric form) to F-actin (fibrillar, polymeric form) can account for the formation and disassembly of the contractile fibrils. During contraction, F-actin is formed. Its association with myosin and release of energy during ATP hydrolysis can account for the contraction phase. Relaxation of the actinomyosin fibrils and solubilization of the actin to form G-actin can account for the relaxation phase.

Role of Calcium and Microfilaments

Current evidence indicates that release of Ca^{2+} from nearby storage sites in intracytoplasmic vesicles may be the trigger that initiates contraction. Likewise, accumulation of calcium into the vesicles by active transport can account for actin disassembly and relaxation. Thus, rhythmic release and resorption of calcium in intracytoplasmic spaces can account for the signal regulating alternating cycles of contraction and expansion in the veins of the slime mold. Stockem (1979) also reports that there may be two kinds of filaments in the veins, those stretching across the vein from one membrane surface to the other, and those forming a peripheral network within the ectoplasm. As with amoeboid protozoa, illumination of *Physarum* cytoplasm alters the contraction process, especially when blue light, or white light with a blue component, is used (Kolodziejczyk and Grebecki 1983). Blue light illumination of the veins produces an increase in force output, but contrarily induces a negative response in the frontal zone. This is particularly marked when the cell is maintained in a uniform red light and then stimulated at the various sites with a beam of blue light. Shading of the frontal zone also produces an increase in the forward streaming time in veins, a higher frequency of their contraction cycles, and a better synchronization between contraction and streaming oscillations (Kolodziejczyk and Grebecki 1982). These data are consistent with the ectoplasmic contraction model of pressure-produced streaming as observed in amoebae, and also accounts for the movement of plasmodia away from illuminated into darkened regions. Illumination of the veins produces an increased net flux of protoplasm into the frontal zone, increasing forward locomotion, whereas illumination of the frontal zone causes resistance to relaxation, a lessening of the pressure gradient, and hence retardation of movement toward the frontal direction. Over time, other advancing fronts develop in areas of lesser illumination, and the organism will locomote toward the darkened region. This remarkably simple yet elegant physiological mechanism for light induced changes in locomotion provides a molecular basis for understanding the ecologically significant responses of amoebae and slime molds to variation in light quality and intensity. Clearly, for bactivorous and detrital-consuming organisms, greater food sources are likely to be located in regions of lesser illumination. Hence, negative phototaxis as explained by the above molecular changes has survival value. These studies provide a unique example of the comprehensive kind of theory building and intellectually intriguing research results that often emerge from interdisciplinary studies in protozoan physiological ecology.

Summative Perspective

Protozoa exhibit varied locomotory behavior and diverse responses to environmental stimuli, but some fundamental cytoskeletal and contractile phenomena mediated by ionic balances increasingly appear to account for many of these processes. Ciliary and flagellar motion are currently understood to be produced by sliding of microtubule doublets, possibly controlled by transient changes in calcium ion concentration. Cytoplasmic flow in amoebae, and shuttle streaming in myxamoebae, are probably the result of differences in hydrostatic pressure within the cell resulting in flowing motion of the cytoplasm. The direction of flow and rhythm of this activity can be explained in part by differences in contractile activity and stability of microfilamentous networks within the cytoplasm and deposited near the plasma membrane. A variety of stimuli including light, calcium ion concentration, and mechanical stimuli alter the contractile patterns of microfilaments. These changes in microfilament tension and contractility may account for some of the behavioral responses of amoeboid organisms which have appeared so enigmatic in the past. The effect of light on microfilament stability and force of contraction may explain the movement of some amoeboid organisms away from illuminated regions. Photoreception and orientation toward illumination by flagellates appear to involve complex interactions of oriented, light-absorbing molecules with the electromagnetic vectors of incident light. The elucidation of the cytomechanics of protozoan motion may have far-reaching implications for our understanding of general processes of locomotion in eukaryotic cells, and contribute to our understanding of the biophysical explanations for normal and pathological states of cellular motion in metazoa.

19
Reproduction

Reproductive Strategies

Documentation of the reproduction and life cycles of protozoa, as for other organisms, was one of the earliest interests of biologists. Substantial progress was made during the early 19th century in clarifying patterns of reproduction and correlating cytological events with reproductive cycles. The diversity of protozoan species, as exemplified by their morphology and physiology, is further illustrated by their varied sexual and reproductive processes. Both asexual and sexual reproductive phases occur in some species, while others reproduce only asexually or sexually. Many amoebae, for example, appear to reproduce only asexually. Some Foraminifera reproduce only sexually, or have alternating sexual and asexual reproductive phases in their life cycles. An additional complication arises in some species due to the separation of sexual processes and reproduction. For example, some ciliates exchange gamete nuclei during conjugation of two mating cells, thus producing zygote nuclei and sharing of genetic material. However, upon separation the conjugants do not immediately undergo reproductive fission. Binary fission typically occurs sometime later. Hence, in these species sexual processes do not coincide with organismic reproduction. The significance of this phenomenon in physiological ecology of protozoa has been discussed conceptually by Fenchel (1986b). It is not entirely clear why there is a temporal separation of gamete exchange and reproduction, but it may be an adaptive mechanism to permit gene recombination and the potential benefits of hybridization without immediately risking hazards due to the potentially delicate stages occurring during cell division. Thus, under less than optimal environmental conditions, gamete exchange occurs and potentially better adapted individuals are produced. A delay in cellular fission may permit locomotion to parts of the environment where the organism is better adapted and offspring are more likely to survive. In other protozoa, such as the planktonic Foraminifer *Globigerinoides sacculifer*, maturation and reproduction occur most rapidly during favorable growth conditions when food is abundant (eg, Bé et al 1981; Anderson 1984). Myriads of flagellated swarmers (ca 10^5) are released by each parent cell, and young are rapidly produced to exploit the favorable environment. During less favorable nutritional conditions, growth is slow and reproduction is delayed, thus potentially providing time for improvement in the environment and increased opportunity for successful reproduction.

Sexual Reproduction

Sexual reproduction in protozoa is varied according to kind of gamete (whether flagellated, amoeboid, or reduced to migrating nuclei as in ciliates). Further distinctions depend on whether the paired gametes are of the same size or of different sizes, and on the basis of how the gametes are liberated or brought into contact to permit syngamy (fusion). Three fertilization modes, described in detail and amply illustrated by Grell (1973), can be distinguished based on these factors: (1) gametogamy, (2) autogamy, and (3) gamontogamy. Gametogamy is perhaps the simplest case. Gametes are released into the water, sometimes in large numbers, although not necessarily so. The individual gametes fuse and form a zygote, followed by reduction division (meiosis) at some point in preparation for the next gamete release. Autogamy occurs when gametes from the same parent (gamont) fuse to form a zygote. Autogamy is a special case of obligate monoecy. Both gametes come from a single (mon) gamont or house (oecy). Whether the gametes or gamete nuclei are sexually different (different mating types) must be determined for each case. Gamontogamy, the third case, occurs when the *gamonts unite*, and either share gametes or produce gamete nuclei that are transferred between the two gamonts. In this case, sexual differentiation (+ or − mating types) can be expressed by the gamonts, gametes, or gamete nuclei. Only pairs of compatible mating types unite or fuse to give rise to zygotes. Some examples of each type of fertilization are given, citing illustrations from as many major groups of protozoa as is appropriate.

Gametogamy

Gametogamy occurs among free-living protozoa in the phytomonads and Foraminifera, and also among endobionts including the Hypermastigida (multiflagellated endocommensals) and some Apicomplexa (including malarial parasites). Among some phytoflagellates, including *Chlamydomonas*, *Dunaliella*, and *Polytoma*, the vegetative cells become differentiated as gametes and fuse directly (hologamy). In other phytoflagellates, however, the vegetative cell first forms a gamont which by successive divisions gives rise to several gametes (merogamy). Some dinoflagellates (eg, *Glenodinium*) produce free-swimming zoogametes that fuse only when they come from parent cells of opposite mating type. The "snow alga," *Chloromonas brevispina*, produces eight zoogametes from each vegetative cell. Syngamy produces a diploid zygote which encysts. Subsequently, meiosis and cell division give rise to haploid biflagellated vegetative cells.

The colonial flagellate *Volvox* exhibits highly specialized sexual reproductive structures as exemplified by *Volvox carteri* (Kochert 1968). Dioecious reproductive strains produce "male" (sperm-producing), and female (egg-producing) colonies. Egg and sperm initials (cells destined to form gametes) are differentiated when the colony matures to a 32-celled stage. In all cases, the reproductive cells are produced by unequal cleavage of the vegetative cell. Female colonies contain about 20 eggs, while male colonies have approximately 50 male initial cells, each of which forms a sperm bundle containing 64 or 128 sperm. The sperm bundles penetrate female colonies and fertilize the eggs. The fertilized egg (zygote) enters a resting stage and then germinates to give rise to a new colony. Usually, the first-formed stage (germling) is a sphere of cells with the flagellar apparatus inwardly oriented. During development the sphere inverts, bringing the flagellar apparatus toward the outside, thus forming an immature, but fully motile vegetative colony. These colonial flagellates exhibit a high degree of differentiation approaching the complexity of sexual reproductive structures found in some simpler metazoa. *Volvox* does not form true tissues and hence is clearly less advanced than metazoa.

Among the Foraminifera, with very elaborate shell structures indicating a highly advanced evolutionary development, definitive sexual reproduction by free-swimming gametes has been established for only a few species. Moreover, benthic Foraminifera have alternation of generations. A diploid agamont stage reproduces asexually and alternates with a haploid gamont stage that reproduces sexually. Fusion of the free-swimming gametes produces the next agamont generation. Autogamy by flagellated gametes has been documented in polythalamic (multichambered) species including *Elphidium*, *Peneroplis*, *Planorbulina*, and *Discorbis*, and in some monothalmic species (*Myxotheca*, *Iridia*, and others). Detailed observations on reproduction and correlated physiological events during asexual reproduction and autogamy in the monothalmic Foraminifera *Allogromia laticollaris* have been presented by Lee and McEnery (1970). The life cycle of many benthic Foraminifera is discussed in detail by Grell (1973). Spinose planktonic Foraminifera including *Globigerinoides sacculifer*, *G ruber*, *Globigerinella aequilateralis* and *Hastigerina pelagica*, among others, produce flagellated gametes (eg, Bé and Anderson 1976; Spindler et al 1978; Anderson 1984). The full life cycle, however, is not known. Radiolaria also produce flagellated swarmers with crystal inclusions contained in large excentric vacuoles (Anderson 1983), but it has not been determined conclusively if these are sexual or asexual swarmers. Inferential evidence from species diversity based on skeletal structure suggests hybridization, and therefore would indicate that at least some species may have sexual reproduction, but the issue is far from settled (Anderson 1983).

In general, it is important not to confuse dispersal swarmers with gametes. For example, in the amoeboflagellates (*Naegleria* and others), the amoeboid stage gives rise to flagellated swarmers, especially upon dilution of the medium with water after anoxia. These swarmers are asexual dispersal stages and should not be categorized as gametes. They do not fuse to form zygotes. They revert to an amoeboid stage when they settle down after swimming. Careful attention must be given, therefore, to the total life history of an organism before making conclusive statements about the reproductive role of flagellated stages. It is also necessary to carefully document the origin of flagellated cells suspected of being reproductive stages. In some cases, parasitic flagellates may swarm from infected cells (as occurs in radiolaria) and may give the impression of being gametes. Likewise dispersal swarmers of amoeboflagellates which may happen to be in a culture with other protozoa can be mistaken for gametes of these cooccurring protozoa or misidentified as free-living zooflagellates. Care should also be taken to clearly differentiate between agamic and gamic fusion of cells. Some protozoa undergo momentary agamic fusion of vegetative cells that can be confused with syngamy. Seravin and Gudkov (1984) have described three main types of agamic cell fusion. The first type (pseudocopulation) is a temporary union of two vegetative individuals to form a united organism. The form and locomotion of the organism is similar to either of the vegetative individuals. Later, the fused cell separates once again into two or sometimes more individuals if mitosis has occurred in the intervening time. There is no evidence of nuclear fusion; hence this is not a case of syngamy. In the second type (plasmodization), several cells fuse to form a multinucleated plasmodium that differs appreciably from the initial cells in volume, morphology, and locomotion. If the plasmodium subsequently reverts by fragmentation into individual cells, this is called deplasmodization. In some cases, however, the plasmodium persists as the main phase of the life cycle and deplasmodization is not observed. The third type of agamic fusion is pseudoconjugation. This is a temporary fusion of individuals by cytoplasmic connections and can be confused with some forms of gamontogamy unless care is taken to document the fate of the nuclei. There is no nuclear fusion in any of the three types of agamic fusion.

Gamontogamy

Gamontogamy occurs when reproducing cells unite during fertilization. For example, in the dinoflagellate *Ceratium* two cells become apposed to each other and establish a conjugation tube (cytoplasmic connection) wherein two protoplasts and their nuclei unite to form a zygote. In many species of polythalamic benthic Foraminifera, sexual reproduction is initiated by union of two or more gamonts to form a clump or *aggregate* (Grell 1973). After formation of gametes, they are released into the common space formed by the united shells where syngamy occurs. In some species the gametes are flagellated and numerous (eg, *Discorbis opercularis, D patelliformis* and others), while in others they are amoeboid and few, sometimes only one per gamont. In the simplest case, two gamonts creep towards one another by rhizopodial locomotion, pair by touching with their ventral surfaces, and exude a cementlike substance which firmly anchors them together. During gamete production, the walls intervening between the two gamonts are dissolved, providing a common space where the gametes can pair. In other species forming aggregates (eg, *Patellina corrugata*) one to 14 gamonts aggregate, produce a common organic membrane that covers them and cements them to the substratum, and thus create a protected space wherein gametes can fuse. Thus far, no instance of gamontogamy has been reported in planktonic Foraminifera. And, little is known about the reproduction of nonspinose planktonic Foraminifera.

In addition to gamete release during gamontogamy, several variations occur where only gamete nuclei are produced and exchanged. These include gamontogamy without gamete formation as found in at least one Oxymonad zooflagellate (*Notila*), and conjugation as occurs among the cilitates. *Notila proteus* produces gamonts that resemble vegetative cells by differentiation of "progamont" stages in the life cycle. Although the morphology of the gamonts is the same as that of nonreproductive cells, sexual differentiation is expressed in the behavior of the gamete nuclei. After fusion of the gamonts of two different mating types, the cytoplasm undergoes reorganization and the nuclei divide meiotically. Two stationary "female" gamete nuclei, each attached to the axostyle, and two motile "male" nuclei are produced. The male nuclei from one cell migrate to and fuse with the female nuclei of the other cell, forming two synkarya within a single cell. The double zygote containing cell then undergoes cytokinesis, forming two diploid daughter cells.

Conjugation in ciliates (Figure 19.1), though fundamentally similar among a wide variety of species, varies in the details of gamete nuclei production, development and fate of the nuclei after syngamy, and form of the gamonts; ie, whether both gamonts are morphologically identical (isogamonty) or of different morphology (anisogamonty). The details of nuclear behavior, temporal duration of different phases of activity during reproduction, and the organization of the cytoplasm accompanying stages of conjugation have been documented for many ciliates beginning in the late 19th century (eg, Bütschli 1873a,b; 1876) and summarized in modern texts (eg, Grell 1973; Giese 1973; Nanney 1980; Wichterman 1986). Only some major features of conjugation in a few representative genera are described here. In isogamonty, the two morphologically similar conjugants lie apposed to one another and become joined at a specialized region known as the conjugation site, usually near the oral region. The membranes do not exhibit the typical cytoskeletal and granular structure of the somatic pellicle (Hufnagel 1986). This conjugation junction contains a dense matrix (ca 500 Å thick) separating the plasma membranes of the two conjugants (Wolfe 1985). At numerous sites, however, the plasma membranes fuse to form membrane-sheathed cytoplasmic strands joining the two cells. Consequently, the adhesion site appears to be perforated by hundreds of microscopic sized pores. This is the region where gamete nuclei will be exchanged. The micronuclei of each conjugant undergo meiotic division in two steps, which eventually produces four haploid gamete nuclei. Three of the nuclei become pycnotic and are resorbed by the cytoplasm. The third nucleus divides to produce two gamete nuclei. This is known as the postmeiotic division. One of the nuclei is migratory; the other is stationary. The conjugants exchange migratory nuclei, and mutual fertilization occurs by fusion of the migratory nucleus from one conjugant with the stationary nucleus of the other. The conjugants separate after syngamy and are known as exconjugants. The synkaryon in each exconjugant divides into two daughter nuclei. One becomes the macronucleus and the other the micronucleus. The original macronucleus disintegrates during the process of conjugation. Variations in timing and coordination of events occur among species. This is particularly so with respect to the number of micronuclei and their behavior during reproduction. The number varies from one to many, depending on the species. When there is more than one micronucleus, all may undergo meiotic division, but only one persists to give rise to a stationary and a migratory nucleus.

Anisogamonty occurs when conjugants are morphologically dissimilar, usually varying in size and sometimes in general form. The stationary nucleus in the larger conjugant (macrogamont) is fertilized by the migratory nucleus of the smaller conjugant (microgamont). The latter is resorbed by the macrogamont, which is the only conjugant to complete fertilization. Subsequent division of the macrogamont, either by binary fission or a variety of multiple fission patterns,

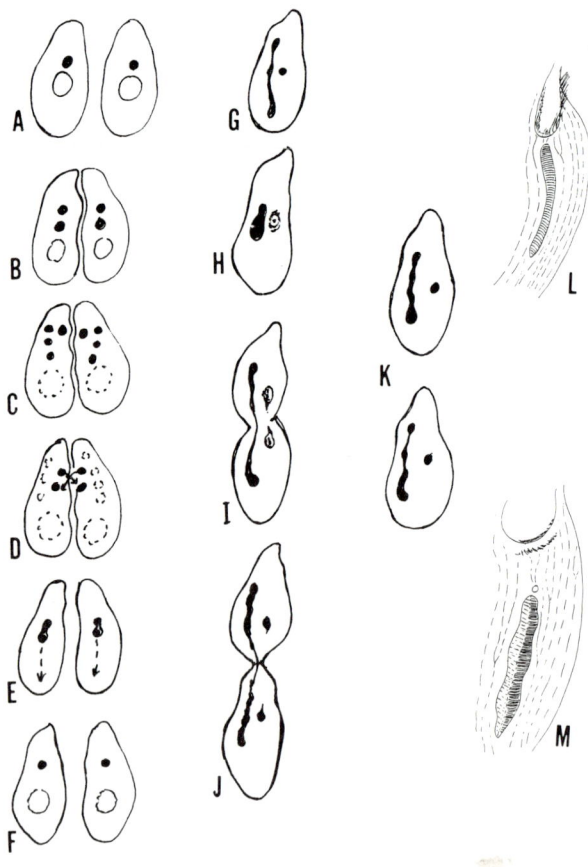

gives rise to daughter cells of unequal size, thus reconstituting the macro- and microgamont forms.

Considerable variation occurs in the form and fate of macronuclei during ciliate reproduction. The shape of the nucleus can vary from a spheroidal form to elongate, scythe-shaped, clubbed or binodal (halteriform), multilobed resembling a string of beads (moniliform), and filamentous (filiform). The variations in nuclear organization, behavior during reproduction, and cytological characteristics have been thoroughly reviewed by Raikov (1982). For example, some primitive species belonging to the karyorelictids possess macronuclei that never divide during reproduction (Raikov 1985). The genera in this group include *Trachelomonas*, *Loxodes*, and *Remanella*. All of the karyorelictids are multinucleate and possess more than one macronucleus. In most species, the macronuclei are more numerous than the micronuclei. The size of the micronuclei are typical of most ciliates (ca 1 to 5 µm), while the macronuclei are decidedly smaller (2.5 to 20 µm). The origin of the macronuclei presents an enigma. Since they do not divide during cell reproduction, it is clear that they cannot be transmitted to both offspring during division. Consequently, the progeny would soon lack macronuclei. The micronuclei, however, do divide during reproduction and proliferate by repeated division in daughter cells after binary fission. Some of the

◄

FIGURE 19.1. Comparison of events during sexual (A–F) and asexual (G–M) reproduction in ciliates. Sexual reproduction occurs by union of sexually compatible conjugants, which join with their oral regions closely apposed and attach by a specialized region of the cell near the anterior end (B). The micronucleus in each cell (solid black) undergoes repeated division, eventually producing only two persistent daughter nuclei. One of these serves as a sperm nucleus that migrates to the reciprocal conjugant (arrows), and the other acts as an egg or stationary nucleus. Fusion of the two nuclei produces a zygote nucleus. Meanwhile, the original macronucleus has disintegrated. The zygote nucleus divides and one daughter nucleus migrates to the posterior of the cell, where it matures into a new macronucleus. During these stages the cells have separated and become exconjugants (E–F). Thus, the exconjugants possess the normal complement of micronucleus and macronucleus. Details vary among species depending in part on the number of micronuclei in the somatic phase of the life cycle. In asexual reproduction (G–K) as exemplified here for *Blepharisma*, the macronucleus remains intact, but first elongates as the cell commences to divide (G), then contracts just before the micronucleus begins to become less dense (H) prior to mitosis (I) of both nuclei and formation of a fission plane through the cell. The daughter cells each receive one daughter nucleus from the macronucleus and micronucleus (K). During division, the oral apparatus of the opisthe (posterior daughter cell) is generated from a specialized region known as the V-region immediately posterior to the oral aperture of the parent cell. The ciliary bases in this region replicate, producing parallel rows of kinetosomes (L). The ciliary bases on the left anatomical side (dense region, M) become more dense and are the precursors of the undulating membrane. The small circle above the V-area is the incipient contractile vacuole of the proter. The opisthe retains the contractile vacuole from the parent cell located at the posterior of the opisthe.

products of micronuclear division differentiate into nondividing macronuclei, thus explaining their continued presence in a reproductive line (Raikov 1985). Moreover, during vegetative growth the macronucleus sometimes degenerates and is replaced by differentiation of a micronucleus. During conjugation, however, the macronuclei apparently never degenerate as in most ciliates. Rather, they persist until later when they degenerate and are replaced by micronuclear differentiation. There are several hypotheses for the origin of this unusual nuclear behavior. Among these is the hypothesis of primitiveness of the macronuclei (Raikov 1985). This explanation assumes that the nondividing macronucleus is a primitive trait that occurred very early in ciliate evolution and gave rise to a phylogenetic lineage quite separate from more "advanced" forms. Thus, the karyorelictids may be a parallel group of evolving ciliates and not a major ancestor to some of the more advanced groups with "typical" macronuclear mitosis. The latter include some of the familiar ciliates such as *Paramecium*, *Tetrahymena*, and others. Examples of conjugation in some of these putatively more advanced ciliates are presented to illustrate the general features of ciliate sexual reproduction.

The general features of conjugation in *Paramecium*, *Tetrahymena*, and *Blepharisma* are compared with some illustrations of variations among species. The major events during conjugation are as presented in the foregoing general description. The first prezygotic division involves an increase in the size of the micronucleus. Thereafter, the nucleus assumes a crescent shape in members of the *P aurelia* complex (*P caudatum*, *P bursaria*, *P multimicronucleatum*, *P calkinsi*, and *P jenningsi*), but not in *P trichium*, *P woodruffi*, or *P wichtermani* (Wichterman 1986). The micronucleus in each conjugant during prezygotic division yields three daughter nuclei; only one persists to divide and yield two pronuclei. One pronucleus remains stationary in each conjugant and the other migrates, uniting with the reciprocal stationary nucleus to form a synkaryon. The fate of the nuclear products formed during the postzygotic divisions of the synkaryon differs among species. In species of the *P aurelia* complex, there are only two postzygotic divisions, not three as in the *P caudatum* group. Among the four nuclei produced by the two divisions, two develop into macronuclear precursors and two become micronuclei. After separation of the conjugants each one undergoes binary fission, contributing one macronuclear anlagen (precursor) and one micronucleus to each daughter cell. The macronucleus matures and the micronucleus divides once to collectively yield two micronuclei and one macronucleus, as is characteristic of the *P aurelia* group. In *P caudatum*, the three postzygotic divisions of the synkaryon yield eight nuclear products. Four of the eight nuclei increase in size to become macronuclear anlagen, one becomes a micronucleus, and the remaining three degenerate. In a sequence of two successive fission steps, the exconjugant distributes the macronuclei among four progeny, and concurrently, the micronucleus divides mitotically to contribute one micronucleus to each individual. Hence, the normal complement of one macronucleus and one micronucleus is restored. *P bursaria* offers some interesting deviations from the general pattern of conjugation. The synkaryon undergoes three postzygotic divisions *before* the conjugants separate. During the first postzygotic division, only one of the nuclear products persists. This enters into the second and third divisions to yield four nuclear products that are present at the time the conjugants begin to separate. In the exconjugant, two of the nuclear products become macronuclei, and the remaining two form micronuclei. Postconjugal binary fission distributes one macronucleus and one micronucleus to each of the two daughter cells. Thus, the normal complement of one macronucleus and one micronucleus is restored. In *P multimicronucleatum*, where there are four micronuclei and one macronucleus in the vegetative cell, a distinctive sequence of steps occurs during the division of the synkaryon (postzygotic division) to restore the nuclear complement (Figure 19.2). The eight products of the postzygotic division of the synkaryon differentiate into four macronuclei and four micronuclei. At the first postconjugal fission, the macronuclei are distributed evenly to the two daughter cells, each containing two macronuclei. The micronuclei all divide mitotically, providing four micronuclei for each of the two progeny. Hence, at this stage they contain two macronuclei and four micronuclei. At the next division, the two macronuclei once again are distributed to give one macronucleus per daughter cell.

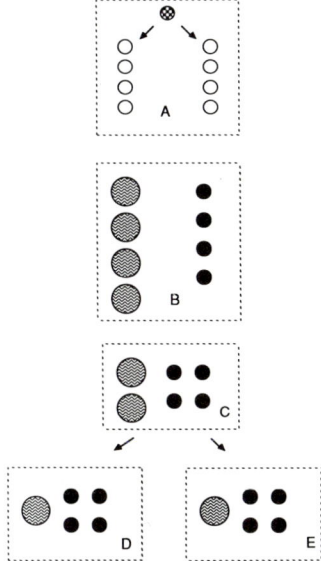

FIGURE 19.2. Production of the macronuclei and micronuclei during the two postzygotic divisions of each exconjugant *Paramecium*. The zygote nucleus (checkered in A) divides meiotically to yield eight daughter nuclei. Four of these (shaded in B) are destined to be macronuclei, and four are micronuclei (solid black). During the first division (B to C), the micronuclei replicate so that the daughter cell receives one-half of the macronuclei (two each) and four micronuclei. In the next (second postzygotic) division (producing D and E), the micronuclei once again divide mitotically, but not the macronuclei. Thus, the daughter cells receive one macronucleus and four micronuclei, as is typical of the species.

The micronuclei divide mitotically, so each daughter cell receives four micronuclei. Hence, at this second postconjugal division, all progeny have a typical complement of one macronucleus and four micronuclei. Normally, conjugants of the *P aurelia* complex remain joined during the sexual process for about 6 hr (25°C), *P caudatum* for 12 to 15 h, and *P bursaria* for ca 24 h (Wichterman 1986). The time required between union and syngamy also varies; ie, 5 to 6 h for *P aurelia* complex, and ca 18 h for *P bursaria*. Further examples of a variety of reproductive patterns and unique characteristics among *Paramecia* spp are summarized by Wichterman (1986).

Conjugation in *Tetrahymena* is similar to the general plan presented for *P aurelia*, but variations occur in the fate of the synkaryon fission products. After union of the two mating cells, the micronucleus elongates to form the crescent stage, followed by two meiotic divisions in the prezygotic phases. Only one of the resulting haploid nuclei persists. This divides once to yield two pronuclei. One of each pair is migratory and by reciprocal exchange fertilizes the stationary nucleus of the other conjugant. The resulting synkaryon divides twice mitotically, producing four nuclear products. Two migrate to the anterior end, and two to the posterior end of the cell. The anterior nuclei mature into macronuclei. Only one of the posterior nuclei matures into a micronucleus. The conjugates separate at about this time. Each one undergoes one binary fission. The two macronuclei are distributed between the daughter cells, and the micronucleus divides mitotically, providing one micronucleus per progeny. Hence, each progeny has one macronucleus and one micronucleus, the normal complement.

In *Blepharisma*, unlike the preceding examples, the macronucleus starts to degenerate early in sexual union immediately after conjugation has begun. During the first few hours following union of the mating pair, the micronuclei migrate from the region of the macronucleus and gather in the midbody of *Blepharisma*. Only about half of the three to 29 micronuclei are functional and undergo the first prezygotic divisions (16 is a typical number). The remaining ones swell and then disintegrate. After the first prezygotic division, only one of the products of each micronucleus is functional. These commence the second prezygotic division, but only one of the products of this division is functional and produces two pronuclei. It is recognizable by its location near the oral aperture, by its larger size, and by the dense enclosing cytoplasmic sphere. The pronuclei differentiate into a stationary nucleus and a migratory nucleus. Reciprocal migration and fusion of the pronuclei between the two conjugants produce a synkaryon in each conjugant. The synkaryon undergoes three successive postzygotic divisions to produce eight nuclei. Six of the eight nuclei are destined to become macronuclei. The other two continue to divide, producing micronuclei. Not all of the macronuclear anlagen survive, and the number persisting varies among species. The number, including additional ones derived from later micronuclear divisions, ranges from one to 13, with six being typical. The micronuclei also multiply until eight or more are produced, varying among species from two to 15.

Following separation of the conjugants, the macronuclear anlagen fuse into two or rarely three masses. Eventually these also fuse, and give rise to one large macronucleus. During the postconjugal binary fission, the macronucleus elongates and divides providing each daughter cell with one macronucleus. Micronuclei are distributed between the daughter cells. Hence, each progeny from the postconjugal division possesses the normal complement of one macronucleus and many micronuclei.

Conjugation is poorly understood in some of the larger spirotrichs such as *Stentor*, even though this organism has been widely cultivated and used extensively in regeneration studies. Skarlato (1982) has examined the fine structure of nuclear events during meiosis. Conjugation requires 15 to 30 h, and during the first 16 to 18 h most micronuclei undergo the first two maturation divisions. These are meiotic. In the first half hour after pairing, the moniliform macronucleus fragments into individual beads. At this time, the micronuclei enter the prophase of the first meiotic division. The nuclear envelope remains intact during the first three meiotic divisions. At telophase of each division, however, the nuclear envelopes are reformed around the daughter nuclei by internal budding of the envelope surrounding the mother nucleus. Intranuclear vesicles are pinched off from these buds and fuse to produce the nuclear envelope of the daughter nuclei. As in other ciliates, only one of the nuclear products of the second maturation division persists to enter the third division and produce the gamete nuclei. The other micronuclei degenerate. Although there is considerable variation in details of events during conjugation among ciliates, the major patterns of gamontogamy are remarkably conserved among major groups, and provide further evidence for phylogenetic affinities among these otherwise highly diverse organisms.

Autogamy

The various forms of gametogamy and gamontogamy in protozoa ensure cross-fertilization. This enhances the probability of new combinations of genes and provides the possibility of genetic variability in breeding populations. Autogamy, however, prevents cross-fertilization. The gametes come from the same parent cell. This process is most likely to be favorable when environmental conditions are rather stable and the individuals are well-adapted to the environment. Under conditions of unpredictable environments, or where there are strong selection pressures and the individuals are not optimally adapted to the environment, gametogamy or gamontogamy should be more advantageous. Likewise, asexual reproduction (producing individuals of identical genotype to the parent cell) is particularly advantageous when the individuals are well-adapted to a relatively stable environment. Since there is no gene recombination and daughter cells contain nuclei with identical chromosomal complement to the parent cells, all of the offspring are genetically identical. If the population is derived from a single parent cell, the individuals are called clones.

Asexual Reproduction

Asexual reproduction in protozoa includes binary fission, multiple fission, and budding. Binary fission produces two usually identical daughter cells by mitotic division. The orientation of the fission plane relative to the long axis of the cell (when one exists) and the geometric relations between major morphogenetic characteristics of the two daughter cells distinguishes two kinds of fission, symmetrogenic and homothetogenic (Figure 19.3).

Symmetrogenic Fission

Symmetrogenic fission (typical of many flagellates, opalinids, and many other nonciliate protozoa) is a mirror image fission. The fission plane typically is coincident with the long axis of the cell. As separation proceeds along the fission plane, two daughter cells are produced with a mirror image relationship. In some cases, as exemplified by *Euglena*, fission commences at the flagellar pole of the cell. The short flagellum within the flagellar pouch elongates to form a second flagellum destined to be the long flagellum of one of the daughter cells. Short flagella must be regenerated from flagellar bases to complete the flagellar apparatus. Duplication of the flagellar apparatus is followed by formation of an anterior cleft yielding two flagellar pouches, fully developed, but still attached posteriorly to the parent cell. Further deepening of the fission cleft results in progressive separation of the two daughter cells, until at the last stage they are connected only by a very thin cytoplasmic bridge. This is eventually pulled to a very fine diameter, concurrent with locomotion of the two developing cells in opposite directions, and eventually it is severed. Thus, the two daughter cells are separated from each other and assume an independent existence.

Homothetogenic Fission

Homothetogenic fission typically occurs in ciliates. It is generally transverse or perikinetal. The parent cell gives rise to two daughter cells by a fission plane that is transverse to the long axis or to the anterior-posterior axis. During initial stages of fission, morphogenetic marker structures develop in topographically identical positions on the two developing daughter cells. For example, in ciliates, the location of the oral apparatus, contractile vacuole (if present), and cytopyge develop simultaneously and in the same topographical location in the proter (anterior developing cell) and the opisthe (posterior developing cell). Eventually, when complete fission occurs, the two daughter cells are typically fully developed and identical in morphology. However, their twinlike development has occurred in tandem to one another, and not parallel to each other as in the symmetrogenic form of fission.

Binary Fission

Binary fission in amoebae usually occurs in a plane transverse to the axis of the nuclear mitotic spindle. A constriction ring gradually bisects the amoeba by producing a cleavage furrow in the midline of the cell, eventually yielding two

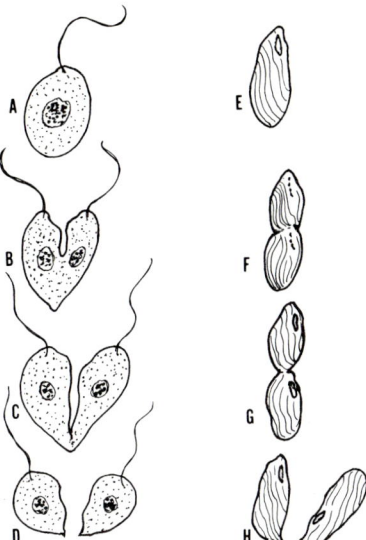

FIGURE 19.3. Comparison of symmetrogenic cell division in a flagellate (A–D) and homothetogenic division in a ciliate (E–H). In symmetrogenic division, the cell divides parallel to the long axis, in this case commencing with the flagellar pole and gradually separating as the division cleft progresses toward the posterior end of the cell. In the last stages, the daughter cells are connected only by a thin cytoplasmic strand which separates. In homothetogenic division, a form of transverse division, the cell divides by a constriction band perpendicular to the long axis of the cell. The oral apparatus of the opisthe (posterior daughter cell) is generated from a specialized region of cilia below the mother cell oral apparatus. Usually, the proter (anterior daughter cell) also resorbs its oral apparatus and regenerates it in unison with the development of the apparatus in the opisthe. The separation of the daughter cells occurs by pinching in two at the site of the constriction ring (G–H).

daughter cells. They frequently exhibit a remarkable phenomenon of locomoting away from one another in paths that are mirror images of one another. This suggests that cellular division is accompanied by some chiral (right and left handedness) organizing property of the intracytoplasmic locomotory system. In testate amoebae, with rather loosely arranged extracellular scales as in *Euglypha*, reserve scales are synthesized in advance of cellular division. The pseudopodia are withdrawn and the intrashell cytoplasm is reorganized, forming a protrusion that bulges from the oral aperture. The bulge grows to approximate the shape of the daughter cell, but is oriented opposite the parent cell. The reserve scales are secreted on the surface of the bulge. At this stage, nuclear division occurs and one of the daugher nuclei migrates into the bulge, which upon further maturation develops pseudopodia and separates from the parent cell. Monothalamic Foraminifera with thin test walls (eg, *Allogromia* sp) reproduce asexually by elongation of the cell and development of granular rhizopodia at each pole (eg, Figure 20.3). The rhizopodia, while under tension, spread outward in opposite directions. A constriction ring develops at the midline of the elongate cell, at first producing an hourglass shape. Further constriction and elongation of the connecting strand of cytoplasm produced by the locomotion and contraction of the rhizopodia at opposing ends of the cell eventually produces two globose cells connected by only a very thin strand of cytoplasm. As further tension is exerted on the strand it finally is severed, releasing the two opposed daughter cells. These immediately contract upon release of the tension, at first forming a wrinkled, globose, cytoplasmic mass with adhering rhizopodia. The cells round up and move apart by rhizopodial extension and contraction.

Multiple Fission

Multiple fission occurs in a wide range of protozoa. This mode of reproduction involves repeated division of the nucleus to produce several to many daughter nuclei that eventually give rise to multiple progeny by repeated cellular fission. Asexual reproduction in some benthic Foraminifera occurs by multiple fission. This is characteristic of the agamont generation in those species exhibiting alternation of generations. The mother cell nucleus undergoes repeated mitotic division, filling the cytoplasm rather uniformly with daughter nuclei. Each one becomes isolated by an envelope of cytoplasm which differentiates into the first developmental stage of the Foraminifer, containing only one or two chambers. The offspring are released, and in some species (eg, *Heterostegina depressa*) the progeny are gently carried away from the parent test by rhizopodial streaming. After a quiescent period, additional chambers are added to the juvenile test and maturation commences.

Multiple Fission in Colonial Volvocida

Colonial Volvocida (suborder: Volvocina) reproduce asexually by specialized modes of multiple fission. This group of organisms includes *Volvox* and its simpler relatives. *Gonium* is a flattened colony containing 16 cells, usually somewhat quadrangular (Figure 19.4). Four-celled daughter colonies are produced by multiple fission and released to develop into 16-celled mature coenobia. The large spherical to ellipsoidal colony formed by *Volvox* reproduces asexually by multiple fission of specialized cells (gonidia). These are larger than the vegetative cells and are also the source of sexual reproductive cells. During asexual reproduction the gonidium sheds its two flagella and divides repeatedly, forming a small, hollow sphere of cells with the flagellar bases and nuclei directed inwards. Asymmetrical division of some of the cells produces gonidia in anticipation of the next reproductive stage. The young colony, still enclosed within the mother sphere, inverts through a pore (phialopore) (analogous to turning a sock inside out), bringing the flagellar bases to the outside. Flagella develop and the young colonies are released by disintegration of the mother colony.

Budding in Ciliates

Budding can be considered to be a specialized type of fission. Nuclear division produces daughter nuclei. Each nucleus migrates into a cytoplasmic bud that is

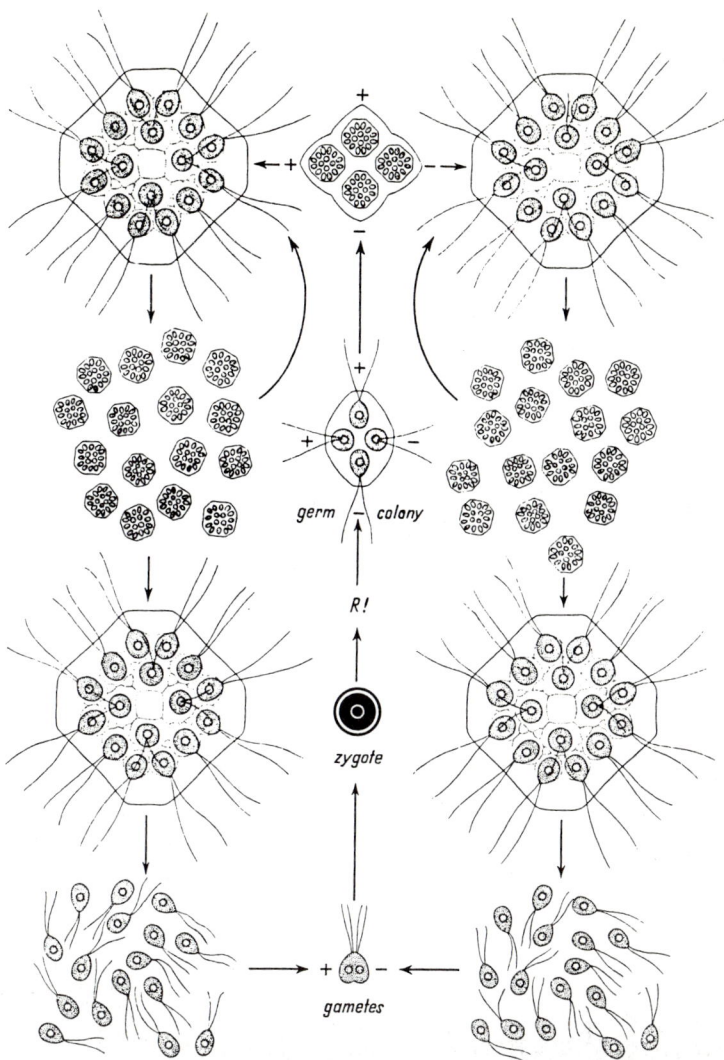

FIGURE 19.4. Reproduction in the colonial flagellate, *Gonium pectorale*, sexual and asexual phases. A four-celled germ colony hatches from the zygote. Of the four gones developed within this colony, two belong to the + sex and two to the − sex. The mature colonies produced by these gones can reproduce asexually as shown by the curved arrows directed upwards. At some point, however, the colony becomes sexually mature and produces gametes (shown in the lower part of the diagram) that fuse to form zygotes, thus completing the reproductive cycle. From Grell (1973); with permission.

390   19. Reproduction

released by cytoplasmic fission. In some cases, the released cell is a motile dispersal stage that migrates from the mother cell and develops into a mature reproductive individual. There are many variations among species, and therefore it is not possible to give a generalized model of reproduction by budding. One of the classic examples, however, is asexual reproduction by budding in the Suctoria (Figure 19.5). Among the various species, buds are produced in three ways: (1) by surface budding; (2) budding initiated within a pouch, but completed at the surface; (3) budding contained entirely within a pouch; and (4) internal differentiation of a bud followed by extrusion through the cell membrane.

During surface budding, one or more protrusions containing daughter nuclei emerge at the anterior end of the stalked suctorian and are eventually released

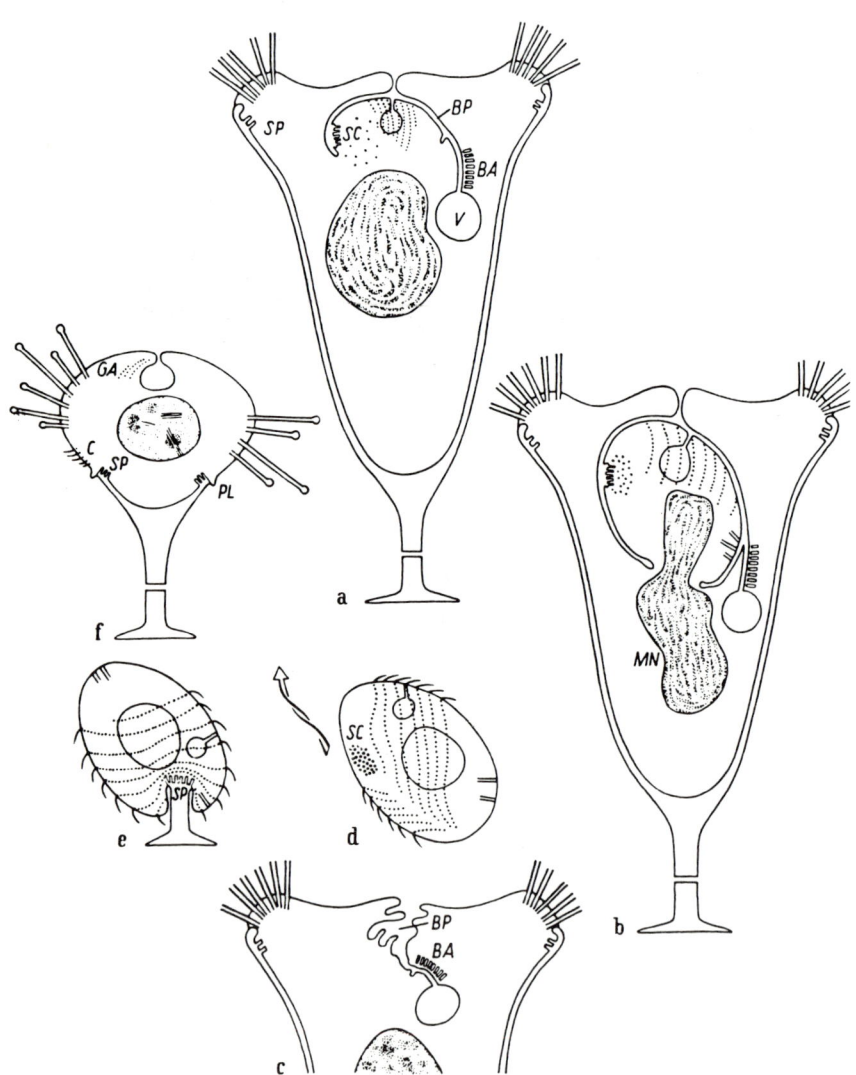

as swarmers by exogenous cytokinesis. For example, in *Ephelota gemmipara*, a large protozoan ca 500 µm long, the mother cell simultaneously produces several ciliated swarmers. As many as four spathate protrusions, each containing a nucleus, form at the anterior end and gradually expand to form a bean-shaped bud. Cilia emerge on the exterior surface of each bud during the last stages of development, when the buds become subtended by a thin stalk. Upon separation from the stalk, the swarmers crawl about using their cilia, eventually become sedentary, secrete a stalk, and develop into a mature suctorian with prey-catching tenacles.

Actineta tuberosa, ca 200 µm in size, produces a swarmer within an intracytoplasmic pouch with endogenous cytokinesis. That is, the swarmer is formed and released from a saclike depression in the cytoplasm of the body of the suctorian. Budding is initiated by invagination of the anterior surface of the cell body to produce a small cavity which enlarges at its lateral surfaces to form a bud enclosed within the enlarging cavity (Figure 19.5). As the bud enlarges, a lobe of the macronucleus protrudes into the bud and is pinched off to form a daughter nucleus within the developing ciliated swarmer. The main part of the mother nucleus remains in the mother cell. During the last stages of swarmer production, the bud is attached within the "brood pouch" by only a very thin strand of cytoplasm. Cilia develop on the surface of the bud in a pattern characteristic of the species, but usually in circumspiral bands in the midregion or at the basal half of the swarmer. Upon maturation, the swarmer is released from the birth cavity and emerges from a small pore on the anterior surface of the cell body. After migrating some distance from the mother cell, the swarmer settles down, attaches to the substratum, secretes a stalk, and develops into a tentacle-bearing, mature form. In those species producing a bud within a pouch followed by exogenous cytokinesis, the fully formed, ciliated bud, still attached to a stalk, is extruded from the pouch onto the outer surface of the cell body before release.

In some lorica-dwelling ciliates, motile dispersal stages are produced by terminal budding. The *folliculinids* are sessile loricate ciliates with terminal ciliated appendages (peristomal wings) that are fluttered near the opening of the lorica to generate water movement for filter feeding. Before fission, the peristomal wings are resorbed, and pigment that previously was dispersed throughout the cytoplasm becomes localized in the anterior half of the cell. A constriction ring forms near the midline of the cell and a vermiform swarmer is pinched off. The

---

FIGURE 19.5. Suctorian reproduction in *Acineta tuberosa*. Asexual reproduction by budding starts with the invagination of a small part of the pellicular area near the opening of the vacuole (shaded area in a). The macronucleus (MN) forms a bud that elongates and invades the cytoplasm of the incipient swarmer. The ciliary bands begin to mature into kineties of the swarmer (dotted lines). Eventually the swarmer is separated as a bud within the brood pouch and released through the pore (d). After swimming for a while, the swarmer settles down and secretes a stalk (e) and matures into a tentacle-bearing feeding stage (f) already showing signs of the brood-pouch vacuole (anterior region) and ciliary anlage. From Bardele (1970a); with permission.

posterior half remains within the ampullate-shaped lorica and produces new peristomal wings. After swimming for a while, the swarmer attaches to a suitable substrate and forms a lorica. Budding is an efficient means of producing offspring while simultaneously conserving the integrity of the mother cell. In the case of the Suctoria, the production of swarmers in a brood pouch permits the mother cell to capture prey and continue feeding during much of the process of budding, thus providing a continuous source of nourishment.

A remarkable example of internal bud formation occurs in the hypotrichous ciliate *Histriculus vorax* (Curds 1966). This ciliate reproduces asexually by binary fission or by internal bud formation. Two sizes of individuals are observed in a culture; larger organisms (190 to 250 µm long) and smaller individuals (70 to 140 µm long). Both size groups undergo binary fission, but only the larger ones produce endogenous buds. Macronuclei divide during bud formation, and the anterior nucleus of each pair migrates to the budding region at the mid- to posterior part of the cell. The nucleus becomes enclosed in an endogenous bud. When mature, the bud is extruded through the body wall of the mother cell. Newly released buds can develop directly into an embryonic form, or if unfavorable conditions prevail, the bud may encyst. Similar embryonic forms, bearing frontal and caudal cirri, are produced either directly from the deposited bud or after excystment of the encysted bud (Curds 1966).

Summative Perspective

The widespread occurrence of protozoa, and their capacity to adapt to a wide range of habitats, is in part a reflection of their varied and environmentally responsive modes of reproduction. Many species reproduce asexually and sexually at different stages in their life cycle or in response to environmental stimuli. Sexual reproduction involves fusion of haploid gametes to form a diploid zygote. However, the kind of gamete produced, whether a flagellated or amoeboid cell, or simply a migratory nucleus as in some ciliates, is characteristic of the species. Furthermore, distinctions are made based on the way the gametes are released and come together, and whether they are from the same or different parents. In the ciliates, especially, sexual processes (syngamy) are separated in time from cellular reproduction. The fertilized cells (exconjugants) undergo fission after reciprocal fertilization and their separation. Asexual reproduction usually produces several to many offspring genetically identical to the parent. Therefore, it is particularly advantageous in stable environments where the species is well-adapted. Gene recombination occurs during sexual reproduction. This provides the possibility of genetic variability and may enhance survival of species that invade new environments, or must adapt to gradual changes in their present habitat. The role of reproduction in the genetics and entire life cycle of protozoa is discussed more fully in the next chapter.

20
Life Cycles and Genetic Continuity

Conceptual Perspective

A substantial literature has accumulated on the details of life cycles among many groups of protozoa and the molecular events that accompany differentiation and transformation of cells during developmental phases. A review of some of these processes can be found in specialized treatises (eg, Giese 1973; Grell 1973; Cox 1980; Nanney 1980; Wichtermann 1986; Hiwatashi and Luporini 1986). A concise conceptual perspective on differentiation in protozoa is presented by Trager (1963), who categorizes two main kinds of morphogenetic events: (1) reproduction, and (2) processes of reorganization. The latter include encystment, excystment, sexual processes (separate from reproduction), structural and physiological adaptations exhibited in the course of complex life cycles, and restorative events elicited by environmental stimuli. Some of these processes, such as the formation of oral apparatus in ciliates during binary fission or during regeneration, offer excellent opportunity to study rapid and extensive cellular differentiation under carefully controlled environmental conditions. Reversible and cyclic transformations in protozoan life histories represent remarkably complex, genetically controlled alternations in form and function. Examples include the amoeba to flagellate transformation among amoebo-flagellates (eg, *Naegleria*), and complex transformations among the various stages in the life cycles of malarial parasites, leishmanias, and trypanosomes. These closely regulated genetic events are not only of intrinsic interest to protozoologists, but may also provide broad insights into mechanisms of genetic control among diverse kinds of eukaryotic cells. Among these highly diverse processes, only some representative examples can be presented within the limited space of this publication. Current knowledge of life cycles and factors regulating stages of reproduction will be discussed for representative species in some of the major phyla.

Flagellates
Reproductive Cycles

Reproductive cycles in some solitary and colonial Volvocidans (eg, Figure 19.4) are discussed as illustrative examples of current research in flagellate life cycle strategies. The solitary Volvocidans *Polytomella* and *Polytoma* have been widely investigated, biochemically and physiologically. The life cycle of *Polytomella caeca* is illustrative of major features of this genus (Lewis et al 1974). Asexual reproduction is by binary fission. The dividing cell cleaves beginning at the posterior end, and the cleavage furrow gradually progresses toward the anterior end. While the two daughter cells are still attached at their anterior ends, the flagella have duplicated and eight flagella extend from the point of union (Figure 20.1). Upon separation, the two daughter cells are fully formed, each bearing four flagella. Sexual reproduction is by fusion of quadriflagellate isogametes that resemble vegetative cells. Gametes engage by intertwining their flagella and gradually merge, beginning with the papilla near the flagellar bases. During fusion, one set of four flagella is resorbed or lost, so that only four remain. Fusion commences at the anterior end and progresses toward the posterior end. Hence, it is the opposite of binary fission. This is followed by a period of zygotic enlargement prior to the fission of the zygote into four daughter cells. Variations in conjugation patterns may occur among species of *Polytomella*. Moore and Cushing (1979) report that syngamy in *P agilis* involves contact in a variety of positions, not only at the apex. The fused cells also do not lose one set of flagella, and the zygote is spherical rather than ovate. Otherwise, the major events are as described for *P caeca*. Sexual reproduction increases in frequency as a population of *Polytomella caeca* becomes more dense. Encystment also occurs during a period of bloom until near the end, approximately 80% of the cells have encysted. The fine structure of encystment has been documented in *Polytomella agilis* (Brown et al 1976). The encysting cells shed their flagella, sink to the bottom of the culture, and form a thick cell wall. The mature cyst wall consists of four layers deposited sequentially next to the plasma membrane. The first layer consists of fine fibrils which are formed partly embedded within the plasma membrane. The remaining layers are thicker and confer protection of the cyst against environmental stress. Endoplasmic reticulum and Golgi bodies proliferate during early phases of encystment, followed by a reduction in size and abundance of these organelles and of plastids during cyst maturation. Microtubules and the flagellar basal bodies dedifferentiate and are not observed in later stages of encystment. Excystment occurs by protrusion of the cytoplasm from one pole of the elongating cyst, and gradual emergence of a fully formed quadriflagellate cell from a porelike rupture in the cyst wall (Figure 20.1).

Morphometric Analysis

Morphometric analysis of changes in the nucleus, chondriome (mitochondrion), and leukoplast during the cell division cycle of *Polytoma papillatum* has shown that the chondriome and leukoplast undergo profound structural changes, while the nucleus exhibits substantial changes in volume (Gaffal et al 1982). The cell volumes during early or midphases of mitosis varied considerably, and seemed to determine the number of subsequent division processes. The chondriome maintained a constant volume ratio to the total cell (ca 8 to 9%) during the whole life cycle, while the ratio of nucleus to cell volume (8 to 10%) correlated only during

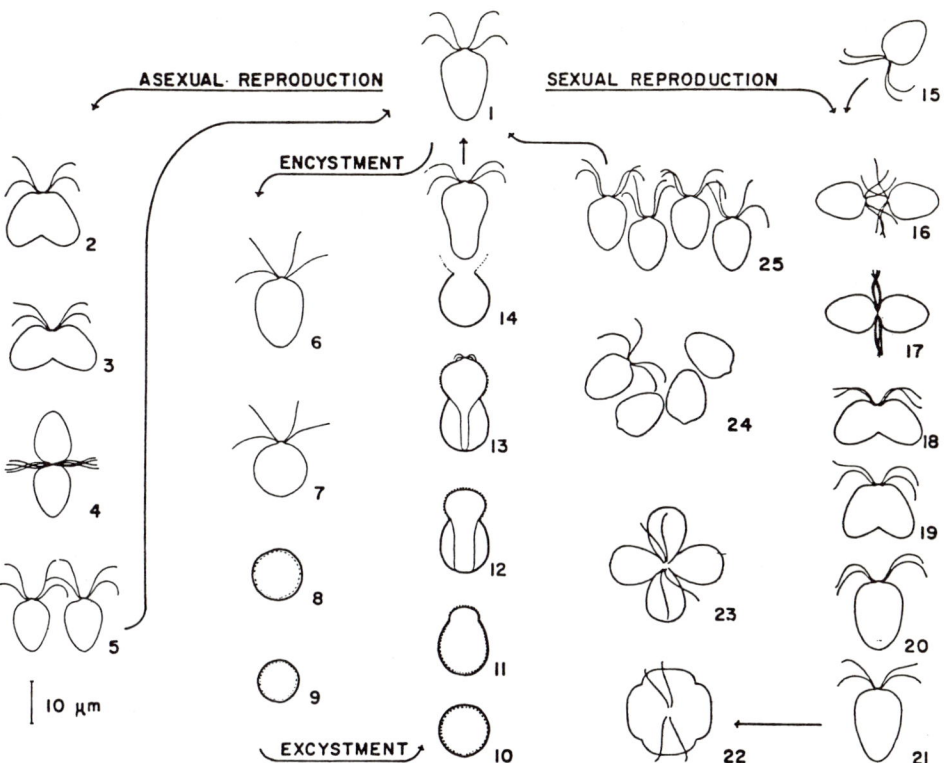

FIGURE 20.1. Life cycle of *Polytomella caeca*. (1) Motile quadriflagellate cell. Stages of asexual reproduction (2–5) show longitudinal division of the cell commencing at the posterior end and ending with separation at the anterior end, where the four flagella have already duplicated prior to separation. Encystment (6–9) occurs by sphering up of the cell, which becomes sedentary and secretes a cyst wall during a process of cell size reduction. Excystment (10–14) occurs by rupture of the cyst wall and emergence of the cell (13) already bearing developing flagella. Sexual reproduction (15–25) occurs by conjugation of compatible individuals which join by their anterior ends and fuse to form a single zygote with four flagella. The zygote enlarges substantially (22) and commences division, forming four longitudinal furrows that deepen gradually, commencing at the posterior end and eventually progressing anteriorly until four daughter cells are released (24). Only one, however, is flagellated. The others subsequently develop flagella. From Lewis et al (1974); with permission.

interphase growth and at the onset of mitosis. At telophase, the nucleus-to-cell-volume ratio was reduced to 2%. This was accompanied by external vesiculation, resulting in a reduction of membrane surface area and loss of neucleoplasm to the surrounding cytoplasm. Nuclear expansion during cytokinesis increased the volume ratio to 4.6% during early cytokinesis, and 6.5% at late cytokinesis. The leukoplast-to-cell-volume ratio (10 to 26%) varied markedly without any relation to the cell cycle. The chondriome exhibits marked changes in structure during the division cycle (Gaffal and Schneider 1980a,b). The number of mitochondria varies from one large reticulate chondriome surrounding the leukoplast at the onset of interphase to numerous (ca 250) smaller mitochondria at the onset of telophase of the first division cycle. The smaller, variously-shaped mitochondria are budded from the reticulate, basket-shaped chondriome. The smaller mitochondria, often localized in groups in the cytoplasm, are of three forms: spherical to ovoid, elongated and poorly branched, and highly reticulated. During the cell division cycle, the specific surface area (outer membrane area/100 μm^3 mitochondrial volume), and the surface-to-volume ratio changed rhythmically. Changes in mitochondrial surface-to-volume ratio (Sc/Vc) were apparently correlated with changes in the diameter (Dc) of the mitochondria. This is approximately represented by the function Sc/Vc = 4/Dc. After cell division, the single mitochondria basket is reorganized by fusion of the smaller mitochondria. During subsequent division cycles, similar periodical changes in the chondriome occur. The leukoplast in early interphase cells is a deep cup-shaped (cyathiform) structure, but undergoes minor changes during later periods of interphase. A thin and less perforated envelope encompassing up to 25% of the external surface, and several double-layered regions, occur sporadically at the tube-shaped part of the cup (anterior end). Alternatively, a bladder-shaped structure consisting of several very thin, concentrically ringed layers occurred at the base of the tube. During early prophase, in anticipation of cell division, the leukoplast becomes transformed into a massive bowl-shaped structure (poculiform). During cell division, the cleavage of the poculiform leukoplast occurred by furrowing which started at early prophase and was completed at the end of cytokinesis. The molecular basis for the rhythmicity and predictable sequence of events during the division cycle in these flagellates remains one of the intriguing questions confronting modern molecular biologists.

Chemical Messengers

Considerable progress has been made in understanding some of the chemical messengers triggering differentiation and development in the colonial flagellate *Volvox* (Starr 1969). Sexual reproduction in dioecious strains occurs by development of sperm packets in "male" colonies and oogonia in female "colonies." For example, in *Volvox aureus* male colonies lack gonidia, but the posterior two-thirds of cells in the colony function as male initials, each of which enlarges, divides, and undergoes rudimentary inversion to form a packet of 32 biflagellate sperm (Darden 1966). Penetration of the sperm into female colonies and fertilization produces zygotes which germinate to form germlings (spheres of cells oriented with their flagellar bases inward). The germling inverts, bringing the flagellar apparatus to the outside, thus forming a young, motile colony. Sexual differentiation of male colonies can be induced by a heat-stable substance secreted into the culture medium of sexually mature colonies. This heat-stable, nondialyzable inducer called "male inducing substance" (MIS) is a large molecular weight compound (> 200,000). Its inducing activity was destroyed by the enzymatic action of trypsin and Pronase, but was unaffected by chymotrypsin. In *Volvox carteri*, male colonies appear spontaneously in the male strain, but female colonies are formed in the female strains only in the presence of a substance produced by male colonies (Kochert 1968). The female-inducing substance is produced primarily by sexually mature male colonies, not by the vegetative stages of growth. The same substance that induces differentiation of female colonies can also induce sexual maturation in vegetative male colonies, as also was found with *V aureus* (Starr 1969). The substance, a glycoprotein, is effective at 6×10^{-17} M, and sufficiently potent that one sexually mature male releases enough inducer to convert the related males and females in a volume of 1000 liters from asexual to sexual reproduction (eg, Kirk and Kirk 1986). This inducing substance is heat labile, nondialyzable, and its activity is destroyed by Pronase, but not by trypsin, chymotripsin, or ribonuclease. Pronase sensitivity is consistent with the interpretation that the inducer is a glycoprotein. The molecular events accompanying induction are not fully determined, but biochemical evidence indicates that the initial binding site for the inducer is in the gelatinous matrix, not the surface of the induced cells (Giles et al 1984). This is in contrast to hormonal activation in higher organisms, where binding to the plasma membrane initiates intracellular changes. Apparently, the binding of the inducer to the matrix macromolecules in quantities as low as 3000 inducer molecules per spheroid triggers a change in composition of phosphoproteins within the matrix surrounding the reproductive cells (Giles and Jaenicke 1984). The signal for this change may be a shift in the matrix cyclic-AMP (cAMP) concentrations triggered by binding of the inducer. The cAMP, acting as a messenger in a signal cascade, could activate many protein kinases (enzymes phosphorylating proteins), thus altering the pattern of phosphoproteins in the matrix. The further influence of these phosphoproteins on the target cells (undifferentiated reproductive cells) is not known. It may involve changes in calcium ion level in the cell and activation of calmodulin. Eventually it is assumed, however, that the signal initiated by the inducing substance (MIS) is transmitted to the nucleus. By action on the chromosomes, this substance could alter the transcription of the genes and thus initiate sexual differentiation of the reproductive cells (Giles et al 1984).

Although it was once thought that the inducing substance was produced only during spermatogenesis and thus was a product of genetic control during sperm production, it is clear from current work by Kirk and Kirk (1986) that environmental factors can also trigger release of the substance. Brief heat shock treatment (42.5°C for 2 h) stimulates production of the inducing substance. Males, including a sterile mutant incapable of producing the substance under normal growth conditions, also produced the inducer in response to heat shock (Kirk and Kirk 1986). This phenomena is probably of significance in the natural life cycle of *Volvox*. During the spring, *Volvox* reproduces asexually in temporary ponds, but at the peak of summer it becomes sexual, and as a result of fertilization produces dormant, overwintering zygotes. These dessication-resistant resting spores, triggered by summer heat and produced before the temporary ponds evaporate, ensure continuity of the species.

Asexual Reproduction

Asexual reproduction in *Volvox* is also entrainable to external stimuli. Cultures become synchronized by a light/dark cycle. Proteins produced at hourly intervals were analyzed by Kirk and Kirk (1983) to determine changes correlated with asexual stages of development. At all stages, marked rapid and reversible changes in the pattern of protein composition occurred when the cells were transferred from light to dark or vice versa. But, this effect was most marked in juvenile spheroids at the end of the dark period accompanying their embryogenesis. The proteins synthesized at the beginning of one light period were not identical to those produced at the end of a preceding light period. These data indicate that there is a selective activation by light of protein synthesis, using mRNA transcripts accumulated during the dark period. This can account for the light-triggered synchronization of cultures entrained to light/dark cycles. An interesting quantitative discussion of factors regulating synchronization of cell division in cultures of protozoa with special emphasis on flagellates and ciliates has been presented by Scherbaum (1962).

Amoebae and Related Organisms

Environmental Influences on Life Cycles

Among asexually reproducing amoebae, environmental variables (eg, temperature and food abundance) influence generation times and reproductive rates. This has been clearly illustrated in *Amoeba proteus* (Rogerson 1980). Temperature has a marked effect on generation time when the amoeba is cultured with *Tetrahymena pyriformis* as food. The length of the cell cycle varied from 44 h at 20°C to 2926 h at 10°C. At three temperature levels of 10°C, 15°C, and 20°C, the generation time of *A proteus* followed a U-shaped curve as a function of prey density. That is, a minimum generation time occurs at intermediate to low prey densities (ca 500 to 2000 prey/500 µl) depending on the temperature. Less prey or much more abundant prey increased the generation times, thus generating the U-shaped curves. In general, the lower the temperature, the lower the optimal density of prey, and the lower the reproductive rate. A decrease in reproductive rate with decreasing temperature has also been reported for other amoebae including *Polychaos fasciculatum* (Baldock and Baker 1980), *Amoeba algonquinensis*, *Saccamoeba limax*, *Vannella* sp (Baldock and Berger 1984; Baldock et al 1980), and *Acanthamoeba* (Heal 1967).

Many free-living and parasitic amoebae include a cyst stage in their life cycles. This is usually correlated with adverse environmental conditions including a reduction in food supply, overcrowding, and/or desiccation. Encystation of the amoeboflagellate *Naegleria gruberi* occurs only if the cells are starving *and* crowded (Chiovetti and Bovee 1982). Experimental evidence suggests that receptor sites on the surface of amoebae respond to cell–cell contact and initiate intracellular reorganization leading to encystment. Starving amoebae did not encyst unless they are critically crowded. Resistance to dessication is particularly significant among terrestrial species subjected to periodic drying of the soil. Encystation enhances survival during adverse environmental conditions. Soil-dwelling species of *Acanthamoeba* have been widely cultivated in the laboratory, and the events accompanying encystment and excystment have been well investigated. During laboratory culture, logarithmic growth is followed by a population growth decline leading into stationary growth. If the culture is not replenished, the cells begin to encyst. Stationary phase of growth typically occurs when cells reach a density of 2 to 3.5×10^6/ml.

Morphometrics

Comparative light and electron microscopic studies of cyst fine structure in *Acanthamoeba castellanii* and *A palestinensis* (Bowers and Korn 1969; Lasman 1977) indicate major similarities in gross features. The cyst is enclosed by two distinct walls, an endocyst and an ectocyst. At places the two walls fuse to form a thin area known as an ostiole, which is closed by a lamina (operculum) that resembles neither the endocyst nor ectocyst in fine structure. The endocyst of *A palestinensis* is nearly rounded, with one to three ostioles scattered on the surface, while in *A castellanii* it is star-shaped with several truncated rays, and the ostioles are located at the edges of the rays. The operculum is thick, arched, and further away from the cell surface. In *A palestinensis*, the operculum lies much closer to the plasma membrane than to the endocyst wall. The exocyst has an irregular shape and in many places is separated from the endocyst by a wide space. This is filled in some regions with a spongy network or with cytoplasmic debris. The inner and outer walls, each 80 to 150 nm thick, are similar in granular texture and staining properties. The cytoplasm is dominated by a large, centrally located nucleus containing a nucleolus surrounded by more electron-dense, spheroidal, nucleolarlike bodies. Bulbous projections protrude from the nuclear envelope and apparently give rise to double-membrane enclosed vesicles in the perinuclear cytoplasm. The cytoplasm contains mitochondria, scattered endoplasmic reticulum, and granular hyaloplasm. As a comparative note, the cysts of *Naegleria* spp lack ostioles, but possess a pore filled with a mucoid plug (Schuster 1975). An electron-dense plaque serves as an additional pore closure. In *N gruberi*, the pore has a pronounced collar and the cyst wall is composed of an inner thick and an outer thin layer. However, only the inner component was present in the smooth-walled cysts of *N fowleri* and *N jadini*.

Induction of Encystment and Changes in Cellular Physiology

Factors promoting encystment in *Acanthamoeba* include glucose and acetate starvation in cultures grown in a chemically defined medium containing, among other nutrients, 21 amino acids (Byers et al 1980). Glucose starvation has also been shown to induce encystment in *Entamoeba invadens* when grown in Diamond's medium (Vazquezdelara-Cisneros and Arroyo-Begovich 1985). Inorganic ions, in conjunction with organics, influence encystment of *Acantamoeba culbertsoni* (Srivastava and Shukla 1983). Magnesium ions serve a key role, while sulfate ions, sodium chloride, and taurine had modulating effects on encystment brought about by Mg ions. The optimum medium for encystment contained an effective balance of Na^+, Mg^{2+}, Cl^-, and SO_4^{2-}, the latter preferably as taurine. An inorganic, nonnutrient medium containing Na_2SO_4 (86 mM) and $MgSO_4$ (15 mM) promoted good encystation of *A culbertsoni* and did not require taurine or other organics.

The physiology of *Acanthamoeba castellanii* and *Naegleria gruberi* during logarithmic growth and population growth deceleration leading to encystment is characterized by marked changes in cell size, macromolecular composition, and oxygen consumption (Weik and John 1977; Byers et al 1969). During transition from log phase growth to population growth deceleration in *Acanthamoeba* cultures, the average cell volume increased approximately 60% and total dry mass increased about 15 to 20% during the later period of deceleration. Although total cell protein remained constant throughout both growth phases, cytochrome oxidase (respiratory enzyme) doubled during growth deceleration. The population growth deceleration begins when oxygen becomes limiting. Hence, the increase in cytochrome oxidase may be an adaptive response to this oxygen deficiency. Increased metabolic activity during encystment is also consistent with the observed increase in cytochrome oxidase activity and the general increase in macromolecular reorganization, especially during early stages of encystment (Griffiths and Hughes 1969). DNA decreases approximately 50%, while RNA increases about 75%. Glycogen decreases 50% during the RNA build-up and then increases to a plateau above the log phase level. Subsequently, glycogen decreases with increasing numbers of cysts formed during late population growth decline. Acid hydrolase (lysosomal enzyme) activities are high in young cultures, but decline to constant low levels in postlog phase cells (Martin and Byers 1976). A differential change in enzyme activity occurs during encystment. For example, ribonuclease decreases most rapidly and acid phosphatase least rapidly. Acid phosphatase is unique, as it exhibits a transient increase in specific activity (activity per unit mass protein) even though its activity per cell is decreasing. Dramatic changes also occur in the quantity of microsomal intracellular membranes (Pauls and Thompson 1978). During transition of the culture into a stationary phase of growth, there is extensive proliferation of cytoplasmic membranes which are subsequently broken-down during cyst formation. The foregoing data suggest that the proliferation of membranes accompanied by an increase in specific lysosomal enzyme activity reflects major reorganization of the cell components toward production of macromolecules required for encystment. The cyst wall, which comprises 37% of the total dry weight of the cell, is rich in proteins (ca 33%) which must be synthesized from existing intracellular pools. The major reorganization of intracellular membranes during cyst formation (as exhibited by electron microscopy) correlates with the biochemical data, indicating a large increase in microsomal membranes prior to encystment. Other factors indicating major changes in intracellular pools of metabolically significant molecules include a net decrease in total adenosine phosphate to about 85%, largely produced by a decrease in the amount of ATP (Jantzen 1974). Intracellular cAMP, however, rises by a factor of two to three from the end of the logarithmic phase to the beginning of the stationary phase. This increase is apparently due to increased activity of the enzyme producing cAMP, and not due to a decrease in its degradation (Gessat and Jantzen 1974). Metabolic studies of *Acanthamoeba castellanii* grown in synchronous culture show that intracellular pools of ATP, ADP, and AMP varied in an oscillating pattern during the cell cycle (7 to 8 h). The fluctuation in concentration of each of these adenylates can be expressed quantitatively by taking the ratio of this maximum value to the minimum value in the cell cycle. This ratio expressed as a percent for ATP, ADP, and AMP was 108%, 194%, and 520%, respectively (Edwards and Lloyd 1978). Adenylate charge values varied between 0.63 and 0.88. The adenylate charge value is a measure of the cell's energy reserve. It is the proportion of ATP or its equivalent (1/2 ADP) to the total adenylate phosphates. Adenylate charge = (ATP + 1/2 ADP)/(ATP + ADP + AMP). Oxygen consumption rates doubled overall during one cell cycle, but rose to seven distinct maxima during this period (Edwards and Lloyd 1978). Respiratory maxima were in phase with maximum ADP levels, but out of phase with maximum values of the ATP/ADP ratio. These data indicate that the overall respiration rates during the cell cycle of the amoeba are regulated by in vivo respiratory controls. When ADP increases, signaling a decline in available ATP, theoretically we expect an increase in aerobic metabolism. This would increase ATP production from ADP by oxidative phosphorylation and bring the cell closer to a favorable energy balance.

Induction of Excystment

Exogenous factors regulating cell cycle and excystment are of particular interest from a physiological ecological perspective, since resumption of active growth of amoebae is significant in accounting for the total productivity and energy flow in microbial communities. Aqueous extracts from soil-dwelling fungi and actively proliferating bacteria (eg, *E coli*) produce good excystment of the soil amoebae *Schizopyrenus russelli* and *Hartmannella rhysodes* (eg, Datta and Kaur 1978). Part of the excystment activity can be attributed to amino acids. In laboratory experiments, pure amino acids and amino acid mixtures can also stimulate excystment of free-living amoebae. During active growth of *Acanthamoeba*, large molecular weight metabolites are also released into the environment (Pigon 1976). Hydrolysis of these compounds yields phosphate, neutral and amino sugars, amino acids, amines, and a small proportion of lipid-soluble compounds. During each cell cycle, the *Acanthamoeba* releases as conditioning material 12 to 37% of its weight. It is not known to what extent this material improves the quality of tne environment for the amoeba and/or acts as positive feedback stimuli to maintain active growth and suppress encystment. It is of interest to note that emetine hydrochloride in fungal extracts was toxic to the cysts of soil amoebae at pH 8.5 (100% mortality), but had no effect at pH 6.2. In the latter case, removal of the emetine in the presence of *E coli* extract produces a fairly large number of excysted amoebae. These data suggest that under conditions where bacterial activity is likely to be high (reduced pH and release of bacterial metabolites), and when fungi are not competing heavily for resources (low emetine concentrations), the soil-dwelling amoebae are induced to excyst. Thus, the survival of the amoeba may be enhanced by this rather interesting set of interacting control factors.

Myxamoebae (Slime Molds)

Although illustrative examples have been taken from free-living, small amoebae, considerable research has been done on the life cycle and biochemistry of plasmodial-forming species of the so-called "slime molds" or myxomycetes. A generalized diagram of the life cycle of a myxamoeba is illustrated (Figure 20.2). A commonly cultured myxomycete is *Physarum polycephalum*. This organism forms large yellow plasmodia, and is easily grown in the laboratory in a Petri dish lined with filter paper, moistened with distilled water or pond water, and sprinkled with rolled oat flakes. No preservatives should be present in the oats. The culture should be kept in the dark to prevent sporulation. The plasmodial phase forms desiccation-resistant resting stages known as sclerotia. These crustose

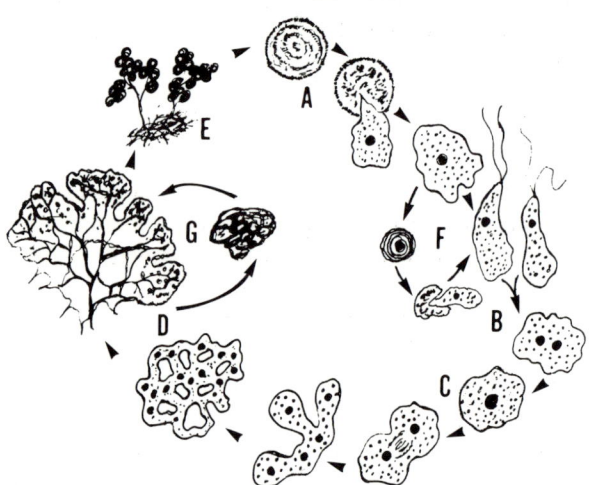

FIGURE 20.2. Life cycle diagram for the slime mold *Physarum polycephalum*. A spore (A) produces a small amoeboid stage that can either form swarmers (B) or enter several cycles of encystation (F). The swarmers fuse to form a zygote (C). Repeated nuclear division and enlargement of the cell produce a multinucleated plasmodium (D) that forms a network of veins in the older parts and an amoeboid advancing edge at the front of the migrating cell mass. Under favorable growth conditions, the plasmodium can form a thickened drought-resistant dried mass known as a sclerotium (G) or produce sporangia (E) containing haploid spores (A) that are dispersed by water or wind, thus completing the life cycle.

masses resume growth under favorable moisture and food conditions. Plasmodia can also become transformed into sporangia. The released spores germinate into swarmers that fuse to form a zygote. The zygote differentiates into a trophozoic plasmodium. Meiosis in myxomycetes occurs just prior to spore cleavage or during spore germination. There is no obligate alternation between asexual and sexual phases of the reproductive cycle. Plasmodial fragmentation and several cycles of sclerotial development and reactivation can occur without an intervening sexual phase.

Foraminifera

As mentioned previously, some benthic Foraminifera regularly exhibit alternation of asexual and sexual phases in their life cycles. An exception, however, is *Allogromia laticollaris* (McEnery and Lee 1976), which has an apogamic life cycle (lacking gamete exchange). During the agamogony phase, the monothalamic asexual cell can undergo repeated alternations between two agamont phases (G1 and G2) labeled I and II (Figure 20.3). In some cases, the agamonts merge into a large plasmodium that fragments and gives rise once again to individual globose agamonts. During the agamogyny cycle, RNA synthesis occurs midway in the G1 phase, followed by nuclear proliferation. This produces a multinucleated agamont which by cell division produces mononucleated progeny agamont II (G2). The G2 cells are characterized by an ameba-form nucleus, shown as a lobed structure. The nucleus undergoes repeated division, and the cell fragments into daughter cells, initiating once again the agamont I phase of the life cycle. Rare instances of gamogony involve the formation of a giant gametocytotomont containing numerous nuclei (B). These form internal gametes which fuse (autogamy) and produce zygotes that emerge from the mother cell to form a G1 agamont cell. The stages giving rise to the giant gamete-producing cell have not been determined.

The life cycle of planktonic Foraminifera remains an enigma, although evidence points toward fusion of gametes to form a zygote that gives rise to a single-chambered stage known as the proloculus. The development of the shell following the proloculus stage has been carefully elucidated by Brummer et al (1986). Five ontogenetic stages are identified in spinose globigerined species as exemplified by *Globigerinoides ruber* (Figure 20.4). The single-chambered proloculus gives rise to a two-chambered deuteroconch which initiates logarithmic growth. The subsequent juvenile stages (Figures 20.4a,b) are characterized by pores along the chamber sutures, sparse, flexible spines, and a marginal aperture (aperture opening on the rim of the shell). The neanic (postjuvenile) stage is characterized by marked changes in shell morphology including more inflated chambers, and larger chamber volume at each increment of growth. Spines become more stout, and the aperture in these later chambers begins to reorient away from the margin of the shell (Figures 20.4c,d). Examination of food vacuoles indicates that the preneanic stages are largely microherbivorous, while later stages become increasingly omnivorous. The stouter and more numerous spines

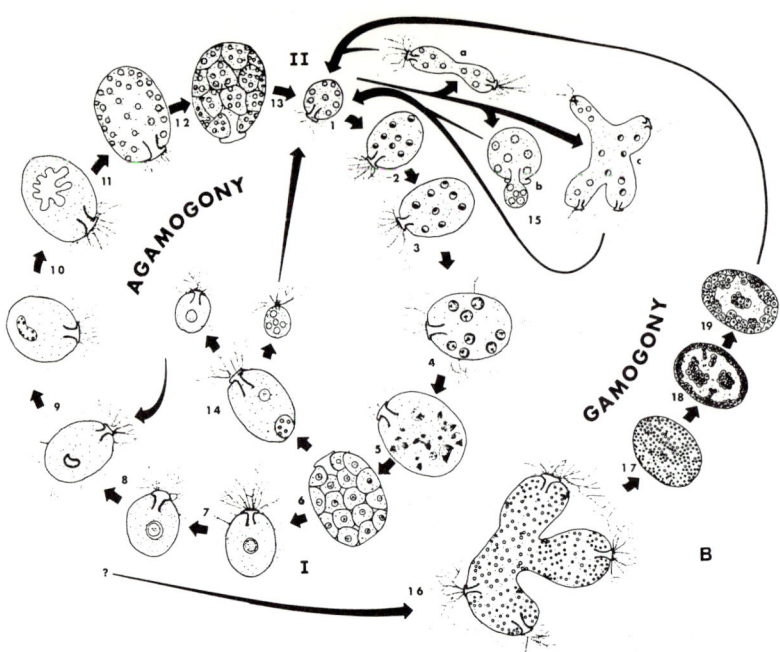

FIGURE 20.3. Life cycle of *Allogromia laticollaris*. (1–13) Agamont phase: (1) juvenile agamont in the agamont stage (II) during growth phase G_1; (2) young agamont during mid growth phase of G_1 (RNA synthesis also occurs in this phase); (3) growing agamont in late G_1 phase of the cell growth cycle; (4) mature agamont–chromosomes in "mushroom-like configuration," RNA accumulated at the periphery; (5) karyokinesis; (6) cytokinesis (schizogony); (7) young agamont in phase (I) of the cycle during S phase of the cell growth cycle (DNA is being synthesized); (9–13) maturation of the agamont (I) stages showing "amoeba-form" nucleus (10), leading to schizogony and production of juvenile agamont (II) individuals. (14) A relatively uncommon life cycle alternate pathway during the agamont (I) phase which includes (a) binary fission, (b) budding, and (c) cytotomy. (16–19) Gamont phase. A giant gametocytotomont (16) gives rise to a multinucleate gamont prior to the formation of gametes. (18) Gamont filled with gametes; (19) gamont with some gametes and zygotes that eventually give rise to stage (1) in phase (II) of agamogony. From McEnery and Lee (1976); with permission.

▶

FIGURE 20.4. Developmental stages of the spinose planktonic foraminifer *Globigerinoides ruber*. (a–b) Juvenile stage (a) spiral view and (b) umbilical view. In this stage, consisting of only a few chambers, the pores are largely restricted to the sutures and there are few spines. The oral aperture opens at the margin. (c–d) Neanic stage (a) spiral view and (b) umbilical view. Chamber number has increased and the size of each new chamber added is greater than in previous stages. Spines begin to develop over the entire shell and the oral aperture is oriented toward the umbilical side. (e) Mature stage showing the porous shell wall with reticulate pattern and large oral aperture oriented with the opening spanning two smaller chambers. Scales = 10 μm. Courtesy of Dr. G. Brummer, from research by Brummer et al (1986).

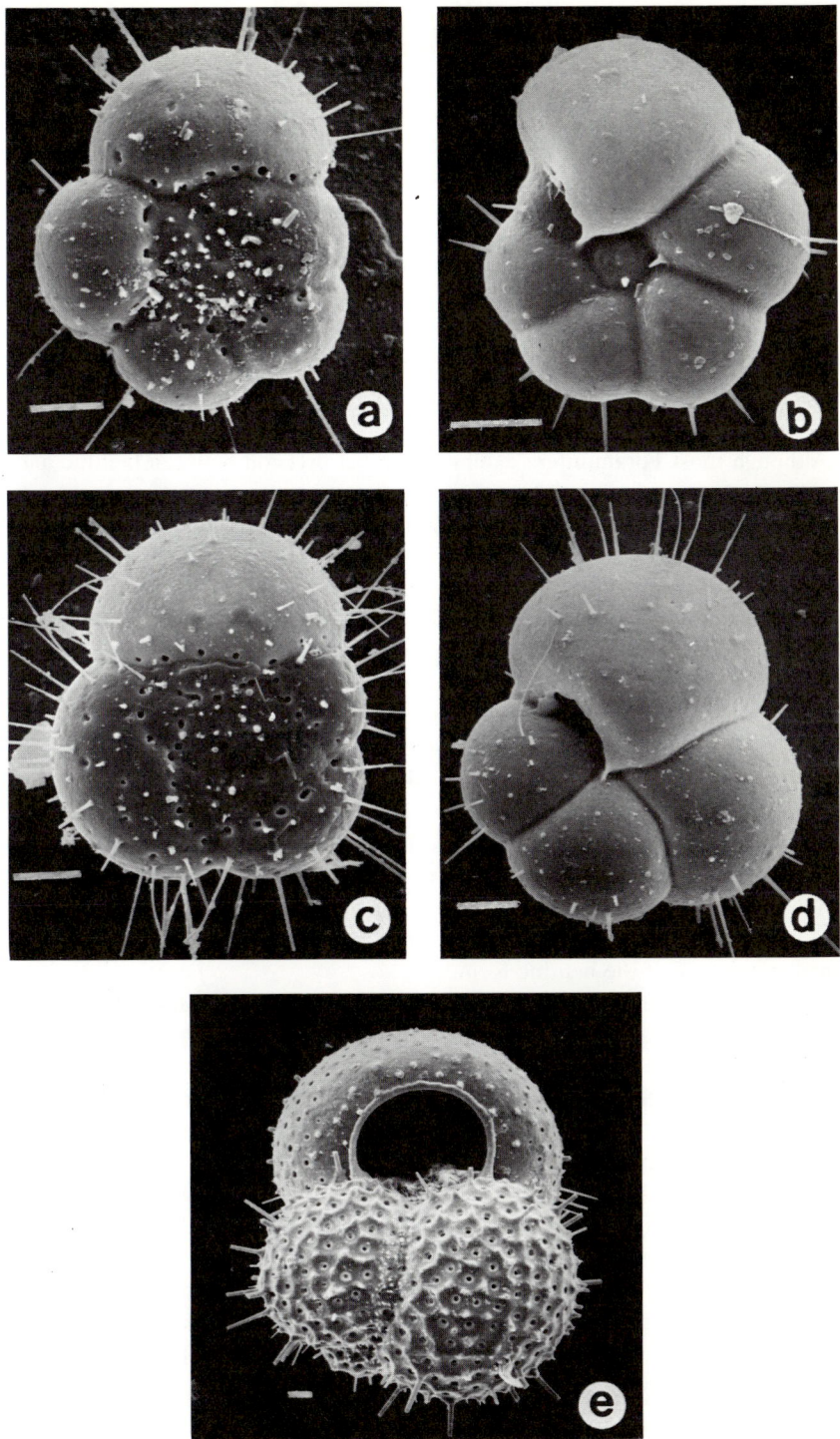

during the neanic stage may potentiate larger prey capture by providing greater mechanical advantage during prey capture. In later stages of the neanic phase, there is a gradual transition toward an adult form. The adult stage is characterized by a variety of morphological features such as development of secondary apertures on the spiral side of the test. The surface of the shell exhibits secondary maturational features such as ridges, thicker spines supported by enlarged spine bases, and pores depressed in pore pits. In the final stages of maturation, preceding release of gametes, additional features may be added including additional thickening of the shell, formation of chambers with aberrant morphology, and in the very last stages, shedding of spines. At reproductive maturity, the cytoplasm is at first fully withdrawn and then protrudes as a bulge, giving rise to myriads of flagellated gametes. These swim vigorously away from the parent cell and presumably fuse to form a zygote, thus initiating a second generation in the reproductive cycle.

Although most Foraminifera exhibit a distinct division between benthic and planktonic modes of existence, some largely benthic species become planktonic during sexual maturity. This involves a dramatic change in the morphology of the shell, as illustrated by *Cymbaloporetta bulloides* (Rückert-Hilbig 1983) and some species of *Rosalina* (Sliter 1965). It is interesting to note that *Rosalina*, maintained in laboratory culture as with *Allogromia laticollaris*, exhibits successive generations of gamonts (apogamic reproduction). During morphogenesis of *C bulloides*, the initially rather flattened immature shell is enlarged by addition to peripheral chambers and at maturity is completed by addition of a spheroidal, floatlike final chamber enclosed by an outer wall bearing numerous pores on the apical surface (Figure 20.5). The internal closed chamber becomes filled with gas (probably carbon dioxide from respiration) and the Foraminifer floats upward in the water column. Gametes are released near the surface. Following syngamy, the zygote either settles directly in the sediments or attaches to surrounding vegetation and forms an amoeboid juvenile stage that crawls toward the sediment. The initial proloculus stage matures by addition of chambers forming the flattened spiral shell of the benthic form.

Environmental factors appear to regulate the onset of sexual reproductive maturity and secretion of the float chamber in many of these species, including *Rosalina globularis* (Sliter 1965). Experiments with laboratory cultures show that induction of the gamontic generation (planktonic stage) occurs when temperature exceeds 18°C. Variations in light intensity had no influence on reproduction, whereas changing the substrate or amount of food influenced motility. Additional evidence indicated that salinity may have a modulating effect on temperature-induced sexual maturation. The stimulation of sexual reproduction during elevated temperatures may have survival value, since offspring produced during warmer periods when primary production may be higher will be better nourished and more likely to mature. Little is known about the molecular biological events accompanying these remarkable physiological and morphological transformations.

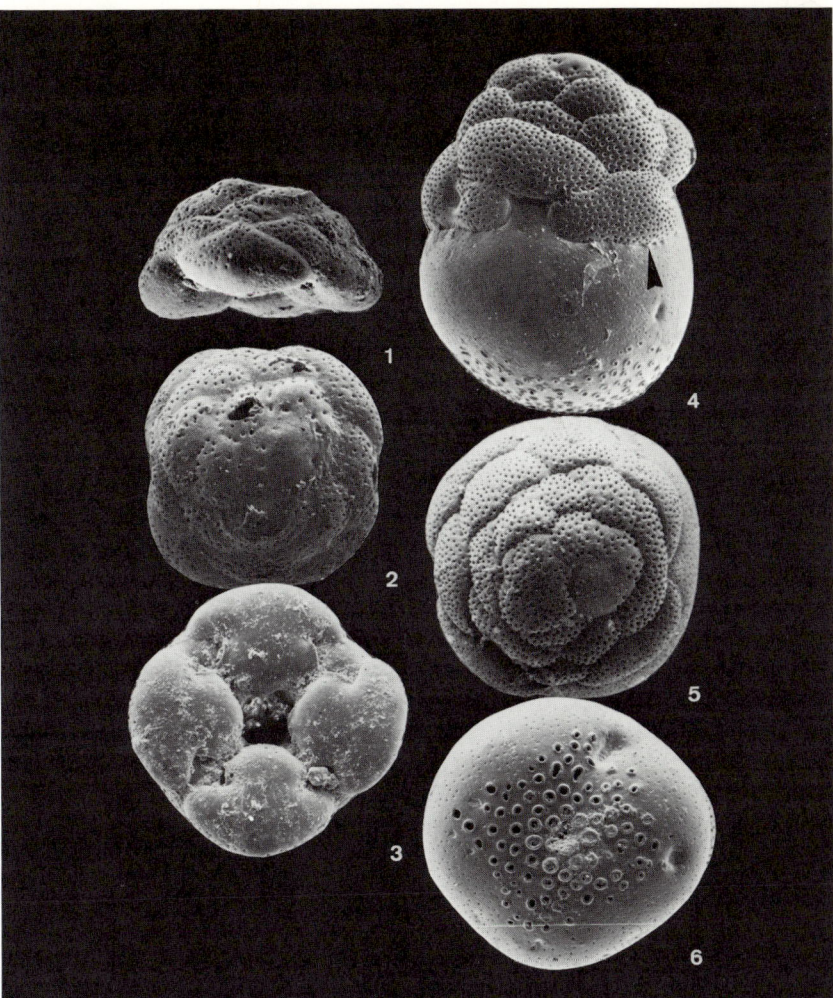

FIGURE 20.5. Growth stages of the benthic foraminifer *Cymbaloporetta bulloides*. Early growth produces a flat spira shell (1) shown in dorsal (2) and ventral (3) views. At the time of gamete production, the organism produces a float chamber (4) which is constructed on the ventral surface, forming a distinct suture (arrow) at the point of attachment. Dorsal (5) and ventral (6) views of the float chamber stage. The float chamber contains gas that provides buoyancy and carries the gamete-producing stage into the surface water where the flagellated gametes are released. From Rückert-Hilbig (1973); with permission.

Ciliates

Among many of the free-living ciliates, conjugation occurs sporadically and is interposed within longer sequences of asexual reproduction. Inception of sexual reproduction may occur due to intrinsic physiological states of the organism, or when food is depleted after longer periods of ample nourishment. Under unfavorable conditions, some ciliates (especially soil-dwelling species) are likely to encyst. Some representative data are presented in the following categories: (1) physiological correlates with life cycle stages, (2) major events during asexual reproduction and cellular regeneration, and (3) some functional aspects of encystment and excystment.

Physiological Correlates of Life Cycles

Quantitative studies of biochemical correlates of growth phases during the life cycle of ciliates requires sufficient cells at each phase of growth to permit chemical analyses. This is achieved by synchronizing the growth of the cultures so that all of the cells divide nearly simultaneously. There are several methods for achieving synchronized cultures. Logarithmically growing cultures can be exposed to a sequence of alternations of light and dark, or subjected to oscillations in temperature at 30 min intervals. For example, alternations in temperature between 34°C and 29°C over several hours induce synchrony in *T pyriformis* (Zeuthen and Rasmussen 1972). The higher temperature arrests differentiation, and the lower temperature (of short duration) permits continued development of the cell, but not sufficiently far to produce insensitivity to the next heat shock. Consequently, the cells become arrested in development with all or most of them held in anaphase. When the culture is returned to a continuous temperature of 29°C, all of the cells begin to divide synchronously for several generations. A similar effect is produced by alternating light and dark periods. This is effective since the division of some ciliates (eg, *Chilodonella*) is light regulated, and the alternating short interval light and dark cycles arrest cytokinesis. When the cultures are placed in continuous light, the cells begin division nearly simultaneously and become growth synchronized. Alternatively, the cultures can be passed through a continuous centrifuge system and cells of a given size selected from the effluent. These cells are typically in the same stage of development and divide synchronously when placed in culture medium.

Cell Cycle Physiology

In synchronous cultures of *Tetrahymena pyriformis*, oxygen uptake increased to maxima three times in each cell cycle and the ATP pool oscillated in phase with respiratory activity (Lloyd et al 1978). There are four major phases to the asexual reproductive cycle in most protozoa, as shown for two hypotrichs (Figure 20.6a,b) and *T pyriformis* (Figure 20.6c). These are: (1) cell division (*D*);

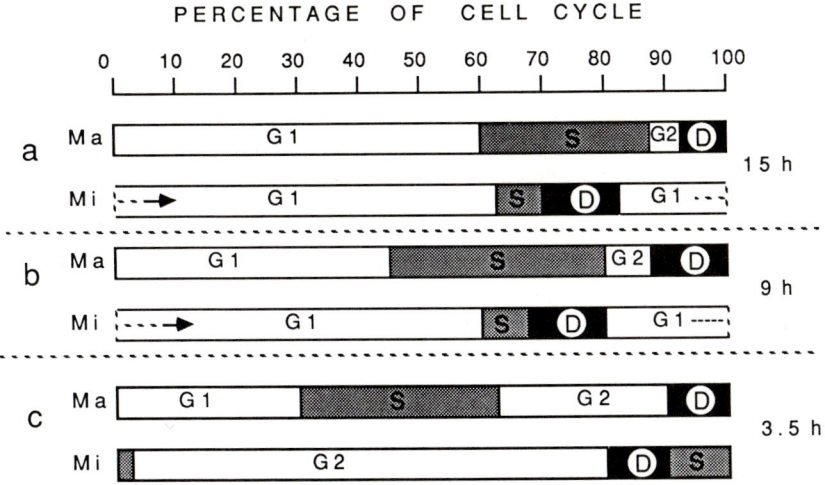

FIGURE 20.6. Representative diagrams of life cycles among ciliates. (a–b) Hypotrichs: (a) *Diophrys scutum* and (b) *Oxytricha bifaria*. (c) Comparative data for *Tetrahymena pyriformis* illustrate differences between these species of ciliates. The cell growth cycle includes (G1) growth following division (D) (usually with substantial increase in protein production); DNA synthesis (S) shown stippled; growth phase (G2) is shown producing additional increase in protein and cell growth. The events for each species are diagrammed for the macronucleus (Ma) and micronucleus (Mi) as a percentage of the total cell cycle. The time for each cell cycle is shown at the right of each species diagram. In some cases, the phase of growth begun during one cycle is continued into the next cycle, as shown by dashed lines and an arrow or stippling. Derived from Dini et al (1975) and Flickinger (1965).

(2) a period of growth and cellular protein synthesis called the first growth phase (*G1*); (3) a period of DNA synthesis in preparation for the next mitotic division labelled (*S*); and (4) a second period of cellular growth (*G2*) preliminary to mitosis and cellular division. The timing and duration of DNA synthesis in the macronucleus and micronucleus usually differ, as shown in Figure 20.6. In *T pyriformis*, macronuclear DNA synthesis occurs in the second hour of the cycle, while micronuclear DNA synthesis occurs immediately after karyokinesis in the few minutes before and after the cell divides. The macronucleus begins swelling in preparation for division at about 70% into the cycle, and division commences at about 80% into the cycle. Micronuclear division occurs at a point about 75% into the cycle. Sensitivity to temperature inhibition of mitosis increases slowly during the cell cycle and reaches a peak at about 80% into the cycle, when the macronucleus and micronucleus are poised for division. Life cycles for the hypotrichs *Diophrys scutum* and *Oxytricha bifaria* are also shown in Figure 20.6a,b (Dini et al 1975). In *Paramecium bursaria*, there is no distinct G1 phase. Macromolecular DNA synthesis (*S* phase) lasts from the first to the 16th h after cell division, and in the micronucleus from the 1st to the 8th h (Just 1973). The total cycle lasts 23 h. RNA

synthesis by the macronucleus rises gradually after cell division, reaching an intermediate broad peak at about the fifth h, and declines to a shallow level until approximately the 19th h, when it rises sharply to a maximum. Protein synthesis reaches an intermediate peak at about the eighth h following the first peak in RNA synthesis, and undoubtedly represents the reponse of the cell to mRNA produced in the macronucleus. It declines slightly after the 10th h and increases slowly, reaching a broad inclined plateau at the 14th h.

The composition of the culture medium and temperature influence the duration of the growth cycle and its substituent phases. For example, at 29°C, *T pyriformis* grown in proteose-peptone medium enriched with liver extract divides every 163 min, and the S phase commences early, at about 20% of the cell cycle. When grown in a chemically defined medium without complex cellular extracts or their hydrolyzates, the length of the cell cycle is increased to 870 min, and the S phase is delayed until 50% into the cell cycle and is nearly doubled in its duration (eg, Elliott 1973). By varying the temperature and chemical composition of the medium, optimal conditions yielding the fastest generation time have been determined for various strains of *T pyriformis*. In the case of strain HSM, optimum growth is at 32.5°C, at pH 7.27 adjusted with NaOH, on a medium relatively free of extra ionized substances (eg, NaCl), and with 1 to 2% proteose peptone enriched with 0.1 to 0.25% liver extract. Under these conditions, the cells divide every 1.83 h (Elliott 1973). The generation time of *Blepharisma* is minimal when grown in an enriched medium containing cerophyl (lyophilized grain leaves) and bacteria as particulate food, with ionic concentration of ca 5 × 10^{-2} osmolar, and a temperature of 30°C (Giese 1973). Generation time for *B japonicum* as a function of temperature varies from 12 h at 30°C to 96 h at 13°C. In laboratory culture, *Blepharisma* tolerates a wide pH range, varying from 5.0 to 8.0. In general, regeneration of cortical structures after transection of the cell is optimal in the same range of conditions as cell division.

Cell Division and Regeneration

The events accompanying cell division and regeneration of cell structures after experimental dissection have been carefully studied in several ciliates including *Tetrahymena*, *Blepharisma*, and *Stentor*. The complex cortical and intracellular structures of many ciliates pose an intriguing question of how the cell reproduces these elaborate organelles during binary fission. Among the several prominent organelles that must be duplicated are the nuclei, contractile vacuole(s), subcortical structures, kineties, oral apparatus, and associated membranelles. The fate of the macronuclei and micronuclei has been carefully documented in many species, and the events during binary fission have been illustrated in Figure 19.1. Attention will be given here to the coordination of cytoplasmic and cortical events during binary fission and regeneration.

Blepharisma

The phases of binary fission in *Blepharisma* are presented and discussed in relation to insights gained from regeneration experiments. More detailed descriptions can be found in Giese (1973). Binary fission occurs every 15 to 20 h at 25°C, and the minimum time required to complete each cell division averages 2.5 h at 20°C. A dividing cell can be recognized by its relatively enlarged and protruding posterior half. Also, the kineties at the periphery of the V-area immediately beneath the oral apparatus are more branched than during interphase. The V-area is a deltoid patch of kineties on the ventral surface of *Blepharisma* posterior to the oral aperture. This region is significant in morphogenesis and cellular regeneration as subsequently explained. The process of cell division and stomatogenesis (origin of the oral apparatus in the daughter cell) is described in four stages. Stage I spans the time from appearance of the adoral anlage (patch of kineties near the posterior edge of the peristome) to macronuclear contraction in preparation for its division. This stage is initiated by the appearance of a crescent-shaped region immediately below the rim of the peristome, where the adoral band terminates. This patch (adoral anlage) will serve as the initiator of a new adoral membranelle in the anterior daughter cell. During this stage, kinetosomes multiply rapidly in the V-area. The proliferating kinetosomes bear minute cilia and are arranged in parallel rows, each consisting of approximately 10 basal granules connected laterally by a slender fibril (Figure 19.1, L and M). The kinetosomes located at the right side, connected by a longitudinal fiber, are somewhat larger and will give rise to the undulating membrane of the posterior daughter cell. The transverse kinetosomes in parallel arrays will form the cilia of the peristomal membranelles. Stage II, commencing with macronuclear contraction and ending with its condensation into a compact mass, is a time of major changes in the V-area. The incipient peristomal area enlarges and the floor of the developing peristome becomes twice the width of the preceding stage. The kinetosomes in the left-hand part of the transverse rows move further to the left, leaving nonciliated, slender fibrils between themselves and the massive, longitudinally aligned granules on the right-hand side. A new contractile vacuole appears on the ventral side of the cell near the anterior end of the developing adoral anlage. This will be the contractile vacuole of the proter (anterior daughter cell). The opsithe (posterior daughter cell) retains the old contractile vacuole, which is located in the posterior half of the parent cell. During Stage III preceding division, when the macronucleus reelongates into a thin strand, the developing peristomal region in the opisthe forms a distinct longitudinal oral groove, and the kinetosomes that had migrated to the left-hand part of the field begin to form the presumptive oral membranelles. Interestingly, the peristomal field and undulating membrane in the anterior half of the cell begin to dedifferentiate. They are resorbed, and a new peristomal apparatus is regenerated from the adoral analge! At this point, the anterior anlage begins to elaborate into a new peristomal field, including development of a new undulating membrane. Consequently, at this point in morphogenesis, both the anterior and posterior peristomal fields are approximately in the same stage of development. Both have a new membranellar band composed of short cilia, but no visible undulating membrane. A visible fission line (constriction ring) begins to form toward the end of stage III. Stage IV extends from the elongation of the macronucleus to completion of cell division. Early in this stage, a zone encircling the cell immediately above the fission line, called a "growing zone," begins to elongate. This produces the posterior region of the proter. The constriction ring becomes more pronounced, clearly demarking the plane of eventual separation of the two cells. The completion of the anterior (reorganizing) and posterior (developing) oral apparatus is synchronous. The new gullets are invaginated and fully formed, their membranellar bands form new terminal spirals at the posterior edge, and a new undulating membrane becomes visible shortly before completion of cell division. As cytoplasmic constriction progresses, the new contractile vacuole migrates toward the constriction plane, eventually situated at the posterior end of the presumptive anterior daughter cell (proter). This sequence of events is clearly visible in vigorous cultures of *Blepharisma*, where dividing individuals can be detected by periodic inspection. The precision and complexity of these events accompanying binary fission and morphogenesis are truly remarkable and aesthetically awe inspiring, given the seemingly "primitive" quality of these unicells. The scientific elucidation of the molecular basis for these celluar events is one of the most intriguing issues that challenges protozoan physiologists.

OTHER CILIATES

In many ciliates, the major events during stomatogenesis are similar. The stomal region of the opisthe is generated by an organizing field of kinetosomes posterior to the parent cell oral region. Maturation of the oral region typically precedes cell division. Variations in detail occur among species. For example, in *Tetrahymena pyriformis*, a loose field of stomatogenic kinetosomes forms posterior to the buccal cavity of the parent cell (Fig. 19.1). Initially, there is no well-defined V-area. Electron microscopic evidence shows that this patch develops by elaboration of kinetosomes between existing cilia. Microtubules orient perpendicular to the pellicle surface and differentiate into basal granules. This initially elongate region expands into an oval field of kinetosomes that differentiate on one side to form the undulating membrane. Peristomal membranelles differentiate from lateral parallel arrays of kinetosomes that become ciliated before invagination of the buccal cavity and development of the deep fiber. In *Dileptus*, bearing a long proboscis, binary fission entails reconstruction of the complex geometry of the oral region and production of the extended proboscis during separation of the opisthe from the proter (Golinska and Doroszewski 1964). Signs of division first appear as a ridgelike protrusion at the site of a ciliary tuft on the midsection of the dorsal side of the cell (opposite the side with the oral aperture). This ridge expands laterally and circumferentially around the cell body. It differentiates on the left side to become the primordium of the cytopharyngeal complex. This appears as a circular depression on the left side of the cell. Meanwhile, the dorsal lip of the ridge begins to elongate, forming the incipient proboscis, so that the organism, in addition to possessing the proboscis of the proter, appears to have an additional one emerging from its middorsal side. In the last stages of division, the proter and opisthe are connected only by a very thin strand of cytoplasm emerging from the left side of the base of the proboscis. If the proboscis of a mature cell is severed, regeneration occurs by resorption of kineties on the dorsal side of the lip followed by emergence of a new proboscis. Cilia are provided by differentiation and proliferation of cortical kinetosomes at the base of the proboscis, which elongates and is reconstituted while the feeding apparatus in the cytopharyngeal pouch is fully active (Golinska and Kink 1976).

Microdissection Studies

Experimental studies using microdissection of interphase cells, or those in various stages of division, have contributed to our understanding of the role of cortical structures in cellular morphogenesis. Some of the historical aspects have been reviewed by Tartar (1967), and further specialized treatises can be found in Elliott (1973) for *Tetrahymena*, Giese (1973) for *Blepharisma*, and Nanney (1980) for *Stentor* and other ciliates. Some generalizations about cortical organizing fields and the role of the macronucleus in various stages of cellular division and regeneration as reported by Giese (1973) for *Blepharisma* are summarized here. The role of the V-area in stomatogenesis has been well-documented; however, the kineties adjacent to the V-area acquire the morphogenetic potential of inducing a new peristome if the entire V-area is removed along with the peristome. The presence of a fully intact oral apparatus inhibits the induction of a peristome by the V-area. Removal of the entire peristome, its anterior three-quarters, or the corresponding anterior part of the cell will remove this inhibition. A new peristome is then produced by the V-area. Simply reducing cellular volume by ablation of other portions of the cell do not have this effect. The oral region and the V-area have a distinct polarity. The oral apparatus is formed as near the anterior end of the cell as is possible, while the main organelles (undulating membrane, adoral spiral or membranellar bands, and gullet) are always formed at the posterior end of the peristome. The posterior half of the cell, especially the anal area, also has an organizing polarity during morphogenesis. A gradient is present, such that induction of a new cytopyge after ablation is more rapid the further posteriorly a cell is transected to remove the cytopyge. The anal area also influences development of the oral region, and certain combinations of dissected pieces joined together in the cell can induce a bipolar peristome in the regenerant. The growing zone immediately above the presumptive fission line is significant in cytokinesis. If the ventral one-third to two-thirds is excised at an early stage of cell division, separation of the cells is not completed even if the presumptive division line is intact. At an earlier stage, however, division is completed even if the fission line is also removed. After the midphase of cell division, the fission line is fully determined, and removal blocks cell division since the potential for regeneration has passed. The macronucleus is essential to development of the peristome. Micronucleate fragments of the cell lacking a macronucleus fail to develop a new peristome. Interestingly, such fragments may develop a new peristome if the macronucleus is removed after the peristomal anlage has fully differentiated. Apparently, gene transcription and sufficient mRNA information have been transmitted to the cytoplasm to permit partial completion of stomatogenesis. However, the peristome is never fully functional and later dedifferentiates. Hence, the macronucleus is essential to coordination of this aspect of cell function. The micronuclei alone are incapable of fulfilling its role. The macronucleus is essential to full development of the V-area and culmination of stomatogenesis during binary fission. Removal of the macronucleus early in division prevents induction of the V-area. Although removal at a later point permits differentiation of the V-area into presumptive peristomal elements, the peristome is never fully functional. These data indicate a progressive role of the macronucleus in sustaining completion of stomatogenesis. This activity, combined with the organizing effects of gradients along the cell cortex, helps to explain the orderly sequence of events accompanying cell development during binary fission. Similar morphogenetic reorganizing events occur during encystment and excystment in cyst-forming species of ciliates.

Encystment

Survival of many ciliates in unpredictable environments clearly depends on their ability to encyst. The adaptive advantages of encystment in relation to other methods of prolonging survival during adverse environmental circumstances have been critically assessed by Jackson and Berger (1985). Among the encysting species they examined, three groups were identified. There are slow growing ones with either short survival times (eg, *Paraurostyla* sp or *Condylostoma vorticella*) or long survival times (eg, *Tokophrya lemnarum*), and those with fast growth and short survival times (eg, *Didinium nasutum*). While cyst formation is advantageous for survival, it also interrupts the normal growth cycle and thus potentially can reduce the competitive advantage the organism has over individuals that may adopt other survival strategies which prolong the growth cycle.

The events accompanying encystment and excystment have been determined in a variety of protozoa, but only some representative examples can be given here. The structure of the cyst and mode of excystment of *Nassula ornata* is shown in Figure 20.7 (Beers 1966). The resting cyst, induced by lack of food, has two wall layers: (1) a faceted, lamellate ectocyst that is colorless and somewhat sticky externally, and (2) a thin, inconspicuous endocyst. The kinetodesmal rows and cytopharynx are retained within the test, and their topographical relationships are unchanged. Hence, the typical polarity of the organism is retained. Excystment was induced effectively by a 0.1% (w/v) aqueous solution of peptone, preinoculated with wild bacteria. Distilled water or plant infusions were relatively ineffective as excystment-inducing agents. Excystment occurs within a 2 to 3.5 h period at 21°C. During excystment, the cytoplasm exhibits energetic cyclosis and the contractile vacuole enlarges (Figure 20.7, 2 through 4). The pressure of the vacuole ruptures the cyst wall. The ectocyst tears first, always at the posterior end of the cyst, and the endocyst ruptures later. Emergence is by cytoplasmic streaming and protrusion of the posterior part of the ciliate through

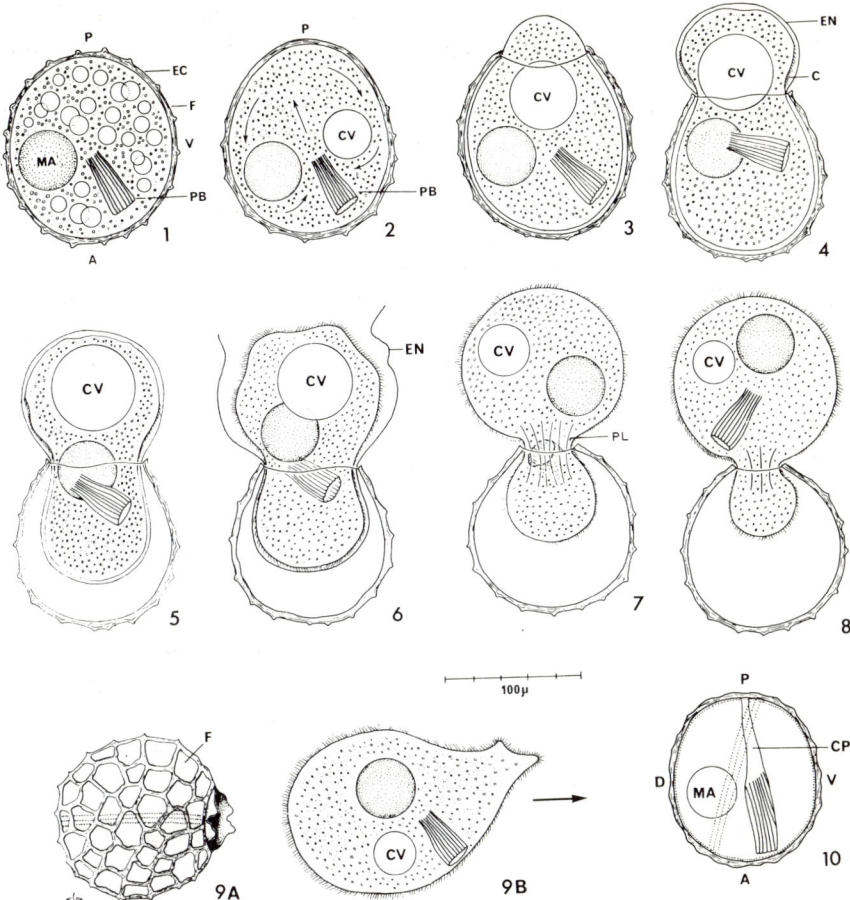

FIGURE 20.7. Excystment in the ciliate *Nassula ornata*. Successive stages of excystment observed in a single cyst at 21 °C. (1) Resting cyst introduced into an excystment medium at 8:00 AM; (2) energetic cyclosis commences by 10:00 AM and rupture of the ectocyst (3) begins at 10:25 AM. (4–5) Emergence by cytoplasmic streaming, leading to rupture of the endocyst (6) at 10:32 AM. (7–8) Emergence of the ciliate, posterior part first, leading to an empty remaining cyst (9A) at 10:40 AM and to the free ciliate (9B). (10) Resting cyst showing the cytopharynx (CP), macronucleus (MA), and three of the kinetodesmal rows (dashed lines). From Beers (1966); with permission.

a small slit in the cyst wall. *Nassula* always emerges posterior end first. The globose, emergent cell exhibits ciliary beating as it emerges, and upon release gradually assumes the pear shape typical of the species.

Encystment and excystment in *Blepharisma stoltei* (Repak 1968) exhibit some interesting contrasts to the foregoing example. Encystment can be divided into four stages: (1) in precystic stages, the normally lancet-shaped cell becomes rounded by overlap of the buccal apparatus with the posterior of the cell; (2) during early encystment, the buccal apparatus is resorbed, the cytoplasm becomes deeply pigmented, and an ectocyst is secreted; (3) an interwall space, endocyst, and plug are produced during late encystment; and (4) disclike structures are secreted on the mature ectocyst, and the macronucleus contains a vacuole.

During early stages of encystment, the organism rotates continuously and ciliary structures are found along the periphery of the ectoplasm. Unbound protein and RNA are present in the cytoplasm throughout the cystic cycle. In later stages, polysaccharides, first formed in the cytoplasm, become deposited in the cyst plug. The plug provides a seal against the environment, but is dissolved to permit exit of the excysting cell. During early excystment, partial kineties are formed. Subsequently, permanent kineties give rise to the anlagen of the buccal apparatus during stomatogenesis. The organism elongates and assumes the characteristic lancet shape of the vegetative phase of growth. Following this late stage of excystment, some cysts divide and the redeveloped organism(s) emerge through the pore. Plug material disappears, permitting emergence, and PAS-positive granules appear in the cytoplasm. The postcystic stage resembles a mature trophozoite except for its small size and lack of pigmentation.

A substantial difference between encystation in *Nassula ornata* and *Blepharisma stoltei* is the extent of cortical reorganization. Substantial resorption of kinetosomes occurs in *Blepharisma* compared to *Nassula*. Among hypotrichs, discriminations can also be made between those species that resorb their kinetosomes during encystation versus those that do not (Walker and Maugel 1980). These distinctions are also augmented by other differences. Kinetosomal resorbing species produce the cyst wall from lamellar precursors which fuse with the cortex at any point on the surface, and the wall is four-layered, mostly fibrillar. Nonresorbing species produce the cyst wall from spherule-filled membranous sacs, and the precursor fuses with the cortex only at restricted regions. The cyst wall is three-layered and amorphous. In the former, the macronuclei fuse, but not in the latter.

The organization of the cyst wall in *Colpoda cucullus*, a commonly cited soil and freshwater-dwelling ciliate, consists of three components: (1) an outer mucous layer, 0.2 to 1.4 µm thick; (2) a middle lamellar layer, 0.2 to 0.4 µm thick; and (3) an inner homogenous layer 0.2 to 0.4 µm thick (Janisch 1980). The inner layer is attached to the pellicle enveloping the cyst. The pellicle is divided into a pattern of irregular polygons and is underlaid by membranes of the subpellicular alveoli. The cytoplasm contains the macronucleus, nuclear fragments composed of chromatin extruded during macronuclear reorganization, the micronucleus, membranes of the endoplasmic reticulum (frequently surrounding mitochondria), lipid droplets, and small vacuoles.

Biochemical analyses of cyst wall production in *Colpoda steinii* (Tibbs and Marshall 1970) show that the cells lose some 30% of their nitrogen during starvation-induced encystment before synthesizing a gluatmic acid-rich protein coat. After 24 h, the coat accounts for 18% of the cyst protein. Settling cells in precyst stages contain ca 29 pg/cell of glutamic acid, while encysted cells contain ca 51 pg/cell. This indicates a substantial amount of glutamic acid and protein synthesis during starvation and subsequent encystment. Some relevant enzymes appeared to decrease as encystment proceeded. Intracellular proteolytic activity was relatively unaltered, but ribonuclease, acid phosphatase, L-alanine:2-oxoglutarate aminotransferase, and L-glutamate:NAPD oxidoreductase were considerably reduced. These activities were assessed on a per cell basis, and more information could have been obtained if the activity had been expressed on a per unit protein basis. The total carbohydrate content of the cell increased during starvation and may represent a food reserve product for the cyst.

Although starvation-induced encystation is common, the quality of feeding prior to starvation can influence capacity to encyst. *Didinium nasutum*, a predator on other ciliates (especially *Paramecium*), maintains maximum binary fission rate and capacity to encyst only if the prey has been well nourished (Butzel and Bolten 1968). Progressive starvation of *Paramecium* prior to their being fed to *Didinia* results in decreased fission rates of the predator, the appearance of abnormal cells, and a loss of ability to encyst. This deficiency is fully reversible by allowing the *Didinia* to feed again on well-nourished paramecia. A minimum of 45 well-fed paramecia is required each day per *Didinium* if maximal fission is to be maintained. Fission and encystment appear to be mutually exclusive processes. Cells grown under optimal conditions for rapid successive fissions, and then deprived of food, failed to encyst. The rate of encystment, however, is positively correlated with capacity to divide. *Didinia* which were fed paramecia sufficient to maintain moderate fission were able to encyst. This may be an adaptation to ensure that only cells with sufficient food reserves to survive prolonged encystation actually form cysts. Excystment of *Didinium* is solely dependent on the concentration of bacteria and not on the presence of *Paramecium*, the main food of the excysted *Didinia* (Butzel and Horwitz 1965). Five types of bacteria including gram-positive, gram-negative, and aerobic and anaerobic species were satisfactory inducers of excystment. This suggests that some general metabolic process or product of the bacteria induces excystment.

The species specific morphology of the cyst wall in ciliates and other protozoa, and the highly ordered events associated with encystment and excystment, suggest that these processes are under tight genetic control. An analysis of the molecular genetics of this process may yield insights into the larger question of how eukaryotic cells regulate events during transformation and development.

Some Perspectives on Protozoan Genetics

The field of protozoan genetics has grown substantially with improved techniques for isolation and cultivation of carefully selected individuals, and the establishment of clonal cultures. These advances have made it possible to perform carefully selected crosses between individuals of known history. Moreover, these innovations coupled with axenic high-yield culture techniques have enhanced electron microscopic and biochemical analyses of parents and progeny, thus bringing together methodologies of classical genetics and modern molecular biology. Some representative data on protozoan genetics are presented as an introduction to the field. Additional insights can be obtained from reviews by Grell (1973), Nanney (1980), Allen and Gibson (1973), and Wichterman (1986).

Expression of Genetic Traits

The expression of genetic traits is the product of genetic information and environmental influences, resulting in observable biological characteristics known as the "reaction norm." Thus, in all living systems, the hereditary information is translated into biological traits through a complex interaction of internal biological processes and external environmental stimuli. The hereditary or genetic characteristics as represented by the gene complement are collectively known as the genotype, and the observable characteristics are termed the phenotype. During asexual reproduction, there is no genetic variability, assuming there are no mutations, since the chromosomes are duplicated with remarkable fidelity. During sexual reproduction, however, gene recombination is possible, as each pair of alleles contains one allele from each of the parent cells. If the zygote undergoes meiosis, producing haploid progeny, then the genetic complement is fully expressed since there is only one allele per gene locus. If, however, the progeny are diploid, then dominance is possible as in heterozygous alleles. The recessive gene is not expressed, since the dominant gene controls metabolic expression. In sexually reproducing species, it is possible to do genetic studies by carefully selecting individuals of a given phenotype and observing the proportion of characteristics in the progeny resulting from the cross. The proportion of offspring of a given phenotype yields evidence about the genetic composition of parents and offspring. Among protozoa, only a limited number of species have been carefully analyzed genetically. The organisms must be capable of sexual reproduction sufficiently regular to permit efficient laboratory manipulation. Three of the major groups of protozoa that have been examined widely in genetic studies will be reviewed here: (1) some Volvocida (eg, *Chlamydomonas* and *Gonium*), (2) myxamoebae (represented by *Physarum polycephalum*), and (3) ciliates (eg, *Tetrahymena*, *Paramecium*, and *Euplotes*).

Genetic Studies of Volvocidans

The small colonial volvocidan *Gonium pectorale* reproduces by dioecious sexual reproduction. Two types of mating colonies, designated (+) and (−), correspond to female and male mating types of higher organisms. Plus and minus type gametes pair and fuse to form a zygote. By analyzing the mating characteristics of the progeny from a single zygote, known as "gones," it is possible to show that the mating type (sex) of the colony is determined by a single pair of alleles. Each zygote differentiates into a germ colony containing four cells. Each of these cells divides repeatedly to yield a 16-celled colony. Thus, there are four daughter colonies produced by each zygote. By culturing the colonies separately and analyzing the gametes produced from each line, it was found that two of the colonies are of the + type and 2 are of the − type. This consistent proportion of 1:1 for the two mating types indicates that one pair of alleles controls sexuality (Stein 1958). That is, during meiosis, two of the cells in the daughter colony receive + genes and the other two receive − genes. Similar studies have been done with *Chlamydomonas*. This solitary flagellate forms haploid gametes of + and − type. The mating type is governed by a single gene designated *mt*. Syngamy occurs only between cells of opposite mating type. In some cases, cells with different amounts of pigment and variations in enzyme complement or type of food storage product can be isolated among the + and − type cells. Thus, it is possible to identify a marker characteristic associated with one of the mating types (for example, a decreased pigment associated with the + type), and trace its appearance in progeny after mating with a "normal" cell of the opposite mating type. The zygote formed by the union of the gametes is diploid, but upon meiotic division gives rise to four haploid progeny. As with *Gonium*, it has been found that the genes segregate in a 1:1 proportion, two of the offspring are + type, and 2 are − type. The identification of different forms of enzymes (isoenzymes) associated with each mating type makes it possible to trace the genes for these enzymes through successive generations. Thus, it is possible to clarify the genetics of this species by tracing the frequency of occurrence of each type of enzyme and also to study the molecular basis for gene expression.

Further studies of this kind by Galloway and Goodenough (1985), using *Chlamydomonas reinhardii*, have shown that a complex set of regulatory genes is associated with the mating-type gene *mt*. These genes apparently act as regulators for the expression of other genes not linked to the *mt* locus. However, since the regulators for these other genes are linked to the *mt* gene, their expression is determined by the *mt* inheritance pattern. For example, Galloway and Goodenough have found three regulatory factors linked to mt^+: (1) a locus designated *sad* functions in sexual adhesion and determines the initial stages of mating; (2) a locus designated *sfu*, that is essential for sexual fusion of the gametes; and (3) a locus symbolized as *upp* (for uniparental plus) that regulates certain aspects of chloroplast inheritance. The clarification that these loci are regulatory in function and not necessarily structural genes helps to explain how genes not linked with the mating-type genes are expressed phenotypically as though they were sex linked.

Genetics of a Myxamoeba

A similar sex-linked marker has been used in crossing experiments with the plasmodia of the myxamoeba *Physarum polycephalum*. This organism produces amoeboid stages as described in the previous section. Plasmodia are normally yellow, but a mutant is white and permits color marking of individuals with genetic traits to be analyzed. For example, Shinnick et al (1983) noted that among populations of *P polycephalum*, most amoeboic stages form plasmodia only by crossing with plasmodia carrying the complementary mating gene; they do not self. The gene controlling this behavior is designated *mat*. However, mutants arise which are capable of selfing (forming plasmodia asexually without fusion) in addition to crossing. Among these "selfing" mutants, a gene separate from the *mat* gene designated as *gad* (symbolizing "greater asexual differentiation") was identified in many strains. The gene promotes plasmodial growth and differentiation in the absence of cross-fertilization. According to standard notation in genetics, gad^- represents the mutant gene and gad^+ represents the "normal" or wild-type gene. In a complex set of crossing experiments, Shinnick et al (1983) showed that the gad^- gene is dominant or semidominant. Strains of the same mating type but differing in the *gad* locus were mixed. In heterozygous plasmodia (ie, those derived from a $gad^- \times gad^+$ cross), the mutant gene dominated and expressed itself, leading to fully differentiated plasmodia. These studies also suggest that the commitment of a cell to differentiate into a plasmodium is under the control of a complex group of genes linked to the *matA* locus.

Genetic Studies of Ciliates

Sexuality has been known in ciliates for a long time, and was clearly documented by Sonneborn (1937) for *Paramecium aurelia*. It had been known that strains of *Paramecium* maintained in culture for many generations never exhibited conjugation, but upon mixing with other strains immediately paired with members of the other strain. By careful analysis of strains isolated from sources around the world, Sonneborn (1937) identified 14 strains, each further divided according to complementary mating type into *E* for even and *O* for odd. Thus, a total of 28 mating types was identified. The degree of compatability among potential conjugants between all possible pairs of mating types has been determined (eg, Wichterman 1986). Some pairs are incompatible and do not conjugate, while other exhibit high probabilities of sexual union. In general, when two potentially compatible cells from different strains are mixed, they clump or pair very promptly and commence conjugation. However, in some cases conjugation does not occur and the cells swim in random patterns, making only occasional contacts, but never uniting. Thus, it is necessary to make several trials to be certain that cells from two strains are compatible. False conjugation can also occur, and it is necessary to clearly document that gamete nuclei have been exchanged. In some cases, pairing results in autogamous fusion of gamete nuclei within each parent cell rather than reciprocal exchange of gametes. Unless care is taken to clearly establish that conjugation has occurred, the hereditary studies can be confounded by self-fertilization or other artifacts. Usually, strains of *Paramecia* isolated from a single parent will remain pure through many generations of asexual reproduction; but occasionally, due to autogamy or other perturbations, variations in mating types will appear in a previously pure strain culture. In some species, moreover, each individual alternates during a 24 h cycle between + and

— mating type. This periodic oscillation is apparently regulated by a biological clock, since the alternation between mating traits occurs at regular intervals. It is believed that the alternation is controlled by the turning on and off of a gene that codes for the alternating mating type. By combining individuals of different mating type and observing the characteristics of their progeny, it is possible to infer the hereditary composition of parents and offspring using classic Mendelian genetic techniques.

Some investigations of sexuality in *Paramecium primaurelia* (Sonneborn 1937; Wichterman 1986) are presented as illustrative examples. When complementary strains conjugate in typical sexual reproduction of *P aurelia*, the karyonides from the first postconjugal fission produce both *E* and *O* type offspring (Figure 20.8). Thereafter, daughter karyonides of each of these progeny will produce progeny of only one type, either *E* or *O*. The appearance of either *E* or *O* types in the first postconjugal division can be explained by the behavior of the macronucleus. Following syngamy, the micronuclei in each conjugate divide to produce two macronuclei. One of the macronuclei is distributed to each daughter cell at the time of postconjugal binary fission. Thus, the mating type of the daughter cell is determined by the mating type of the macronucleus. This pattern of reproduction is called "two-type" reproduction (governed by gene *mat*-2), since the progeny can be of either mating type *E* and *O*. In another strain, however, progeny are consistently only of the *O* mating type. These strains are known as "one-type" (*mat-1*), since the progeny are all of one kind (as diagrammed in Figure 20.8). When complementary conjugants of the one-type strain are crossed with those of the two-type strain, the progeny in the first postconjugal division are mixed, producing both *E* and *O* mating types. In other words, the two-type trait is dominant (Nanney 1980). When the progeny undergo autogamy, however, the genes segregate and the offspring are one-half "one-type" and one-half "two-type," as shown in the lower half of Figure 20.8.

Application of Mendelian genetic techniques to the study of sexual inheritance in the hypotrich *Euplotes crassus* has demonstrated that the capacity for autogamy is controlled by a single dominant gene (Dini and Luporini 1980). When individuals capable of reproducing by autogamy (selfing), and carrying the gene (a^+), are crossed with those that are incapable of autogamy (a^-), all of the offspring are capable of autogamy, indicating a simple dominance relationship as follows:

Parent genotypes	a^+/a^+	×	a^-/a^-
phenotypes	autogamous		nonautogamous
F_1 Generation			
genotype		a^+/a^+	
phenotype		autogamous	

Moreover, when a classic backcrossing experiment was done in which the heterozygous F_1 offspring carrying a^+/a^- alleles were backcrossed to the homozygous

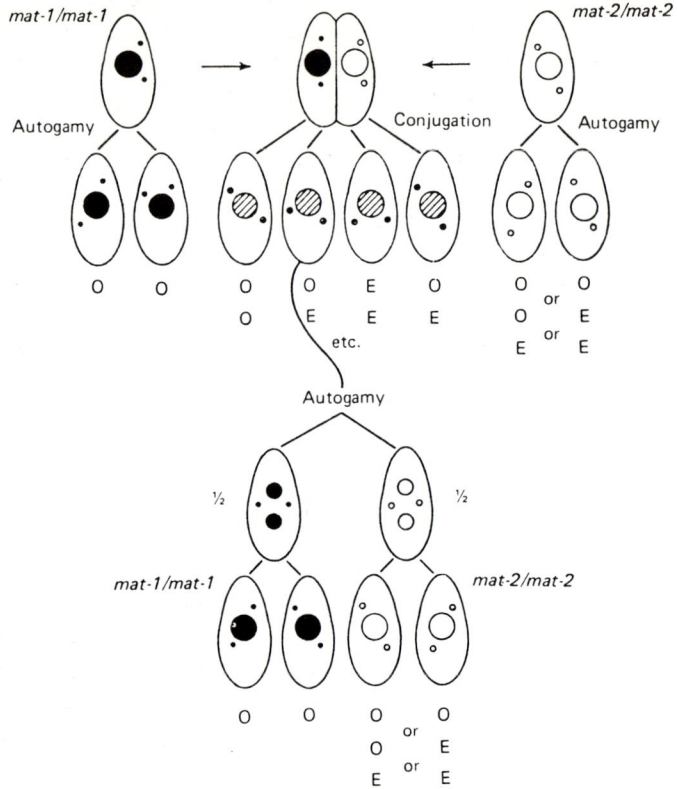

FIGURE 20.8. Genetic control of mating type potentialities in *Paramecium primaurelia*. Crosses between a one-type strain able to produce only Odd-type (O) progeny and a normal two-type strain producing both Even (E) and Odd (O) progeny. The one-type strain is homozygous for an allel, *mat-1*, which limits its autogamous progeny to type Odd. The two-type strain has the *mat-2* allele and produces karyonides randomly determined for mating type (O and E). The F_1 between the strains has both kinds of karyonides, demonstrating the dominance of the *mat-2* allele. The F_2 by autogamy consists of one-half restricted to the Odd mating type and one-half showing random karyonidal distribution. From Nanney (1980) and Wichterman (1986); with permission.

recessive parent with a^+/a^- alleles, the proportion of offspring was approximately 1:1 for autogamy to nonautogamy phenotypes:

Back Cross
    genotype a^+/a^- × a^-/a^-
    phenotype   autogamous   ×   nonautogamous
Progeny
    genotype a^+/a^- + a^-/a^-
    phenotype   autogamous   :   nonautogamous
Ratio   1   :   1

This further confirmed a dominance relationship and suggested that the trait is governed by a single gene locus.

Application of classic genetic techniques and modern physiological methods of identifying phenotypic characteristics has begun to clarify the molecular genetic basis for irregular swimming traits in *Paramecium tetraurelia*. A mutant, known as "dancer," shows prolonged backward swimming when exposed to K^+ or Ca^{2+} in its medium. By crossing dancer mutants with normal individuals, Hinrichsen et al (1984) demonstrated that the dancer gene is dominant or semidominant. Nearly all of the progeny in the F_1 generation exhibited the dancer trait. However, the phenotypic expression of the trait appeared only after the progeny divided asexually for several cycles. This suggested two hypotheses for the gene action: (1) some deficiency in a cytoplasmic factor is produced by the gene and several fissions are required for the deficiency to develop, or (2) the gene codes for an abnormal membrane transport protein that must accumulate in sufficient amount to produce the dancer trait. Since injection of normal cell cytoplasm into the F_1 progeny did not reverse the abnormality, it was concluded that abnormal membrane ion channels may account for the mutant effect. Further inheritance studies have clarified the genetic basis for absence of backward swimming behavior. Some mutants, designated "atlanta," are incapable of swimming backwards and gyrate wildly when stimulated. Research by Hinrichsen and Kung (1984) has demonstrated that this mutant gene is recessive. When mutants are crossed with wild-type (normal) individuals, the progeny are normal. The abnormality appears to be produced by "leaky" membrane ion channels that do not properly regulate Ca^{2+} diffusion across the membrane. Apparently, based on the crossing studies, the wild type produces a normal protein that ameliorates the error introduced by the mutant gene.

Summative Perspective

The life cycles of protozoa involve complex variations in form and physiology, often including alternations of generations. Cyst stages or other stress resistant phases of life are often found among species subjected to unpredictable environments. In some cases, these phases may occur after sexual reproduction. But, in other species, environmentally resistant resting stages occur during periods of stress or in response to chemical signals. Incystment can be induced by factors such as crowding, depletion of nutrients, or excessive amounts of waste products. Excystment in resting stages may be initiated by intrinsic physiological variables or by environmental inducers. In soil-dwelling amoebae, for example, exudates of potential food induce excystment. The nuclear and cytoplasmic events accompanying the cell cycle have been carefully documented in several groups of protozoa, using synchronized cell cultures. This experimental technique promises to be a useful approach in clarifying the fundamentals of eukaryotic cell physiology. Modern molecular genetics techniques have also begun to explain the genetic bases for protozoan behavior and specificity of reproductive responses. The discovery of chemical inducers which trigger sexual maturation in some colonial Volvicida, and their relationship to survival during the annual growth cycle, constitute a promising field of ecological and genetic research. Since protozoa are eukaryotes, and as some have been carefully studied both physiologically and biochemically in axenic cultures, they are likely candidates for more extensive molecular genetic research, using modern genetic engineering techniques. For example, De Lozanne and Spudich (1987) have demonstrated that the gene for production of heavy chain myosin, mediating motility in the slime mold *Dictyostelium*, can be altered by insertion of a gene fragment for myosin, resulting in defective myosin production. These altered cells are defective in cytokinesis (cell division) and become large and multinucleate. However, in spite of the absence of native myosin, these cells exhibit many forms of normal cell movement, including membrane ruffling, phagocytosis, and chemotaxis. These results provide genetic proof that the intact myosin molecule is required for cytokinesis, but not for karyokinesis (nuclear division). Research of this kind with protozoa can enhance the generalizability of molecular genetic studies among eukaryotes, and may provide new model systems for clarifying the mechanisms of genetic control among diverse types of cells.

References

Aaronson, S. Descriptive biochemistry and physiology of the chrysophyceae (with some comparisons to prymnesiophyceae). In: Levandowsky, M., Hutner S.H., eds. Biochemistry and Physiology of Protozoa; vol. 3. New York: Academic Press; 1980: pp. 117–169.

Aaronson, S; Baker, H. A comparative biochemical study of two species of Ochromonas. J. Protozool. 6:282–284; 1959.

Abrams, P. Some comments on measuring niche overlap. Ecology. 61:44–49; 1980.

Adam, K. The amino acid requirement of *Acanthamoeba* sp. Neff. J. Protozool. 11:98–100; 1964.

Adam, K.M.G.; Blewett, D.A. Carbohydrate utilization by the soil amoeba *Hartmanella castellanii*. J. Protozool. 14:277–282; 1967.

Albach, R.A.; Booden, T. Amoebae. In: Krier, A.P., ed. Parasitic Protozoa; vol. 2. New York: Academic Press; 1978: pp. 455–506.

Alexander, M. Introduction to Soil Microbiology. 2nd ed. New York: John Wiley and Sons; 1977: pp. 89–102.

Alexander, M. Why microbial predators and parasites do not eliminate their prey and hosts. Ann. Rev. Microbiol. 35:113–133; 1981.

Alexopoulos, C.J. The Myxomycetes II. Bot. Rev. 29:1–78; 1963.

Aley, S.B.; Cohn, Z.A.; Scott, W.A. Endocytosis in *Entamoeba histolytica*: evidence for a unique non-acidified compartment. J. Exp. Med. 160:724–737; 1984.

Allen, R.D. A new theory of amoeboid movement and protoplasmic streaming. Exp. Cell Res. Suppl. 8:17–31; 1961.

Allen, R.D. Fine structure, reconstitution and possible function of components of the cortex of *Tetrahymena pyriformis*. J. Protozool. 14:553–565; 1967.

Allen, R.D. *Paramecium* phagosome membrane: from oral region to cytoproct and back again. J. Protozool. 31:1–6; 1984.

Allen, R.D.; Fok, A.K. Stages of digestive vacuoles in *Paramecium*: membrane surface differences and location. European J. Cell Biol. 35:149–155; 1984.

Allen, R.D.; Francis, D.; Zeh, R. Direct test of the positive pressure gradient theory of pseudopod extension and retraction in amoebae. Science. 174:1237–1240; 1971.

Allen, S.; Gibson, I. Genetics of *Tetrahymena*. In: Elliott, A.M., ed. Biology of Tetrahymena. Stroudsburg, PA: Dowden, Hutchinson & Ross; 1973: pp. 307–373.

Anderson, O.R. Cytoplasmic fine structure of nassellarian Radiolaria. Marine Micropaleontol. 2:251–264; 1977a.

Anderson, O.R. Fine structure of a marine ameba associated with a blue-green alga in the Sargasso Sea. J. Protozool. 24:370–376; 1977b.

Anderson, O.R. Light and electron microscopic observations of feeding behavior, nutrition, and reproduction in laboratory cultures of *Thalassicolla nucleata*. Tissue and Cell. 10:401–412; 1978.

Anderson, O.R. Radiolaria. New York: Springer-Verlag. 1983; 335 pp.

Anderson, O.R. Cellular specialization and reproduction in planktonic foraminifera and radiolaria. In: Steidinger, K.A.; Walker, L.M., eds. Marine Plankton Life Cycle Strategies. Boca Raton: Chemical Rubber Co. Press; 1984: pp. 35–66.

Anderson, O.R. Silicification in radiolaria–deposition and ontongenetic origins of form. In: Leadbeater, B.S.; Riding, R., eds. Biomineralization in Lower Plants and Animals. Oxford: Clarendon Press; 1986: pp. 375–391.

Anderson, O.R. The fine structure of a silica-biomineralizing testate amoeba, *Netzelia tuberculata*. J. Protozool. 34:302–309; 1987.

Anderson, O.R.; Bé, A.W.H. The ultrastructure of a planktonic foraminifer *Globigerinoides sacculifer* (Brady) and its symbiotic dinoflagellates. J. Foraminiferal Res. 6:1–21; 1976.

Anderson, O.R.; Bé, A.W.H. Recent advances in foraminiferal fine structure research. In: Hedley, R.H.; Adams, C.G., eds. *Foraminifera*. vol. 3. London: Academic Press; 1978: pp. 121–202.

Anderson, O.R.; Moss, M.L.; Skalak, R. The cytoskeletal and biomineralized supportive structures in radiolaria. In: Bereiter-Hahn, J.; Anderson, O.R.; Reif, W., eds. *Cytomechanics*. Heidelberg: Springer-Verlag. 1987.

Anderson, O.R.; Roels, O.A. Myelin-like configurations in *Ochromonas malhamensis*. J. Ultrastruc. Res. 20:127–139; 1967.

Anderson, O.R.; Spindler, M.; Bé, A.W.H.; Hemleben, C. Trophic activity of planktonic foraminifera. J. Mar. Biol. Assoc., U.K. 59:791–799; 1979.

Anderson, O.R.; Swanberg, N.R.; Bennett, P. Laboratory studies of the ecological significance of host–algal nutritional associations in solitary and colonial radiolaria. J. Mar. Biol. Assoc., U.K. 65:263–272; 1985.

Anderson, O.R.; Tuntivate-Choy, S. Cytochemical evidence of peroxisomes in planktonic foraminifera. J. Foraminiferal Res. 14:203–205; 1984.

Andrus, W.DeW.; Giese, A.C. Mechanisms of sodium and potassium regulation in *Tetrahymena pyriformis* strain W. J. Cell. Comp. Physiol. 61:17–30; 1963.

Antia, N.J. Nutritional physiology and biochemistry of marine cryptomonads and chrysomonads. In: Levandowsky, M.; Hutner, S.H., eds. Biochemistry and Physiology of Protozoa. 2nd ed., vol. 3. New York: Academic Press; 1980: pp. 67–115.

Antia, N.J.; Cheng, J.Y.; Fogle, R.A.J.; Percival, E. Marine cryptomonad starch from autolysis of glycerol-grown *Chroomonas salina*. J. Phycol. 15:57–62; 1979.

Arnold, Z.M. Field and laboratory techniques for the study of living foraminifera. In: Hedley, R.H.; Adams, C.G., eds. Foraminifera. vol. 1. London: Academic Press; 1974: pp. 153–206.

Austin, M.P. Continuum concept, ordination methods, and niche theory. Ann. Rev. Ecol. Syst. 16:39–61; 1985.

Baker, H.; Frank, O.; Khalil, F.; DeAngelis, B.; Hutner, S.H. Determination of metabolically active B_{12} and inactive B_{12} analog titers in human blood using several microbial reagents and a radiodilution assay. J. Amer. College of Nutr. 5:467–475; 1986.

Baker, J.R. Parasitic Protozoa. Hutchinson, PA. 1973: 176 pp.

Baker, J.R. Systematics of parasitic protozoa. In: Krier, J.P., ed. Parasitic Protozoa. New York: Academic Press; 1977: pp. 35–56.

Baker, W.B.; Buetow, D.E. Hydrolytic enzymes of *Euglena gracilis*: characterization and activity as a function of culture age and carbon deprivation. J. Protozool. 23:167–176; 1976.

Baldock, B.M.; Baker, J.H. The occurrence and growth rates of *Polychaos fasciculatum*, a rediscovered amoeba. Protistologica. 16:79–83; 1980.

Baldock, B.M.; Baker, J.H.; Sleigh, M.A. Laboratory growth rates of six species of freshwater Gymnamoebia. Oecologia. 47:156–159; 1980.

Baldock, B.M.; Berger, J. The effects of low temperature on the growth of four freshwater amoebae (Protozoa: Gymnamoebia). Trans. Am. Microsc. Soc. 103:233–239; 1984.

Bamforth, S.S. Ecological studies on the planktonic Protozoa of a small artificial pond. Limnol. and Oceanogr. 3:398–412; 1958.

Bamforth, S.S. The numbers and proportions of testacea and ciliates in litters and soils. J. Protozool. 18:24–28; 1971.

Bamforth, S.S. Population dynamics of soil and vegetation protozoa. J. Exp. Zool. 13:171–176; 1973.

Bamforth, S.S. Ecology of protozoa. In: Lee, J.J.; Hutner, S.H.; Bovee, E.C., eds. Illustrated Guide to the Protozoa. Lawrence, Kansas: Society of Protozoologists, 1985a.

Bamforth, S.S. Symposium on "Protozoan ecology": The role of protozoa in litters and soils. J. of Protozool. 32:404–409; 1985b.

Band, R.N. The amino acid requirements of the soil amoeba, *Hartmannella rhysodes* Singh. J. Protozool. 9:377–379; 1962.

Bardele, C.F. Budding and metamorphosis in *Acineta tuberosa*. An electron microscopic study on morphogenesis in Suctoria. J. Protozool. 17:51–70; 1970a.

Bardele, C.F. Comparative ultrastructural studies on Centrohelida. J. Protozool. 17 (Suppl.), abstr. 17; 1970b.

Bardele, C.F. A microtubule model for ingestion and transport in the suctorian tentacle. Zeitschrift für Zellforschung und Mikroskopische Anatomie. 126:116–134; 1972.

Bardele, C.F. Transport of materials in the suctorian tentacle. In: Sleigh, M.A.; Jennings, D.H., eds. Transport at the Cellular Level. Society of Experimental Biology Symposium. Cambridge: Cambridge University Press; 1974: no. 28, pp. 191–208.

Bardele, C.F. The fine structure of the centrohelidian heliozoan *Heterophyrs marina*. Cell Tissue Res. 161:85–102; 1975.

Bardele, C.F. The imprints of ciliate phylogeny revealed by comparative freeze-fracture study of the ciliary membrane. In: Schwemmles, W.; Schenk, H.E.A., eds. Endocytobiology. vol. 1. Hawthorne, NY: DeGruyter; 1980: pp. 51–61.

Bardele, C.F. Functional and phylogenetic aspects of the ciliary membrane: a comparative freeze-fracture study. Biosystems. 14:403–421; 1981.

Barras, D.R.; Stone, B.A. The chemical composition of the pellicle of *Euglena gracilis* var. *bacillaris*. Biochem. Journ. 97:14–15; 1965.

Barry, S.C. Utilization of glucose by *Astasia longa*. J. Protozool. 9:394–400; 1962.

Barsdate, R.J.; Prentki, R.T.; Fenchel, T. Phosphorous cycle of model ecosystems: Significance for decomposer food chains and effect on bacterial grazers. Oikos. 25:239–251; 1974.

Bates, R.C.; Hurlbert, R.E. The effect of acetate on *Euglena gracilis* var. *bacillaris* as a function of environmental conditions. J. Protozool. 17:134–138; 1970.

Bé, A.W.H. An ecological, zoogeographic and taxonomic review of recent planktonic foraminifera. In: Ramsay, A.T.S., ed. Oceanic Micropaleontology. London: Academic Press; 1977: 1:1–100.

Bé, A.W.H.; Anderson, O.R. Gametogenesis in planktonic foraminifera. Science. 192:890–892; 1976.

Bé, A.W.H.; Caron, D.A.; Anderson, O.R. Effects of feeding frequency on life processes of the planktonic foraminifer *Globigerinoides sacculifer* in laboratory culture. J. Mar. Biol. Ass., U.K. 61:257–277; 1981.

Bé, A.W.H.; Hemleben, C.; Anderson, O.R.; Spindler, M.; Hacunda, J.; Tuntivate-Choy, S. Laboratory and field observations of living planktonic foraminifera. Micropaleontology. 23:155–179; 1977.

Bé, A.W.H.; Tolderlund, D.S. Distribution and ecology of living planktonic Foraminifera in surface waters of the Atlantic and Indian Oceans. In: Funnell, B.M.; Riedel, W.R., eds. The Micropaleontology of the Oceans. London: Cambridge University Press; 1971: pp. 105–149.

Bean, B. Geotactic behavior of *Chlamydomonas*. J. Protozool. 24:394–401; 1977.

Beaver, J.R.; Crisman, T.L. The trophic response of ciliated protozoans in freshwater lakes. Limnol. Oceanogr. 27:246–253; 1982.

Beers, C.D. The excystment process in the ciliate *Nassula ornata* Ehrb. J. Protozool. 13:79–83; 1966.

Beezley, B.B.; Gruber, P.J.; Frederick, S.E. Cytochemical localization of glycolate dehydrogenase in mitochondria of *Chlamydomonas*. Pl. Physiol. 58:315–319; 1976.

Belcher, J.H. The fine structure of the loricate colourless flagellate *Bicoeca planctonica* Kisselew. Arch. Protistenk. 117:78–84; 1975.

Belsky, M.M.; Schultz, J. Partial characterization of hexokinase from *Euglena gracilis* var. *bacillaris*. J. Protozool. 9:195–200; 1962.

Bereiter-Hahn, J.; Anderson, O.R.; Reif, W., eds. Cytomechanics. Heidelberg: Springer-Verlag, 1987.

Bereiter-Hahn, J.; Strohmeier, R. Hydrostatic pressure in metazoan cells in culture and its involvement in locomotion and shape generation. In: Bereiter-Hahn, J.; Anderson, O.R.; Reif, W, eds. Cytomechanics. Heidelberg: Springer-Verlag; 1987.

Bernstein, B.B.; Hessler, R.R.; Smith, R.; Jumars, P.A. Spatial dispersion of benthic foraminifera in the abyssal central North Pacific. Limnol. Oceanogr. 23:401–416; 1978.

Bick, H. Population dynamics of protozoa associated with the decay of organic materials in fresh water. Amer. Zool. 13:149–160; 1973.

Bird, D.F.; Kalff, J. Bacterial grazing by planktonic lake algae. Science. 231:493–495; 1986.

Birdsey, E.C.; Lynch, V.H. Utilization of nitrogen compounds by unicellular algae. Science. 137:763–764; 1962.

Blum, J.J.; Sommer, J.R.; Kahn, V. Some biochemical, cytological and morphogenetic comparisons between *Astasia longa* and a bleached *Euglena gracilis*. J. Protozool. 12:202–209; 1965.

Bohatier, J. Structure et ultrastructure de *Lacrymaria olor* (O.F.M. 1786). Protistologica. 6:331–342; 1970.

Boltovskoy, E.; Wright, R. Recent Foraminifera. The Hague: W. Junk, 1976.

Borowitz, M.J.; Stein, R.B.; Blum, J.J. Quantitative analysis of the change of metabolite fluxes along the pentose phosphate and glycolytic pathways in *Tetrahymena* in response to carbohydrates. J. Biol. Chem. 252:1589–1605; 1977.

Borror, A.C. Spatial distribution of marine ciliates: Micro-ecologic and biogeographic aspects of protozoan ecology. J. Protozool. 27:10–13; 1980.

Bovee, E.C. An ecological study of amebas from a small stream in northern Florida. Hydrobiologia. 25:69–87; 1965.

Bovee, E.C. Class Lobosea Carpenter. In: Lee, J.J.; Hutner, S.H.; Bovee, E.C., eds. Illustrated Guide to the Protozoa. Lawrence, Kansas: Society of Protozoologists; 1985: pp. 158–211.

Bovee, E.C.; Jahn, T.L. Locomotion and behavior. In: Jeon, K.W., ed. The Biology of Amoeba. New York: Academic Press; 1973: pp. 52–56.

Bowers, B. Comparison of pinocytosis and phagocytosis in *Acanthamoeba castellanii*. Exp. Cell Res. 110:409–418; 1977.

Bowers, B.; Korn, E.D. The fine structure of *Acanthamoeba castellanii*. I. trophozoite. J. Cell Biol. 39:95–111; 1968.

Bowers, B.; Korn, E.D. The fine structure of *Acanthamoeba castellanii* (Neff) strain. J. Cell Biol. 41:786–805; 1969.

Bowers, B.; Olszewski, T. *Acanthamoeba castellanii* discriminates internally between digestible and indigestible particles. J. Cell Biol. 97:317–322; 1983.

Bowser, S.S. Invasive activity of *Allogromia* pseudopodial networks: skyllocytosis of a gelatin/agar gel. J. Protozool. 32:9–12; 1985.

Bowser, S.S.; Israel, H.A.; McGee-Russell, S.M.; Reider, C.L. Surface transport properties of reticulopodia: do intracellular and extracellular motility share a common mechanism? Cell Biol. Int. Rep. 8:1051–1063; 1984.

Bowser, S.S.; McGee-Russell, S.M.; Rieder, C.L. Digestion of prey in Foraminifera is not anomalous: a correlation of light microscopic, cytochemical, and HVEM technics to study phagotrophy in two allogromiids. Tissue and Cell. 17:823–839; 1985.

Bowser, S.S.; Rieder, C.L. Evidence that cell surface motility in *Allogromia* is mediated by cytoplasmic microtubules. Can. J. Biochem. Cell Biol. 63:608–620; 1985.

Boyce, M.S. Restitution of r- and K- selection as a model of density-dependent natural selection. Ann. Rev. Ecol. Syst. 15:427–447; 1984.

Braatz-Schade, K.; Stockem, W. Pinocytosis and locomotion of amoebae. IX. The influence of different cations on the moving behavior of *Amoeba proteus*. Arch. Protistenk. 114:272–290; 1972.

Bradbury, P.C. The life cycle and morphology of the apostomatous ciliate, *Hyalophysa chattoni* n.g., n. sp. J. Protozool. 13:209–225; 1966.

Bragg, P.D.; Reeves, R.E. Pathways of glucose dissimilation in the Laredo strain of *Entamoeba histolytica*. Exp. Parasitol. 12:393–400; 1962.

Bray, D.F.; Nakamura, K.; Costerton, J.W.; Wagenaar, E.B. Ultrastructure of *Chlamydomonas eugametous* as revealed by freeze-etching: cell wall, plasmalemma and chloroplast membrane. J. Ultrastruc. Res. 47:125–141; 1974.

Bretthauer, R. Laboratory aquatic microcosms. In: Giesy, J. P., Jr., ed. Microcosms in Ecological Research. Technical Information Center, U. S. Dept. of Energy, CONF-781101; 1980: pp. 416–445.

Briand, J.; Calvayrac, R. Paramylon synthesis in heterotrophic and phototrophic *Euglena gracilis* (Euglenophyceae). J. Phycol. 16:234–239; 1980.

Broad, T.E.; Dawson, R.M.C. Role of choline in the nutrition of the rumen protozoan *Entodinium caudatum*. J. Gen. Microbiol. 92:391–397; 1976.

Brock, T.D. Lower pH limit for the existence of blue-green algae: evolutionary and ecological implications. Science. 179:480–483; 1973.

Brooker, B.; Preston, T.M. The cytostome in trypanosomes and allied flagellates. J. Protozool. 14 (Suppl.): 41; 1967.
Brown, D.L.; Massalski, A.; Leppard, G.G. Fine structure of encystment of the quadriflagellate alga, *Polytomella agilis*. Protoplasma 90:139–154; 1976.
Bruce, V.G. The biological clock in *Chlamydomonas reinhardi*. J. Protozool. 17:328–334; 1970.
Brugerolle, G. Etude ultrastructurale du flagellé *Protrichomonas legeri* (Léger 1905), parasite de l'estomac des bogues (*Box Boops*). Protistologica. 16:353–358.
Brugerolle, G. Des trichocystes chez les bodonides, un caractére phylkogénétique supplémentaire entre Kinteoplastida et Euglenida. Protistologica 16:353–358; 1985.
Brugerolle, G.; Lom, J.; Nohynkova, E.; Joyon, L. Comparison et evolution des structures cellulaires chez plusiers espéces de bodonidés et cryptobiidés appartenant aux genres *Bodo*, *Cryptobia*, et *Trypanoplasma* (Kinetoplastida, Mastigophora). Protistologica 15:197–221; 1979.
Brummer, G-J.A.; Hemleben, C.; Spindler, M. Planktonic foraminiferal ontogeny and new perspectives for micropaleontology. Nature. 319:50–52; 1986.
Buechler, D.G., Dillon, R.D. Phosphorus regeneration in fresh-water *Paramecia*. J. Protozool. 21:339–343; 1974.
Buetow, D.E. Amino acids as nitrogen sources for the growth of *Euglena gracilis* and *Astasia longa*. J. Protozool. 13:585–587; 1966.
Buetow, D.E., ed. The Biology of Euglena. New York: Academic Press; 1982.
Buetow, D.E.; Padilla, G.M. Growth of *Astasia longa* on ethanol, I. Effects of ethanol on generation time, population density, and biochemical profile. J. Protozool. 10:121–123; 1963.
Buetow, D.E.; Schuit, K.E. Phosphorus and the growth of *Euglena gracilis*. J. Protozool. 15:770–773; 1968.
Butschli, O. Vorlaufige Mittheilung einiger Resultate von Studien über die Conjugation der Infusorien und die Zelltheilung. Z. Wiss. Zool. 25:426–441; 1873a.
Butschli, O. Eineges über Infusorien. Arch. Mikrosk. 9:657–678; 1873b.
Butschli, O. Studien über die ersten Entwicklungsvorgange der Eizelle, die Zellteilung und der Konjugation der Infusorien. Abh. Senckenb. Naturforsch. Ges. 10:1–250; 1876.
Butzel, H.M., Jr.; Horwitz, H. Excystment of *Didinium nasutum*. J. Protozool. 12:413–416; 1965.
Butzel, H.M., Jr.; Bolten, A.B. The relationship of the nutritive state of the prey organisms *Paramecium aurelia* to the growth and encystment of *Didinium nasutum*. J. Protozool. 15:256–258; 1968.
Byers, T.J.; Akins, R.A.; Maynard, B.J.; Lefken, R.A.; Martin, S.M. Rapid growth of *Acanthamoeba* in defined media: induction of encystment by glucose-acetate starvation. J. Protozool. 27:216–219; 1980.
Byers, T.J.; Rudick, V.L.; Rudick, M.J. Cell size, macromolecule composition, nuclear number, oxygen consumption and cyst formation during two phases in unagitated cultures of *Acanthamoeba castellanii*. J. Protozool. 16:693–699; 1969.
Cachon, J.; Cachon, M. L' infrastructure des axopodes chez les Radiolaires Sphaerellaires Periaxoplastidies. C.R. Acad. Sci. (Paris). 261:1388–1391; 1965.
Cachon, J.; Cachon, M. Le systeme axopodial des Radiolaires Nassellaires. Arch. Protistenkd. 113:80–97; 1971.
Cain, J. Nitrogen utilization in 38 freshwater chlamydomonad algae. Can. J. Bot. 43:1367–1378; 1965.

Cairns, J., Jr. The chemical environment of common fresh-water Protozoa. Not. Nat. Acad. Nat. Sci. Philadelphia. 365:1–16; 1964.

Cairns, J., Jr.; Dahlberg, M.L.; Dickson, K.L.; Smith, N.; Waller, W.T. Rate of species diversity restoration following stress in protozoan communities. Univ. Kansas Sci. Bull. 48:209–224; 1969.

Cairns, J., Jr.; Dickson, K.L.; Yongue, H., Jr. The consequences of nonselective periodic removal of portions of fresh-water protozoan communities. Trans. Amer. Micro. Soc. 90:71–80; 1971.

Cairns, J., Jr.; Ruthven, J. The relation between artificial substrate area and the number of fresh-water protozoan species. Trans. Amer. Micro. Soc. 89:100–109; 1970.

Cairns, J., Jr.; Yongue, W.H., Jr. Factors affecting the number of species in freshwater protozoan communities. In: Cairns, J.J., Jr., ed. Aquatic Microbial Communities. New York & London: Garland Publishing, Inc.; 1977: pp. 257–303.

Calow, P. Conversion efficiencies in heterotrophic organisms. Biol. Rev. 52:385–409; 1977.

Canning, E.U. Microsporida. In: Kreier, J.P., ed. Parasitic Protozoa. New York: Academic Press; 1978: vol. 4, pp. 155–196.

Cantor, M.H.; Withroe, J. Cellular events during metabolic adaptation in *Polytomella agilis*. Am. Soc. Zool. 19:514; 1970.

Capriulo, G.M. Feeding of field collected Tintinnid micro-zooplankton on natural food. Marine Biology. 71:73–86; 1982.

Capriulo, G.M.; Carpenter, E.J. Grazing by 35 to 202 µm micro-zooplankton in Long Island Sound. Marine Biology. 56:319–326; 1980.

Casey, R.E. Distribution of polycystine radiolaria in the oceans in relation to physical and chemical conditions. In: Funnell, B.M.; Riedel, W.R., eds. The Micropaleontology of Oceans. Cambridge: Cambridge University Press; 1971: pp. 151–159.

Casey, R.E. The ecology and distribution of recent radiolaria. In: Ramsay, A.T.S., ed. The Micropaleontology of Oceans. Cambridge: Cambridge University Press; 1977: pp. 331–349.

Cassin, P.E. Isolation, growth, and physiology of acidophilic chlamydomonads. J. Phycol. 10:439–447; 1974.

Chapman, L.F.; Cirillo, V.P.; Jahn, T.L. Permeability to sugars and fatty acids in *Polytoma obtusum*. J. Protozool. 12:47–51; 1965.

Chen, Y.T. Investigations of the biology of *Peranema trichophorum* (Euglenineae). Quart. J. Microscop. Sci. 91:279–308; 1950.

Chiovetti, R., Jr.; Bovee, E.C. Initiation of encystment by multiple contacts among starving amoebas, *Naegleria gruberi*. Acta Protozool. 21:149–156; 1982.

Cirillo, V.P. Long-term adaptation to fatty acids by the phytoflagellate *Polytoma uvella*. J. Protozool. 4:60–62; 1957.

Cleveland, L.R. *Tritrichomonas fecalis* nov. sp. of man; its ability to grow and multiply indefinitely in faeces diluted with tap water and in frogs and tadpoles. Am. J. Hyg. 8:232–255; 1928.

Cleveland, L.R. Origin and development of the achromatic figure. Biol. Bull. 74:41–55; 1938.

Cole, G.T.; Wynne, M.J. Endocystosis of *Microcystis aeruginosa* by *Ochromonas danica*. J. Phycol. 10:397–410; 1974.

Coleman, G.S. Rumen ciliated protozoa. In: Levandowsky, M.; Hutner, S.H., eds. Biochemistry and Physiology of Protozoa. 2nd ed., vol. 2. New York: Academic Press; 1979: pp. 381–408.

Collyard, K.J.; Danforth, W.F. The relation of phosphate metabolism to oxidative assimilation of acetate by *Euglena gracilis* var. *bacillaris*. Effects of 2,4-Dinitrophenol. J. Protozool. 17:334–340; 1970.

Cook, J.R. Photo-inhibition of cell division and growth in euglenoid flagellates. Journal of Cell Physiology. 71:177–184; 1968.

Coombs, G.H. In: Klein, R.A.; Miller, P.G.G. Alternate metabolic pathways in protozoan energy metabolism. Parasitology 82:7–9; 1981.

Corliss, J.O. Comments on the systematics and phylogeny of the Protozoa. Systematic Zoology. 8:169–190; 1960.

Corliss, J.O. The ciliate Protozoa and other organisms: some unresolved questions of major phylogenetic significance. American Zoologist. 12:739–753; 1972.

Corliss, J.O. The Ciliated Protozoa: Characterization, Classification, and Guide to the Literature. 2nd ed. New York: Pergamon Press; 1979: 455 pp.

Corliss, J.O. Concept, definition, prevalence, and host-interactions of xenosomes (cytoplasmic and nuclear endosymbionts). J. Protozool. 32:373–376; 1985.

Costas, M.; Griffiths, A.J. Taxonomic significance of the fatty acid composition of *Acanthamoeba* (Amoebida, acanthamoebidae). Protistologica. 20:27–31; 1984.

Costas, M.; Griffiths, A.J. Physiological characterizations of *Acanthamoeba* strains. J. Protozool. 33:304–309; 1986.

Cowles, R.P.; Schwitalia, A.M. The hydrogen ion concentration of a creek, its water fall, swamp, and pond. Ecol. 4:402–416; 1923.

Cowling, A.J. Respiration and growth of terrestrial protozoa from the maritime antarctic. Great Britain: Council for National Academic Awards; 1983; 280 pp. Ph.D. Thesis.

Cowling, A.J. Application of gradient diver micro-respirometry to protozoan studies. Br. Antarct. Surv. Bull. 65:91–107; 1984.

Cowling, A.J.; Smith, H.G. Protozoa in the microbial communities of maritime Antarctic fellfields. *Les Ecosystemes Terrestres Subantarctiques: Contraintes selectives sur les organismes et les communautes*. Abstracts, p. 34. Comite National Francais des Recherches Antarctiques, Universite de Rennes I, Rennes, 1986.

Cox, D. Prolonged survival of *Tetrahymena* at 0–5°C in citrated, lecithinized, defined media. J. Protozool. 17:150–152; 1970.

Cox, E.R. Phytoflagellates: the organisms and the taxonomic categories. In: Cox, E.R., ed. Phytoflagellates. vol. 2. New York: Elsevier; 1980: pp. 1–4.

Creutz, C.; Diehn, B. Motor response to polarized light and gravity sensing in *Euglena*. J. Protozool. 23:552–556; 1976.

Curds, C.R. Endogenous bud formation in *Histriculus vorax*, a new asexual reproductive process in the Hypotrichida. J. Protozool. 13:155–164; 1966.

Curds, C.R. The role of protozoa in activated-sludge process. Amer. Zool. 13:161–169; 1973.

Curds, C.R. British and other freshwater ciliated protozoa. Part I. Ciliophora: Kinetofragminophora. Cambridge: Cambridge University Press; 1982: 474 pp.

Curds, C.; Bazin, M.J. Protozoan predation in batch and continuous cultures. In: Droop, M.R.; Jannasch, H.W., eds. Advances in Aquatic Microbiology. Symposium Series no. 6. Academic Press: 1977: pp. 115–176.

Curds, C.R.; Bamforth, S.S.; Finlay, B.J. Report on the freshwater workshop in Kisumu, Kenya (30 June–5 July 1985). Insect Sci. Applic. 7:447–449; 1986.

Curds, C.R.; Cockburn, A. Studies on the growth and feeding of *Tetrahymena pyriformis* in axenic and monoxenic culture. J. Gen. Microbiol. 54:343–358; 1968.

Curds, C.R.; Gates, M.A.; Roberts, D.M. British and other freshwater ciliated protozoa. Part 2. Ciliophora: Oligohymenophora and Polyhymenophora. Cambridge: Cambridge University Press; 1983: 474 pp.

Curds, C.R.; Vandyke, J.M. The feeding habits and growth rates of some fresh-water ciliates found in activated-sludge plants. J. Appl. Ecol. 3:127–137; 1966.

Current, W.L. *Cryptobia* sp. in the snail *Triadopsis multilineata* (Say): Fine structure of attached flagellates and their mode of attachment to the spermatheca. J. Protozool. 27:278–287; 1980.

Cutler, D.W.; Bal, D.V. The influence of protozoa on the process of nitrogen fixation by *Azotobacter chroococcum*. Ann. Appl. Biol. 13:516; 1926.

Dahlgren, L. On the ultrastructure of the gamontic nucleus and the adjacent cytoplasm of the monothalamous foraminifer *Ovamina opaca* Dahlgren. Zool. Bidrag, Uppsala. 37:77–112; 1967.

Daley, R.J.; Morris, G.P.; Brown, S.R. Phagotrophic ingestion of a blue-green alga by *Ochromonas*. J. Protozool. 20:58–61; 1973.

Danforth, W.F.; Wilson, B.W. Adaptive changes in the acetate metabolism of *Euglena*. J. Protozool. 4:52–55; 1957.

Daniels, E.W. Ultrastructure. In: Jeon, K.W., ed. The Biology of Amoeba. New York: Academic Press; 1973: pp. 125–170.

Darden, W.H., Jr. Sexual differentiation in *Volvox aureus*. J. Protozool. 13:239–255; 1966.

Datta, T.; Kaur, J. Effect of fungal extracts on excystment of soil amoebae. Protistologica. 14:121–123; 1978.

Davis, P.G.; Caron, D.A.; Sieburth, J.McN. Oceanic amoebae from the North Atlantic: culture, distribution, and taxonomy. Trans. Amer. Micros. Soc. 97:73–88; 1978.

Davis, P.G.; Sieburth, J.McN. Estuarine and oceanic microflagellate predation of actively growing bacteria: estimation by frequency of dividing-divided bacteria. Mar. Ecol. Progr. Series. 19:237–246; 1984.

Dawkins, R. The Extended Phenotype. Oxford: Freeman; 307 pp.

Dehority, B. Protozoa of the digestive tract of herbivorous mammals. Insect Sci. Applic. (Special Issue, Progess in Protozoology). 7:279–296; 1986.

de la Cruz, V.F.; Gittleson, S.M. The genus *Polytomella*: A review of classification, morphology, life cycle, metabolism, and motility. Arch. Protistenk. 124:1–28; 1981.

De Lozanne, A.; Spudich, J.A. Disruption of the *Dictyostelium* myosin heavy chain gene by homologous recombination. Science. 236:1086–1091; 1987.

Dewey, V.C.; Heinrich, M.R.; Kidder, G.W. Evidence for the absence of the urea cycle in *Tetrahymena*. J. Protozool. 4:211–219; 1957.

Dewey, V.C.; Kidder, G.W. Growth studies on ciliates. VI. Diagnosis, sterilization, and growth characteristics of *Perspira ovum*. Biol. Bull. 79:255–271; 1940.

Dewey, V.C.; Kidder, G.W. Serine synthesis in *Tetrahymena* from non-amino acid sources; compounds derived from serine. J. Gen. Microbiol. 22:79–92; 1960.

Diamond, L.S. Axenic cultivation of *Entamoeba histolytica*. Science. 134:336–337; 1961.

Diamond, L.S.; Harlow, D.R.; Cunnick, C.C. A new medium for the axenic cultivation of *Entamoeba histolytica* and other *Entamoeba*. Trans. R. Soc. Trop. Med. Hyg. 72:431–432; 1978.

Diehn, B. Phototaxis and sensory transduction in *Euglena*. Science. 181:1009–1015; 1973.

Dillon, R.D.; Bierle, D.A. Microbiocoenoses in an antarctic pond. In: Giesy, J.P., Jr., ed. Microcosms in Ecological Research. Technical Information Center, U.S. Dept. of Energy, CONF-781101; 1980: pp. 446-457.

Dini, F.; Bracchi, P.; Luporini, P. Cellular cycle in two ciliate Hypotrichs. Acta Protozoologica. 14:59-66; 1975.

Dini, F.; Luporini, P. Genic determination of the autogamy trait in the hypotrich ciliate, *Euplotes crassus*. Genet. Res. Camb. 35:107-119; 1980.

Docampo, R.; Cruz, F.S.; Leon, W.; Schmunis, G.A. Acetate oxidation by bloodstream forms of *Trypanosoma cruzi*. J. Protozool. 26:301-303; 1979.

Dodge, J.D. The Fine Structure of Algal Cells. London: Academic Press; 1973: 261 pp.

Dodge, J.D. The fine structure of *Trachelomonas* (Euglenophyceae). Arch. Protistenk. 117:65-77; 1975.

Dodge, J.D. Atlas of Dinoflagellates—A Scanning Electron Microscope Survey. London: Farrand Press; 1985: 119 pp.

Dodge, J.D.; Crawford, R.M. The fine structure of the dinoflagellate *Oxyrrhis marina*. 1 & 2. Protistologica. 7:295-304; 399-409; 1971.

Dolphin, W.D. Effect of glucose on glycine requirements of *Acanthamoeba castellanii*. J. Protozool. 23:455-457; 1976.

Doran, D.J. Studies on Trichomonads. III. Inhibitors, acid production and substrate utilization by 4 strains of *Tritrichomonas foetus*. J. Protozool. 6:177-182; 1959.

Doughty, M.J.; Diehn, B. Photosensory transduction in the flagellated alga, *Euglena gracilis*. I. Action of divalent cations, Ca^{2+} antagonists and Ca^{2+} ionophore on motility and photobehavior. Biochim. et Biophys. Acta. 588:148-168; 1979.

Doughty, M.J.; Diehn, B. Anion sensitivity of motility and step-down photophobic responses of *Euglena gracilis*. Arch. Microbiol. 138:329-332; 1984.

Doughty, M.J.; Grieser R.; Diehn, B. Photosensory transduction in the flagellated alga, *Euglena gracilis*. II. Evidence that blue light effects alteration in Na^+/K^+ permeability of the photoreceptor membrane. Biochim. Biophys. Acta. 602:10-23; 1980.

Drainville, G.; Cagnon, A. Osmoregulation in *Acanthamoeba castellanii*. I. Variations of the concentrations of free intracellular amino acids and of the water content. Comp. Biochem. Physiol. 45A:379-388; 1973.

Droop, M.R. Heterotrophy of carbon. In: Stewart, W.D.P., ed. Algal Physiology and Biochemistry. Berkeley: University of California Press; 1974: pp. 530-582.

Duguay, L.; Taylor, D. Primary production and calcification by the soritid Foraminifer *Archaias angulatus* (Fichtell & Moll). J. Protozool. 25:356; 1978.

Dunham, P.B.; Child, F.M. Ion regulation in *Tetrahymena*. Biol. Bull. 121:129-140; 1961.

Dunham, P.B.; Kropp, D.L. In: Elliott, A.M., ed. Biology of Tetrahymena. Stroudsburg: Dowden, Hutchinson and Ross; 1973: pp. 165-198.

Dunstan, W.M. A comparison of the photosynthesis-light intensity relationship in photosynthetically different marine microalgae. J. Exp. Mar. Biol. Ecol. 13:181-187; 1973.

Dute, R.; Kung, C. Ultrastructure of the proximal region of somatic cilia in *Paramecium tetraurelia*. J. Cell Biol. 78:451-464; 1978.

Edwards, S.W.; Doulah, F.A. Elevation of AMP levels during phagocytosis in *Acanthamoeba castellanii*. J. Gen. Microbiol. 128:2919-2926; 1982.

Edwards, S.W.; Lloyd, D. Oscillations of respiration and adenine nucleotides in synchronous cultures of *Acanthamoeba castellanii*: Mitochondrial respiratory control in vivo. J. Gen. Microbiol. 108:197-204; 1978.

Eisler, K. Licht–und elektronenmikroskopische untersuchungen zur corticalen Morphologie und Morphogenese nassulider Ciliaten. Tübingen, W. Germany: Tübingen Univ.; 1986. 224 pp. Dissertation.

Elliott, A.M. The fine structure of *Teterahymena pyriformis* during mitosis. In: Levine, L., ed. The Cell in Mitosis. New York: Academic Press, 1963: pp. 107–121.

Elliott, A.M. Biology of *Tetrahymena*. Stroudsburg: Dowden, Hutchinson, and Ross, Inc.; 1973: 508 pp.

Ellis, B.F.; Messina, A.R. Catalogue of Foraminifera. New York: American Museum of Natural History; 1940. 270 pp. 30 vols.

Englemann G. Über Sauerstoffausscheidung von Pflanzenzellen in Microspectrum. Bot. Ztg. 40:419–426; 1882.

Esteve, J.-C. Pertubation du comportement alimentaire par atteinte du revetement cellulaire chez le cilié *Dileptus*. Protistologica. 17:479–488; 1981.

Ettienne, E.M. Control of contractility in *Spirostomum* by dissociated calcium ions. J. Gen. Physiol. 56:168–179; 1970.

Eyden, B.P. Light and electron microscope study of *Dunaliella primolecta* Butcher (Volvocida). J. Protozool. 22:336–344; 1975.

Fabczak, S. Measurements of intracellular potassium-activity in *Stentor coeruleus* with the use of an ion-selective microelectrode. Acta Protozool. 22:175–182; 1983.

Fairlamb, A.H. In: Klein, R.A.; Miller, P.G.G. Alternate metabolic pathways in protozoan energy metabolism. Parasitology. 82:1–3; 1981.

Farina, M.; Attias, M.; Souto-Padron, T.; DeSouza, W. Further studies on the organization of the paraxial rod of trypanosomatids. J. Protozool. 33:552–557; 1986.

Febvre-Chevalier, C. Organisation ultrastructurale des axopodes chez un héliozoaire. Protistologica. 9:35–43; 1972.

Febvre-Chevalier, C. (with contributions by E.C. Bovee and D. J. Patterson). IV. Class Heliozoea Haeckel 1866. In: Lee, J.J.; Hutner, S.H.; Bovee, E.C., eds. Illustrated Guide to the Protozoa. Lawrence, Kansas: Society of Protozoologists; 1985: pp. 302–317.

Febvre-Chevalier, C.; Bilbaut, A.; Bone, Q.; Febvre, J. Sodium-calcium action potential associated with contraction in the heliozoan *Actinocoryne contractilis*. J. Exp. Biol. 122:177–192; 1986.

Feinleib, M. E.; Curry, G.M. The nature of the phototactic receptor in phototaxis. In: Lowenstein, W.R., ed. Principles of Receptor Physiology. Heidelberg: Springer-Verlag; 1971a: pp. 366–395.

Feinleib, M.E.; Curry, G.M. The relation between stimulus intensity and oriented phototactic response (topotaxis) in *Chlamydomonas*. Physiol. Plant. 25:346–352; 1971b.

Fenchel, T. The ecology of marine microbenthos. I. The quantitative importance of ciliates as compared with metazoans in various types of sediments. Ophelia. 4:121–137; 1967.

Fenchel, T. The ecology of marine microbenthos II. The food of marine Benthic Ciliates. Ophelia. 5:73–121; 1968a.

Fenchel, T. The ecology of marine microbenthos III. The reproductive potential of ciliates. Ophelia. 5:123–136; 1968b.

Fenchel, T. The ecology of marine microbenthos IV. Structure and function of the benthic ecosystem, its chemical and physical factors and the microfauna communities with special reference to the ciliated protozoa. Ophelia. 6:1–182; 1969.

Fenchel, T. Intrinsic rate of natural increase: the relationship with body size. Oecologia (Berl.). 14:317–326; 1974.

Fenchel, T. Suspension feeding in ciliated protozoa: structure and function of feeding organelles. Archiv. Protistenkd. 123:239–260; 1980.

Fenchel, T. The bioenergetics of a heterotrophic microflagellate. Ann. Inst. Oceanogr. Paris. 58:55–60; 1982.

Fenchel, T. Suspended marine bacteria as food source. In: Fasham, M.J., ed. Energy and Materials in Marine Ecosystems. New York: Plenum Press; 1984: pp. 301–315.

Fenchel, T. The ecology of heterotrophic flagellates. In: Marshall, K.C., ed. Advances in Microbial Ecology. vol. 9. New York: Plenum Press: 1986a: pp. 57–97.

Fenchel, T. The Ecology of Protozoa. Madison, Wisconsin: Science Tech. Publishers; 1986b: 196 pp.

Fenchel, T.; Finlay, B.J. Geotaxis in the ciliated protozoon *Loxodes*. J. Exp. Biol. 110:17–33; 1984.

Fenchel, T.; Finlay, B.J. The structure and function of Müller vesicles in loxodid ciliates. J. Protozool. 33:69–76; 1986.

Fenchel, T.; Perry, T.; Thane, A. Anaerobiosis and symbiosis with bacteria in free-living ciliates. J. Protozool. 24:154–163; 1977.

Findenegg, I. Untersuchungen uber die Okologie und die Produktionsverhaltnisse des Planktons im Karntner Seegebiete. Internat. Rev. Ges. Hydrobiol. Hydrogr. 43:368–429; 1943.

Finlay, B.J. The dependence of reproductive rate on cell size and temperature in freshwater ciliated protozoa. Oecologia (Berl.). 30:75–81; 1977.

Finlay, B.J. Temporal and vertical distribution of ciliphoran communities in the benthos of a small eutrophic loch with particular reference to the redox profile. Freshwater Biol. 10:15–34; 1980.

Finlay, B.J. Oxygen availability and seasonal migrations of ciliated protozoa in a freshwater lake. J. Gen. Microbiol. 123:173–178; 1981.

Finlay, B.J. Effects of seasonal anoxia on the community of benthic ciliated protozoa in a productive lake. Arch. Protistenk. 125:215–222; 1982.

Finlay, B.J.; Curds, C.R.; Bamforth, S.S.; Bafort, J.M. Ciliated protozoa and other microorganisms from two African soda lakes (L. Nakuru & L. Simbi, Kenya). Arch. Protistenk. 133:81–91; 1987.

Finlay, B.J.; Fenchel, T. Physiological ecology of the ciliated protozoon *Loxodes*. Rep. Freshwat. Assn. 54:73–96; 1986.

Finley, H.E.; McLaughlin, D. Ecological studies on peritrichs. Am. Zool. 2:523, 1962.

Fischer-Defoy, D.; Hausmann, K. Untersuchungen zur phagocytose bei *Climacostomum virens*. Protistologica. 13:459–476; 1977.

Flickinger, C.J. The fine structure of the nuclei of *Tetrahymena pyriformis* throughout the cell cycle. J. Cell Biol. 27:519–529; 1965.

Flickinger, C.J. The fine structure of four "species" of *Amoeba*. J. Protozool. 21:59–68; 1974.

Foissner, W. The silverline system of ciliates: facts, hypotheses, problems. Mikroskopie. 38:16–26; 1981.

Foissner, W. Soil protozoa: fundamental problems, ecological significance, adaptation, indicators of environmental quality, guide to the literature. In: Corliss, J.O.; Patterson, D.J., eds. Progress in Protistology. 2:69–212; 1987.

Fok, A.K.; Lee, Y.; Allen, R.D. The correlation of digestive vacuole pH and size with the digestive cycle in *Paramecium caudatum*. J. Protozool. 29:409–414; 1982.

Fulton, J.D. Metabolism and pathogenic mechanisms of parasitic protozoa. In: Chen, T.-T., ed. Research in Protozoology. vol. 3. Oxford: Pergamon Press; 1969: pp. 389–504.

Gaffal, K.P.; Gaffal, S.I.; Schneider, G.J. Morphometric analysis of several intracellular events occurring during the vegetative life cycle of the unicellular alga *Polytoma papillatum*. Protoplasma. 110:185–195; 1982.

Gaffal, K.P.; Schneider, G.J. Morphogenesis of the plastidome and the flagellar apparatus during the vegetative life cycle of the colorless phytoflagellate *Polytoma papillatum*. Cytobios. 27:43–61; 1980a.

Gaffal, K.P.; Schneider, S. Numerical, morphological and topographical heterogeneity of the chondriome during the vegetative life cycle of *Polytoma papillatum*. J. Cell Sci. 46:299–312; 1980b.

Galloway, R.E.; Goodenough, U.W. Genetic analysis of mating locus linked mutations in *Chlamydomonas reinhardii*. Genetics. 111:447–461; 1985.

Gantt, E. Photosynthetic cryptophytes. In: Cox, E.R., Phytoflagellates. vol. 2. Amsterdam: Elsevier; 1980: pp. 381–405.

Gause, G.F. The Struggle for Existence. Baltimore: Williams and Wilkins; 1934: p. 96, et seq.

Gawlitta, W.; Stockem, W. Pinocytosis and locomotion of amoebae: 15. Visualization of calcium-dynamics by chlortetracycline fluorescence during induced pinocytosis in living *Amoeba proteus*. Cell Tissue Res. 213:9–20; 1980.

Gessat, M.; Jantzen, H. Die Bedeutung von Adenosin-3′,5′-monophosphat fur die Entwicklung von *Acanthamoeba castellanii*. Arch. Microbiol. 99:155–166; 1974.

Gibbs, S.P. The chloroplasts of *Euglena* may have evolved from symbiotic green algae. Can. Journ. Bot. 56:2883–2889; 1978.

Gibbs, S.P.; Mak, R.; Ng, R.; Slankis, T. The chloroplast nucleoid in *Ochromonas danica*. J. Cell Sci. 16:579–591; 1974.

Gibson, T.G.; Buzas, M. Species diversity: patterns in modern and Miocene foraminifera of the eastern margin of North America. Bull. Geol. Soc. Am. 84:217–238; 1973.

Giese, A.C. Blepharisma: The Biology of a Light-sensitive Protozoan. Palo Alto, California: Stanford University Press; 1973: 366 pp.

Giles, R.; Jaenicke, L. Extracellular phosphoproteins are involved in sexual differentiation of *Volvox carteri*. Hoppe-Seyle's Zeitschr. Physiol. Chemie. 365:990; 1984.

Giles, R.; Giles, C.; Jaenicke, L. Pheromone-binding and matrix-mediated events in sexual induction of *Volvox carteri*. Z. Naturforsch. 39c:584–592; 1984.

Gill, D.E. Intrinsic rates of increase, saturation densities, and competitive ability. I. An experiment with *Paramecium*. Amer. Nat. 106:461–471; 1972.

Gill, D.E.; Hairston, N.G. The dynamics of a natural population of *Paramecium* and the role of interspecific competition in community structure. J. Animal Ecol. 41:137–151; 1972.

Gnekow, M.A. Observations on the biology and ultrastructure of the moss-dwelling thecamoeba *Nebela tincta* (Rhizopoda). Arch. Protistenk. 124:36–69; 1981.

Gold, K.; Baren, C.F. Growth requirements of *Gyrodinium cohnii*. J. Protozool. 13:255–257; 1966.

Gold, K.; Morales, E.A. Seasonal changes in lorica sizes and the species of Tintinnida in the New York Bight. J. Protozool. 22:520–528; 1975.

Goldacre, R.J. The action of general anesthetics on amoebae and the response to touch. Symp. Soc. Exp. Biol. 6:128–144; 1952.

Goldman, J.C. A conceptual role for microaggregates in pelagic water. Bull. Mar. Sci. 35:462; 1984.

Goldman, J.C.; Caron, D.A. Experimental studies on an omnivorous microflagellate: implications for grazing and nutrient regeneration in the marine microbial food chain. Deep Sea Res. 32:899–915; 1985.

Golinska, K.; Doroszewski, M. The cell shape of *Dileptus* in the course of division and regeneration. Acta Protozoologica. 2:59–67; 1964.

Golinska, K.; Kink, J. The regrowth of oral structures in *Dileptus cygnus* after partial excision. Acta Protozoologica. 15:143–163; 1976.

Goulder, R. The vertical distribution of some ciliated protozoa in the plankton of a eutrophic pond during summer stratification. Freshwater Biol. 2:162–176; 1972.

Goulder, R. The seasonal and spatial distribution of some benthic ciliated protozoa in Esthwaite water. Freshwater Biol. 4:127–147; 1974.

Graves, L.B., Jr.; Becker, W.M. Beta-oxidation in glyoxysomes from *Euglena*. J. Protozool. 21:771–774; 1974.

Graves, L.B., Jr.; Trelease, R.N.; Grill, A.; Beckerd, W.E. Localization of glyoxylate cycle enzymes in glyoxysomes in *Euglena*. J. Protozool. 19:527–532; 1972.

Gray, W.D.; Alexopoulos, C.J. Biology of the Myxomycetes. New York: Ronald Press; 1968; 269 pp.

Grebecki, A. Organization of motory functions in amoebae and in slime moulds plasmodia. Acta Protozoologica. 18:43–58; 1979.

Grebecki, A. Behavior of *Amoeba proteus* exposed to light-shade difference. Protistologica. 16:103–116; 1980.

Grebecki, A. Effects of localized photic stimulation on amoeboid movement and their theoretical implications. Eur. J. Cell Biol. 24:163–175; 1981.

Grebecki, A. Relationship between cytoskeleton and motility. Insect Sci. Applic. Special Issue on Progress in Protozoology. 7:379–386; 1986.

Grebecki, A.; Cieslawski, M. Motive force generation site in plasmodium of *Physarum polycephalum*, a dissection study. Acta Protozoologica. 23:123–134; 1984.

Grebecki, A.; Grebecka, L. Morphodynamic types of *Amoeba proteus*: A terminological proposal. Protistologica. 14:349–358; 1978.

Grebecki, A.; Kolodziejczyk, J. Contraction and streaming relations recorded simultaneously at two points along the plasmodial veins and frontal channels of *Physasrum polycephalum*. Acta Protozoologica. 22:1–18; 1983.

Green, J.C.; Pienaar, R.N. The taxonomy of the order Isochrysidales (Prymnesiophyceae) with special reference to the genera *Isochrysis* Parke, *Dicrateria* Parke and *Imantonia* Reynolds. J. Mar. Biol. Assn., U. K. 57:7–17; 1977.

Grell, K.G. Protozoology. Berlin: Springer-Verlag; 1973: 554 pp.

Grell, K.; Benwitz, G. Die Zellhulle von *Paramoeba eilhardi* Schaudinn. Z. Naturforsch. 21b:600–601; 1966.

Grell, K.; Benwitz, G. Ultrastruktur mariner Amöben. I. *Paramoeba eilhardi* Schaudinn. Arch Protistenk. 112:119–137; 1970.

Griffin, J.L. Pathogenic freeliving amoebae. In: Krier, J.P., ed. Parasitic Protozoa. vol. 2. New York: Academic Press; 1978: pp. 507–549.

Griffiths, A.J.; Hughes, D.E. The physiology of *Hartmanella castellanii*. J. Protozool. 16:93–99; 1969.

Grim, J.N. Fine structure of the surface and infraciliature of *Gastrostyla steinii*. J. Protozool. 19:113–126; 1972.

Grunewald, J. The hydro-chemical and physical conditions of the environment of the immature stages of some species of the simulium (Edward-sellum) damnosum complex (Diptera). Tropenmedizin und Parasitologie. 27:438–454; 1976.

Gutteridge, W.E. In: Klein, R.A.; Miller, P.G.G., eds. Alternate metabolic pathways in protozoan energy metabolism. Parasitology. 82:6–7; 1981.

Guttman, H.N. First defined media for *Leptomonas* spp. from insects. J. Protozool. 13:390–392; 1966.

Haga, N.; Forte, M.; Ramanathan, R.; Hennessey, T.; Takahashi, M.; Kung, C. Characterization and purification of a soluble protein controlling calcium-channel activity in *Paramecium*. Cell 39:71–78; 1984.

Haines, T.H. Halogen- and sulfur-containing lipids of *Ochromonas*. Annu. Rev. Microbiol. 27:403–411; 1973.

Hairston, N.G. Studies on the limitation of a natural population of *Paramecium aurelia*. Ecology. 48:904–910; 1967.

Hairston, N.G.; Allan, J.D.; Colwell, R.K.; Futuyma, D.J.; Howell, J.; Lubin, M.D.; Mathias, J.; Vandermeer, J.H. The relationship between species diversity and stability: An experimental approach with protozoa and bacteria. Ecology. 49:1091–1100; 1968.

Hairston, N.G.; Smith, F.E.; Slobodkin, L.B. The relationship between species diversity and stability: An experimental approach with protozoa and bacteria. Ecology. 49:1091–1101; 1968.

Halevy, S.; Finkelstein, S. Lipid composition of soil amoebae. J. Protozool. 12:250–252; 1965.

Hall, R.P. Nutrition and growth of protozoa. In: Chen, T.-T., ed. Research in Protozoology. New York: Pergamon Press; 1967: pp. 338–404.

Hamilton, R.D.; Preslan, J.E. Cultural characteristics of a pelagic marine hymenostome ciliate, *Uronema* sp. J. Exp. Mar. Biol. Ecol. 4:90–99; 1969.

Hanson, W.L.; McGhee, R.B.; Blake, J.D. Experimental infection of various latex plants of the family Asclepiadaceae with *Phytomonas elmassiani*. J. Protozool. 13:324–327; 1966.

Harrison, F.W.; Dunkelberger, D.; Watabe, N.; Stump, A.B. The cytology of the testaceous rhizopod *Lesquereusia spiralis* (Ehrenberg) Penard. I. Ultrastructure and shell formation. J. Morph. 150:343–358; 1976.

Hauser, D.C.; Levandowsky, R.M.; Hutner, S.H.; Chunosoff, L.; Hollwertz, J.S. Chemosensory responses by the heterotrophic marine dinoflagellate *Crypthecodinium cohnii*. Microb. Ecol. 1:246; 1975.

Hausman, L.A. Observations on the ecology of protozoa. Amer. Natur. 51:157–172; 1917.

Hausmann, K.; Mignot, J.P. Untersuchungen und den mucocysten von *Euglena splendens* Dangeard 1901. Protistologica. 13:213–217; 1977.

Hausmann, K.; Mulisch, M. Das Epiplasma des Ciliaten *Pseudomicrothorax dubius*, ein Cytoskelett. Arch. Protistenk. 124:410–416; 1981.

Hausmann, K.; Peck, R.K. Microtubules and microfilaments as major components of a phagocytic apparatus: the cytopharyngeal basket of the ciliate *Pseudomicrothorax dubius*. Differentiation. 11:157–167; 1978.

Hausmann, K.; Peck, R.K. The mode of function of the pharyngeal basket of the ciliate *Pseudomicrothorax dubius*. Differentiation. 14:147–158; 1979.

Haynes, J.R. Foraminifera. New York: Wiley and Sons; 1981: 453 pp.

Heal, O.W. Quantitative studies on soil amoeba. In: Graff, O.; Satchell, J.E., eds. Progress in Soil Biology. Braunschweig: F. Vieweg; 1967: pp. 120–126.

Hecky, R.E.; Kling, H.J. The phytoplankton and protozooplankton of the euphotic zone of Lake Tanganyika: Species composition, biomass, chlorophyll content, and spatio-temporal distribution. Limnol. and Oceanogr. 26:548–564; 1981.

Hedley, R.H.; Ogden, C.G. Biology and fine structure of *Euglypha rotunda* (Testacea: Protozoa). Bull. Brit. Mus. (Nat. Hist.), Zoology. 25:121–137; 1973.

Hedley, R.H.; Ogden, C.G. Observations on *Trinema lineare* Penard (Testacea: Protozoa). Bull. Brit. Mus. (Nat. Hist.), Zoology. 26:187–199; 1974.

Hedley, R.H.; Wakefield, J.St.J. Fine structure of *Gromia oviformis* (Rhizopodea: Protozoa). Bull. Brit. Mus. (Nat. Hist.), Zoology. 18:5–89; 1969.

Heinbokel, J.F. Studies on the functional role of tintinnids in the Southern California Bight. I. Grazing and growth rates in laboratory cultures. Mar. Biol. 47:177–189; 1978.

Heinrich, B.; Cook, J.R. Studies on the respiratory physiology of *Euglena gracilis* cultured on acetate or glucose. J. Protozool. 14:548–553; 1967.

Heiple, J.M.; Taylor, D.L. pH Changes in pinosomes and phagosomes in amoeba, *Chaos carolinensis*. J. Cell Biol. 94:143–149; 1982.

Helenski, L.L.; Walne, P.L. Ultrastructure of mucocysts in *Peranema trichophorum* (Euglenophyceae). J. Protozool. 30:491–496; 1983.

Hemleben, C.; Anderson O.R.; Berthold, W.; Spindler, M. Calcification and chamber formation in Foraminifera – a brief overview. In: Leadbeater, B. S. C.; Riding, R., eds. Biomineralization in Lower Plants and Animals. Oxford: Clarendon Press; 1986: pp. 237–249.

Hemleben, C.; Bé, A.W.H.; Anderson, O.R.; Tuntivate, S. Test morphology, organic layers and chamber formation of the planktonic foraminifer *Globorotalia menardii* (D'Orbigny). J. Foram. Res. 7:1–25; 1977.

Hennessey, T.; Machemer, H.; Nelson, D.L. Injected cyclic AMP increases ciliary beat frequency in conjunction with membrane hyperpolarization. European J. Cell Biol. 36:153–156; 1985.

Hibberd, D.J. Prymnesiophytes (= Haptophytes). In: Cox, E.R., ed. Phytoflagellates. vol. 2. New York: Elsevier; 1980: pp. 273–317.

Hibberd, D.; Leedale, G. Order 4. Chrysomonadida. In: Lee, J.J.; Hutner, S.; Bovee, E.C., eds. Illustrated Guide to the Protozoa. Lawrence, Kansas: Soc. of Protozoologists; 1985: pp. 54–88.

Hills, G.J.; Gurney-Smith, M.; Roberts, K. Structure, composition and morphogenesis of the cell wall of *Chlamydomonas reinhardtii*. II. Electron microscopy and optical diffraction analysis. J. Ultrastruc. Res. 43:179–192; 1973.

Hinrichsen, R.D.; Kung, C. Genetic analysis of axonemal mutants in *Paramecium tetraurelia* defective in their response to calcium. Genet. Res. Camb. 43:11–20; 1984.

Hinrichsen, R.D.; Saimi, Y.; Kung, C. Mutants with altered Ca^{2+}-channel properties in *Paramecium tetraurelia*: isolation, characterization and genetic analysis. Genetics. 108:545–558; 1984.

Hirschberg, R.; Stavis, R. Phototaxis mutants of *Chlamydomonas reinhardtii*. J. Bacteriol. 129:803–808; 1977.

Hitchen, E.T. The fine structure of the colonial kinetoplastid flagellate *Cephalothamnium cyclopum* Stein. J. Protozool. 21:221–231; 1974.

Hiwatashi, K.; Luporini, P. Cell interactions in sexual phemonena. Insect. Sci. Applic. Special Edition: Progress in Protozoology. 7:433–439; 1986.

Ho, T.S.-S.; Alexander, M. The feeding of amebae on algae in culture. J. Phycol. 10:95–100; 1974.

Hoffmann, C.; Bouck, G.B. Immunological and structural evidence for patterned intussusceptive surface growth in a unicellular organism. J. Cell Biol. 69:693–715; 1976.

Hogg, J.F. Peroxisomes in *Tetrahymena* and their relation to gluconeogenesis. In: Hogg, J.F., ed. The Nature and Function of Peroxisomes (Microbodies, Glyoxysomes). Ann. N. Y. Acad. Sci. 168:281–291; 1969.

Hoham, R.W. Unicellular Chlorophytes—Snow Algae. In: Cox, E.R., ed. Phytoflagellates. New York: Elsevier; 1980: pp. 61–84.

Hollande, A. Identification du parasome (Nebenkern) de *Janickina pigmentifera* a un symionted (*Perkinsiella amoebae* nov. gen-nov. sp.) apparente aux flagelles kinetoplastidies. Protistologica. 17:147–154; 1980.

Holz, G.G., Jr. The nutrition of *Tetrahymena*: Essential nutrients, feeding, and digestion. In: Elliott, A.M., ed. Biology of Tetrahymena. Stroudsburg: Dowden, Hutchinson & Ross; 1973: pp. 89–98.

Holz, G.G., Jr.; Conner, R.L. The composition, metabolism, and roles of lipids in *Tetrahymena*. In: Elliott, A.M., ed. Biology of Tetrahymena. Stroudsburg: Dowden, Hutchinson & Ross; 1973: pp. 99–122.

Holz, G.G.; Wagner, B.; Erwin, J.; Kessler, D. The nutrition of *Glaucoma chattoni* A. J. Protozool. 8:192–199; 1961.

Honigberg, B.M. Trichomonads of importance in human medicine. In: Kreier, J.P., ed. Parasitic Protozoa. vol. 2. New York: Academic Press; 1978a: pp. 275–454.

Honigberg, B.M. Trichomonads of veterinary importance. In: Krier, J.P., ed. Parasitic Protozoa. New York: Academic Press; 1978b: pp. 220–231.

Honigberg, B.M.; King, V.M. Structure of *Trichomonas vaginalis* Donné. J. Parasitol. 50:345–364; 1964.

Honigberg, B.M.; Mattern, C.F.T.; Daniel, W.A. Fine structure of the mastigont system in *Tritrichomonas foetus* (Riedmüller). J. Protozool. 18:183–198; 1971.

Hosotani, K.; Yokota, A.; Nakano, Y.; Kitaoka, S. The metabolic pathway of propionate in *Euglena gracilis* Z grown under illumination. Agric. Biol. Chem. 44:1097–1104; 1980.

Hottinger, L. Comparative anatomy of elementary shell structures in selected larger Foraminifera. In: Hedley, R.H.; Adams, C.G., eds. Foraminifera. vol. 3. London: Academic Press; 1978: pp. 203–266.

Hottinger, L.; Dreher, D. Differentiation of protoplasm in Nummulitidae (Foraminifera) from Elat, Red Sea. Mar. Biol. 25:41–61; 1974.

Hottinger, L.; Leutenegger, S. The structure of calcarinid Foraminifera. Schweiz Palaontol. Abhandl. 101:115–151; 1980.

Huang, B.; Pitelka, D.R. The contractile process in the ciliate *Stentor coeruleus*. I. The role of microtubules and microfilaments. J. Cell Biol. 57:704–728; 1973.

Hudock, G.A.; Hudock, M.O. Phototaxis: Isolation of mutant strains of *Chlamydomonas reinhardi* with reversed sign of response. J. Protozool. 20:139–140; 1973.

Hufnagel, L.A. The cell cortex during ciliate morphogenesis and ciliogenesis. Insect Sci. Applic. Special Edition on Progress in Protozoology. 7:249–260; 1986.

Hulsmann, N., Haberey, M. Phenomena of amoeboid movement. Behavior of the cell surface in *Hyalodiscus simplex* Wohlforth-Bottermann. Acta Protozoologica. 12:1171–1182; 1973.

Hunter, F.R.; Lee, J.W. On the metabolism of *Astasia longa* (Jahn). J. Protozool. 9:74–78; 1962.

Hurlbert, R.E.; Bates, R.C. Glucose utilization by *Euglena gracilis* var. *bacillaris* at higher pH. J. Protozool. 18:298–306; 1971.

Hurlbert, S.H. The nonconcept of species diversity: a critique and alternative parameters. Ecology. 52:577–586; 1971.

Hurlbert, S.H. The measurement of niche overlap and some relatives. Ecology. 59:67–77; 1978.

Hutchinson, G.E. A treatise on limnology. vol. 1, Geography, Physics, and Chemistry. New York: Wiley; 1957.
Hutner, S.H. The nutritional requirements of two species of *Euglena*. Arch. Protistenk. 88:93–106; 1936.
Hutner, S.H. Induced enzyme synthesis in the phytoflagellate, *Polytoma*. J. Protozool. 3:69–74; 1956.
Hutner, S.H.; Bacchi, C.J.; Baker, H. Nutrition of the Kinetoplastida. In: Lumsden, W.H.R.; Evans, D.A., eds. Biology of the Kinetoplastida. vol. 2. London: Academic Press; 1979: pp. 653–692.
Hutner, S.H.; Bach, M.K.; Ross, G.I.M. A sugar-containing medium for vitamin B_{12} assay with *Euglena*, application to body fluids. J. Protozool. 3:101–112; 1956.
Hutner, S.H.; Baker, H.; Aaronson, S.; Nathan, H.A.; Rodriguez, E.; Lockwood S.; Sanders, M.; Petersen, R.A. Growing *Ochromonas malhamensis* above 35°C. J. Protozool. 4:259–269; 1957.
Hutner, S.H.; Baker, H.; Frank, O.; Cox, D. Nutrition and metabolism in protozoa. Ch. 5. In: Fiennes, R.N., ed. International Encyclopedia of Food and Nutrition. vol. 18. New York: Pergamon Press; 1972: pp. 85–117.
Hutner, S.H.; Provasoli, L. The phytoflagellates. In: Lwoff, A., ed. Biochemistry and Physiology of Protozoa. New York: Academic Press; 1951: pp. 27–128.
Hutner, S.H.; Provasoli, L. Comparative biochemistry of flagellates. In: Hutner, S.H.; Lwoff, A., eds. Biochemistry and Physiology of Protozoa. II. New York: Academic Press; 1953: pp. 17–43.
Ingalls, C.G.; Brent, M.M. Defined minimal growth medium for *Acanthamoeba polyphaga*. J. Protozool. 30:606–608; 1983.
Jackson, K.M.; Berger, J. Life history attributes of some ciliated protozoa. Trans. Am. Microsc. Soc. 104:52–63; 1985.
Jacobs, M.H. Some aspects of cell permeability to electrolytes. Cold Spring Harbor Sympos. Quant. Biol. 16; 1940.
Jacobson, D.M. The ecology, feeding biology of thecate heterotrophic dinoflagellates. Woods Hole Oceanographic Institution, M.I.T. Joint Program; 1987, 210 pp. Thesis.
Jacobsen, D.M.; Anderson D.M. Thecate heterotrophic dinoflagellate feeding behavior and mechanism. J. Phycol. 22:249–258; 1986.
Janisch, R.A freeze-etch study of the ultrastructure of *Colpoda culcullus* protective cysts. Acta Protozool. 19:239–246; 1980.
Jantzen, H. Das Adenosinphosphat-System wahrend Wachstum und Entwicklung von *Acanthamoeba castellanii*. Arch. Microbiol. 101:391–399; 1974.
Jennings, H.S. Behaviour of Lower Organisms. Bloomington: Indiana Press; 1906.
Jensen, A.L.; Ball, R.C. Variation in the availability of food as a cause of fluctuations in predator and prey population densities. Ecology. 51:517–520; 1970.
Jeon, K.W.; Jeon, M.S. Scanning electron microscope observations of *Amoeba proteus* during phagocytosis. J. Protozool. 23:83–86; 1976.
Johansson, P.; Josefsson, J.-O. Selective inhibition of calcium-stimulated cation-induced pinocytosis by starvation and inhibitors of protein synthesis in *Amoeba proteus*. Exp. Cell Res. 154:367–375; 1984.
Johnson, W.H.; Brennan, M.D.; Berard, D.K.; Morrow, J.H.; Hudson, K.D. Proteins and polysaccharides in the nutrition of *Paramecium multimicronucleatum*. J. Protozool. 5 (Suppl.):14; 1980.
Jones, T.W.; Galloway, R.A. Effect of light quality and intensity on glycerol content in *Dunaliella teritolecta* (Chlorophyceae) and the relationship to cell growth/osmoregula-

tion. J. Phycol. 15:101–106; 1979.

Josefsson, J.-O. Studies on the mechanism of induction of pinocytosis in *Amoeba proteus*. Acta Physiol. Scand. Suppl. 432:6–65; 1975.

Jurand, A.; Selman, G.G. The Anatomy of Paramecium Aurelia. London: Macmillan; New York: St. Martin's Press; 1969: 218 pp.

Just, E. Untersuchungen zur zeitlichen Ordnung der Syntheseleistungen in der Interphase von *Paramecium busaria*. Arch. Protistenk. 115:22–68; 1973.

Kaneshiro, E.S.; Holz, G.G.; Dunham, P.B. Osmoregulation in a marine ciliate *Miamiensis avidus* 2. Regulation of intracellular free amino acids. Biol. Bull. 137:161–169; 1969.

Kaneshiro, E.S.; Holz, G.G., Jr. Observations on the ultrastructure of *Uronema* spp., marine scuticociliates. J. Protozool. 23:503–517; 1976.

Kaneshiro, E.S.; Beischel, L.S.; Merkel, S.J.; Rhoads, D.E. The fatty acid composition of *Paramecium aurelia* cells and cilia: changes with culture age. J. Protozool. 26:147–158; 1979.

Karpenko, A.A.; Railkin, A.I.; Seravin, L.N. Feeding behavior of unicellular animals. II. The role of prey mobility in the feeding behaviour of protozoa. Acta Protozoologica. 16:333–344; 1977.

Kauss, H. Biochemistry of osmotic regulation. Int. Rev. Biochem. 13:120–140; 1977.

Keller, S.E.; Hutner, S.H.; Keller, D.E. Rearing the colorless marine dinoflagellate *Cryptothecodinium cohnii* for use as a biochemical tool. J. Protozool. 15:792–795; 1968.

Kempner, E.S.; Miller, J.H. The molecular biology of *Euglena gracilis* III. General carbon metabolism. Biochem. 4:2735–2739; 1965.

Kempner, E.S.; Miller, J.H. The molecular biology of *Euglena gracilis* VII. Inorganic requirements for a minimal culture medium. J. Protozool. 19:343–346; 1972.

Kennedy, J.R., Jr. The morphology of *Blepharisma undulans* Stein. J. Protozool. 12:542–561; 1965.

Kidder, G.W.; Dewey, V.C. The biochemistry of ciliates in pure culture. In: Lwoff, A., ed. Biochemistry and Physiology of Protozoa. vol. 1. New York: Academic Press; 1951: pp. 323–400.

Kidder, G.W. Nitrogen: distribution, nutrition, and metabolism. In: Florkin, M.; Scheer, B.T.; Kidder, G.W., eds. Chemical Zoology vol. 1. Ch. 4. New York: Academic Press; 1967: pp. 93–159.

Kiefer, D.A.; Mitchell, B.G. A simple, steady state description of phytoplankton growth based on absorption cross section and quantum efficiency. Limnol. Oceanogr. 28:770–776; 1983.

Kirk, D.L.; Kirk, M.M. Protein synthetic patterns during the asexual life cycle of *Volvox carteri*. Dev. Biol. 96:493–506; 1983.

Kirk, D.L.; Kirk, M.M. Heat shock elicits production of sexual inducer in *Volvox*. Science. 231:51–54; 1986.

Kitching, J.A. Contractile vacuoles, ionic regulation and secretion. In: Chen, T.-T., ed. Research in Protozoology vol. 1. New York: Pergamon Press; 1967: pp. 307–336.

Klausener, C. Die Blutseen der Hochalpen. Internat. Rev. Ges. Hydrobiol. Hydrogr. 1:359–424; 1908.

Klein, R.A. In: Klein, R.A.; Miller, P.G.G., eds. Alternate metabolic pathways in protozoan energy metabolism. Parasitology. 82:3–5; 1981.

Klein, R.A.; Miller, P.G.G. Workshop no. 1. Alternate metabolic pathways in protozoan energy metabolism. Parasitology. 82:1–30; 1981.

Klopocka, W. Intracellular factors influencing the direction of new pseudopodia produced by *Amoeba proteus*. Protistologica. 18:389–396; 1982.

Knight-Jones, E.W. Relations between metachronism and the direction of ciliary beat in Metazoa. Q. Jl. Microsc. Sci. 95:503–521; 1954.

Kochert, G. Differentiation of reproductive cells in *Volvox carteri*. J. Protozool. 15:438–452; 1968.

Kolkwitz, R.; Marsson, M. Ökologie der pflanzlichen saprobien. Ber. Deutsch. Bot. Ges. (A). 26:505–519; 1908.

Kolodziejczyk, J.; Grebecki, A. Further studies on the relation between contraction and streaming oscillations in the plasmodial veins of *Physarum polycephalum*. Acta Protozoologica. 21:37–53; 1982.

Kolodziejczyk, J.; Grebecki, A. Effects of white-red illumination changes on the coordination of some motor functions in plasmodia of *Physarum polycephalum*. Acta Protozoologica. 22:19–31; 1983.

Korn, E.D. Biosynthesis of unsaturated fatty acids in *Acanthamoeba* sp. J. Biol. Chem. 239:396–400; 1964.

Korohoda, W. Experimental induction of locomotion in enucleated fragments of *Amoeba proteus* and its bearing on the theories of amoeboid movement. Cytobiologie. 14:338–349; 1977.

Kuhlmann, H.-W.; Heckmann, K. Interspecific morphogens regulating prey-predator relationships in protozoa. Science. 227:1347–1349; 1985.

Kyle, D.E.; Noblet, G.P. Seasonal distribution of thermotolerant free-living amoebae. I. Willard's Pond. J. Protozool. 33:422–434; 1985.

Kyle, D.E.; Noblet, G.P. Seasonal distribution of thermotolerant free-living amoebae. II. Lake Issaqueena. J. Protozool. 34:10–15; 1987.

Landis, W.G. The spatial and temporal distribution of *Paramecium bursaria* in the littoral zone. J. Protozool. 29:159–161; 1982.

Lang, N.J. Electron-microscopic demonstration of plastids in *Polytoma*. J. Protozool. 10:333–339; 1963.

Langdon, C. On the causes of interspecific differences in the growth:irradiance relationship for phytoplankton. J. Plank. Res. 9:459–482; 1987.

Langreth, S.G.; Balber, A.L. Protein uptake and digestion in bloodstream and culture forms of *Trypanosoma brucei*. J. Protozool. 22:40–53; 1975.

Lasman, M. Light and electron microscopic observations on encystment of *Acanthamoeba palestinensis*, Reich. J. Protozool. 24:244–248; 1977.

Latorella, A.H.; Vadas, R.L. Salinity adaptation by *Dunaliella tertiolecta*. I. Increases in carbonic anhydrase activity and evidence for light-dependent Na^+/H^+ exchange. J. Phycol. 9:273–277; 1973.

Laval-Peuto, M. Cortex périlemme et réticulum vésiculeux de *Cyttarocylis brandti* (cilié tintinnide). Les ciliés a Périlemme. Prostistologica. 11:83–98; 1975.

Laverde, A.V.; Brent, M.M. Simplified soluble media for the axenic cultivation of *Naegleria*. Protistologica. 16:11–15; 1980.

Laybourn, J.E.M. Respiratory energy losses in *Stentor coeruleus* Ehrenberg (Ciliophora). Oecologia (Berlin). 21:273–278; 1975.

Laybourn, J.E.M. Energy budgets for *Stentor coeruleus* Ehrenberg (Ciliophora). Oecologia (Berl.). 22:431–437; 1976.

Laybourn, J.E.M.; Finlay, B.J. Respiratory energy losses related to cell weight and temperature in ciliated protozoa. Oecologia (Berl.). 24:349–355; 1976.

Laybourn, J.E.M.; Stewart, J.M. Studies on consumption and growth in the ciliate *Col-

pidium campylum Stokes. J. Anim. Ecol. 44:165–174; 1975.
Leadbeater, B.S.C. Distribution and chemistry of microfilaments in choanoflagellates, with special reference to the collar and other tentacle systems. Protistologica. 19:157–166; 1983.
Leadbeater, B.S.C.; Riding, R., eds. Biomineralization in Lower Plants and Animals. Oxford: Clarendon Press; 1986: 401 pp.
Lee, J.J. Towards understanding the niche of Foraminifera. In: Hedley, R.H.; Adams, C.G., eds. *Foraminifera*. vol. 1. London: Academic Press; 1974: pp. 201–260.
Lee, J.J. Nutrition and physiology of the Foraminifera. In: Levandowsky, M.; Hutner, S.H., eds. Biochemistry and Physiology of Protozoa. New York: Academic Press; 1980a: p. 45.
Lee, J.J. Informational energy flow as an aspect of protozoan nutrition. J. Protozool. 27:5–9; 1980b.
Lee, J.J. Trichomonadida. In: Lee, J.J.; Hutner, S.H.; Bovee, E.C., eds. Illustrated Guide to the Protozoa. Lawrence, Kansas: Society of Protozoologists; 1985: pp. 119–127.
Lee, J.J.; Hallock, P. Algal symbiosis as the driving force in the evolution of larger Foraminifera. Endocytobiology III. In: Proceedings of the New York Academy of Sciences, 1987: pp. 330–347.
Lee, J.J.; McEnery, M.E. Autogamy in *Allogromia laticollaris* (Foraminifera). J. Protozool. 17:184–195; 1970.
Lee, J.J.; Muller, W.A. Trophic dynamics and niches of salt marsh Foraminifera. Amer. Zool. 13:215–223; 1973.
Lee, J.J.; Pierce, S. Growth and physiology of Foraminifera in the laboratory; Pt. IV – Monoxenic culture of an allogromiid with notes on its morphology. J. Protozool. 10:404–411; 1963.
Lee, J.J.; Hutner, S.H.; Bovee, E.C., eds. Illustrated Guide to the Protozoa. Lawrence, Kansas: Society of Protozoologists; 1985a: pp. 16–17.
Lee, J.J.; Lee, M.J.; Weis, D.S. Possible adaptive value of endosymbionts to their protozoan hosts. J. Protozool. 32:380–382; 1985b.
Lee, R.E. Formation of scales in *Paraphysomonas vestita* and the inhibition of growth by germanium oxide. J. Protozool. 25:163–166; 1978.
Leedale, G.F.; Hibberd, D.J. Observations on the cytology and fine structure of the euglenoid genera *Menoidium* Perty and *Rhabdomonas* Fresnius. Arch. Protistenk. 116:319–345; 1974.
Legner, M. Experimental approaches to the role of protozoa in aquatic ecosystems. Amer. Zool. 13:177–192; 1973.
Leick, V.; Hellung-Larsen, P. Chemosensory response in *Tetrahymena*: involvement of peptides and other signal substances. J. Protozool. 32:550–553; 1985.
Lembi, C.A. Unicellular chlorophytes. In: Cox, E.R., ed. Phytoflagellates. vol. 2. New York: Elsevier; 1980: pp. 5–59.
Lembi, C.A.; Lang, N.J. Electron microscopy of *Carteria* and *Chlamydomonas*. Am. J. Bot. 52:464–477; 1965.
Leslie, P.H. An analysis of the data for some experiments carried out by Gause with populations of the protozoa, *Paramecium aurelia* and *Paramecium caudatum*. Biometrika. 44:314–327; 1957.
Levandowsky, M.; Hutner, S. Introduction. In: Levandowsky, M.; Hutner, S., eds. Biochemistry and Physiology of Protozoa. 2nd ed. vol. 2. New York: Academic Press; 1979: p. 6.

Levandowsky, M.; Hutner, S., eds. Biochemistry and Physiology of Protozoa. vol. 3. New York: Academic Press; 1980: 446 pp.

Levine, N.D.; Corliss, J.O.; Cox, F.E.G.; Deroux, G.; Grain, J.; Honigberg, B.M.; Leedale, G.F.; Loeblich, A.R.; Lom, J.; Lynn, D.; Merinfeld, E.G.; Page, F.C.; Poljansky, G.; Sprague, V.; Vavra, J.; Wallace, F.G. A newly revised classification of the protozoa. J. Protozool. 27:37–58; 1980.

Lewis, E.; Munger, G.; Watson, R.; Wise, D. Life cycle of *Polytomella caeca* (Phytomonadida, Polyblepharidae). J. Protozool. 21:647–649; 1974.

Lindberg, R.E.; Bovee, E.C. Induction of phagocytosis and "cannibalism" by the giant ameba *Chaos carolinensis*. J. Protozool. 23:333–336; 1976.

Lindblom, G.P. Carbohydrate metabolism of trichomonads: growth, respiration, and enzyme activity in four species. J. Protozool. 8:139–150; 1961.

Lindholm, T. *Mesodinium rubrum*—a unique photosynthetic ciliate. Adv. in Aquat. Microbiol. 3:1–48; 1985.

Lindmark, D.G. Energy metabolism of the anaerobic protozoan *Giardia lamblia*. Mol. Biochem. Parasitol. 1:1–12; 1980.

Linnebach, M.; Hausmann, K.; Patterson, D.J. Ultrastructural studies on the food vacuole cycle of a heliozoan (feeding by *Actinophrys sol*): 3. Protoplasma 115:43–51; 1983.

Lloyd, D.; Cantor, M.H. Subcellular structure and function in acetate flagellates. In: Levandowsky, M.; Hutner, S.H., eds. Biochemistry and Physiology of Protozoa. 2nd ed. vol. 2. New York: Academic Press; 1979: pp. 9–63.

Lloyd, D.; Phillips, C.A.; Statham, M. Oscillations of respiration, adenine nucleotide levels and heat evolution in synchronous cultures of *Tetrahymena pyriformis* ST prepared by continuous-flow selection. J. Gen. Microbiol. 106:19–26; 1978.

Loeblich, A. R.; Tappan, H. Sarcodina, chiefly "Thecamoebians" and Foraminiferida: treatise on invertebrate paleontology, part C, Protista 2. vols. 1–2. Geological Society of America and University of Kansas Press; 1964: 900 pp.

Loeblich, L. Growth limitation of *Dunaliella salina* by CO_2 at high salinity. J. Phycol. 6 (Suppl.):9; 1970.

Lousier, J.D. Response of soil testacea to soil moisture fluctuations. Soil Biol. Biochem. 6:235–239; 1974.

Lousier, J.D. Colonization of decomposing deciduous leaf litter by Testacea (Protozoa, Rhizopoda): species succession, abundance and biomass. Oecologia. 52:381–388; 1982.

Lousier, J.D. Population dynamics and production studies of species of Nebelidae (Testacea, Rhizopoda) in an aspen woodland soil. Acta Protozoologica. 23:145–159; 1984a.

Lousier, J.D. Population dynamics and production studies of species of Euglyphidae (Testacea, Rhizopoda, Protozoa) in aspen woodland soil. Pedobiologia. 26:309–330; 1984b.

Lousier, J.D. Population dynamics and production studies of *Phryganella acropodia* and *Difflugiella oviformis* (Testacea, Rhizopoda, Protozoa) in aspen woodland soil. Pedobiologia. 26:331–347; 1984c.

Lousier, J.D. Annual population dynamics and production ecology of testacea (Protozoa, Rhizopoda) in an aspen woodland soil. Soil Biol. Biochem. 16:103–114; 1984d.

Lousier, J.D. Population dynamics and production studies of species of Centropyxidae (Testacea, Rhizopoda) in an aspen woodland soil. Arch. Protistenkd. 130:165–178; 1985.

Luckinbill, L.S.; Fenton, M.M. Regulation and environmental variability in experimental populations of protozoa. Ecology. 59:1271–1276; 1978.

Luckinbill, L.S. Regulation and diversity in a model experimental microcosm. Ecology. 60:1098–1102; 1979.

Lui, N.S.T.; Roels, O.A. Nitrogen metabolism of aquatic organisms. I. The assimilation and formation of urea in *Ochromonas malhamensis*. Arch. Biochem. Biophys. 139:269–277; 1970.

Luksas, A.J.; Erwin, J.A. The lipid requirement of the ciliated protozoon *Tetrahymena patula*. J. Protozool. 30:8–13; 1983.

Lynn, D.H. Comparative ultrastructure and systematics of the Colpodida. Structural conservatism hypothesis and a description of *Colpoda steinii* Maupas. J. Protozool. 23:302–314; 1976.

Lynn, D.H. Dimensionality and contractile vacuole function in ciliated protozoa. J. Exp. Zool. 223:219–230; 1982.

MacArthur, R.; Wilson, E.O. An equilibrium theory of insular zoogeography. Evolution. 17:373–387; 1963.

MacArthur, R.; Wilson, E.O. The Theory of Island Biogeography. Princeton: Princeton University Press; 1967: 199 pp.

Maguire, B. Community structure of protozoans and algae with particular emphasis on recently colonized bodies of water. In: Cairns, J., Jr., ed. Aquatic Microbial Communities. New York and London: Garland Publishing, Inc.; 1977: pp. 357–397.

Maihle, N.J.; Dedman, J.R.; Means, A.R.; Chafouleas, J.G.; Satir, B.H. Presence and indirect immunofluorescent localization of calmodulin in *Paramecium tetraurelia*. J. Cell Biol. 89:695–699; 1981.

Mansfield, J.M. Nonpathogenic trypanosomes of mammals. In: Krier, A.P., ed. Parasitic Protozoa. vol. 6. New York: Academic Press; 1978: pp. 297–327.

Manton, I. Some phyletic implications of flagellar structure in plants. In: Preston, R.D., ed. Advances in Botanical Research. vol. 2. London and New York: Academic Press; 1965: pp. 1–34.

Manton, I. Functional parallels between calcified and uncalcified periplasts. In: Leadbeater, B.; Ryding, R., eds. Biomineralization in Lower Plants and Animals. Oxford: Clarendon Press; 1986: pp. 157–172.

Manton, I.; Parke, M. Further observations on small green flagellates with special reference to possible relatives of *Chromulina pusilla* Butcher. J. Mar. Biol. Assn. U. K. 39:275–298; 1960a.

Manton, I.; Parke, M. Observations on the fine structure of two species of *Platymonas* with special reference to flagellar scales and the mode of origin of the theca. J. Mar. Biol. Assn., U.K. 45:743–754; 1960b.

Margulis, L. Origin of Eukaryotic Cells. New Haven: Yale University Press; 1970.

Margulis, L. The classification and evolution of prokaryotes and eukaryotes. In: King, R.C., ed. Handbook of Genetics. vol. 1. New York: Plenum; 1974: pp. 1–41.

Martin, S.M.; Byers, T.J. Acid hydrolase activity during growth and encystment in *Acanthamoeba castellanii*. J. Protozool. 23:608–613; 1976.

Mast, S.O. Structure, movement, locomotion and stimulation in amoeba. J. Morph. 41:347–425; 1925.

Matthews, H.M.; Daly, J.J. Metabolic changes in *Trichomonas gallinae* resulting from growth in various carbohydrates. J. Protozool. 22:139–145; 1975.

Mavrides, C. Regulation of glyconeogenesis from amino acids by acetate in *Tetrahymena pyriformis*. Can. J. Biochem. 51:323–333; 1973.

May, L.; Andrson, O.R.; Hogg, J.F. Cellular reorganization during glucoenogenesis in *Tetrahymena pyriformis*. J. Ultrastruc. Res. 81:271–289; 1982.

McEnery, M.E.; Lee, J.J. *Allogromia laticollaris*: A forminiferan with an unusual apogamic metagenic life cycle. J. Protozool. 23:94–108; 1976.

McGee-Russell, S.M. Dynamic activities and labile microtubules in cytoplasmic transport in the marine foraminiferan *Allogromia*. In: Transport at the Cellular Level, Symposia of the Society for Experimental Biology XXVIII. Cambridge: Cambridge University Press; 1974: pp. 157–189.

McGhee, R.B.; McGhee, A. H. The relation of migration of *Oncopeltus fasciatus* to distribution of *Phytomonas elmassiani* in the eastern United States. J. Protozool. 18:344–352; 1971.

McLaughlin, J.; Aley, S. The biochemistry and functional morphology of the *Entamoeba*. J. Protozool. 32:221–240; 1985.

McNeil, P.I.; Tanasugarn, L.; Meigs, J.; Taylor, D.L. Acidification of phagosomes is initiated before lysosomal enzyme activity is detected. J. Cell Biol. 97:692–702; 1983.

Meiklejohn, J. The effect of *Colpidium* on ammonia production by soil bacteria. Ann. Appl. Biol. 19:584; 1932.

Meis, J.F. G.M.; Verhave, J.P.; Wirtz, P.; Meuwissen, J.H.E.T. Histochemical observations on the exoerythrocytic malaria parasite *Plasmodium berghei* in rat liver. Histochemistry 81:417–425; 1984.

Meskill, V.P. Factors influencing the growth of *Glaucoma chattoni* in a chemically defined medium. J. Protozool. 17:104–107; 1970.

Mignot, J.-P. Quelques particulaires de l'ultrastructure d'*Entosiphon sulcatum*. C.R. Acad. Sci. Paris. 257:2500–2503; 1963.

Mignot, J.-P. Étude ultrastructurale des *Bicoeca*, protistes flagellés. Protistologica 10:543–565; 1974.

Miller, L.H.; Howard, R.J.; Carter, R.; Good, M.F.; Nussenzweig, V.; Nussenzweig, R.S. Research toward malaria vaccines. Science. 234:1349–1356; 1986.

Miller, R., Deane, M.P. The cytostome of *Trypanosoma cruzi* and *T. conorhini*. J. Protozool. 16:730–737; 1969.

Miller, S. The predatory behavior of *Dileptus anser*. J. Protozool. 15:313–319; 1968.

Moestrup, O.; Thomsen, H.A. Fine structural studies on the flagellate genus *Bicoeca* I – *Bicoeca maris* with particular emphasis on the flagellar apparatus. Protistologica 12:101–120; 1976.

Moore, J.; Cantor, M.H.; Sheeler, P.; Kahn, W. The ultrastructure of *Polytomella agilis*. J. Protozool. 17:671–676; 1970.

Moore, J.; Cushing, S.D. Conjugation in *Polytomella agilis* Aragao (Volvicida, Polyblepharidae). J. Protozool. 26:63–65; 1979.

Moraczewski, J. Taxocenoses des *Testacea* de quelques petits bassins de terrains inondables de la Narew. Acta Protozoologica. 3:189–213; 1965.

Moraczewski, J. Formation des taxocenoses des *Testacea* dans le lac de Zegrze. Acta Protozoologica. 4:327–342; 1967.

Morris, I. Nitrogen assimilation and protein synthesis. In: Stewart, W.D., ed. Algal Physiology and Biochemistry. Berkeley: University of California Press; 1973: pp. 583–609.

Morris, I. In: Stewart, W.D., ed. Algal Physiology and Biochemistry. vol. 10. Berkeley: University of California Press; 1974: pp. 584–585.

Mosser, J.L.; Mosser, A.G.; Brock, T.D. Photosynthesis in the snow: the alga *Chlamydomonas nivalis* (Chlorophyceae). J. Phycol. 13:22–27; 1977.

Müller, M. Peroxisomes of protozoa. In: Hogg, J.F., ed. The Nature and Function of Peroxisomes (Microbodies, Glyoxysomes). vol. 168. Ann. N. Y. Acad. Sci. 1969. 168 pp.

Müller, M. The hydrogenosome. In: Gooday, G.W.; Lloyd, D.; Trinei, A.P.J., eds. The Eukaryotic Microbial Cell. Cambridge: Cambridge University Press; 1980.

Müller, M. In: Klein, R.A.; Miller, P.G.G., eds. Alternate metabolic pathways in protozoan energy metabolism. Parasitology. 82:5; 1981.

Müller, M. Search for cell organelles in protozoa. J. Protozool. 32:559–563; 1985.

Müller, M.; Hogg, J.F.; de Duve, C. Distribution of tricarboxylic acid cycle enzymes and of glyoxylate cycle enzymes between mitochondria and peroxisomes in *Tetrahymena pyriformis*. J. Biol. Chem. 243:5385–5395; 1968.

Muller, W.A. Graphic representation of niche width and its application to salt marsh littoral foraminifera. New York: City University of New York. 1972. Thesis.

Muller, W.A. Competition for food and other niche-related studies of three species of salt-marsh foraminifera. Mar. Biol. 31:339–351; 1975.

Muller, W.A.; Lee, J.J. Apparent indispensability of bacteria in foraminiferan nutrition. J. Protozool. 16:471–478; 1969.

Muller, W.A.; Lee, J.J. Biological interactions and the realized niche of *Euplotes vannus* from the salt marsh Aufwuchs. J. Protozool. 24:523–527; 1977.

Munoz, M. DeL.; Rojkind, M.; Calderon, J.; Tanimoto, M.; Arias-Negrete, S.; Martinez-Palomo, A. *Entamoeba histolytica*: collagenolytic activity and virulence. J. Protozool. 31:468–470; 1984.

Murray, D.R.; Giovanelli, J.; Smillie, R.M. Photoassimilation of glycolate, glycine and serine by *Euglena gracilis*. J. Protozool. 17:99–104; 1970.

Murray, J.W. Comparative studies of living and dead benthic foraminiferal distributions. In: Hedley, R.H.; Adams, C.G., eds. *Foraminifera*. vol. 2. London: Academic Press; 1976: pp. 47–102.

Naitoh, Y.; Eckert, R. Ionic mechanisms controlling behavioral responses of *Paramecium* to mechanical stimulation. Science. 164:963–965; 1969.

Naitoh, Y.; Eckert, R. Sensory mechanisms in *Paramecium*. II. Ionic basis of the hyperpolarizing mechanoreceptor potential. J. Exper. Biol. 59:53–65; 1973.

Nanney, D.L. Experimental Ciliatology. New York: Wiley and Sons; 1980: 304 pp.

Neale, P.J.; Marra, J. Short-term variation of P_{max} under natural irradiance conditions: a model and its implications. Mar. Ecol. Prog. Ser. 26:113–124; 1985.

Neff, R.J. Purification, axenic cultivation, and description of a soil amoeaba, *Acanthamoeba* sp. J. Protozool. 4:176–182; 1957.

Nerad, T.H.; Visvesvara, G.; Daggett, P.M. Chemically defined media for the cultivation of *Naegleria*: pathogenic and high temperature tolerant species. J. Protozool. 30:383–387; 1983.

Netzel, H. Gehäusewandbildung des Gehäuses bei *Difflugia oviformis* (Rhizopoda, Testacea). Arch. Protistenkd. 118:321–339; 1983.

Netzel, H.; Dürr, G. Dinoflagellate cell cortex. Dinoflagellates. New York: Academic Press; 1984: Ch. 3, pp. 43–105.

Nicholls, K.H.; Lynn, D.H. *Lepidotrachelophyllum fornicis*, n.g., n.sp., a ciliate with an external layer of organic scales (ciliophora, litostomatea, haptoria). J. Protozool. 31(3):413–419; 1984.

Nisbet, B. An ultrastructural study of the feeding apparatus of *Peranema trichophorum*. J. Protozool. 21:39–48; 1974.

Nisbet, B. Nutrition and Feeding Strategies in Protozoa. London: Croom-Helm; 1984: 280 pp.

Noirot-Timothée, C. Recherches sur l'ultrastructure d'*Opalina ranarum*. Ann. Sci. Nat. Biol. 12:265–281; 1959.

Noland, L.E. Factors influencing the distribution of freshwater ciliates. Ecology. 6:437–452; 1925.

Norris, R.E. Prasinophytes. In: Cox, E.R, ed. Phytoflagellates. vol. 2. New York: Elsevier; 1980: pp. 85–145.

Nozawa, Y.; Thompson, G.A., Jr. Lipids and membrane organization in *Tetrahymena*. In: Levandowsky, M.; Hutner, S.H., eds. Biochemistry and Physiology of Protozoa. 2nd ed. vol. 2. New York: Academic Press; 1979: pp. 275–338.

Nozawa, Y.; Nagao, S. Functional aspects of calmodulin in protozoa. Insect Sci. Appl. Special Issue, Protozoa. 7:267–277; 1986.

O'Dell, W.D. Isolation, enumeration, and identification of amebae from a Nebraska lake. J. Protozool. 26:265–269; 1979.

Ogden, C.G. Siliceous structures secreted by members of the subclass Lobosia (Rhizopodea: Protozoa). Bull. Brit. Mus. (Nat. Hist.) Zoology. 36:203–207; 1979.

Ogden, C.G. Aspects of shell structure in the genus *Difflugia* (Rhizopodea). J. Protozool. 27:57A; 1980.

Ogden, C.G.; Hedley, R.H. An Atlas of Freshwater Testate Amoebae. British Museum (Natural History) and Oxford Univ. Press, Oxford; 1980: 222 pp.

Ogura, A.; Machemer, H. Distribution of mechanoreceptor channels in the *Paramecium caudatum* surface membrane. J. Comp. Physiol. and Sens. Neural Behav. Physiol. 135:233–242; 1980.

Old, K., Chakraborty, S. Mycophagous soil amoebae: their biology and significance in the ecology of soil-borne plant pathogens. In: Corliss, J.O.; Patterson, D.J., eds. Progress in Protistology. vol. 1. Bristol: Biopress, Ltd.; 1986: pp. 163–194.

Opas, M. Studies on locomotion of *Amoeba proteus* I. Response to hydrogen ion concentration in the medium. Acta Protozoologica. 13:285–294; 1975.

Opas, M. Effects of induction of endocytosis on adhesiveness of *Amoeba proteus*. Protoplasma. 107:161–170; 1981.

Opperdoes, F.R. In: Klein, R.A.; Miller, P.G.G., eds. Alternate metabolic pathways in protozoan energy metabolism. Parasitology. 82:10–13; 1981.

Organ, A.E.; Bovee, E.C.; Jahn, T.L.; Wigg, D.; Fonseca, J.R. The mechanism of the nephridial apparatus of *Paramecium multimicronucleatum*. I. Expulsion of water from the vesicle. J. Cell Biol. 37:139–145; 1968.

Orlovskaja, E.E.; Karvanen, L.N.; Seravin, L.N. Susceptibility of food chemoreceptors in carnivorous *Protozoa*. Acta Protozoologica. 23:197–211; 1984.

Paasche, E. Effect of ammonia and nitrate on growth, photosynthesis, and ribulosediphosphate carboxylase content of *Dunaliella tertiolecta*. Physiologia Pl. 25:294–299; 1971.

Page, F.C. An Illustrated Key to Freshwater and Soil Amoebae. Ambelside: Freshwater Biological Association; 1976a.

Page, F.C. A revised classification of the Gymnamoebia (Protozoa: Sarcodina). Zool. J. Linn. Soc. 58:61–67; 1976b.

Page, F.C. Some comparative notes on the occurrence of Gymnamoebia (Protozoa: Sarcodina) in British and American habitats. Trans. Amer. Micros. Soc. 95:385–394; 1976c.

Page, F.C. An electron microscopical study of *Thecamoeba proteroides* (Gymnamoebia), intermediate between Thecamoebidae and Amoeidae. Protistologica. 14:77–85; 1978.

Page, F.C. Fine structure of some marine strains of *Platyamoeba* (Gymnamoebia, Thecamoebidae). Protistologica. 16:605–612; 1980.

Page, F.C. Marine Gymnamoebae. Cambridge: Institute of Terrestrial Ecology; 1983: 54 pp.

Palmer, E.G.; Togasaki, R.K. Acetate metabolism by an obligate phototrophic strain of *Pandorina morum*. J. Protozool. 18:640–644; 1971.

Park, B.; Hirotani, A.; Nakano, Y.; Kitaoka, S. The physiological role and catabolism of arginine in *Euglena gracilis*. Agric. Biol. Chem. 47:2561–2568; 1983.

Patterson, D.J. On the behavior of contractile vacuoles and associated structures of *Paramecium caudatum* (Ehrbg). Protistologica. 13:205–212; 1977.

Patterson, D.J. The fine structure of *Opalina ranarum* (Family Opalinidae): Opalinid phylogeny and classification. Protistologica. 21:413–428; 1985.

Patterson, D.J.; Dürrschmidt, M. Siliceous structures formed by heliozoa and heliozoon-like amoebae. In: Leadbeater, B.S.C.; Riding, R., eds. Biomineralization in Lower Plants and Animals. Oxford: Clarendon Press; 1986: pp. 361–374.

Patterson, D.J.; Hausmann, K. Feeding by *Actinophrys sol* (Protista, Heliozoa): 1 light microscopy. Microbios. 31:39–55; 1981.

Paul, J.H.; Cooksey, K.E. Regulation of L-asparaginase (E.C. 3.5.1.1) in *Chlamydomonas* species in response to ambient concentrations of combined nitrogen. J. Bacteriol. 147:9–12; 1981.

Paulin, J.J.; Krascheninnikow, S. An electron microscopic study of *Balantidium caviae*. Acta Protistologica. 12:97–104; 1973.

Pauls, K.P.; Thompson, J.E. Growth- and differentiation-related enzyme changes in cytoplasmic membranes of *Acanthamoeba castellanii*. J. Gen. Microbiol. 107:147–153; 1978.

Peak, J.G.; Peak, M.J. Heterotrophic carbon dioxide fixation by *Euglena*, function of phosphoenolpyruvate carboxylase. Biochim. et Biophys. Acta. 677:390–396; 1981.

Peck, R.K. Feeding behavior in the ciliate *Pseudomicrothorax dubius* is a series of morphologically and physiologically distinct events. J. Protozool. 32:492–501; 1985.

Peck, R.K.; Hausmann, K. Primary lysosomes of the ciliate *Pseudomicrothorax dubius*: Cytochemical identification and role in phagocytosis. J. Protozool. 27:401–409; 1980.

Pennick, N.C. Comparative ultrastructure and occurrence of scales in *Pyramimonas* (Chlorophyta, Prasinophyceae). Arch. Protistenk. 128:3–11; 1984.

Petraitis, P.S. Likelihood measures of niche breadth and overlap. Ecology. 60:703–710; 1979.

Petrova, M.A.; Smirnova, T.P. Ecology of planktonic infusoria in a secondary oligotrophic lake. Hydrobiologica J. 10:17–21; 1974.

Petruschevskya, M.G. On the natural system of polycystine Radiolaria (Class Sarcodina). In: Farinacci, A., ed. Proceedings of the II Planktonic Conference Roma 1970. Rome: Edizioni Tecnoscienza; 1971: pp. 981–992.

dePeyer, J.E.; Machemer, H. Hyperpolarizing and depolarizing mechanoreceptor potentials in *Stylonichia*. J. Comp. Physiol. 127:255–266; 1978.

Pfeffer, W. Über die chemotaktischen Bewegungen von Bakterien, Flagellaten und Volvocineen. Untersuch. Bot. Inst. Tübingen. 2:582–661; 1888.

Philips, R.S. Malaria. London: E. Arnold; 1983: 58 pp.

Picken, L.E.R. The structure of some protozoan communities. J. Ecol. 25:368–384; 1937.

Pienar, R. Chrysophytes. In: Cox, E.R., ed. Phytoflagellates. vol. 2. New York: Elsevier; 1980: pp. 213–242.

Pigon, A. Cell surface and medium conditioning in *Acanthamoeba* culture. Cytobiologie. 13:107–117; 1976.

Pitelka, D.R. Electron-Microscopic Structure of Protozoa. New York: Pergamon Press; 1963: 269 pp.

Pollard, T.D.; Aebi, U.; Cooper, J.A.; Fowler, W.E.; Kiehart, D.P.; Smith, P.R.; Tseng, P.C. Actin and myosin function in *Acanthamoeba*. Philos. Trans. R. Soc. Lond. B. Biol. Sci. 299:237–246; 1982.

Pollero, R.J.; Brenner, R.R.; Dumm, C.G. Comparative biosynthesis of polyethylenic fatty acids in *Acanthamoeba castelanii* and *Ochromonas danica*. Acta Physiol. Lat. Am. 25:412–424; 1976.

Porter, K.G.; Sherr, E.B.; Sherr, B.F.; Pace, M.; Sanders, R.W. Protozoa in planktonic food webs. J. Protozool. 32:409–415; 1985.

Pratt, J.R.; Cairns, J., Jr. Functional groups in the protozoa: Roles in differing ecosystems. J. Protozool. 32:415–423; 1985.

Precali, R.; Falkowski, P.G. Incorporation of carbon-14-labeled glutamate into proteins and chlorophylls in *Dunaliella tertiolecta*, a marine chlorophyte. Biol. Plant. 25:187–195; 1983.

Pringsheim, E.G. Kulturversuche mit chlorophyll fuhren-den Microorganismen. II. Zur physiologie der *Euglena gracilis*. Beitr. Biol. Pflanz. 12:1–47; 1912.

Proper, G.; Garver, J.C. Mass culture of the protozoa *Colpoda steinii*. Biotechnol. Bioeng. 8:287–296; 1966.

Provasoli, L. Nutrition and ecology of protozoa and algae. Ann. Rev. Microbiol. 12:279–308; 1958.

Provasoli, L.; Carlucci, A.F. Vitamins and growth regulators. In: Stewart, W.D.P., ed. Algal Physiology and Biochemistry. Berkeley: University of California Press; 1974: pp. 741–787.

Provasoli, L.; Pintner, I.J. Ecological implications of in vitro nutritional requirements of algal flagellates. Ann. New York Acad. Sci. 56:839–851; 1953.

Prusch, R.D. Active calcium extrusion by *Amoeba proteus*. J. Exp. Zool. 212:475–477; 1980a.

Prusch, R.D. Endocytotic sucrose uptake in *Amoeba proteus* induced with the calcium ionophore A23187. Science 209:691–692; 1980b.

Prusch, R.D. Bulk solute extrusion as a mechanism conferring solute uptake specificity by pinocytosis in *Amoeba proteus*. Science. 213:668–670; 1981.

Prusch, R.D.; Hannifin, J.A. Sucrose uptake by pinocytosis in *Amoeba proteus* and the influence of external calcium. J. Gen. Physiol. 74:523–535; 1979.

de Raadt, P.; Seed, J.R. Trypanosome causing disease in man in Africa. In: Krier, J.P., ed. Parasitic Protozoa. vol. 1. New York: Academic Press; 1977: pp. 175–237.

Raikov, I.B. The Protozoan Nucleus. Morphology and Evolution. (Cell Biol. Mongr. 9). Wien and New York: Springer-Verlag; 1982.

Raikov, I.B. Primitive never-dividing macronuclei of some lower ciliates. Int. Rev. Cytol. 95:267–325; 1985.

Raikov, I.B.; Gerassimova-Matvejeva, Z.P.; de Puytorac, P. Cytoplasmic fine structure of the marine psammobiotic ciliate *Tracheloraphis dogieli* Raikov. I. Somatic infraciliature and cortical organelles. Acta Protozoologica. 14:17–42; 1975.

Raikov, I.B.; Kovaleva, V.G. Fine structure of the circumoral ciliature and pharyngeal apparatus of the marine ciliate *Helicoprorodon gigas* (Gymnostomata). Protistologica. 16:177–198; 1980.

Randall, J.T.; Jackson, S.F. Fine structure and function in *Stentor polymorphus*. J. Biophys. Biochem. Cytol. 4:807–830; 1958.

Rankin, J.C.; Davenport, J. Animal Osmoregulation. Glasgow: Blackie and Son, Ltd.; 1981: 202 pp.

Rassoulzadegan, F. Dependence of grazing rate, gross growth efficiency, and food size range on temperature in a pelagic, oligotrichous ciliate *Lobmanniella spiralis* Leeg., fed on naturally occurring particulate matter. Ann. Inst. Océanogr. Paris. 58:177–184; 1982.

Rassoulzadegan, F., Etienne, M. Grazing rate of the tintinnid *Stenosemella ventricosa* (Clap. & Lachm.) Jörg. on the spectrum of the naturally occurring particulate matter from a Mediterranean neritic area. Limnol. Oceanogr. 26:258–270; 1981.

Rausch, M.; Grunewald, J. Light and electron microscopic observations on some microsporidian parasites (Cnidosporidia: Microsporidia) of blackfly larvae (Diptera: Simuliidae). Parasitenkunde. 63:1–11; 1980.

Reeves, R.E. Phosphopyruvate carboxylase from *Entamoeba histolytica*. Biochim. Biophys. Acta. 220:346–349; 1970.

Reeves, R.E.; Warren, L.G.; Susskind, B.; Lo, H. An energy-conserving pyruvate-to-acetate pathway in *Entamoeba histolytica*. J. Biol. Chem. 252:726–731; 1977.

Reisser, W. Endosymbiotic associations of freshwater protozoa and algae. In: Corliss, J.O.; Patterson, D.J., eds. Progress in Protistology. vol. 1. Bristol: Biopress, Ltd.; 1986: pp. 195–214.

Repak, A.J. Encystment and excystment of the heterotrichous ciliate *Blepharisma stoltei* Isquith. J. Protozool. 15:407–412; 1968.

Reynoso, G.T.; DeGamboa, B.A. Salt tolerance in the freshwater algae *Chlamydomonas reinhardii*: Effect of proline and taurine. Comp. Biochem. Physiol. 73A:95–99; 1982.

Rheinheimer, G. Aquatic Microbiology. 3rd ed. Chichester: John Wiley and Sons; 1985: 257 pp.

Rieckmann, K.H.; Silverman, P.H. Plasmodia of man. In: Kreier, J.P., ed. Parasitic Protozoa. vol. 3. New York: Academic Press; 1977: pp. 493–527.

Rivkin, R.B.; Voytek, M.A.; Seliger, H.H. Phytoplankton division rate in light-limited environments: two adaptations. Science. 215:1123–1125; 1982.

Rivkin, R.B.; Swift, E.; Biggley, W.H.; Voytek, M.A. Growth and carbon uptake by natural populations of oceanic dinoflaggelates *Pyrocystis noctiluca* and *Pyrocystis fusiformis*. Deep Sea Res. 31:353–367; 1984.

Roberts, K. Crystalline glycoprotein cell walls of algae: their structure, composition and assembly. Phil. Trans. Roy. Soc. Lond. B. 268:129–146; 1974.

Roberts, K.; Gurney-Smith, M.; Hills, G.J. Structure, composition, and morphogenesis of the cell wall of *Chlamydomonas reinhardtii*. I. Ultrastructure and preliminary chemical analysis. J. Ultrastruc. Res. 40:599–613; 1972.

Rogers, T.D.; Scholes, V.E.; Schlichting, H.E. An ultrastructural comparison of *Euglena gracilis* Klebs, bleached *Euglena*, and *Astasia longa* Pringsheim. J. Protozool. 19:133–139; 1972.

Rogerson, A. Generation times and reproductive rates of *Amoeba proteus* (Leidy) as influenced by temperature and food concentration. Can. J. Zool. 58:543–548; 1980.

Roitman, C.; Roitman, I., Azevedo, H.P. Growth of an insect trypanosomatid at 37°C in a defined medium. J. Protozool. 19:346–349; 1972.

Rosenberg, H.; Munk, N. Transport phenomena associated with the deposition and disappearance of pyrophosphate granules in *Tetrahymena pyriformis*. Biochim. Biophys. Acta. 184:191–197; 1969.

Rossman, D.A. Protozoan fauna of a limestone sinkhole in north-central Florida. Quart. J. Fla. Acad. Sci. 22:121–124; 1959.

Roth, J.S.; Eichel, H.J. Studies on the metabolism of L-Phenylalanine by *Tetrahymena pyriformis* W. J. Protozool. 8:69–71; 1961.

Roth, J.S.; Eichel, H.J.; Ginter, E. The oxidation of amino acids by *Tetrahymena pyriformis* W. Arch. Biochem. and Biophys. 48:112–119; 1954.

Routledge, L.M.; Amos, W.B.; Gupta, B.L.; Hall, T.A.; Weis-Fogh, T. Microprobe measurements of calcium-binding in the contractile spasmoneme of a vorticellid. J. Cell Sci. 19:195–201; 1975.

Rückert-Hilbig, A. Megalospheric gamonts of *Rosalina globularis*, *Cymbaloporetta bulloides*, and *Cymbaloporetta miletti* (Foraminifera) with differently constructed swimming-apparatus. Institut und Museum fur Geologie und Palaontologie der Universitat Tübingen; 1983; 69 pp. Thesis.

Rudzinska, M.A. The fine structure and function of the tentacle of *Tokophrya infusionum*. J. Cell Biol. 25:459–477; 1965.

Ruttner, F. Fundamentals of Limnology. Translated by: Frey, D.G.; Frey, F.E.G. Toronto: University of Toronto Press; 1953: 242 pp.

Ryley, J. F. Studies on the metabolism of protozoa 3. Metabolism of the ciliate *Tetrahymena pyriformis* (*Glaucoma pyriformis*). Biochem. J. 52:483–492; 1952.

Saliba, R.; Dive, D., Devis, R. An inexpensive and practical medium for rearing mass cultures of *Tetrahymena*. Protistologica. 19:417–421; 1983.

Salt, G.W. Changes in cell volume of *Didinium nasutum* during population increase. J. Protozool. 22:112–114; 1975.

Santore, U.J.; Leedale, G.F. Order 1. Cryptomonadida Senn, 1900. In: Lee, J.J.; Hutner, S.H.; Bovee, E.C., eds. Illustrated Guide to the Protozoa. Lawrence, Kansas: Society of Protozoologists; 1985: pp. 19–22.

Satir, P. The generation of ciliary motion. J. Protozool. 31:8–12; 1984.

Schafer, C.T. Sampling and spatial distribution of benthonic foraminifera. Limnol. Oceanogr. 16:944–951; 1971.

Scherbaum, O.H. A comparison of synchronized cell division in protozoa. J. Protozool. 9:61–64; 1962.

Schliwa, M. The role of divalent cations in the regulation of microtubule-assembly. In vivo studies on microtubules of the heliozoan axopodium using ionophore A 23187. J. Cell Biol. 70:527–540; 1976.

Schonborn, W. Production studies on Protozoa. Oecologia. 27:171–184; 1977.

Schonefeld, A.; Alfermann, W.; Schultz, J.E. Economic mass cultivation of *Paramecium tetraulia* on a 200-liter scale. J. Protozool. 33:222–225; 1986.

Schuster, F.L. Ultrastructure of cysts of *Naegleria* spp.: A comparative study. J. Protozool. 22:352–359; 1975.

Schuster, F.L.; Hershenov, B. *Khawkinea quartana*, a colorless euglenoid flagellate. I. Ultrastructure. J. Protozool. 21:33–39; 1974.

Segura, J.J.; Mata, B.D. Cellular behavior of axenic amoebae cultured in the presence of different carbohydrates. Arch. Invest. Med. 11 (Suppl. 1):115–122; 1980.

Seravin, L.N.; Gudakov, A.V. The main types and forms of agamic fusion in Protozoa. Tristologiya. 26:123–131; 1984.

Shinnick, T.M.; Anderson, R.W.; Holt, C.E. Map and function of *gad* mutations in *Physarum polycephalum*. Genet. Res. Camb. 42:41–57; 1983.

Shrago, E.; Elson, C. Intermediary metabolism of *Tetrahymena*. In: Levandowsky, M.; Hutner, S. H., eds. Biochemistry and Physiology of Protozoa. 2nd ed. vol. 3. New York: Academic Press; 1980: pp. 287–312.

Sibbald, M.J.; Albright, L.J.; Sibbald, P.R. Chemosensory responses of a heterotrophic microflagellate to bacteria and several nitrogen compounds. Mar. Ecol. Prog. Ser. 36:201–204; 1987.

Siddiqui, W.A.; Rudzinska, M.A. The fine structure of axenically grown trophozoites of *Entamoeba invadens* with special reference to the nucleus and helical ribonucleoprotein bodies. J. Protozool. 12:448–458; 1965.

Sieburth, J.McN. Sea Microbes. New York: Oxford Univ. Press; 1979: 491 pp.

Sinden, R.E. Report on the satellite meeting on malaria. Insect Sci. Applic. 7:313–318; 1986.

Skarlato, S.O. Electron microscope study of the micronuclei of the ciliate *Stentor coeruleus* during meiosis. Protistologica. 18:281–288; 1982.

Sladecek, V.V. Four metasaprobic communities of colourless flagellates. Arch. Protistenk. 114:245–248; 1972.

Sleigh, M.A. Cilia and Flagella. London: Academic Press; 1973: 500 pp.

Sleigh, M.A. Radiation of the eukaryotic Protista. In: House, M.R., ed. The Origin of Major Invertebrate Groups. London and New York: Academic Press; 1979: pp. 23–53.

Sleigh, M.A. The integrated activity of cilia: function and coordination. J. Protozool. 31:16–21; 1984a.

Sleigh, M.A. III. Motile systems of ciliates. Protistologica. 20:299–305; 1984b.

Sliter, W.V. Laboratory experiments on the life cycle and ecological controls of *Rosalina globularis* d'Orbigny. J. Protozool. 12:210–215; 1965.

Small, E.B. A study of ciliate protozoa from a small polluted stream in east-central Illinois. Amer. Zool. 13:225–230; 1973.

Small, E.B.; Lynn, D.H. Phylum Ciliophora. In: Lee, J.J.; Hutner, S.; Bovee, E.C., eds. Illustrated Guide to the Protozoa. Lawrence, Kansas: Soc. of Protozoologists; 1985: pp. 393–567.

Small, E.B.; Lynn, D.H. Phylum Ciliophora Doflein, 1901. In: Lee, J.J.; Hutner, S.H.; Bovee, E.C., eds. Illustrated Guide to the Protozoa. Lawrence, Kansas: Society of Protozoologists; 1985: pp. 393–575.

Smith, W.O.; Barber, R.T. Carbon budget for the autotrophic ciliate *Mesodinium rubrum*. J. Phycol. 15:27–33; 1979.

Smyth, R.D.; Ebersold, W.T. A *Chlamydomonas* mutant with altered phototactic response. Genetics. 64:s62; 1970.

Sobota, A.; Burovina, I.V.; Pogorelov, A.G.; Solus, A.A. Correlation between potassium and phosphorus content and their nonuniform distribution in *Acanthamoeba castellanii*. Histochemistry. 81:201–204; 1984.

Soldo, A.T.; Merlin, J. The nutrition of *Parauronema acutum*. J. Protozool. 24:556–562; 1977.

Soldon, A.T.; van Wagtendonk, W.J. The nutrition of *Paramecium aurelia*, Stock 299. J. Protozool. 16:500–506; 1969.

Sols, A. Regulation of carbohydrate transport and metabolism in yeast. In: Mills, A.K.; Krebs, H., eds. Aspects of Yeast Metabolism. Philadelphia: F. A. Davis; 1967: pp. 47–66.

Sonneborn, T.M. Sex, sex inheritance and sex determination in *Paramecium aurelia*. Proc. Natl. Acad. Sci. U.S.A. 23:378–385; 1937.

Spero, H.J. Phagotrophy in *Gymnodinium fungiforme* (Pyrrhophyta): the peduncle as an organelle of ingestion. J. Phycol. 18:356–360; 1982.

Spero, H.J. Chemosensory capabilities in the phagotrophic dinoflagellate *Gymnodinium fungiforme*. J. Phycol. 21:181–184; 1985.

Spindler, M.; Anderson, O.R.; Hemleben, C.; Bé, A.W.H. Light and electron microscopic observations of gametogenesis in *Hastigerina pelagica* (Foraminifera). J. Protozool. 25:427–433; 1978.

Spindler, M.; Hemleben, C. Formation and possible function of annulate lamellae in a planktonic foraminifer. J. Ultrastruc. Res. 81:341–350; 1982.

Srivastava, D.K.; Shukla, O.P. Encystment of *Acanthamoeba culbertsoni* in non-nutrient inorganic media. Indian J. Exp. Biol. 21:440–443; 1983.

Stabenau, H.; Beevers, H. Isolation and characterization of microbodies from the alga *Chlorogonium elongatum*. Pl. Physiol. 53:866–869; 1974.

Starr, R.C. Structure, reproduction, and differentiation in *Volvox carteri* f. *nagariensis* Iyengar, Strain HK 9 & 10. Arch. Protistenk. 111:204–222; 1969.

Stavis, R.L.; Hirschberg, R. Phototaxis in *Chlamydomonas reinhardtii*. J. Cell Biol. 59:367–377; 1973.

Stebbings, H.; Hyams, J.S. Cell Motility. London: Longman; 1979: pp. 177–188.

Steele, J.H. The Structure of Marine Ecosystems. Cambridge: Harvard University Press; 1974: 128 pp.

Steiger, R.F.; Steiger, E. Cultivation of *Leishmania braziliensis* in defined media: nutritional requirements. J. Protozool. 24:437–441; 1977.

Stein, J.R. A morphological and genetic study of *Gonium pectorale*. Amer. J. Bot. 45:664–672; 1958.

Steinbuchel, A.; Müller, M. Anaerobic pyruvate metabolism of *Tritrichomonas vaginalis* hydrogenosomes. Mol. Biochem. Parasitol. 20:57–65; 1986.

Stevens, A.R.; Prescott, D.M. Reformation of nucleolus-like bodies in the absence of post-mitotic RNA synthesis. J. Cell Biol. 48:443–454; 1971.

Stewart, W.D., ed. Algal physiology and biochemistry. vol. 10. Berkeley and Los Angeles: University of California Press; 1974: 987 pp.

Stockem, W. Membrane-turnover during locomotion of *Amoeba proteus*. Acta Protozoologica. 11:83–93; 1972.

Stockem, W. Cell surface morphology and activity in *Amoeba proteus* and *Physarum polycephalum*. Acta Protozoologica. 18:33–41; 1979.

Stockem, W.; Hoffman, H.U.; Gawlitta, W. Spatial organization and fine structure of the cortical filament layer in normal locomoting *Amoeba proteus*. Cell Tiss. Res. 221:505–519; 1982.

Stoecker, D.; Guillard, R.R.L.; Kavee, R.M. Selective predation by *Favella ehrenbergii* (Tintinnia) on and among dinoflagellates. Biol. Bull. 160:136–145; 1981.

Stoltze, H.J.; Lui, N.; Anderson, O.R.; Roels, O.A. The influence of the mode of nutrition on the digestive system of *Ochromonas malhamensis*. J. Cell Biol. 43:296–409; 1969.

Stoner, L.C.; Dunham, P.B. Regulation of cellular osmolarity and volume in *Tetrahymena*. J. Exp. Biol. 53:391–399; 1970.

Stout, J.D. The relation between protozoan populations and biological activity in soils. In: Progress in Protozoology. Amsterdam: Excerpta Medica Found.; 1965: p. 119.

Stout, J.D. The relationship between protozoan populations and biological activity in soils. J. Exp. Zool. 13:193–201; 1973.

Stout, J.D. The role of protozoa in nutrient cycling and energy flow. Adv. Microb. Ecol. 4:1–50; 1980.

Stout, J.D.; Bamforth, S.S.; Lousier, J.D. Protozoa. In: Black, C.A., ed. Methods of Soil Analysis, Part 2. Chemical and Microbiological Properties—Agronomy Monograph

no. 9 (2nd ed.). Madison, Wisconsin: American Society of Agronomy; 1982: pp. 1103-1120.
Suckow, R. Schwefelmikrobengesellschaften der See—un Boddengewasser von Hiddensee. Z. Allg. Mikrobiol. 6:309-315; 1966.
Suzaki, T.; Williamson, R.E. Pellicular ultrastructure and euglenoid movement in *Euglena ehrenbergii* Klebs and *Euglena oxyuris* Schmarda. J. Protozool. 33:165-171; 1986a.
Suzaki, T.; Williamson, R.E. Ultrastructure and sliding of pellicular structures during euglenoid movement in *Astasia longa* Pringsheim (Sarcomastigophora, Euglinida). J. Protozool. 33:179-184; 1986b.
Swanberg, N.R.; Bjørklund, K.R. The radiolarian fauna of western Norwegian fjords: patterns of abundance in the plankton. Mar. Micropal. 11:231-241; 1986.
Swanberg, N.R.; Bennett, P.; Lindsey, J.L.; Anderson, O.R. A comparative study of predation in two Caribbean radiolarian populations. Mar. Microbiol. Food Webs. 1:105-118; 1986.
Takeuchi, A.; Weinbach, E.C.; Diamond, L.S. Pyruvate oxidase (CoA acetylating) in *Entamoeba histolytica*. Biochem. Biophys. Res. Commun. 65:591-596; 1975.
Tamburro, K.; Hutner, S.H. Carbohydrate-free media for *Crithidia*. J. Protozool. 18:667-672; 1971.
Tartar, V. Morphogenesis in protozoa. In: Chen, T.-T., ed. Research in Protozoology. vol. 2. New York: Pergamon Press; 1967: pp. 1-116.
Tatchell, E.C. An ultrastructural study of extension and contraction in *Lacrymaria olor* (O.F. Muller). Protistologica. 16:167-175; 1980.
Tatchell, E.C. An ultrastructural study of prey capture and ingestion in *Lacrymaria olor* (O.F.M. 1786). J. Protozool. 17:59-66; 1981.
Taub, F.B. A continuous gnotobiotic (species defined) ecoystem. In: Cairns, J., Jr., ed. Structure and Function of Freshwater Microbial Communities. Blacksburg: Virginia Polytechnic and State University; 1971: pp. 101-120.
Taylor, W.D. Maximum growth rate, size and commonness in a community of bactivorous ciliates. Oecologia (Berl.) 36:263-272; 1978.
Taylor, W.D. Sampling data on the bactivorous ciliates of a small pond compared to neutral models of community structure. Ecology. 60:876-883; 1979.
Taylor, W.D.; Berger, J. Growth response of cohabiting ciliate protozoa to various prey bacteria. Can. J. Zool. 54:1111-1114; 1976.
Taylor, W.D.; Berger, J. Microspatial heterogeneity in the distribution of ciliates in a small pond. Microb. Ecol. 6:27-34; 1980.
Taylor, D.L.; Wang, Y.-L.; Heiple, J.M. Contractile basis of amoeboid movement: 7. The distribution of fluorescently labeled actin in living amoebas. J. Cell Biol. 86:590-598; 1980.
Tendal, O.S.; Swinbanks, D.D.; Shirayama, Y. A new infaunal xenophyophore (Xenophyophorea, Protozoa) with notes on its ecology and possible trace fossil analogues. Oceanol. Acta. 5:325-329; 1982.
Tibbs, J.; Marshall, B.J. Cyst wall protein synthesis and some enzyme changes on starvation and encystment in *Colpoda steinii*. J. Protozool. 17:125-128; 1970.
Tolmashova, G.T. Adaptive modification in the flagellate *Astasia longa* (Protozoa) to changes of calcium (II) content in the medium. Zool. Zh. 60:653-660; 1981.
Trager, W. Differentiation in protozoa. J. Protozool. 10:1-6; 1963.
Travis, J.L.; Kenealy, J.F.X.; Allen, R.D. Studies on the motility of the Foraminifera: 2. The dynamic microtubular cytoskeleton of the reticulopodial network of *Allogromia*

laticollaris. J. Cell Biol. 97:1668-1676; 1983.

Triemer, R.E.; Fritz, L. Structure and operation of the feeding apparatus in a colorless euglenoid, *Entosiphon sulcatum*. J. Protozool. 34:39-47; 1987.

Umbach, J.A. Changes in intracellular pH affect calcium currents in *Paramecium caudatum*. Proc. R. Soc. Lond. Biol. Sci. 216:209-224; 1982.

Van Der Wal, P. Calcification in two species of coccolithophorid algae. GUA Papers of Geol. Series 1. 20:1-112; 1984.

Vandermeer, J.H. The competitive structure of communities: An experimental approach with protozoa. Ecology. 50:362-371; 1969.

Vandermeer, J.H. Niche theory. Ann Rev. Ecol. System. 3:107-133; 1972.

Van Houten, J. A mutant of *Paramecium* defective in chemotaxis. Science. 198:746-748; 1977.

Van Wagtendonk, W.J. Nutrition of *Paramecium*. In: Van Wagtendonk, W.J., ed. *Paramecium*: A Current Survey. Amsterdam: Elsevier; 1974: pp. 339-376.

Vazquezdelara-Cisneros, L.G.; Arroyo-Begovich, A. Induction of encystation of *Entamoeba invadens* by removal of glucose from the culture medium. J. Parasitol. 70:629-633; 1985.

Verni, F.; Rosati, G. Preliminary survey of the morphology and cytochemistry of polysaccharide reserves in ciliates. Protistologica. 16:427-434; 1980.

Vickerman, K. The fine structure of *Trypanosoma congolense* in its bloodstream phase. J. Protozool. 16:54-69; 1969.

Vickerman, K. The mode of attachment of *Trypanosoma vivax* in the proboscis of the tsetse fly *Glossina fuscipes*: An ultrastructural study of the epimastigote stage of the trypanosome. J. Protozool. 20:394-404; 1973.

Vickerman, K. The diversity of the kinetoplastid flagellates. In: Lumsden, W.H.R.; Evans, D.A., eds. Biology of the Kinetoplastida. vol. 1. London and New York: Academic Press: 1976: pp. 1-34.

Vickerman, K.; Preston, T.M. Comparative cell biology of the kinetoplastid flagellates. In: Lumsden, W.H.R.; Evans, D.A., eds. Biology of the Kinetoplastida. vol. 1. London and New York: Academic Press; 1976: pp. 35-130.

Votta, J.J.; Jahn, T.L.; Levedahl, B.H. The mechanism of onset of the stationary phase in *Euglena gracilis* grown with 10 mM succinate: Intracellular pH values. J. Protozool. 18:166-170; 1971.

Waidyasekera, P.L.D.; Kitching, J.A. Oxygen consumption and Krebs cycle activity of *Acanthamoeba castellanii* at high hydrostatic pressure. Arch. Protistenk. 117:314-319; 1975.

Walker, G.K. Observations on a unique cortical network in the hypotrich ciliate *Euplotes vannus*. Protistologica. 11:275-278; 1975.

Walker, G.K.; Maugel, T.K. Encystment and excystment in hypotrich ciliates II. *Diophrys scutum* and remarks on comparative features. Protistologica. 16:525-531; 1980.

Walne, P. Euglenoid flagellates. In: Cox, E.R., ed. Phytoflagellates. vol. 2. New York: Elsevier; 1980: pp. 165-212.

Walne, P.L.; Arnott, H.J. The comparative ultrastructure and possible function of eyespots: *Euglena granulata* and *Chlamydomonas eugametos*. Planta. 77:325-353; 1967.

Walne, P.L.; Moestrup, O.; Norris, R.E.; Ettl, H. Light and electron microscopical studies of *Eutreptiella eupharyngea* sp. nov. (Euglenophyceae) from Danish and American waters. Phycologia. 25:109-126; 1986.

Walter, M.F.; Schultz, J.E. Calcium receptor protein calmodulin isolated from cilia and cells of *Paramecium tetraurelia*. Eur. J. Cell Biol. 24:97–100; 1981.

Wang, Y.-L.; Lanni, F.; McNeil, P.L.; Ware, R.; Taylor, D.L. Mobility of cytoplasmic and membrane-associated actin in living cells. Proc. Natl. Acad. Sci. U.S.A. 76:4660–4664; 1982.

Wangersky, P.J. Lotka-Volterra population models. Ann. Rev. Ecol. Syst. 9:189–218; 1978.

Warner, A.C.I. Proteolysis by rumen microorganisms. J. Gen. Microbiol. 14:749–762; 1956.

Weik, R.R.; John, D.T. Cell size, macromolecular composition, and O_2 consumption during agitated cultivation of *Naegleria gruberi*. J. Protozool. 24:196–200; 1977.

Weinbach, E.C.; Diamond, L.S. *Entamoeba histolytica*. 1. Aerobic metabolism. Exp. Parasitol. 35:232–243; 1974.

Weiser, J. Phylum Microspora Sprague. In: Lee, J.J.; Hutner, S.H.; Bovee, E.C., eds. Illustrated Guide to the Protozoa. Lawrence, Kansas: Society of Protozoologists; 1985: pp. 375–383.

Werries, E.; Nebinger, P. The occurrence of beta-amylase (EC 3.2.1.2) in *Entamoeba histolytica*. Mol. Biochem. Parasitol. 11:329–336; 1984.

Wessenberg, H.; Antipa, G. Capture and ingestion of *Paramecium* by *Didinium nasutum*. J. Protozool. 17:250–270; 1970.

Westbroek, P.; van der Wal, P.; van Emburg, P.R.; de Vrind-deJong, E.W.; de Bruijn, W.C. Calcification in the coccolithophorids *Emiliania huxleyi* and *Pleurochrysis carterae*. I. Ultrastructural aspects. In: Leadbeater, B.; Ryding, R., eds. Biomineralization in Lower Plants and Animals. Oxford: Clarendon Press; 1986: pp. 189–203.

Whittaker, R.H. New concepts of kingdoms of organisms. Science. 163:150–160; 1969.

Whittaker, R.H. Broad classification: the kingdoms and the protozoans. In: Krier, J.P. Parasitic Protozoa. New York: Academic Press; 1977: pp. 1–34.

Wichterman, R. The Biology of Paramecium. 2nd ed. New York and London: Plenum Press; 1986: 599 pp.

Williams, N.E. The nature and organization of filaments in the oral apparatus of *Tetrahymena*. J. Protozool. 33:352–358; 1986.

Williams, N.E.; Bakowska, J. Scanning electron microscopy of cytoskeletal elements in the oral apparatus of *Tetrahymena*. J. Protozool. 29:382–389; 1982.

Wilton, D.C. Effect of cholesterol on ubiquinone and tetrahymanol biosynthesis in *Tetrahymena pyriformis*. Biochem. J. 216:203–206; 1983.

Winet, H.; Jahn, T.L. Geotaxis in protozoa. I. A propulsion-gravity model for *Tetrahymena* (Ciliata). J. Theor. Biol. 46:449–465; 1974.

Wise, D.L. Carbon sources for *Polytomella caeca*. J. Protozool. 2:156–158; 1955.

Wise, D.L. Carbon nutrition and metabolism of *Polytomella caeca*. J. Protozool. 6:19–23; 1959.

Wohlfarth-Bottermann, K.D. Contraction phenomena in *Physarum*: New results. Acta Protozoologica. 18:59–71; 1979.

Wolfe, J. Cytoskeletal reorganization and plasma membrane fusion in conjugating *Tetrahymena*. J. Cell Sci. 73:69–85; 1985.

Wolken, J.J. *Euglena*. New York: Appleton-Century-Crofts; 1967; 204 pp.

Wood, D.C. Membrane permeabilities determining resting, action and mechanoreceptor potentials in *Stentor coeruleus*. J. Comp. Physiol. 146:537–540; 1982.

Woodruff, L.I. Observations on the origin and sequence of the protozoan fauna of hay infusions. J. Exper. Zool. 12:205–264; 1912.

Yagiu, R.; Shigenaka, Y. Electron microscopy of the longitudinal fibrillar bundle and the contractile fibrillar system in *Spirostomum ambiguum*. J. Protozool. 10:364–369; 1963.

Yamaoka, I.; Kawamura, N.; Mizuno, M.; Nagatani, Y. Scale formation in an Amoeba, *Cochliopodium* sp. J. Protozool. 31:267–272; 1984.

Yarlett, N.; Orpin, C.G.; Munn, E.A.; Yarlett, N.C.; Greenwood, C.A. Hydrogenosomes in the rumen fungus *Neocallimastix patriciarum*. Biochem. J. 226:729–739; 1986.

Yates, G.T. How microorganisms move through water. Amer. Scient. 74:358–365; 1986.

Yongue, W.H., Jr.; Cairns, J., Jr. Micro-habitat pH differences from those of the surrounding water. Hydrobiologia. 38:453–461; 1971.

Yongue, W.H., Jr.; Cairns, J., Jr.; Boatin, H., Jr. A comparison of fresh-water protozoan communities in geographically proximate but chemically dissimilar bodies of water. Arch. Protistenk. 115:154–161; 1973.

Yoshida, N. A macromolecule-free partially defined medium for *Trypanosoma cruzi*. J. Protozool. 22:128–130; 1975.

Zeuthen, E.; Rasmussen, L. Synchronized cell division in Protozoa. In: Chen, T.-T., ed. Research in Protozoa. Oxford: Pergamon Press; 1972: pp. 9–145.

Zuckerman, A.; Lainson, R. Leishmania. In: Kreier, J.P., ed. Parasitic Protozoa. vol. 6. New York: Academic Press; 1978: pp. 57–133.

Subject Index

A

Abundance of species
  aquatic, 78, 79–84, 92, 121–129, 137–138, 144–147
  terrestrial, 36, 78–79, 85–86, 148–149
Acetate (acetic acid), 276, 280–282, 283, 286–287, 290–291, 296, 299, 315, 317, 328
Acetate flagellates, 3, 37–38, 40, 52, 286–289, 320, 337
Acetic-thiokinase, 286
Acetyl-CoA, 276, 278–281, 283, 293, 295–297
Acetyl-CoA-synthetase, 293
Acid hydrolases (digestive or lytic enzymes), 305–306, 311, 314, 335
Acid phosphatase, 314, 400, 417
Adaptations
  anoxic, 31, 121, 129, 133, 139–140, 148, 150, 153, 166
  facultative heterotrophy, 28, 38, 144
  geographic, 36, 79, 86–88, 125–126, 141–149, 165
  osmotic, 36, 87–88, 91, 93, 342–348
  parasitic, 5, 38–39, 150–152, 159, 164, 169–173
  Photo-, 23, 36–37, 111, 185, 208, 210, 216
  thermal, 39, 75, 77, 80–81, 82, 87–88, 91, 112, 114, 119, 121–123, 143
Adenylate charge, 401
Adoral anlage, 411
Alpha-hydroxy acid oxidases, 294
Alpha linolenic acid, 303, 304

Amino acids
  metabolites, 276, 278–280, 285–288, 292, 295–297, 300–301
  nutrients, 285, 314–318, 320, 323–324, 328–329, 331, 335
  osmotic regulators, 242–243
  respiratory substrates, 339
Amino acid metabolism (*see also* Physiology), 278–279, 296, 314–317, 323, 328, 339, 343
Aminotransferase, 417
Amoebae (*see also* Sarcodina)
  nontestate
    free-living, 6, 7, 9, 11, 13, 58–62, 77–78, 135, 139, 141, 147, 185–186, 191, 231–233, 291–292, 304–305, 323–327, 340, 343, 345, 347, 366–373
    parasitic, 152–156, 291, 293, 295, 301, 302, 305, 327
  testate, 58, 63–65, 64, 78–86, 233, 246–247, 324, 339, 387
Amoebic dysentery, 153
Amoebic meningoencephalitis, 154
  compared to bacterial disease, 155
Amoebo-flagellates, 6, 7, 9, 59, 60, 155, 292, 324, 369, 377, 378, 393
Amoeboid locomotion, 366–371
  ion and pH effects, 370–371
  light effects, 368
AMP (adenosine monophosphate), 355, 397, 401
Amylase, 324
Anisogamonty, 379

Anthropozoonosis, 165
Apical complex, 170, 171
Apical ring, 171
Apicomplexa, 14, 151, 169–173
Arginine deiminase, 328
L-Asparaginase, 286
ATP, 179, 183, 192, 276–278, 282–287, 283, 293, 295, 297, 360, 401, 408
ATP/ADP ratio, 401
ATPase (*see also* Na⁺/K⁺ ATPase), 358–360
Autogamy, 385, 422
Axoneme, 187, 200, 241
Axostyle, 158, 223

B

Balantidial dysentery, 173
Balantidosis, 173
Basal body, 16, 181, 202–203, 411
Benthic Foraminifera, 65–69, 67, 68, 86–87, 247–248, 249, 378, 388, 403–407
Beta keto-oxidation, 192, 278–279, 288–289
Beta-galactosidase, 314
Beta-glucosidase, 314
Binary fission, 386–388
Biochemistry (*see also* Physiology and Metabolism), 275–306
Biological clock, 363, 408–410, 421
Brood pouch, 390–391
Budding, 388–392

C

Calcium
  and cellular regulation, 325, 352–355, 361–362, 365–366, 371–373
  metabolism, 345–348
  nutrition, 316, 325
Calcium gated channels, 353–354, 423
Calmodulins, 284, 354–355, 397
Capitulum, 158–159
Capping protein, 371
Catalase, 192, 288, 294
Cathepsin D, 314
Cell cycle, 408–410

Cell enclosures (shells, loricae, tests)
  ciliates, 97, 104, 105, 253–263
  flagellates
    mineralized, 21, 24, 26, 217–219
    organic, 22–26, 166, 208, 210, 213, 214–216, 220
    scales, 24, 25, 200
  sarcodines
    mineralized, 13, 56, 63, 65, 66, 236, 247–251, 403–407
    organic, 57, 63, 65, 66, 233, 238, 244–246, 399, 416
    scales, 13, 58, 191, 233, 245–246, 387
Cell membrane (*see* Plasma membrane)
Cellulases, 305
Centrioles, 178, 182, 184, 186–187
Centrosphere, 186
Cerophyl, 410
Chaga's disease, 167–168
Chemical messengers (signals), 396–397
Chemoautotrophic bacteria, 131
Chloroplasts, 25, 27, 28, 185, 206, 209, 219
Choanomastigote, 160, 164
Chondriome (*see* Mitochondrion)
Chromatin
  euchromatin, 207, 209, 239
  heterochromatin (condensed), 216, 266
Cilia (*see also* Undulipodia), 97–99, 254–256
Ciliary bud and necklace, 180
Ciliary rosette, 180
Ciliata (Ciliates), 7, 9, 11, 14, 95–100, 108–130, 173, 180, 253–271, 294–299, 303–305, 311–314, 328–333, 339–340, 345–348, 352–357, 408–417, 420–424
Ciliophora (*see also* Ciliata), 14
Circadian rhythms (*see also* Biological clock), 363
Circular DNA, 185
Cirrus, 94, 102, 262
Citrate (citric acid), 276, 277, 281, 287, 315, 328
Coccolithophorids, 36, 54, 217–219, 301
Collecting tubules, 193, 194

Colonies, 1, 13, 17, 71, 377, 389, 396–398
Condensing enzyme, 287
Conjugation, 379–385
Conjugation tube, 378
Contractile vacuole, 26, 97, 102, 107, 173, 192–195, 193, 194, 222, 226, 238, 269, 341–342, 410, 411, 414
Cortex, 97, 253–263, 255, 257
Cryptomonadida, 9, 11, 18, 21–22, 27, 44, 139, 144, 216, 308
Culture media (*see* Nutrient media)
Cultures, 54, 72, 113, 314–330, 399, 402, 408, 410, 414
Cutaneous leishmaniasis, 159, 161
Cyanobacteria, 129, 139, 142, 146, 147, 313
Cyrtophora, 15
Cyrtos (pharyngeal basket), 95, 102, 106, 264, 266–267, 313
Cysts, 21, 26, 30, 59, 61, 62, 79, 85, 97, 120–121, 136, 152, 154, 156, 173, 207–208, 305, 323, 394–396, 399–402, 414–417
Cytochrome oxidase, 285, 400
Cytokalymma, 250
Cytomechanics, 355–360, 366–369
Cytopharynx
  ciliates, 263–266, 312, 330, 415
  flagellates, 31–33, 225, 228
Cytopyge (cytoproct), 100, 270, 312, 413
Cytoskeleton, 179–181, 205, 212, 228, 235, 241, 240–242, 254, 360–362, 366–368
Cytostomal pouch
  ciliates, 265, 330
  flagellates, 209, 211, 212, 226–228, 309

D

Decorated tubules, 194
Dehydrogenase, 283
  lactate, 285, 294, 298
Density dependent effects, 75
Desaturase, 303–304
Diastole, 193, 341–342

Digenetic parasites, 150, 157, 162, 164, 165, 170, 172
Digestive vacuole, 178, 183, 209, 221, 223, 227, 232, 237, 308, 310, 312, 321–323, 327
Distribution of species
  spatially, 35–36, 54–55, 61, 78, 136, 148, 165
  temporally, 49–53, 136–141, 144–149
Diversity, 3, 57
  coefficients, 47–48
  and community stability, 118
  of ciliates, 117–118
  of flagellates, 47–53
  of sarcodines, 57, 83, 86, 89, 90
DNA, 15, 183, 185, 243, 400, 409
DNase (deoxyribonuclease), 314, 333

E

E_h (*see* Redox potential)
E and O mating types, 420–422
Ecology
  abiotic factors, 35, 75, 109–111, 133–134, 144–150, 401–402
  biotic factors, 35, 45, 75, 118–129, 138–144, 401–402
  community stability, 35, 49, 53, 117, 134–138
  community structure, 51, 77, 89, 92, 108, 120–129, 132–134, 141–149
  and environmental variables, 120, 126–129
  competition, 76, 77, 112, 118–120
  competitive exclusion, 117
  competitive strategies, 90, 119–120
  cycles
    carbon, 111, 131, 133, 137, 143, 149
    mineral, 44, 111, 133, 134, 149
  density-dependent factors, 75, 90
  density-independent factors, 75
  niches, 53, 74–78
    and population growth, 119
    overlap, 75–76
  population
    density regulation, 114, 116
    equilibrium, 110
    growth, 108, 112–117

Ecology (cont.)
   population (cont.)
      r/K-selected, 109
      stability, 49–53, 115, 117–118, 132, 134–135
      predator-prey interactions, 52–53, 109–110, 135, 330–333, 417
      resources
         patchiness, 121
         utilization, 113, 116, 131–136
      succession, 49–52, 91, 138–141, 145, 147
      trophic interactions, 4, 8, 20, 31, 37–38, 71, 76–78, 90–91, 111, 114–117, 122–124, 131–135, 328, 333, 417
Ectoplasmic tube contraction model, 367
Embden-Meyerhof (glycolytic) pathway, 276–278, 285, 292, 294, 297, 299
Emetine hydrochloride, 401
Endoplasmic reticulum (microsomes), 178, 182, 204, 207, 216, 225, 229, 231, 235, 238, 240, 242, 253, 269, 394, 400
   rough, 178, 190, 207
   smooth, 178, 204, 212, 213, 216
Endosome, 154, 186, 231, 232
Endosymbionts, 188, 189–191, 190, 226, 227, 231, 232, 233, 237–238, 242–243, 333–336
Estuarine communities, 86–89, 125, 129
Euglendia (see euglenoids under flagellates), 23
Eukaryotes, 10
Eutrophy (see Saprobic environments)
Evolution, 11–12
   ciliates, 8, 107, 263, 271
   endosymbiosis hypothesis, 8, 291
   flagellates, 8, 20, 189, 212, 222, 227–228, 361
   from ancestral flagellates, 8
   heterotroph hypothesis, 8
   metazoa, 8
   parasites, 150
   polyphyletic, 8
   sarcodines, 8, 57, 66, 70, 73
Eyespot (ocelli & stigma), 26, 206, 209, 210, 216, 359, 364

F

Feeding behavior, 23, 28, 32, 37–39, 56, 91, 99–100, 113–117, 127, 128, 134, 138, 153, 171, 183, 200, 211, 220, 226, 228, 263–267, 308–314, 321–323, 327, 330–333, 417
Fermentation (see Physiology, anaerobic metabolism)
Fine structure, 178–196
   amoebae, 231–247
      nontestate, 231–233
      testate, 233, 236, 246–247
   cell model, 178
   centrioles, 182, 184
   ciliates, 253–271
   cilium, 180, 253–263
      membrane particles, 180
   contractile vacuole, 192–195, 269
   cysts, 394, 399
   cytoplasm, 178, 182–185, 191–195, 205, 253, 263–271
   cytoplasmic organelles, 182–185, 190, 238–244
   cytoskeleton, 177–181, 205, 212, 228, 235, 241, 240–242, 254, 256, 257, 261, 262, 263–264, 267, 268
   digestive vacuole, 178, 182–183, 209, 221, 232, 237, 254
   endosymbionts, 190, 232, 237, 242–244
   flagellates
      phyto-, 205–222
      zoo-, 222–229
   flagellum, 181, 198–203, 258, 357–360
      basal body, 181, 202–203
      mastigonemes, 198
      paraflagellar rods, 200, 201
   Foraminifera, 233–235, 247–250
   Golgi apparatus, 178, 182–183, 190, 206, 219, 221, 227, 229
   haptonema, 199, 203–204
   heliozoa, 234, 238, 241–242, 245
   hydrogenosome, 190, 223
   kinetosome (basal body), 202–203, 253, 258
   nucleus, 178, 184, 185–187, 190, 206, 209, 232, 234, 239–240, 254, 256, 266

Peroxisome (microbody, glyoxysome), 190, 192
plasma membrane, 191, 222, 253, 254
plastid, 185, 206, 207, 209
radiolaria, 234, 235-238, 250-251
striated fibers (rootlets), 181, 202-203, 223, 224, 256
Five-kingdom classification, 5
Flagellar pouch, 195, 203, 211, 212-213, 309
Flagellates, 9, 13, 16-34, 157-169, 197-230, 280
  phytoflagellates, 21-28
    chrysomonads, 9, 11, 18, 24, 27, 38, 40, 52, 139, 140, 142, 144, 203-205, 220, 286, 289, 302, 304, 321, 323, 343-344
    cryptomonads, 9, 11, 18, 21-22, 27, 44, 139, 144, 216, 308
    dinoflagellates, 9, 11, 18, 22-23, 27, 37, 45, 55, 144, 145, 185, 199, 214-216, 308, 321, 350-351, 378
    euglenoids, 4, 11, 18, 20, 23-24, 27, 37, 38, 41, 44, 55, 131, 135, 139, 141, 142, 145, 181, 195, 199, 209-214, 286-287, 288, 301, 309, 314-318, 359, 364-366, 359, 363-366
    Prasinomonads, 25-27
    prymnesiomonads, 24-25, 27, 191, 197, 217-219, 289, 290, 301
    silicoflagellates, 26, 27
    volvocidans, 1, 9, 11, 18, 25, 27, 36, 38-41, 50, 51, 140, 146, 186, 202-203, 205-208, 286-288, 300-301, 314-317, 319-320, 342-343, 363, 388-389, 394-396, 418-419
  zooflagellates, 28-33
    cercomonads, 18, 19, 29, 33
    choanoflagellates, 11, 29
    diplomonads, 18, 30-31, 33, 159, 291
    hypermastigida, 7, 31, 33, 187, 291
    Kinetoplastida, 9, 11, 19, 31, 33, 36, 51, 52, 139, 144, 148, 159-169, 189, 195, 225-229, 350, 351
    opalinids, 7, 13, 20, 33, 257, 258
    trichomonads, 9, 20, 30, 33, 157-159, 192, 222-225, 291, 320
    trypanosomes, 31, 162-166, 228-229, 285-286, 290, 320
Foraminifera (*see* Benthic foraminifera or Planktonic foraminifera)
Frontal contraction model, 367
Fusules, 72, 237

G

G-actin, 368, 371
Gametocytotomont, 403, 404
Gametogamy, 376-378
Gamma-linolenic acid, 303, 304
Gamontogamy, 378-385, 403, 404, 406
Genes
  a, 421
  atlanta, 423
  ciliate, 420-423
  dancer, 423
  E and O type, 421
  flagellate, 418-419
  gad, 420
  mat, 421
  mt, 419
  pawn, 329, 354
  sad, 419
  sarcodine, 419-420
  sfu, 419
  upp, 419
Genetic markers, 418-419
Genetics, 417-424
  Mendelian, 417-423
  molecular, 12, 300, 423, 424
Genotype, 418, 421, 422
Geotaxis, 95, 269, 270
Giardial diarrhea, 159
Glucose metabolism (*see* Physiology)
Glucose-6-phosphate, 275-276
Glutamate (glutamic acid), 276, 288, 295, 297, 315, 324, 329, 335
Glutamine, 287, 297, 335
Glutarate (alpha-keto), 287, 288, 297, 335
Glyceraldehyde phosphate, 276, 277, 278, 298

Glycerophosphate (phosphoglycerate), 276, 277, 285, 289, 296
Glycerophosphate oxidase, 285
Glycocalyx, 57, 311, 369, 397
Glycolate dehydrogenase, 288, 290
Glycolate oxidase, 294
Glycolysis (*see* Embden-Meyerhof pathway)
Glycoprotein, 397
Glycosynthesome, 295
Glyoxylate cycle, 192
Glyoxylate oxidase, 294
Glyoxysome, 192, 280
GMP (guanosine monophosphate), 355
Golgi apparatus (body), 17, 173, 178, 182, 190, 205, 217, 222, 226, 327, 394
Gones, 418
Granuloplasm, 57, 231
Growth
  density factors, 75, 108
  efficiency, 134
  equation of, 108
  inhibition
    chemical, 37
    pH, 36, 286
    photo, 37, 145, 153
  light effects, 398
  population stability, 115
  species characteristics, 109
  specific coefficient (μ or r), 45–46, 108
  temperature response, 112, 153, 321, 397–398, 406, 408, 411

H

H equation, 47
Habitats
  ciliates, 122–129
  colonization, 137–138, 148
  flagellates, 22, 25, 31–34, 35–39, 50–53, 141–147, 165
  freshwater, 42–43, 50, 79–84, 120–124, 141–148
  high temperature, 39, 40
  low temperature, 39, 78, 88, 125, 143
  marine
    open ocean, 53, 90–93
    tidal zone, 86–89, 125
  sarcodines, 57, 65, 77–78, 84, 147–149, 152–156
  terrestrial, 36–37, 79, 125, 148–149
Haemoflagellates, 159–169, 172–173
Haemozoin, 171
Haptonema, 24, 197–204, 199
Heat shock, 397, 408
Heliozoans, 11, 13, 68, 72–73, 187, 193, 234, 238, 241–242, 244, 245–246, 327, 351
Heredity (*see* Genetics)
Holotrich, 94, 146, 298–299
Homothetogenic fission, 386
Hyaloplasm, 57, 231
Hydrogenosome, 150, 183, 190, 192, 222, 223, 225, 282, 283, 291, 294, 299
Hypotrich, 94, 102, 125, 408–409

I

Intestinal protozoa
  parasites, 152–155, 159, 173, 291–294, 295, 302, 327
  symbionts, 34, 153, 298–299, 305
Isocitrate lyase, 295
Isofloridoside, 344
Isogamonty, 379

K

K (population maximum), 108, 119
Kala-azar, 161
Karyomastigont, 30–31
Killer strains, 119
Kinase (*see also* Phosphorylase), 397
Kineties, 97, 98, 100, 103, 105, 253–263, 411, 416
Kinetoplast, 31, 160, 166, 189, 226–229
Kinetosome (*see* Basal body)

L

Labyrinthomorpha, 13
Lactate (lactic acid), 288, 294, 295, 296, 299, 337
Lactate dehydrogenase, 285, 294, 298
Laredo-type strain, 153

Leishman-Donovan (L.D.) bodies, 160–161
Leishmaniasis, 159–162
Life cycle, 62, 154–155, 160, 162, 164, 167–168, 170, 173, 208–209, 246, 393–417
Lipid synthesis, 302–305
Locomotion (*see also* Sensation), 350–374
 amoeboid, 17, 56, 59, 61, 154, 360–362, 366–373
 cilia, 94, 97, 104, 253–259, 355–360
 contraction, 259–261, 351–353, 360–362
 euglenoid (metaboly), 23, 214
 oar-like organelles, 72
 undulipodial, 16–17, 357–360
Logistic equation, 108–109

M

mRNA, 178, 186, 300, 410, 414
Macrogametocyte, 171
Macrogamont, 380
Malaria, 151, 169–173, 285
Malate (malic acid), 276, 280–281, 283, 287, 293, 295, 298, 315
Malate dehydrogenase, 293
Malate synthase, 289, 295
Male-inducing substance (MIS), 396–397
Malignant malaria, 169
Mastigoneme, 21, 198, 199
Mastigophora, 6, 13, 17
Mating types, 418, 420–422
Membranelles, 263–266, 330
Merozoite, 171
Mesokaryote, 10, 185, 190
Metabolism (*see also* Physiology), 275–279
Microanatomy (*see* Fine structure)
Microbody (peroxisome-glyoxysome), 178, 190, 192, 253, 270, 280
Microdissection studies, 413–414
Microfilaments, 178, 179, 184, 186, 259, 260, 269, 360–362, 367–368, 372–373
Microflagellates, 19, 52, 114, 139, 143, 144, 149, 323, 344, 351
Microgametocyte, 171

Microgamont, 380
Micromastigote, 160, 161, 164
Micronemes, 171
Microspora, 14
Microtubules, 179, 180, 181, 182, 187, 199, 204, 213, 234, 235–238, 245, 254–263, 267, 268, 353, 356, 358–360, 394
Mitochondrion (chondriome), 17, 178, 183, 189, 190, 205, 227, 233, 234, 238, 253–257, 269, 277, 278, 280, 285, 287, 288, 290, 294, 297, 298, 394–396
Mitosis
 closed, 9, 184
 open, 9, 184
Molecular genetics (*see* Genetics)
Monogenetic parasites, 157
Morphogenetic polarity, 413
Morphology
 anatomical right and left, 99–100
 ciliates, 7, 94–100, 253–263
 comparative, 5, 7, 9, 177, 232–238
 flagellate, 6, 16, 197–200
 sarcodine, 8, 56–57, 59, 152, 232–252
Morphometric analysis, 394–396
Motility (*see* Locomotion)
MTOC, 182, 238, 241
Mucocutaneous leishmaniasis, 159–161
Mucocysts, 97, 253
Mucoprotein (mucin), 177, 210, 253
 glycocalyx, 57, 311, 369, 397
Müller vesicles, 269, 270, 362
Multiple fission, 388
Mutants, 317, 329, 354, 423
Mycetozoea, 62, 240, 305, 371–373, 402–403, 419–420
Myoneme, 261, 361
Myosin, 361, 367, 371, 372, 424
Myxozoa, 14, 151

N

Na^+/H^+ ATPase, 342
Na^+/K^+ ATPase, 179, 283, 347
Nagana, 165
Nebenkörper (*see* Parasome)
Nematodesmata, 102, 266, 268, 313
Nitrate reductase, 300

Nitrite reductase, 300
Nuclei
  dikaryotic (macro and micro), 94, 173, 187, 266
  filiform, 381
  fine structure, 185–187, 231, 235, 239–240, 254–256, 266, 394–396
  halteriform, 381
  macro-, 14, 94, 95, 96, 173, 381, 383, 384–385, 413–414, 416
  micro-, 14, 94, 96, 173, 379, 383, 384–385
  mitosis of, 59, 184, 185–186, 380, 403–404, 408
  moniliform, 381
  multiple, 13, 31, 32, 62, 73, 187, 403
Nucleoid of microbody, 190, 192
Nucleolus (see also Endosome), 186, 239, 266, 399
Nutrient media, 314–321, 323–324, 329–330
  Aaronson and Baker (*O. danica*), 315–316
  comparative composition, 315–316
  Diamond's medium (*E. histolytica*), 291, 324
  Hutner and Provasoli (*O. malhamensis*), 315–316
Nutrition
  amoebae, 323–327
  autotrophy, 4, 20, 131, 335, 337
  auxotrophy, 35, 110
  bactivory, 52, 77, 113, 116, 117, 118, 323, 329, 373
  carnivory, 124, 132, 331
  ciliates, 113, 117, 122–123, 328–333, 417
  flagellates, 38, 46, 50, 52, 308–309, 314–323
  herbivory, 77, 124–125, 132, 321, 403
  heterotrophy, 20, 131, 308
  myxotrophy, 12
  nutrient requirements
    carbon, 45, 317, 320
    minerals, 44, 314
    nitrogen, 41, 318
    phosphorus, 314–317, 329, 347
    vitamins, 314–317
  omnivory, 124, 403

osmotrophy, 37, 172, 325
phagotrophy, 23, 37, 99, 153, 171, 212, 312, 321–323, 324–327, 330–333
saprophagy, 131
sarcodines, 66, 69, 70, 77, 91–93, 309–311, 323–327

O

Oleic acid, 303, 304–305
Oleoyl-CoA, 304
Oocyst, 172
Oogonia, 396
Ookinete, 170, 172
Opalinata (*see also* Flagellates), 13
Opisthe, 411
Opisthomastigote, 160, 164
Oral apparatus
  ciliates, 97, 99, 102, 104, 106, 107, 263–266, 311–313
  flagellates, 211, 212–213, 226–227, 321–323
Osmoregulation (*see also* Contractile vacuole), 340–348
  salinity tolerance, 36, 345–346
Oxidoreductase, 283, 417

P

Paralabial organelle, 180
Parasitophorus vacuole, 171
Pébrine (silkworm disease), 173
Pelta, 158, 223, 224
pH, 38, 40, 42–44, 74, 79–82, 122, 125, 130, 288–289, 311, 314, 317, 318–319, 321, 324, 326, 327, 353, 370–371, 402, 410
pK, 40, 319
Parasites
  of algae, 14
  amoebae, 152–156, 291–294, 295, 302, 327
  of animals, 14, 165, 168–169, 173
  Apicomplexa, 14, 151, 169–173, 285
  ciliates, 173, 261–262
  digenetic, 32, 150, 157, 162, 165, 175
  flagellates, 157–169, 201, 222–225, 285, 290–291, 320–321, 340

microspora, 14, 173
myxozoa, 14, 151
of plants, 36
Parasome, 61, 243
Peduncle, 23
Pellicle, 98, 209, 213, 214, 254–255, 262
Peltar-axostylar complex, 159, 224
Peniculus, 263–265
Perialgal vacuole, 190, 233, 335
Peritrich, 103, 130, 330
Peroxisome (*see* Microbody)
Phenotype, 418, 421, 422
Phosphoenolpyruvate, 276, 293, 295, 296, 297, 298
Phosphoenolpyruvate carboxykinase, 286
Phosphoenolpyruvate carboxylase (PEPC), 286, 289, 293, 295, 297
Phosphorylase, 280
Photophobic response, 365
Photoreceptor, 364–365
Photosynthesis
  compensation intensity, 36–37
  efficiency, 37, 111
  of symbionts, 335
Phylogenetic relationships (*see also* Evolution), 8–12
Physiology, 275–306, 338–349
  acetate metabolism, 37–38, 40, 52, 276, 278, 286–291, 299
  aerobic metabolism, 276, 277–278, 285, 290–291, 401, 408–410
  amino acid metabolism, 276, 278–279, 281, 286, 289, 292, 295, 296, 297, 300–301, 318, 343, 398, 409
  amoebae, etc., 291–294, 309–311, 323–327, 339–340, 343, 345, 347, 370–371, 398–402
  anaerobic metabolism, 277, 290, 292, 295, 298–299, 301
  beta keto-oxidation, 278–279, 288–289
  ciliates, 269, 294–299, 301–305, 311–314, 328–333, 339–342, 345–348, 353–355, 408–410
  2-carbon metabolism, 278, 279, 281
  chemical messengers, 396–397
  cysts and resting stages, 305, 323, 394–396, 399–402, 414–417
  energetics, 275, 282–284, 287, 297, 338–339, 399–401
  fatty acid metabolism, 278, 279, 281
  feedback regulation, 282–284, 297, 341, 347, 401
  flagellates, 285–291, 295, 299, 300–302, 308–309, 314–323, 342–345, 362–366
  gluconeogenesis, 279–282, 289, 292, 294–295
  glucose metabolism, 275–278, 281, 285, 289, 292, 294, 297, 298–299, 305, 320, 323, 344, 399
  glycolysis, 275–278, 285, 287, 292, 297
  hydrogen-generating metabolism, 282, 283, 290, 299
  lipid metabolism, 278, 279, 296, 302–305, 328–329
  metabolism
    carbon, 275–284, 286–300, 317–318
    mineral, 283–284, 315–316, 345–348
    nitrogen, 41, 300–302, 318, 323, 335
    nucleic acid, 285, 296, 301–302, 328
  nutrient assimilation, 41–42, 275, 279, 285, 289, 294, 297, 300–301, 308, 319, 335
  nutrition
    amoebae, 280, 291–294, 305–306, 309–311, 323–327, 399
    ciliates, 280, 294–299, 301, 305–306, 311–314, 328–333, 410
    flagellates, 280, 283, 285–291, 300, 305–306, 314–323
  osmotic balance, 97
  parasites, 283, 285–286, 290–294, 295, 301, 305–306, 320–321, 327
  Pasteur effect, 281
  protein metabolism, 294, 317, 325, 398, 409, 410, 416
  Q_{10}, 113
  respiration, 275–278
  sarcodines, 291–294
  symbionts, 298–299, 305, 333
  TCA cycle, 276, 277–278, 285–286, 289–291, 297
Phytomastigophorea (phytoflagellates), 16, 21, 205–223, 286–291, 300–302, 342–344
Pigments
  biliproteins, 22

Pigments (cont.)
  blepharismin, 105
  carotenoids, 22, 24–26, 210
  chlorophylls, 185
    a, 22, 24–26
    b, 24, 25
    c, 22, 24–26
  cytoplasmic, 105, 391, 402, 416, 419
  lutein, 26
  peridinin, 23
  red, 22, 37
  xanthophylls, 26
Planktonic Foraminifera
  ecology, 90–92
  microanatomy, 233–234
  nonspinose, 65–66, 69
  reproduction, 375, 377, 403–405
  spinose, 65–66, 69, 310, 334, 403–405
Plasma membrane, 177, 178, 191, 286, 302, 317
  action potential (depolarization), 350–354
  extrinsic proteins, 177
  hyperpolarization, 352
  intrinsic proteins, 177
  porters (and active transport), 178–179, 282–284, 286, 297, 308, 319, 325, 335, 340–342, 345–348, 366
  resting potential, 351
  surface coat, 166, 178, 191, 210, 216–219, 245–248, 311, 369
Plasmodization, 378
Plastids, 17, 22, 24–27
  chromoplasts (see Chloroplasts)
  leukoplasts (nonpigmented), 21, 205, 207, 396
Podite, 107
Polar filament, 173
Polarized light, 359, 364–365
Populations (see Ecology)
Population variance, 114–118
Postciliary ribbon (microtubules), 255–262
Postciliodesmatophora, 14
Posteronuclear form (trypanosome), 166
Posterosome, 173
P.I.E. coefficient, 48
Pr (primary production coefficient), 111

Prey
  algal, 71, 77, 91–92, 114, 122–123, 125, 127, 140, 142, 146, 321
  bacterial, 52, 71, 89, 91, 113, 117, 121, 122–123, 124, 132, 134, 140, 142, 144, 153, 226, 237, 323, 329, 410, 417
  other protozoa, 128, 132, 134, 142, 211, 223, 331, 417
Productivity, 45, 47, 78, 84, 111, 134–135
Prokaryote, 10, 185, 192
Prolifin, 371
Promastigote, 160, 164
Pronase, 333, 397
Protein metabolism (see Physiology)
Proter, 411
Protista, 5
Protozoa
  animal properties, 4
  defined, 3, 6
  phylogenetic advancement, 12
  in research, 3, 12, 15, 54–55, 73, 93, 149, 173–174, 187–188, 229–230, 252, 271, 306, 336–337, 373–374, 423–424
  taxonomy of, 5, 12–15, 17, 57, 95
Pseudoconjugation, 378
Pseudocopulation, 378
Pseudopodia
  axo- or actinopodia, 57, 70, 235, 241
  filopodia, 57, 58, 65
  lobopodia, 56, 58
  morphology, 56, 59, 369–370
  reticulo- (rhizo) podia, 65, 69, 70, 233, 310–311
Pseudostome, 63
Pusule, 26, 216
Pyrophosphatase, 347–348
Pyruvate (pyruvic acid), 276–277, 280, 283, 285–286, 293, 297, 301, 318
Pyruvate oxidase, 293
Pyruvate phosphate dikinase (PPDK), 293
Pyruvate synthase, 293

Q

Quartan malaria, 169

R

r (growth rate), 108–112
r-K selection, 109–110
Radiolaria, 8, 11, 12, 13, 71–72, 91–93, 235–238, 310, 377
Redox potential (E_h), 42, 74, 111, 125–127, 129
  of water column, 43
Regulatory genes, 419
Reproduction
  asexual, 386–392, 394, 398
  ciliates, 113, 121, 124, 379–385, 388–392
  flagellates, 3, 376, 386–388
  karyorelictids, 381–382
  sarcodines, 246, 377
  sexual, 376–386
Reproductive cycles
  ciliates, 408–417
  flagellates, 393–398
  sarcodines, 398–407
Reserve substance (*see* Storage products)
Reservoir (gullet) flagellar pocket, 181, 209, 210–212, 227
Respiration, 339–340, 401, 408
  balanced with photosynthesis, 36
Rhabdophora, 14
Rhoptries, 171
RNase (ribonuclease), 314, 333, 397, 400, 417
Ribosomes, 178, 182, 186
RNA, 15, 178, 186, 239, 266, 400, 403, 409–410, 416
Rod organ, 28, 209, 211, 212–214
Romanovsky stains, 31

S

Salivaria, 165
Saprobic (organic-enriched) environments, 38, 44, 50, 61, 86, 121
  eutrophy, 114
  oligotrophy, 114
Saprobien system, 44
Sarcodina, 13, 57–73, 75–93, 147–149, 231–252, 291–294, 309–311, 323–328, 366–373, 398–408, 419–420
Sarcomastigophora (*see also* Flagellates), 13
Secretory vesicles, 182–183, 312, 323
Sensation, 350–352
  avoidance responses, 97, 105
  chemo (chemical), 105, 331–332
  geotactic, 95, 269, 270, 362–363, 368
  photo (light), 23, 208, 210, 216, 363–366
Shells (*see* Cell enclosures)
Shuttle streaming, 371–372
Silicoflagellates, 21
Skeletons (*see* Cell enclosures)
Sleeping sickness, 165–167
Slime molds (*see* Mycetozoea)
Spasmoneme, 361–362
Sphaeromastigote, 160, 164
Spindle granules, 187
Spirotrich, 94, 105, 259–260, 329, 330
Sporoblastoid body, 172
Sporozoite, 169, 170, 171
Starch hydrolysis, 280, 281, 298, 324
Stearic acid, 303, 304–305
Stearoyl-CoA, 304
Stercoraria, 165, 167
Sterols, 177, 302–303, 328
Storage products
  Food reserves, 2, 276–281, 299
    amylose (starch), 22, 23, 25, 207, 209
    glycogen (animal starch), 21, 153, 178, 184, 254, 256, 276, 280, 289, 294–295, 299, 342–345, 400
    leucosin, 24, 344
    lipids (oils), 23, 24, 233, 235, 253, 289
    paramylon, 24, 28, 212, 287, 290
  metabolic products
    metaphosphates, 24, 347
    organic, 235, 276, 281, 282, 285–305
Striated rootlets, 181, 202–203, 223, 256
Sublittoral sand communities, 126–128
Succinate (succinic acid), 276, 281, 283, 287–288, 295, 296, 315, 318–319, 321
Succinyl-CoA, 283
Sulphuretum, 129
Suture, 98, 99

Symbioses
  bacterial, 226, 227, 231, 232, 333
  ecto-, 107
  endo (*see* Endosymbionts)
  dinoflagellate, 22, 232, 237
  Janickina, 243
  Paramoeba, 243
  Perkinsiella, 243
  prasinomonad, 25
  with ciliates, 125, 333–336
  with flagellates, 226, 227
  with ruminants, 298–299, 305
  with sarcodines, 70, 73, 231, 232, 237
  with xylophagous insects, 34, 153
Symmetrogenic fission, 386
Synchronized reproduction (cultures), 401, 408
Synthetase (glycogen), 280
Systole, 193

T
Taxonomy, 5
TCA (tricarboxylic acid cycle) (*see also* Physiology), 192
Tentacles, 107
Termite gut symbiont, 34, 299, 305
Tertian malaria, 169
Tetrahymanol, 302
Thylakoids, 185
Tolerance (*see also* Adaptations)
  pH, 38, 283, 318–319, 321
  salinity, 36, 342–345
  thermal, 39, 304
Tomite, 107
Transformations
  amoeboflagellate, 7, 59, 62, 393
  autotroph to heterotroph, 4
  cell cycle, 32, 62, 393
  life cycle, 32, 62, 79, 154, 164, 169–172, 207, 246, 394–417

Trichomonads, 9, 20, 33, 157–159, 192, 222–225, 290–291, 320, 340
Trophic relations (*see* Nutrition)
Trophozoite, 171
Trypomastigote, 160, 164, 285
Trypanosomes, 36, 228–229, 285, 290, 320–321
Tubulin, 179

U
UDP, 280, 344
Undulating membrane
  ciliate, 264–265, 411, 413
  flagellate, 157, 162, 164, 167, 224
Undulipodia, 16, 179, 180–181, 195, 197–203, 258
Urease, 301
Uroid, 61, 153
UTP/UDP, 280, 344

V
V-area, 380, 411–414
Veneral disease, 158
Visceral leishmaniasis, 159, 161

W
Water expulsion vacuole, 192, 193 (*see also* Contractile vacuole)
Water purification, 139

X
Xenophyophores, 5

Z
Zoomastigophorea, 17 (*see also* flagellates, zooflagellates)
Zoonoses, 162, 165, 167, 170

Genera and Species Index

A

Acanthamoeba, 80, 134, 154, 186, 291, 304–305, 315–316, 339, 371, 398–402
  castellanii, 78, 292, 305, 340, 343, 345, 369, 399–401
  culbertsoni, 155, 400
  glebae, 78
  palestinensis, 399
  polyphaga, 60, 78, 147
Acanthocystis, 244
  erinaceoides, 245
  turfacea, 241
Acineta, 101, 107
  tuberosa, 390, 391
Acrasis rosea, 60, 62
Actinocoryne, 239, 241, 242, 352
  contractilis, 351
Actinomonas mirabilis, 46
Actinophrys, 241
  sol, 68, 72, 141, 145, 193, 235
Actinosphaerium, 239, 241
  eichhorni, 73
Aerobacter aerogenes, 117, 124
Alcaligenes faecalis, 124
Allantion, 36
Allogromia, 66, 239, 247, 387
  laticollaris, 89–91, 125, 377, 403, 404, 406
Amastigomonas debruynei, 19
Ammonia beccarii, 69, 87, 88, 89–91
Ammotium salsum, 68, 87, 89
Amoeba, 7, 368
  algonauinensis, 398
  amazonas, 239, 244
  discoides, 77, 80, 239, 244
  dubia, 239, 244
  limax, 141
  navina, 59
  proteus, 13, 56, 59, 77, 80, 82, 141, 142, 239, 240, 243, 345, 347, 368, 369, 371, 398
Amphidinium, 215
  carterae, 36, 37, 45, 46
Amphileptus
  claparedei, 122
Amphorellopsis, 65
Amphileptus, 124
Amphisorus hemprichii, 67
Amphistegina, 67
Anacanthotermes ochraceus, 31
Arcella, 63, 86
  discoides, 79, 81, 82
  gibosa, 83
  mitrata, 141
  vulgaris, 58, 64, 81, 82, 83, 84, 141
Arenoparrella mexicana, 87
Ascoglena, 210
Aspidisca, 126
  angulata, 112
  costata, 122
Assulina
  muscorum, 79
  seminulum, 79
Astasia, 4, 141, 200, 286
  klebsii, 24
  longa, 37, 46, 148, 211, 213, 214, 286, 287, 315–316, 348

Genera and Species Index

Astrorhiza, 66
Astylozoon, 104

B

Balantidium, 101
  coli, 173, 261
Bacillus, 124
  cereus, 114
  subtilis, 113
Barbulanympha, 187
Beggiatoa, 129
Bicoeca, 220, 221, 222, 358, 359
  kepneri, 220
  lacustris, 220
Blepharisma, 7, 101, 105, 118, 126, 259, 260, 329, 346–347, 380–381, 384, 410–414, 415–416
  clarissimum, 129
  japonicum, 299, 410
  lateritia, 122
  salinarum, 129
  stoltei, 415
Bodo, 36, 52, 139, 144, 200, 225, 227, 228, 350
  caudatus, 19, 227
  edax, 19, 51
  globosus, 51
  minimus, 51, 148
  putrinus, 51
  saltans, 19, 31, 32, 226
Bodomorpha minima, 19
Bolivina
  doniezi, 91
  lowmani, 88
Box boops, 223
Brachiomonas, 208
Buccella firgida, 87
Buliminella basendorfensis, 88
Bursaria, 101
  truncatella, 96, 104

C

Caenomorpha, 146
  medusula, 121
Calcarina calcar, 67
Calcituba polymorpha, 91, 249
Camptonema, 239
Carchesium, 104, 124, 141
  polypinum, 96
Cardiostomella vermiforme, 127
Carex, 134
Carteria, 202, 203, 207, 208
Centropyxis, 63, 81
  aerophila, 79
  ecornis, 83
  cassis, 83
  constricta, 83, 148
  discoides, 83
  hirsuta, 64, 79
Cephalothamnium, 228
Ceramium, 89
Ceratium, 215, 378
  hirundinella, 40
Cercobodo (= *Cercomonas*), 19, 36, 52, 139
  agilis, 51
  grandis, 51
  varians, 51
Cercomastix, 139
Cercomonas, 18, 29, 36
Chaos, 59
  carolinense, 79, 80, 82, 371
  chaos, 345
Chilodon, 142
  cucullus, 121, 122
  uncinatus, 122
Chilodonella, 101, 140, 263, 408
  caudata, 148
  cucullus, 95, 106, 146
Chilomonas, 22, 141
  paramecium, 39, 40, 46, 77
Chlamydomonas, 25, 38, 140, 144, 145, 146, 186, 199, 202, 203, 205, 207, 286, 300, 333, 363, 366, 376, 418–419
  acidophila, 289
  gloeopara, 300
  microsphaerella, 300
  mundana, 46
  nivalis, 39
  peterfi, 300
  pulsatilla, 46
  reinhardtii, 46, 208, 288, 343, 363, 419

Chlorella, 334
Chlorogonium, 208, 288
  elongatum, 330
Chloromonas brevispina, 376
Chroomonas, 22, 46
  mesostigmatica, 18, 21
  salina, 290
Chromulina, 139
Chrysochromulina, 197
Chrysomonas ovalis, 140
Ciliofaurea, 127
Ciliophrys, 242
Cinetochilum margaritaceum, 122
Chathrulina, 244
  elegans, 72
Climacostomum virens, 330
Clydonella vivax, 78
Cochliopodium, 186
  bilimbosum, 81
Codium, 90
Coleps, 97, 101, 106, 126, 127, 141
  hirtus, 121, 122, 146
  tesselatus, 125
Colonympha grassii, 7
Colpidium, 140–142, 257
  campylum, 95, 103, 113, 114–116, 118, 120, 328, 331–332, 339
  colpoda, 121–122
  glaucoma, 120
Colpoda, 101, 138, 141, 148, 149, 416
  cucullus, 95, 104, 416
  steinii, 134, 339, 416
Colponema
  loxodes, 19
  agitans, 19
Condylostoma, 128, 259, 260, 414
  patulum, 112
Corallomyxa, 62, 78
  chattoni, 60
Corythion, 65
  dubium, 64, 79, 339
Cruzella marina, 19
Cymbaloporetta bulloides, 406, 407
Crypthecodinium cohnii, 315–316, 351
Cryptobia, 32, 200, 223, 227
  branchialis, 227, 228
  dahli, 227
  helicis, 227, 228
  intestinalis, 227
Cryptodifflugia, 65
  oviformis, 58, 64
Cryptomonas, 139, 144
  erosa, 40
Cyclidopsis acus, 24
Cyclidium, 114, 140, 148
  glaucoma, 120, 122
Cyclogyra involvens, 88
Cyclopyxis
  arcelloides, 83
  eurystoma, 79
Cyphoderia ampulla, 81
Cyprinus carpo, 227
Cyrtolophosis, 149
Cyttarocylis brandti, 262

D
Dactylamoeba bulla, 60, 61, 191
Dasytricha, 299
Dictyocha, 26
Dictyostelium, 305, 424
  discoides, 305
Didinium, 101, 114, 138
  nasutum, 96, 106, 113, 117, 118, 146, 331–332, 414, 417
Dientamoeba, 16, 223, 229
  fragilis, 30
Difflugia, 63, 86, 142
  acuminata, 81, 83, 84, 141
  corona, 83
  elegans, 81, 83
  globulosa, 79, 81, 82, 83, 84
  gramen, 58, 83
  lacustris, 64
  lobostoma, 81
  oblonga, 81, 82, 83, 84
  oviformis, 81
  tuberculata, 81
  urceolata, 81, 83, 84
Dileptus, 106, 114, 126, 128, 412
  anser, 96, 331, 333
  gigas, 121, 122
Dinobryon
  bavaricum, 40
  divergens, 40
  sociale, 40

476   Genera and Species Index

Diophrys scutum, 112, 299, 409
Discorbis, 69, 91, 377
  opercularis, 378
  patelliformis, 378
Dunaliella, 25, 39, 205, 206, 207, 376
  salina, 18, 36, 343
  teriolecta, 46, 300–301, 342
Dysteria
  marina, 125
  monostyla, 125

E

Echinamoeba, 147
Echinosphaerium, 239
  nucleolfilum, 73
Eggerella advena, 87, 88
Elphidium, 248, 310, 377
  advenum, 89
  articulatum, 91
  clavatum, 87
  crispum, 67, 91
  excavatum, 88, 91
  galvestonense, 89
  gunteri, 87, 89
  incertum/clavatum, 69, 88, 89
Emiliana huxleyi, 46, 217, 218
Endamoeba, 153
Endolimax nana, 154
Entamoeba, 152, 291
  coli, 152, 154
  gingivalis, 152
  hartmanni, 154
  invadens, 154, 239
  histolytica, 152–154, 156, 239, 243, 292–294, 302, 305
Enteromorpha intestinalis, 89
Entodinium caudatum, 134, 298, 302
Entosiphon, 24, 36, 148, 200, 212, 227, 228, 309
  sulcatum, 28, 209
Ephelota gemmipara, 391
Epistylis plicatilis, 114
Escherichia
  coli, 114, 134, 140, 402
  freundi, 124
Eudiplodinium maggii, 305
Euglena, 36, 38, 131, 139, 186, 198, 200, 210–212, 214, 290, 359, 362, 386

  acus, 20, 23, 24
  ehrenbergii, 214
  gracilis, 4, 18, 23, 24, 28, 37, 41, 46, 214, 286–289, 300–302, 315–316, 363–366
  granulata, 209
  haematodes, 37
  oxyuris, 213, 214
  sanguinea, 37
  splendens, 210
  viridis, 28
Euglypha, 63, 65, 142, 387
  acanthophora, 83
  loevis, 83
  phryganella, 79
  rotundata, 58, 83
  strigosa, 79
Euplotes, 101, 126, 129, 142, 333, 418
  bisulcatus, 125
  crassus, 209, 421
  eurystomus, 299
  octocarinatus, 333
  patella, 122
  quinquicarinatus, 125
  vannus, 112, 125, 263
Eutreptiella eupharyngea, 210

F

Fabrea, 101
Favella
  arcuata, 114
  ehrenbergii, 114
Flabellula
  mira, 80
  simplex, 80
Frontonia, 101, 126, 129, 141–143, 146
  acuminata, 122
  arenaria, 126
  leucas, 146, 180, 339
  marina, 129
Furgasonia, 263

G

Gastrocirrhus, 126
Gastrostyla steinii, 262, 299
Geleia, 127, 259
Geophryocapsa oceanica, 218

Giardia, 157, 291
　intestinalis (= *lamblia*), 18, 159
Gigantomonas, 30
Glaucoma, 101, 142, 257
　chattoni, 328
　scintillans, 95, 103, 121, 122
Glenodinium, 144, 145, 215, 376
Gloeocystis
　gigas, 300
　maxima, 300
Globigerinella aequilateralis, 70, 91, 377
Globigerinoides
　ruber, 377, 403–405
　sacculifer, 70, 92, 239, 375, 377
Globorotalia, 69
　menardii, 249
　hirsuta, 91
　truncatulinoides, 91
Glossina, 163
　moristans, 165
　palpalis, 165
Gocevia, 186
Gonium, 25, 145, 388, 418
　pectorale, 389, 418–419
Gonyaulax
　grindleyi, 199
　polyedra, 37, 199
　tamarensis, 37
Gromia, 65, 81
　oviformis, 244, 245
Gymnodinium, 23, 215
　breve, 22
　dogieli, 18
　fungiforme, 351
　helveticum, 40
Gymnosphaera, 239, 241, 242
Gyrodinium cohnii, 46

H
Haematococcus, 208
　pluvialis, 46
Halteria, 140
　grandinella, 120, 121, 122, 146
Hartmanella, 78, 80, 147, 305
　rhysodes, 401
Hastigerina pelagica, 68, 69, 70, 91, 92, 377
Hedraiophyrys, 239, 241, 242
Hedriocystis reticulata, 72

Helicoprorodon gigas, 128, 266, 268
Helicostomella subulata, 114
Hemiophrys
　filum, 128
　pleurosigma, 124
Hemiselmis, 22
　virescens, 300
Herpetomonas
　megaseliae, 201
Heterocapsa, 215
Heterophrys, 242, 244
　elati, 241
　marina, 241
Heteronema acus, 28
Heteromita, 36
　globosa, 19, 143
Heterostegina, 248
　depressa, 388
Heterophrys
　glabrescens, 145
　marina, 234
Hexamitus, 30, 50, 139
　crassus, 51
　fissus, 51
　fusiformis, 51
　inflatus, 50, 51
　pusillus, 51
Histomonas, 223
Histriculus vorax, 113, 392
Holophrya, 145, 146
Holosticha, 148
Homalozoon, 126
　caudatum, 127
Hyalophacus, 210

I
Icthyobodo necator, 19
Iodamoeba butschlii, 154
Iridia, 66, 377
Isonema, 227
Isotricha, 299

J
Janickina, 243

K
Karotomorpha, 30
Katodinium, 215

Kentrophorus, 127, 259
Keronopsis
  gracilis, 96
  rubra, 112
Khawkinea quartana, 211, 212
Klebsiella aerogenes, 114, 134

L

Lacrymaria, 101, 106, 126, 127, 128, 129, 260, 266
  marina, 112, 126
  olar, 122, 332
Leishmania, 31, 160, 164, 225, 285
  braziliensis, 161, 162
  chagasi, 161, 162
  donovani, 161
  infantum, 162
  major, 162
  tropica, 161, 162
Lembadion lucens, 333
Lepidotrachelophyllum, 107
  fornicis, 97
Lesquereusia, 63, 81, 83, 239
  spiralis, 64
Lionotus
  anser, 121, 122, 142
  fasciola, 122, 146
Litonotus, 124
  lamella, 112, 128
Loxodes, 101, 105, 133, 144, 257, 259, 269, 270, 271, 300, 362, 381
  rostrum, 95, 121, 122
Loxophyllum, 126, 128, 129
Lycnophora, 125

M

Macacus, 163
Mallomonas
  acaioides, 40
  alpina, 40
  epithelattia, 289
Marginopora ammonoides, 67
Mastigella polymastix, 7
Mayorella, 78, 305
  bigemma, 77
  cultura, 80

  dofleini, 77, 80, 82
  vespertilio, 77
Menoidium bibacillatum, 214
Mesodinium, 126
  pulex, 121, 122
  pupula, 127
  rubrum, 126, 334
Metacoronympha, 30, 31, 34
Metacylis
  angulata, 114
  annulifera, 114
Metarotaliella, 67
Metopus, 126, 129, 146, 148
  sigmoides, 121, 122
Miamiensis avidus, 343
Microchlamys, 86
  patella, 79
Microglena arenicola, 289
Microthorax, 101, 102, 140
  pusillus, 123
  sulcatus, 121, 123
Miliammina fusca, 87, 91
Miliolina, 247
Monas (= *Spumella*), 36, 51, 139, 141, 148
  arhabdomonas, 51, 52
  vulgaris, 51, 52
Monocercomonas, 291
Monochrysis lutheri, 46
Mycterothrix tuamotuensis, 104
Myrionecta (see *Mesodinium*)
Myxotheca, 66, 377

N

Naegleria, 147, 377, 399
  fowleri, 154, 155, 228, 399
  gruberi, 60, 61, 77, 155, 228, 292, 369, 399–400
  Jardini, 399
Nassula, 101, 102, 106, 263, 264
  aurea, 95
  ornata, 414–416, 415
Nebela, 63, 79, 86
  bigibbosa, 79
  collaris, 79
  dentistoma, 64
  tincta, 79

Nemogullima, 66
Netzelia, 246
  tuberculata, 247
Nonionella opima, 88
Nosema bombycis, 173
Notila proteus, 379

O

Ochromonas, 18, 52, 140, 144, 198, 200, 286
  danica, 24, 46, 219, 301, 315–316
  malhamensis, 24, 38, 219, 222, 289, 301–302, 315–316, 343–344
  sociabilis, 140, 302
  Oicomonas, 139
  socialis, 51, 52
Oikomonas, 36, 141, 148
Olisthodiscus luteus, 37
Opalina, 257
  ranarum, 7, 20, 32, 258
Opercularia coarctata, 114
Operculina ammonoides, 68
Ophrydium, 104
  versatile, 121
Ophryoglena, 126
Ophryoscolex purkinjei, 180
Opisthonecta, 104
Orbulina universa, 68, 91, 92
Oscillatoria, 147
Ovamina opaca, 240
Oxyrrhis, 202, 215, 216
Oxytricha
  bifaria, 299, 409
  falax, 299
  ferruginea, 121, 123
  pellionella, 121, 123

P

Palmerinella gardensis-landensis, 87
Pandorina, 25
  morum, 289
Parabodo nitrophilus, 19
Parahistomonas, 223
Paramecium, 7, 15, 97, 100, 101, 112, 118, 138, 140, 262–264, 265, 270, 305, 328, 331, 333, 342, 346, 352, 354–355, 365, 417, 418, 420–423

aurelia, 102, 117, 119, 121, 124, 253, 255, 329, 339, 382, 384, 420
bursaria, 12, 102, 121, 123, 124, 299, 329, 334, 382, 384, 409
calkinsi, 382
caudatum, 77, 113, 117, 123, 146, 180, 312, 331, 341–342, 352, 382, 384
jenningsi, 382
multimicronucleatum, 329, 342, 382–383
primaurelia, 115, 116, 421, 422
tetraurelia, 115, 329, 350, 354, 423
trichium, 382
wichtermani, 382
woodruffi, 382
Paramoeba, 243
  aesturina, 78
  pemaquidensis, 60, 78
Paranassula microstoma, 95
Paraphysomonas
  imperforata, 46
  vestita, 344
Paraspathidium, 126
  fuscum, 127
Paratetramitus, 147
Parauronema acutum, 329
Paraurostyla, 414
Paascherella tetras, 289
Patellina corrugata, 88, 91, 378
Pavlova (see also *Monochrysis*)
  gyrans, 289
Pelomyxa, 11
  carolinensis, 239, 240, 243
  illinoisensis, 243
  limicola, 80
  lucens, 80
  palustria, 60, 61, 77, 239, 242
  villosa, 80
Peneroplis, 69, 248, 377
  planatus, 67
Pentatrichomonas hominis, 157
Peranema, 38, 142, 200, 202, 212, 228, 309
  trichophorum, 28, 46, 210, 211, 315–316
Peridinium
  cinctum, 141, 199, 215

willei, 40
Perkinsiella amoebae, 243
Perspira ovum, 113
Phacus, 200, 210
  triqueter, 24
Phascolodon vorticella, 145
Phryganella nidulus, 79
Phyllomitus
  amylophagus, 19
  undulans, 19
Physa, 125
Physarum, 240
  polycephalum, 62, 371–373, 402, 419–420, 418–420
Physematium muelleri, 92
Phytomonas, 36
Plagiopogon, 126
Plagiopyla, 126
  frontata, 129
Plagiopyxis, 79
Planorbulina, 377
Plasmodium, 170
  berghei, 285
  falciparum, 169
  malariae, 169
  ovale, 169
  vivax, 169
Platyamoeba, 78
  bursella, 61
Platymonas, 203
Pleurochrysis carterae, 217, 219
Pleuromonas, 139, 148
  jaculans, 19
Pleuronema, 126, 141
  chrysalis, 123
  marinum, 127
Podophrya, 107
Polychaos
  dubium, 77, 80
  fasciculatum, 398
Polygonum ambiguum, 82
Polyplastron multivesiculatum, 305
Polysiphonia, 89, 90
Polytoma, 52, 203, 207, 208, 286, 288, 376, 394
  fusiformis, 51
  ocellatum, 41, 51
  papillatum, 394
  uvella, 50, 51

Polytomella, 38, 203, 286, 287, 288, 394
  agilis, 207
  caeca, 18, 25, 315–316, 394, 395
Polytrichum formosum, 79
Potamogeton, 82
Poteriochromonas stipitata, 330
Prorocentrum, 215
Protelphidium tisburyensis, 89
Proteromonas lacertae, 18, 30
Protrichomonas, 222, 223
  legeri, 222
Prymnesium, 198
  parvum, 290
Pseudobodo tremulans, 351
Pseudomicrothorax, 101, 102, 263, 361
Pseudomonas, 77, 143
  fluorescens, 113, 140
Pseudourostyla
  weissei, 180
Pyramimonas, 144
Pyrocystis
  fusiformis, 37, 70
  noctiluca, 45, 70

Q

Quadrullela, 63
  symmetrica, 64
Quinqueloculina, 69
  lata, 89
  seminulum, 89
  striata, 67
  weisneri, 87

R

Rabdiophrys annulifera, 73
Raphidiocystis lemani, 73
Raphidiophrys, 242, 244
  pallida, 73
Remanella, 101, 126, 127, 269, 271, 381
Rhabdomonas, 200
Rhizobium, 52
Rhizonympha, 31
Rhodomonas
  lacustris, 40
  minuta, 145
Rhyncobodo, 227

Rhynchomonas, 139, 225, 228
   nasuta, 19, 31, 32, 227
Rosalina, 88, 406
   golobularis, 406
   leei, 89, 90, 91
Rotaliella heterocaryotica, 91

S

Saccamoeba, 77, 78
   limax, 398
Sainouran, 36
Salpingoeca amphoroideum, 18
Schizopyrenus russelli, 401
Sonderia, 126, 129
Sorites, 69, 248
Spartina, 125
Spathidium, 114, 146
Spirillina, 67
   vivipara, 88
Spirogyra, 77
Spirolocullina, 69, 247
   hyalina, 67, 90, 91
Spiromonas, 36
   angusta, 19
Spirostomum, 101, 142, 260, 361
   ambiguum, 96, 121, 123, 259, 331
   minus, 146
   teres, 121, 123, 146, 339
Spongomonas, 36
Spumella hovassei, 7
Stenosemella, 114
   ventricosa, 333
Stentor, 14, 105, 107, 141, 259, 260, 361, 385, 410
   coeruleus, 134, 339, 346
   igneus, 123
   polymorpha, 146
   roeseli, 96, 101, 123
Sticholonche zanclea, 72, 239, 241
Stokesia vernalis, 146
Streptococcus faecalis, 114
Strichotricha, 125
Strobilidium, 114
Strombomonas, 210
Strombidium, 114, 126, 145
   kahli, 125
   purpureum, 125
   viride, 146, 147

Stylonychia, 101, 140, 141, 299, 332, 352
   mytilus, 331, 333
Synura, 141, 142, 199
Syracosphaera carterae, 289

T

Tetrahymena, 77, 101, 103, 255, 260, 262, 264, 269, 294, 298, 301–305, 384, 408–410, 412, 418
   corlissi Th-X, 303
   paravorax RP, 302
   patula, 95, 103
   pyriformis, 95, 103, 134, 253, 254, 256, 257, 301–305, 328–329, 339, 343, 345–348, 354, 398, 408, 409, 410
   setifera HZ-1, 303
   thermophila, 265, 350
   vorax, 140
Tetramitus, 36
   rostratus, 7
Teteraselmis
   convolutae, 25
   cordiformis, 26
Thalassiosira weissflogii, 111
Thecamoeba, 78
   striata, 60, 61
   lanceolata, 81
   papyracea, 77
   sphaeronucleolus, 77, 81
   striata, 81
   terricola, 77
Tintinnida, 114
Tintinnidium, 114
Tintinnopsis levigata, 114
Tiphotrocha comprimata, 87
Tokophrya, 107
   lemnarum, 414
Trachelocerca, 105
Trachelomonas, 199, 210, 381
   abrupta, 41
   grandis, 24
   hispida, 40, 145
   pertyi, 41
   pyrum, 41
   volvocina, 40, 145
Trachelonema, 259

Trachelophyllum, 146
  pusillum, 123
Tracheloraphis, 101, 105, 126, 127, 129, 259
  dogieli, 260, 261
  kahli, 129
Trepomonas, 30
Triatoma, 165
Trichamoeba
  myakka, 77
  osseosaccus, 77
  villosa, 77, 80
Trichodesmium (= *Oscillatoria*), 78
Trichomitus fecalis, 158, 159
Trichomonas, 30, 225
  gallinae, 340
  tenax, 157, 158
  vaginalis, 20, 157, 158, 291
Trichonympha, 151
Trigonomonas, 52
Triloculinella obliquinoda, 87
Trinema, 65
  camplanatum, 79
  enchelys, 79
  lineare, 81, 83, 236
Tritrichomonas, 30
  foetus, 224, 291
  muris, 18
Trochammina
  inflata, 87, 89
  ochracea, 88
Trochilia palustris, 121
Truncatulinoides, 69
Trypanoplasma borreli, 227, 228
Trypanosoma, 31, 162, 164, 225
  brucei, 165, 285, 286
  congolense, 163, 229
  conorhini, 228
  cruzi, 165, 167, 228, 285, 286, 290
  cyclops, 163
  dionisii, 163
  gambiense, 165-167
  grayi, 163
  lewisi, 168
  mega, 163
  raiae, 163
  rangeli, 163
  rhodesiense, 165-167
  rotatorium, 163, 168
  vivax, 163, 228

U

Ulva lactuca, 89, 90
Urocentrum turbo, 121, 123, 146
Uroleptus piscis, 123
Uronema, 113, 114, 140, 269
  filificum, 125
  marina, 112
  turbo, 120
Uronychia, 129
Urostyla grandis, 333
Urotricha, 145

V

Vahlkampfia, 77, 78, 147
Vexellifera, 78
  telma, 80
Vanella, 77, 398
Volvox, 25, 377, 388, 396-398
  aureus, 396-397
  carteri, 377, 397
Vorticella, 101, 103, 138, 140-142, 147, 260, 361, 414
  campanula, 121, 123, 146
  longifilum, 146
  microstoma, 113, 339
  nebulifera, 96, 123, 146

W

Woloszynskia, 215
Woodruffia metabolica, 117, 118

Z

Zoothamnium, 104, 312
  geniculatum, 362
Zostera marina, 89, 90